CHAOS THEORY

Modeling, Simulation and Applications

Selected Papers from the 3rd Chaotic Modeling and Simulation International Conference (CHAOS2010)

CHAOS THEORY

Modeling, Simulation and Applications

Selected Papers from the 3rd Chaotic Modeling and Simulation International Conference (CHAOS2010)

Chania, Crete, Greece 1 – 4 June 2010

Christos H Skiadas
Technical University of Crete, Greece

Ioannis Dimotikalis
Technological Educational Institute of Crete, Greece

Charilaos Skiadas
Hanover College, Indiana, USA

editors

World Scientific

NEW JERSEY • LONDON • SINGAPORE • BEIJING • SHANGHAI • HONG KONG • TAIPEI • CHENNAI

Published by

World Scientific Publishing Co. Pte. Ltd.
5 Toh Tuck Link, Singapore 596224
USA office: 27 Warren Street, Suite 401-402, Hackensack, NJ 07601
UK office: 57 Shelton Street, Covent Garden, London WC2H 9HE

British Library Cataloguing-in-Publication Data
A catalogue record for this book is available from the British Library.

CHAOS THEORY
Modeling, Simulation and Applications
Selected Papers from the 3rd Chaotic Modeling and Simulation International Conference (CHAOS2010)

ISBN-13 978-981-4350-33-4
ISBN-10 981-4350-33-8

Printed in Singapore by World Scientific Printers.

Preface

This book includes a collection of the best papers presented in the 3rd International Conference (CHAOS2010) on Chaotic Modeling, Simulation and Applications, Chania, Crete, Greece, June 1-4, 2010. The first part of the book contains the papers addressed by the keynote and plenary speakers. The conference continues the tradition of the previous conferences on chaotic modeling and simulation that is to invite and bring together people working in the main topics of nonlinear and dynamical systems and chaotic analysis and simulation.

Interesting papers on various important topics of Chaotic Modeling and Simulation are presented as:

- Scattering by many small inhomogeneities and applications, Exploring the process of fibre breaking in NOL samples of composite during quasi-static process of fracture.
- Classical versus Quantum Dynamical Chaos: Sensitivity to external perturbations, and reversibility.
- Nonlinearity of Earth: Astonishing diversity and wide prospects, Optimizing nonlinear projective noise reduction for the detection of Planets in mean-motion resonances in transit light curves, Chaos game technique as a tool for the analysis of natural geo-morphological features.
- Lagrangian approach to chaotic transport in the Japan Sea, Theory of turbulence and the Kolmogorov constant, Simulations of steady isotropic turbulence.
- Dynamics of a rubbing Jeffcott rotor with three blades, Classifying periodic orbits, Dynamics of steel turning by recurrence plots.
- Socio-Economic and financial chaos, Modeling recent economic debates, Chaoticity in the time evolution of foreign currency exchange rates, Importance of chaos for computational processes of collective intelligence in social structures.
- New enciphering algorithm based on chaotic Generalized Hénon map, Stability on logistic-like iterative maps, Time variant chaos encryption, A highly chaotic attractor for a dual-channel single-attractor private communication system, Modified chaotic shift keying using indirect coupled chaotic synchronization for secure digital communication, Rendering statistical significance of information flow measures.
- Bifurcation problems, Analysis of homoclinic bifurcation in Duffing oscillator under two-frequency excitation, Predicting chaos with second

method of Lyapunov, Characteristic relations and reinjection probability densities of Type-II and II Intermittencies.

- Beta(p,q)-Cantor sets — determinism and randomness, Complexity theory: From microscopic to macroscopic level, concepts and applications, Regular variation, Paretian distributions, and the interplay of light and heavy tails in the fractality of asymptotic models.
- Chaotic music composition, Aesthetic considerations in algorithmic and generative composition, Pre-fractal patterns in Iannis Xenakis' algorithmic composition, Computer aided composition, Algorithmic sound composition using coupled cellular automata, Chaos as compositional order, Composing chaotic music from the letter m, On the timbre of chaotic algorithmic sounds, The Rainbow Effect on composing chaotic algorithmic music.
- Solitons, Dissipative solitons: Perturbations and chaos formation, Non Hamiltonian chaos from Nambu dynamics of surfaces.
- Plasma physics, Acoustic emission within an atmospheric Helium corona discharge jet, The stabilization of a chaotic plasma turbulence, Manifestation of chaos in collective models of nuclei.
- Data analysis synchronization and control, Dynamical principles of prognosis and control, Efficient large-scale forcing in finite-difference, Symbolic dynamics and chaotic synchronization, Scale invariance in chaotic time series: Classical and quantum examples, Investigation of the cross-correlation function and the enhancement factor for graphs with and without time reversal symmetry, Approximation of Markov chains by a solution of a stochastic differential equation.

We thank all the contributors to the success of the CHAOS 2010 International Conference, the committees, the plenary speakers, the reviewers and especially the authors of this volume. Special thanks to the Conference Secretary Dr. Anthi Katsirikou for her work and assistance. Finally, we would like to thank Mary Karadima, Aggeliki Oikonomou, Aris Meletiou and George Matalliotakis for their valuable support.

November 30, 2010

Christos H. Skiadas, Technical University of Crete, Greece
Ioannis Dimotikalis, Technological Educational Institute of Crete, Greece
Charilaos Skiadas, Hanover College, Indiana, USA

Honorary Committee

David Ruelle
Academie des Sciences de Paris
Honorary Professor at the Institut des Hautes Etudes Scientifiques of Bures-sur-Yvette, France

Leon O. Chua
EECS Department, University of California, Berkeley, USA
Editor of the International Journal of Bifurcation and Chaos

Ji-Huan He
Donghua University, Shanghai, China
Editor of Int. Journal of Nonlinear Sciences and Numerical Simulation

Gennady A. Leonov
Dean of Mathematics and Mechanics Faculty, Saint-Petersburg State University, Russia. Member (corresponding) of Russian Academy of Science

Ferdinand Verhulst
Mathematics Faculty, Utrecht, The Netherlands

International Scientific Committee

C. H. Skiadas (Technical University of Crete, Chania, Greece), Chair
H. Adeli (The Ohio State University, USA)
J.-O. Aidanpää (Div. of Solid Mechanics, Lulea University of Technology, Sweden)
N. Akhmediev (Australian National University, Australia)
M. Amabili (McGill University, Montreal, Canada)
J. Awrejcewicz (Technical University of Lodz, Poland)
J. M. Balthazar (UNESP-Rio Claro, State University of Sao Paulo, Brasil)
S. Bishop (University College London, UK)
T. Bountis (University of Patras, Greece)
Y. S. Boutalis (Democritus University of Thrace, Greece)
C. Chandre (Centre de Physique Theorique, Marseille, France)
M. Christodoulou (Technical University of Crete, Chania, Crete, Greece)
P. Commendatore (University of Napoli 'Federico II', Italy)
D. Dhar (Tata Institute of Fundamental Research, India)
I. Dimotikalis (Technological Educational Institute, Crete, Greece)
B. Epureanu (University of Michigan, Ann Arbor, MI, USA)
G. Fagiolo (Sant'Anna School of Advanced Studies, Pisa, Italy)
V. Grigoras (University of Iasi, Romania)
K. Hagan (University of Limerick, Ireland)
L. Hong (Xi'an Jiaotong University, Xi'an, Shaanxi, China)

Keynote Talks

Gennady Leonov

Member (corresponding) of Russian Academy of Science
Dean of Mathematics and Mechanics Faculty
Saint-Petersburg State University, Russia

Attractors, limit cycles and homoclinic orbits of low dimensional quadratic systems

Sergey V. Prants

Laboratory of Nonlinear Dynamical Systems
Pacific Oceanological Institute of the Russian Academy of Sciences
Vladivostok, Russia

De Broglie-wave chaos

Alexander G. Ramm

Mathematics Department, Kansas State University
Manhattan, KS 66506-2602, USA
http://www.math.ksu.edu/~ramm

Scattering by many small inhomogeneities

Valentin V. Sokolov

Budker Institute of Nuclear Physics and
Novosibirsk Technical University
Novosibirsk, Russia

Classical versus quantum dynamical chaos: Sensitivity to external perturbations, stability and reversibility

Contents

Preface v

Honorary Committee and International Scientific Committee vii

Keynote Talks ix

Part I. Plenary and Keynote Papers 1

Lagrangian approach to chaotic transport and mixing in the Japan sea
M. V. Budyansky, V. I. Ponomarev, P. A. Fyman, M. Yu. Uleysky and S. V. Prants 3

Nonlinearity of Earth: Astonishing diversity and wide prospects
O. B. Khavroshkin and V. V. Tsyplakov 14

Dynamical priciples of prognosis and control
Gennady A. Leonov 21

On a problem of approximation of Markov chains by a solution of a stochastic differential equation
Gabriel V. Orman 30

Scattering by many small inhomogeneities and applications
Alexander G. Ramm 41

Modeling recent economic debates
Christos H. Skiadas 53

Classical versus quantum dynamical chaos: Sensitivity to external perturbations, stability and reversibility
Valentin V. Sokolov, Oleg V. Zhirov and Yaroslav A. Kharkov 59

On logistic-like iterative maps
Dimitrios A. Sotiropoulos 77

Part II. Invited and Contributed Papers 87

Improved ε expansion in theory of turbulence: Calculation of Kolmogorov constant and skewness factor
L. Ts. Adzhemyan, M. Hnatich and J. Honkonen 89

Dynamics of a rubbing Jeffcott rotor with three blades
Jan-Olov Aidanpää and Göran Lindkvist 97

Exploring process of fibre breaking in tube samples of composite during quasi-static process of fracture
Dorota Aniszewska and Marek Rybaczuk 105

Non Hamiltonian chaos from Nambu dynamics of surfaces
Minos Axenides 110

A methodology for classifying periodic orbits
Jayanta K. Bhattacharjee, Sagar Chakraborty and Amartya Sarkar 120

Chaoticity in the time evolution of foreign currency exchange rates in Turkey
O. Cakar, O. O. Aybar, A. S. Hacinliyan and I. Kusbeyzi 127

Symbolic dynamics and chaotic synchronization
Acilina Caneco, Clara Grácio and J. Leonel Rocha 135

New enciphering algorithm based on chaotic Generalized Hénon Map
Octaviana Datcu, Jean-Pierre Barbot and Adriana Vlad 143

Noise influence on the characteristic relations and reinjection probability densities of type-II and type-III intermittencies
Ezequiel Del Rio, Sergio Elaskar, Jose M. Donoso and Luis Conde 151

Multifractal and wavelet analysis of epileptic seizures
Olga E. Dick and Irina A. Mochovikova 159

Fractal based curves in musical cretivity: A critical annotation
Anastasia Georgaki and Christos Tsolakis 167

Time variant chaos encryption
Victor Grigoras and Carmen Grigoras 175

Aesthetic considerations in algorithmic and generative composition
Kerry L. Hagan 183

Optimizing nonlinear projective noise reduction for the detection of planets in mean-motion resonances in transit light curves
N. Jevtic, J. S. Schweitzer and P. Stine 191

Dissipative solitons: Perturbations and chaos formation
Vladimir L. Kalashnikov 199

Modified chaotic shift keying using indirect coupled chaotic synchronization for secure digital communication
Rupak Kharel, Krishna Busawon and Z. Ghassemlooy 207

Chaos problems in observer's mathematics
Boris Khots and Dmitriy Khots 215

Freedom and necessity in computer aided composition: A thinking framework and its application
Johannes Kretz 223

A predator-prey model with the nonlinear self interaction coupling $x^k y$
I. Kusbeyzi, O. O. Aybar and A. S. Hacinliyan 231

Evidence for deterministic chaos in aperiodic oscillations of acute lymphoblastic leukemia cells in long-term culture
George I. Lambrou, Aristotelis Chatziioannou, Spiros Vlahopoulos, Maria Moschovi and George P. Chrousos 239

Scale invariance in chaotic time series: Classical and quantum examples
Emmanuel Landa, Irving O. Morales, Pavel Stránský, Rubén Fossion, Victor Velàzquez, J. C. López Vieyra and Alejandro Frank 247

Acoustic emission within an atmospheric helium discharge jet
V. J. Law, C. E. Nwankire, D. P. Dowling and S. Daniels 255

Experimental investigation of the enhancement factor for irregular undirected and directed microwave graphs
Michał Ławniczak, Szymon Bauch, Oleh Hul and Leszek Sirko 265

Algorithmic sound composition using coupled cellular automata
Jaime Serquera and Eduardo R. Miranda 273

Efficient large-scale forcing in finite-difference simulations of steady isotropic turbulence
Ryo Onishi, Yuya Baba and Keiko Takahashi 281

Rendering statistical significance of information flow measures
Angeliki Papana and Dimitris Kugiumtzis 289

Complexity theory and physical unification: From microscopic to macroscopic level
G. P. Pavlos, A. C. Iliopoulos, L. P. Karakatsanis, V. G. Tsoutsouras and E. G. Pavlos 297

Regular variation, Paretian distributions, and the interplay of light and heavy tails in the fractality of asymptotic models
Dinis D. Pestana, Sandra M. Aleixo and J. Leonel Rocha 309

Tools for investigation of dynamics of DC-DC converters within Matlab/Simulink
Dmitry Pikulin 317

Chaos as compositional order
Eleri Angharad Pound 325

Beta(*p*,*q*)-Cantor sets — Determinism and randomness
J. Leonel Rocha, Sandra M. Aleixo and Dinis D. Pestana 333

Predicting chaos with second method of Lyapunov
Vladimir B. Ryabov 341

Analysis of homoclinic bifurcation in Duffing oscillator under two-frequency excitation: Peculiarity of using Melnikov method in combination with averaging technique
Vladimir Ryabov and Kenta Fukushima 349

Exploring life expectancy limits: First exit time modeling, parameter analysis and forecasts
Christos H. Skiadas and Charilaos Skiadas 357

Composing chaotic music from the letter m
Anastasios D. Sotiropoulos 369

On the timbre of chaotic algorithmic sounds
Dimitrios A. Sotiropoulos, Anastasios D. Sotiropoulos and Vaggelis D. Sotiropoulos 379

The rainbow effect on composing chaotic algorithmic music
Vaggelis D. Sotiropoulos 388

A highly chaotic attractor for a dual-channel single-attractor, private communication system
Banlue Srisuchinwong and Buncha Munmuangsaen 399

Manifestation of chaos in collective models of nuclei
Pavel Stránský, Michal Macek, Pavel Cejnar, Alejandro Frank, Ruben Fossion and Emmanuel Landa 406

Importance of the chaos for computational processes of collective intelligence in social structures
Tadeusz (Ted) Szuba 414

Complex signal generators based on capacitors and on piezoelectric loads
Horia-Nicolai L. Teodorescu and Victor P. Cojocaru 423

Drift waves' synchronization by using an external signal. The stabilization of a chaotic plasma turbulence
C. L. Xaplanteris and E. Filippaki 431

Chaos game technique as a tool for the analysis of natural geomorphological features
G. Žibret and T. Verbovšek 439

Dynamics of a steel turning process
Grzegorz Litak and Rafał Rusinek 445

Author Index 449

PART I

Plenary and Keynote Papers

Lagrangian approach to chaotic transport and mixing in the Japan Sea

M. V. Budyansky, V. I. Ponomarev, P. A. Fyman, M. Yu. Uleysky, and S. V. Prants

Pacific Oceanological Institute of the Russian Academy of Sciences,
690041 Vladivostok, Russia
Email: prants@poi.dvo.ru

Abstract: We use the Lagrangian approach to study surface transport and mixing of water masses in a selected region of the Japan Sea using velocity fields generated by a numerical MHI multi-level eddy-resolved sea-circulation model. Evolution of patches with a large number of tracers, chosen in different parts of the selected region, is computed. The pictures obtained demonstrate clearly regions of strong mixing and stagnation zones coexisting with each other. Computing finite-time Lyapunov exponents for a long period of time, we plot a Lyapunov synoptic map quantifying surface transport and mixing and revealing Lagrangian coherent structures.
Keywords: chaotic mixing, Lyapunov synoptic map, Japan Sea, mesoscale dynamics.

1 Introduction

Surface transport and mixing processes play a crucial role in the ocean dynamics. Water masses of different origins interchange their contents of heat, salinity, other physical and chemical characteristics and of biological nutrients with a profound impact on the ocean and atmospheric weather. Understanding and quantifying these processes are important for addressing some practical problems as well. To list a few, we mention plankton blooms, anthropogenic pollution, fishering quotas, etc.

Strong currents, streamers (quasistationary jets of different scales), and mesoscale eddies (with the size of the order 30 – 300 km) are the ocean features playing important role in transport and mixing processes. Currents and jets are transport barriers because it is difficult for water particles to

Chaos Theory: Modeling, Simulation and Applications
C. H. Skiadas, I. Dimotikalis and C. Skiadas (Eds)

cross them, and waters on both sides of a jet may have different temperatures and distinct contents of salt, nutrients, and chemicals. Eddies are regions enclosed by streamfunction contours within which water maintains its properties for a long time being trapped and transported within the eddy. Ocean eddies may travel hundreds to thousands kilometers and live from a few months to years. Eddy cores remain coherent for some time but stirring of the surrounding water provides eventually mixing.

Search for order in an apparent disorder of the ocean motion is a hard problem. The Lagrangian approach is the most effective in studying transport and mixing phenomena. This approach does not aim at studying individual trajectories of fluid particles but at searching for and identifying spatial structures organizing the whole flow and known in theory of dynamical systems as invariant manifolds. In this theory they are smooth subspaces in an abstract phase space. In fact, it is an application and elaboration of the old Poincarés idea of searching for geometrical structures in the phase space of a dynamical system. In two-dimensional fluid flows they are material curves, consisting of fluid particles and corresponding to transport barriers, lines of maximal stretching and convergence. Intersections of stable and unstable manifolds are regions of strong mixing with typical filamentary and convoluted structures visible in laboratory experiments on chaotic mixing and in satellite images of surface temperature and chlorophyll in the ocean.

Methods of dynamical systems theory have been successfully used for the last two decades in studying surface transport and mixing in the ocean (for recent reviews see [2,7,3]). Starting with simplified kinematic models, researchers have then assimilated dynamically consistent models of ocean currents and eddies. Now numerical circulation models are used with the aim of elucidating water mixing in different seas. In this work, we use velocity data from the Japan Sea circulation model [8] to characterize surface transport and mixing in the active region comprising the Primorskoye (Liman) current which is known to generate mesoscale eddies [6].

2 Lagrangian approach in studying surface transport and mixing in the ocean

In Lagrangian approach, a fluid particle is advected by a two-dimensional Eulerian velocity field

$$\frac{dx}{dt} = u(x, y, t), \quad \frac{dy}{dt} = v(x, y, t), \tag{1}$$

where (x, y) is the location of the particle, and u and v are the zonal and meridional components of its velocity at the location (x, y). Even if the velocity field is fully deterministic, the Lagrangian trajectories may be very complicated and practically unpredictable. It means that a distance between two initially nearby particles grows exponentially in time

$$\|\delta \boldsymbol{r}(t)\| = \|\delta \boldsymbol{r}(0)\| e^{\lambda t}, \tag{2}$$

where λ is a positive number, known as a Lyapunov exponent, which characterizes asymptotically (at $t \to \infty$) the average rate of the particle dispersion, and $\| \cdot \|$ is a norm of the vector $\boldsymbol{r} = (x, y)$. It immediately follows from (2) that we unable to forecast the fate of the particles beyond the so-called predictability horizon

$$T_p \simeq \frac{1}{\lambda} \ln \frac{\|\Delta\Gamma\|}{\|\Delta\Gamma(0)\|}, \tag{3}$$

where $\|\Delta\Gamma\|$ is a confidence interval of the particle location and $\|\Delta\Gamma(0)\|$ is a practically inevitable inaccuracy in specifying the initial location. The deterministic dynamical system (1) with a positive maximal Lyapunov exponent for almost all vectors $\delta\boldsymbol{r}(0)$ (in the sense of nonzero measure) is called chaotic. It should be stressed that the dependence of the predictability horizon T_p on the lack of our knowledge of exact location is logarithmic, i.e., it is much weaker than on the measure of dynamical instability quantified by λ. Simply speaking, with any reasonable degree of accuracy on specifying initial conditions there is a time interval beyond which the forecast is impossible, and that time may be rather small for chaotic systems.

In the last two decades, the interest rapidly grows in application of chaos theory and dynamical systems approach to transport and mixing processes in the ocean [2,7,3]. Since the phase plane of a two-dimensional dynamical system (1) is a physical space for fluid particles, many abstract mathematical objects from dynamical systems theory are material surfaces, points and curves in fluid flows. Say, a stagnation point in a steady flow is a fluid particle with zero velocity. Besides "trivial" elliptic stagnation points, the motion around which is stable, there are hyperbolic (saddle) stagnation points which organize fluid motion in their neighborhood in a specific way. There are two opposite directions (for each saddle point) along which nearby trajectories approach the point at an exponential rate and two other directions along which nearby trajectories move away from it at an exponential rate which are known as stable, W_s, and unstable, W_u, invariant manifolds.

Hyperbolic stagnation points under a periodic perturbation of steady flows become hyperbolic (unstable) periodic trajectories whose invariant manifolds, in general, intersect each other transversally forming a complex structure known as a homoclinic (heteroclinic) tangle. Water parcels in that tangle experience stretching and folding at progressively small scales in course of time providing effective mixing in unsteady flows known as chaotic advection [3].

Stable and unstable manifolds are useful tools in studying realistic flows modeling the ocean. In aperiodic flows it is possible to identify aperiodically moving hyperbolic points with stable and unstable effective manifolds [2]. Unlike the manifolds in steady and periodic flows, defined in the infinite time limit, the "effective" manifolds of aperiodic hyperbolic trajectories have a finite lifetime just like the very trajectories. The point is that they may play the same role in organizing oceanic flows as do invariant manifolds in simpler flows. The effective manifolds in course of their life undergo stretching and folding at progressively small scales and intersect each other in the homoclinic points in the vicinity of which fluid particles move chaotically. Trajectories of initially nearby fluid particles diverge rapidly in these regions, and particles from other regions appear these. It is a mechanism for effective transport and mixing of water masses. Moreover, stable and unstable effective manifolds constitute Lagrangian transport barriers between different regions because they are material invariant curves that cannot be crossed by purely advective processes.

Any preselected region in a circulation basin is an open system in the sense that fluid particles come into the region from outside and soon or later leave it. So, we deal with a scattering problem that can be illustrated with an unsteady deterministic open flow as follows. Passive particles are advected by the incoming flow into a mixing region, where their motion may be chaotic, and then most of them are washed away from that region. It is known in the theory that these exists the chaotic invariant set consisting of an infinite number of hyperbolic particle trajectories that never leave the mixing region [5,4]. If a particle belongs to the set at an initial moment, then it remains in the mixing region forever. Most of particles soon or later leave the mixing region, but their behavior is strongly influenced by the presence of the chaotic invariant set. Each trajectory in the set and therefore, the whole set possesses stable and unstable manifolds. Theoretically, these manifolds have infinite spatial extent, and the tracer, belonging to the stable manifold, is advected by the incoming flow into the mixing region and remains there forever. The corresponding initial conditions make up a set of zero measure. However, the particles that are initially close to those in W_s follow them for a long time and eventually deviate from them, and leave the mixing region along the unstable manifold W_u.

3 MHI Japan Sea circulation model

The Japan Sea is a deep marginal sea with shallow straits connected with the East China Sea, Okhotsk Sea, and North Pacific. The typical large scale circulation over the Northwestern Japan Sea includes two cyclonic gyres, cold Primorskoye (Liman) Current streamed southwestward along the continental slope of the Japan Basin, and warm northern current along the slope of Japanese Islands. The southwestern cyclonic gyre over southern and central areas of the Japan Basin is simulated in the model domain as a large scale circulation.

The mesoscale dynamics over the shelf and steep continental slope includes the jet currents, streamers, and eddies being controlled by synoptic scale wind forcing and sea baroclinicity. According to satellite images, the anticyclonic mesoscale eddies (clockwise rotating) of relatively small scale are observed in the northwestern marginal area directly over the steep continental slope and anticyclonic mesoscale eddies of larger scale is clearly seen in the southern marginal area of the slope and shelf of the Peter the Great Bay. According to the theory, the mesoscale dynamics over the continental slope could be associated with the coastal Kelvin waves propagated downstream of Primorskoe Current to the southwest and catched by the wide shelf of the Peter the Great Bay. The integration of diagnostic 3D hydrodynamic model using observed temperature and salinity profiles from oceanographic R/V surveys shows the eddies along the continental slope and in the Peter the Great Bay [1].

The MHI (Marine Hydrophysical Institute, Sebastopol, Ukraine) ocean circulation model [8] is 3D primitive equations under the hydrostatic and Boussinesq approaches with free surface boundary condition. The MHI model of the Japan Sea is also described briefly in [6]. It belongs to a class of layered models in which the sea consists of a number of quasi-isopycnal layers. Interfacial surfaces between layers can freely move up and down and layers can deform, physically vanish and restore. Equations of the MHI model, vertically integrated within layers, are formulated at the beta-plane, with the x-axis directed from the west to east and the y-axis directed from the south to north.

We simulate the nonlinear mesoscale eddy dynamics over the shelf, continental slope, and Japan Basin taking into account realistic bottom topography and daily mean external atmospheric forcing. The present study is focused on simulation of mesoscale dynamics over the continental slope and shelf in the closed sea area of the cyclonic gyre occupied southern and central area of the Japan Basin. The sea domain extends from 39° to 44° N and from 129° to 134° E with horizontal grid steps $1'.44$ along latitude and $1'.92$ along longitude. It is practically the same resolution along x and

y in kilometers. The total number of the grid points is 210×280. We set 10 quasi-isopicnal layers including upper mixed layer. The bottom topography is adopted from navigation maps.

The near-surface daily meteorological conditions were set from the NCEP/ NCAR Reanalysis. It includes short wave radiation flux, wind stress, wind speed, air temperature, and precipitation. The numerical experiments with minimized coefficients of the horizontal and vertical viscosity show the intensive mesoscale dynamics, particularly, synoptic scale variability of anticyclonic/ cyclonic eddies and streamers over the shelf and continental slope. The anticyclonic eddies generated over the shelf break and continental slope are usually moving southwestward along the slope like the topographic Kelvin waves with prevailing phase velocity of about 6-8 cm/s. The spatial scale of the anticyclonic eddies is usually increased near the Peter the Great Bay shelf where it exceeds significantly the baroclinic Rossby deformation radius.

4 Results

Satellite image of the surface temperature in the part of this region in the infrared range is shown in Fig. 1 (left panel). White and dark colors in the figure correspond to low and high temperatures, respectively. Mesoscale eddies are visible along the coast of the Primorye region (Russia). Snapshot of the color-coded vorticity field rot $\boldsymbol{v}$ at the 15th day of integration (Fig. 1, right panel) demonstrates a complex picture of mixing in that region with a number of mainly anticyclonic and cyclonic eddies of different sizes with negative and positive vorticity, respectively.

To study the surface transport and mixing in the first horizontal layer we select a few rectangulars of smaller size with $9 \cdot 10^4$ homogeneously distributed particles in each of them and compute the particle trajectories for 50 days using the surface velocity field generated by the numerical model. The particles are assumed to be of infinitesimally small size, neutrally buoyant, and they do not affect the dynamics of the fluid. To obtain the velocity at any point between the grid points we use a bicubic interpolation in space. Temporal resolution of the data is 24 hours. To obtain an approximation of the velocity field between discrete days we use an interpolation with third-order Lagrangian polynomials. Tracking evolution of patches of ocean surface is important for a few practical transport problems like oil and other pollutant spills, harmful algal blooms, etc. It is necessary to know the directions of motion and the form of the evolved patch not details of trajectories of the particles inside the patch.

Fig. 2a shows locations of the patches 1 and 2 which were chosen to be rather close to each other initially. However, the evolution of those

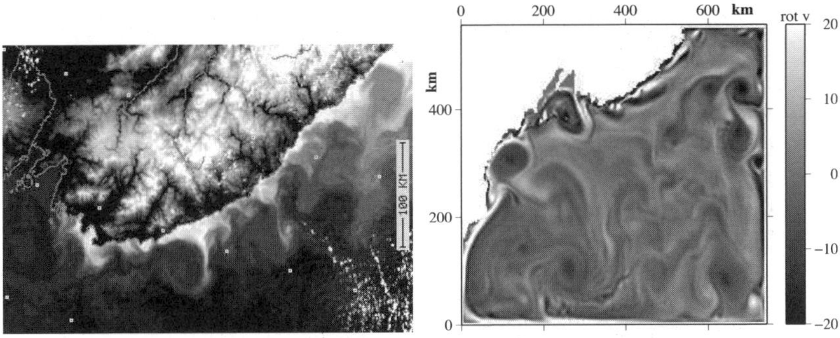

Fig. 1. Left: Satellite image of the water surface temperature in the selected region of the Japan Sea in the infrared range (NOAA AVHRR data, 15. 09. 1997). White and dark colors correspond to low and high temperatures, respectively. Mesoscale eddies are visible along the coast of the Primorye region (Russia). Right: Snapshot of the vorticity field in the preselected region of the Japan Sea at the 15th day of integration plotted against initial positions. Color modulates the values of the vorticity field rot $\boldsymbol{v}$. The land is in white.

patches (see Fig. 2b, c, and d) is very different. The patch 2 is deformed slightly being rather compact after 50 days of integration. Its latitude practically have not been changed during a long period of time, and it only extended in the longitude direction. The evolution of the patch 1 is cardinally different because it was chosen initially to cover an anticyclonic mesoscale eddy of the Primorskoye current. The eddy moves downstream to the south-west (see Fig. 2b). Then a group of particles is pinched off from the eddy being involved in a counter-propagating current (see Fig. 2c) forming a complicated structure with long filaments. After 50 days, we see the eddy transported along the North Korean continental slope and a long filamentary-like tail (Fig. 2d).

Fig. 3a shows initial positions of the other two patches 3 and 4 which were chosen in the eastern part of the selected region. The patch 3, chosen in the region of the Primorskoye current, evolves in a complicated way forming patterns with strong mixing and moves mainly downstream. Whereas the patch 4 is situated in the stagnation region of the sea. Its evolution is much more simple.

In theory of dynamical systems Lyapunov exponents λ's are known to be quantitative criteria of chaotic motion in the asymptotic limit. In practice, one is forced to compute Lyapunov exponents for a finite time. The finite-time Lyapunov exponent (FTLE) is a finite-time average of the maximal separation rate for a pair of neighboring advected particles. The FTLE at

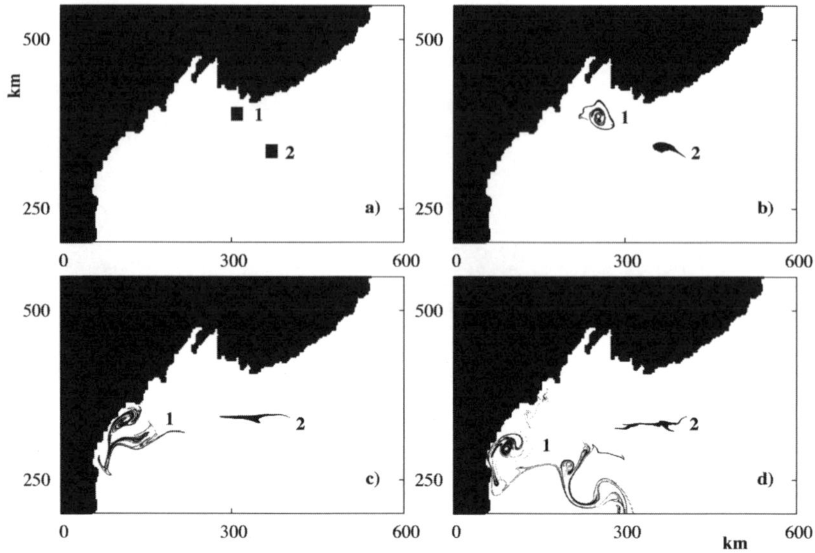

Fig. 2. Snapshot of evolution of the patches 1 and 2 at (a) $t = 0$, (b) $t = 15$, (c) $t = 30$, and (d) $t = 50$ days. The land is in dark.

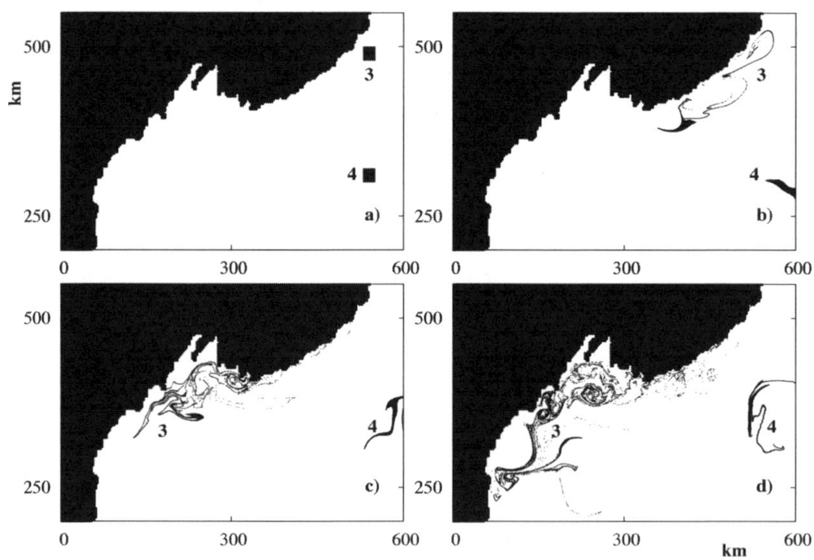

Fig. 3. Snapshot of evolution of the patches 3 and 4 at (a) $t = 0$, (b) $t = 15$, (c) $t = 30$, and (d) $t = 50$ days.

position $\boldsymbol{r}$ at time τ is given by

$$\lambda(\boldsymbol{r}(t)) \equiv \frac{1}{\tau} \ln \sigma(G(t)), \tag{4}$$

where τ is an integration time, and $\sigma(G(t))$ denotes the largest singular value of the evolution matrix $G(t)$ which governs evolution of small displacements in linearized advection equations.

Computing FTLEs is useful in oceanography because they are mathematical analogues of drifter launching in the ocean and characterize quantitatively dispersion of water masses. Moreover, they enable to reveal Lagrangian coherent structures hidden in the velocity field including stable and unstable manifolds of finite-time hyperbolic trajectories, large-scale transport barriers, and eddies.

A uniform grid of 1000×1000 particles is advected by the numerically generated velocity field. After 50 days (starting on September, 15), the FTLEs are computed using Eq. (4). Spatial distribution of the FTLEs, plotted against initial positions in Fig. 4, may be called a Lyapunov synoptic map. This map shows that there is a large range of λ values from 0.01 to 0.3 days^{-1} which corresponds to mixing times (e-folding times) from 100 to 3 days.

The Lyapunov synoptic map in Fig. 4 reveals a number of structures. There are eddies along the continental slope which are easily visible. The eddy cores are characterized by low values of the Lyapunov exponents. The particles inside the cores tend to stay therein for a long time. There are filaments that wind up around the eddies in spirals which correspond an ejection of water out off the eddy. This process is visualized with the patch 1 in Fig. 2b. The filaments around the eddies may act as transport barriers, the water from outside cannot enter the eddy core. Moreover, there are very long filaments, correspond to the largest Lyapunov exponents, which are not associated with any eddies. Filamentary structures in Figs. 2, 3, and 4 align along stretching directions of the velocity field.

It is evident that in numerically generated aperiodic velocity fields hyperbolic trajectories are of a transient nature, they may appear and exist for a while and then disappear. Transient hyperbolic trajectories induce the corresponding finite-time stable and unstable invariant manifolds. Finite-time stable and unstable invariant manifolds of a hyperbolic trajectory $\gamma(t)$ on the interval $[t_0, T]$ is a set of initial particle positions that approach to $\gamma(t)$ forwards and backwards in time, respectively, as long as $t \in [t_0, T]$. They are material lines changing in time and space. Like any material line they provide transport barriers because particles cannot cross them. However, they are very special material lines forming a kind of the front separating waters with different characteristics. Isolines of maximal val-

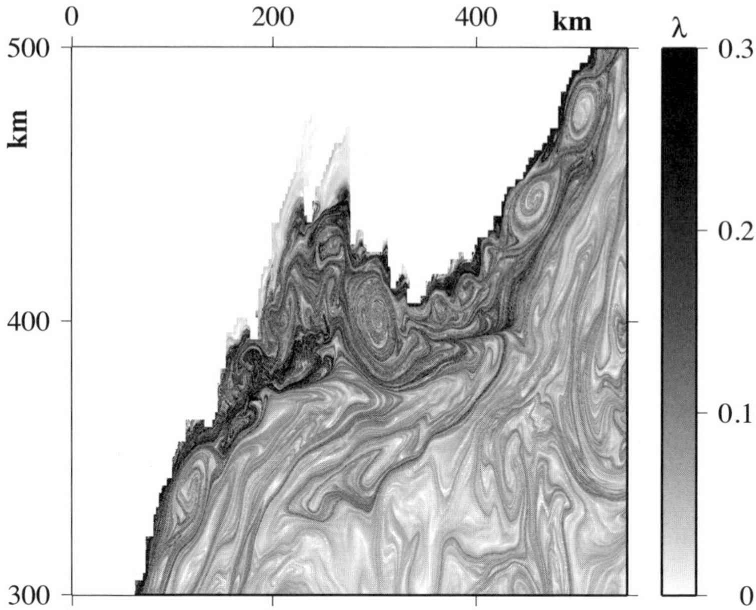

Fig. 4. Lyapunov synoptic map of the selected region shows the maximal finite-time Lyapunov exponents λ vs initial particle positions. λ is in units days^{-1}. Integration time is 50 days.

ues of λ, computed by integration of advection equations forwards in time, provide good approximation of stable manifolds of finite-time hyperbolic trajectories. They are repelling material lines of different sizes and geometry that are clearly visible in Fig. 4. Isolines of maximal values of λ, computed by integration of advection equations backwards in time, provide in turn good approximation of unstable manifolds of finite-time hyperbolic trajectories (not shown in the paper). They are attracting material lines. Both the manifolds and the corresponding ridges in the λ maps organize the transport processes in the basin.

5 Conclusion

Using the Lagrangian approach, we have studied surface transport and mixing processes in a selected region of the Japan Sea from velocity fields generated by the numerical circulation model. Methods of dynamical systems theory have been applied to identify and quantify dynamical structures in realistic flows provided by the model. We have shown that mesoscale eddies generated in the Primorskoye current play an important role in the mixing and transport of water masses with their physical, chemical, and biological characteristics. Computing finite-time Lyapunov maximal exponents, we plotted a Lyapunov synoptic map of the selected region for 50 days which enables to quantify mixing processes and reveal coherent structures in the flow like eddies and jets.

The ideas and methods of theory of dynamical systems and chaos seem to be useful to solve a general oceanographic problem of large-scale transport and mixing. We hope that the results obtained are of interest to the oceanographic community.

References

1.P.A. Fyman and V.I. Ponomarev. Diagnostic simulation of sea currents in the peter the great bay based on ferhri oceanographic surveys. *Pacific Oceanography*, 4:56–64, 2008.
2.G. Haller. Distinguished material surfaces and coherent structures in 3d fluid flows. *Physica D*, 149:248–277, 2001.
3.K.V. Koshel and S.V. Prants. Chaotic advection in the ocean. *Physics – Uspekhi*, 49:1151–1178, 2006.
4.S.V. Prants M.V. Budyansky, M.Yu. Uleysky. Hamiltonian fractals and chaotic scattering of passive particles by a topographical vortex and an alternating current. *Physica D*, 195:369–378, 2004.
5.E. Ott. Chaos in dynamical systems. page 381, Cambridge, 1993. Cambridge University Press.
6.V. Ponomarev and O. Trusenkova. Circulation patterns of the japan sea. *La Mer*, 38:189–198, 2000.
7.R.M. Samelson and S. Wiggins. Lagrangian transport in geophysical jets and waves. page 147, New York, 2006. Springer.
8.N.B. Shapiro. Formation of the black sea circulation under forcing of wind with stochastic component. *Marine Hydrophysical Journal*, 6:26–40, 1998.

Nonlinearity of Earth: Astonishing diversity and wide prospects

O. B. Khavroshkin and V. V. Tsyplakov

Schmidt Institute of the Earth Physics, RAS, Moscow, Russia
Email: khavole@ifz.ru

Preface. Astonishing diversity of nonlinearity of seismic waves, fields and processes really have many peculiarities which in common are similary nonlinear effects of other scientific division. Only a seismic acoustic emission and the modulation of high frequency seismic noise are belonging for seismology. Therefore description of its is general and other direction will be shortly mention.

The prevalence of nonlinear physics at the present stage of development of natural sciences ensures in the theoretical plan operation of many models of nonlinear seismology in every respect. Therefore shaping of last from experiment becomes preferable. Let's present a first stage of experimental search, investigation and analysis of nonlinear seismic effects, processes and medium as formative structure of nonlinear seismology. Basic elements of such structure: (1) seismic acoustic emission as component of regional high-frequency seismic noise radiated by geological medium of lithosphere of the Earth and governed characteristically by level of energy action on medium of processes, deforming the Earth, and structural geological features of medium, thus such emission component is the integrated response to every strains, simultaneously acting on region; (2) the statistical characteristics of seismic acoustic emission can be contained hidden periodicities which correspond rhythms of processes deforming geological structure of tested region (the lunar-solar tides, the Earth.s proper oscillations, technique effects etc.), that is, the regional noises are modulated by changes of emission component; (3) seismic wave fields, wave packets and waves from the deterministic sources (vibrators, hydroelectric power stations, the explosions etc.) show properties, which are also defined by concepts of wave dynamics and theory of open systems: three-frequency interactions, radiative forces, elements of chaos and self-organizing, action of a signal on a noise field.

The modulation of high frequency seismic noise (15-300 Hz) by long-time deformation processes of the Earth are being studied experimentally from the very moment of its discovery in 1975 till present. A method of a narrow band filtration and singling out an envelope curve for recording some noise characteristics has been first grounded and applied by Khavroshkin and Tsyplakov. According to definition and characteristics of an enveloping curve variations of the accidental process at the output of narrow-band filter the data of registration of envelope amplitude give information about process intensity and its low-frequency changes. The relation of these variations (a modulation effect) of regional noise level to the processes which deform the Earth's lithosphere: the lunar-solar tides, the Earth's proper oscillations, microseism storms and wave packets from earthquakes and explosions has been found and studied [1-3]. A qualitative mechanism of generation of a part of high frequency noise has been considered. A model of a local distraction and/or reconstruction of various scale defects of the deformed stressed geophysical media has been used. The concept of a seismic acoustic emission (SAE), analogue of acoustic emission has been

Chaos Theory: Modeling, Simulation and Applications
C. H. Skiadas, I. Dimotikalis and C. Skiadas (Eds)
 (pp. 14 - 20)

introduced. Long-duration research revealed that usually the anomalous variations of SAE relate tectonic activity growth (earthquakes) in specific form [4].

These SAE anomalies were found to exceed considerably the variations resulting from the other known regional noise effects (like tides, changes of meteorological and fluid-dynamic conditions) [1].

1.Tectonic activity of region is adequately represented by SAE envelope.
2. Seismic self-oscillations; self-chaotic and self-order of vibroseismic signals.
3. Solitary sign and peculiarity of seismic waves and fields.
4. Seismic waves interraction; conversion of seismic wave front.
5. Applied and fundamental using of seismic nonlinearity.
6. Cosmogonic nonlinearity.

The prevalence of nonlinear physics at the present stage of development of natural sciences ensures in the theoretical plan operation of many models of nonlinear seismology in cvcry respect. Therefore shaping of last from experiment becomes preferable. Let's present a first stage of experimental search, investigation and analysis of nonlinear seismic effects, processes and medium as formative structure of nonlinear seismology. That is, it is effort to understand the development of this branch of geophysics on the basis of possibilities of instrumental and methodical provision of research. Basic elements of such structure need to assign to the following.

Seismic Acoustic Emission

Seismic acoustic emission as component of regional high-frequency seismic noise is radiated by geological medium of lithosphere of the Earth and governed characteristically by level of energy action on medium of processes, deforming the Earth, and structural geological features of medium, thus such emission component is the integrated response to every strains, simultaneously acting on region.
Seismic emission may be observed experimentally as increasing level of envelope of the noise in narrow band at seismic active region with a rise of tectonic stresses (Fig. 1), in the moment of passing seismic waves (Fig. 2) and with other occasional changing energetic state of region.

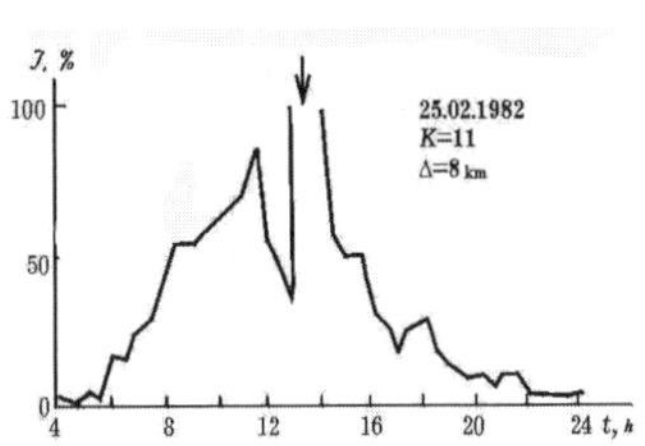

Fig. 1

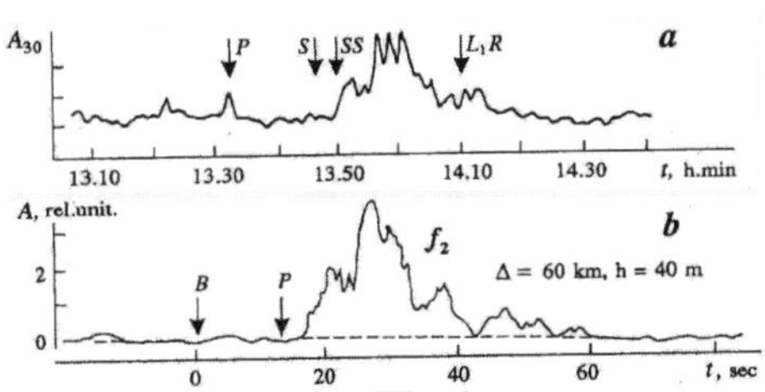

Fig. 2

Fig. 1. Seismic acoustic emission (SAE) and seismicity. Turkmenistan's region. Local earthquakes of K classes, K - the value which is linearly dependent on M (K=1.8M+4.0); *J* - intensity of SAE under Δf=0.1 Hz: f = 30 Hz. The arrows mark the earthquakes of local seismicity of tested region.

Fig. 2. Seismic noise modulation by wave tides from teleseismic events in Obninsk (a) and from remote explosions in Turkmenia (b); *a* — Earthquake in the Pacific Ocean on 18.12.78: *M=6,0,* A° ≈ 110, A_{30} in the f=30 Hz, t - current time (local), hour, min; *b* — monitoring of an explosion with a capacity of 800 kg at a distance of 60 km, the instrumentation 40 m deep, a record of the noise enveloping curve in the frequencies f_2=20 Hz, *t* - current time, sec, *A* -amplitude, rel. un., *B* - an explosion moment, *P* - P-waves arrival

Modulation of High Frequency Microseism and Noise

Variations of any statistical parameters of regional high frequency microseism or noise that correlate with temporal variations of processes deforming the examined region and a geophysical medium should be understood as a modulation effect discovered in 1975.

Statistical characteristics of seismic acoustic emission can be contained hidden periodicities which correspond rhythms of processes deforming geological structure of tested region (the lunar-solar tides, the Earth's proper oscillations, technique effects etc.), that is, the regional noises are modulated by changes of emission component. The other form of definition takes into consideration the original recording methods.

Structural elements of such a medium of local zones mechanical stress concentration type, boards of loaded fissures in a near critical condition are active macroscopic systems, and a spectrum of a wave (noise) field, being formed by a radiation of an ensemble of elements is one of the parameters of the medium condition.

A method of a narrow band filtration and singling out an enveloping curve for recording some stochastic noise characteristics is grounded and applied; variations of natural regional high frequency noise in time and at long intervals of observations are investigated. A relation of these variations (a modulation effect) of regional noise level with processes deforming the Earth's lithosphere: the lunar-solar tides, the Earth's

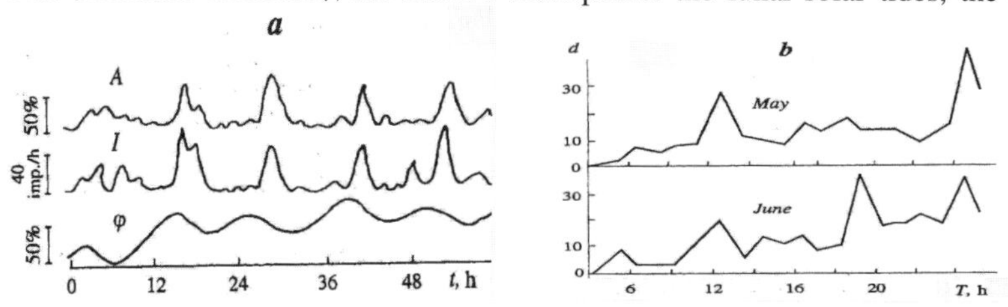

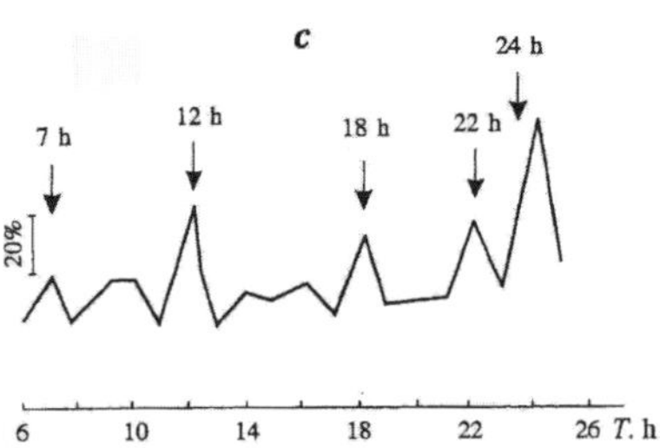

Fig. 3. High frequency seismic noise modulation with periodicity of lunar-solar tides, Ashkhabad region, 1983 *a* — in an obvious form: an example of compressed section of record (5-7 May), noise parameters' temporal run *(A* - averaged amplitudes of the enveloping curve at the frequency f=*46* Hz; I - intensity of noise, a number of impulses in an hour) and tidal tilt of the Earth's surface (direction B-3); *t* - current time; b — in a non obvious form: periodograms of temporal graphs I-noise f=46 Hz; *d* - relative intensity of a period; *T* – periodicity in an hour, c – in a non-obvious form, modulation complex periodicity:periodogram for April-June, 1983

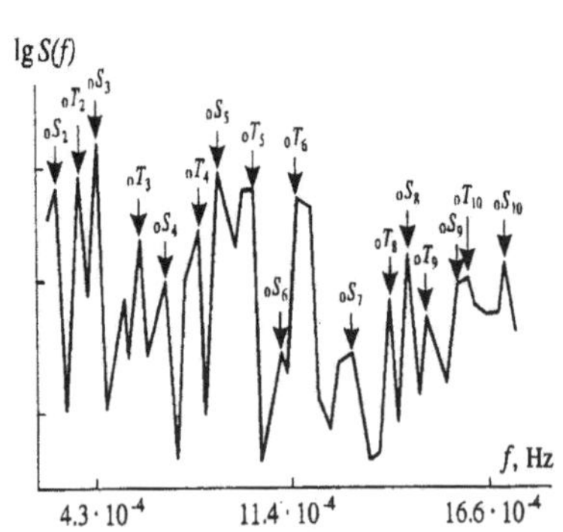

Fig. 4. Seismic noise modulation with a periodicity of he Earth's proper oscillations, in a non-obvious form; energy spectrum of temporal variation of the noise enveloping curve, obtained in 10 hours after the earth-quake in Alaska on 1.03.79, 00h26m30s, M=7,2; Dbninsk, *S(f)* - spectral density

proper oscillations, microseismic storms and wave packets from earthquakes and explosions. The modulation of seismic noises by deformational waves from lunar-solar tides and free oscillations of the Earth has been researched in more detail (Fig. 3, 4)

Soliton Properties of Seismic Wave Fields, Wave Packets etc

Seismic wave fields, wave packets and waves from the deterministic sources (vibrators, hydroelectric power stations, the explosions etc.) show properties, which are also defined by concepts of wave dynamics and theory of open systems: three-frequency interactions, radiative forces, elements of chaos and self-organizing, action of a signal on a noise field. The manifestations of soliton features in nonlinear seismology are of a great variety. Effects are observed both to velocities of solitary waves resulted from underground nuclear explosions (Fig. 5) and to envelopes of wave packets of microseismic field and to distant earthquakes (Fig. 6). Effect depends on signal amplitude.

Effects of chaos and self-organisation have been found and researched in case of harmonic vibrosignals (Fig. 7) and deformational tides waves and seismic actions The self-maintained seismic processes at a zone of seismically active foils attach by their unusual. The foregoing does not reflects completely all features of research of nonlinear seismology.

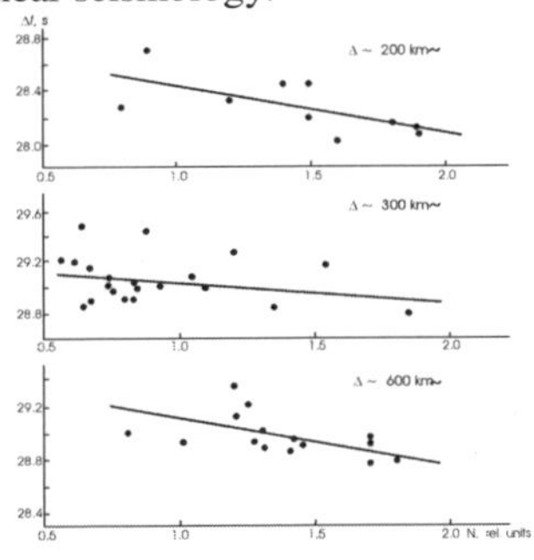

Fig. 5

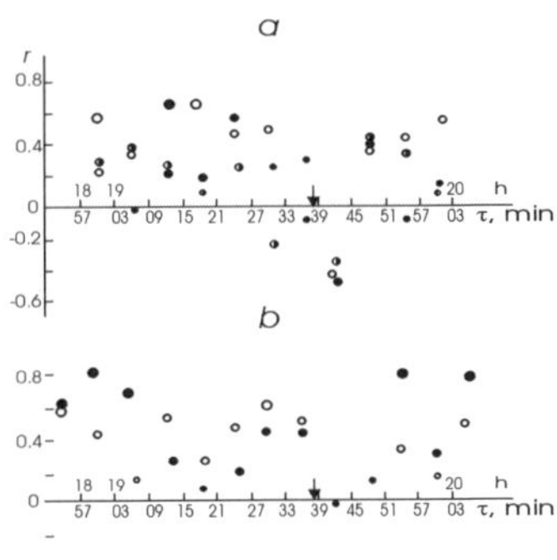

Fig. 6

Fig. 5. Decrease of transit time Δt with increasing explosive power N
Fig. 6. Correlation between width (half-width) of microseism trains of waves and their amplitudes at Hoodat-Dagestan profile before, in moment and after Iran earthquake: a – point 1; b – point 2; distance between points ~ 25 km; r – correlation; τ – local current time; ↓ - moment of first wave event; 1 – data of Z – component, 2 – data of X – component, 3 – data of Y – component; $P < 0.9$; $P \geq 0.99$; $P \geq 0.999$

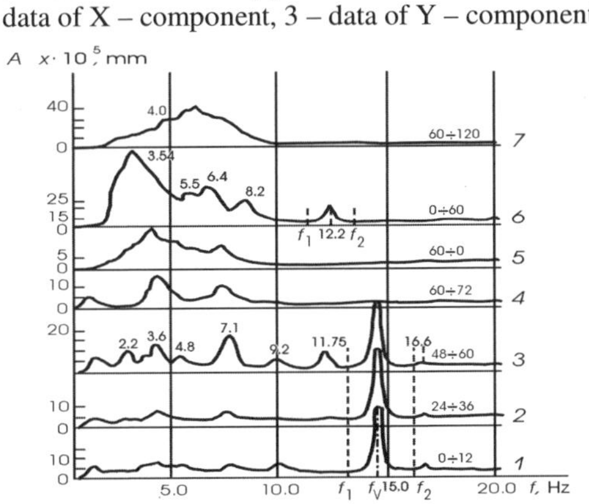

Figure 7. Chaotization of vibrosignal in time for 14 Hz (spectra 1-3) and 12.2 Hz (spectrum 6); curves 4, 5, 7 are spectra of microseism noise; digits above curves are durations of analyzed records in current time (min), the time origin – vibrator switch-on; A – amplitude in relative units; f – frequency

Seismic Wave Fields and Anharmonic Quasi-Steady Radiative Stresses

Preface. Radiative stress is proportional to coefficient of nonlinearity n and quadrate of Mach number (M). The record of the stresses was made using radiation of the vibrator for different experimental conditions. During the intense excitation directly under the plate of vibrator the observed values of quasi-steady radiative stresses (QSRS) (100-200 kPa) correspond to unloading of preliminary loading of the soil. The initiation of quasi-steady deformations of day surface under intense excitation by vibrator is recognised in experiments on measuring a inclination of day surface near working vibrator. This additional inclination disappears at lockout of vibrator. The nonlinear seismic effects are found by Aptikaev according to the Feynman equation describing nonlinear response of absorbing medium. In geological media irrespective of the mechanism of absorption the sufficiently strong seismic waves leads to production of the QSRS.

INTRODUCTION

According to the nonlinear theory of waves propagation radiative stress $\delta_{\text{рад}}$ is proportional to coefficient of nonlinearity *n* and quadrate of Mach number (*M*) and is defined as follows:

$$\sigma_{\text{рад}} = n\rho_0\overline{v}_0^2 = n\rho_0 c_0^2 \mathrm{M}^2,$$

where ρ_0 - medium density, v_0 - amplitude of oscillatory velocity (velocity of displacement of a soil), c_0 - velocity of a wave propagation. It is known for seismic exploration vibrator that under sufficient power $M=10^{-3} - 10^{-4}$ and $n= 10^3 - 10^4$ in nonwave and a near wave zones of vibrational field in soft soil.

The initiation of a quasi-steady component of stress in an epicentral band of strong earthquake (radiative force) was first to consider Nikolaev and Aptikaev.

EXPERIMENTS

The experiments were field conducted in Byelorussia, Yaroslavl and Krasnodar areas. The tensometric system of Yu.I.Vasiliev and M.N.Shcherbo was used. The record of the stresses was made using monochromatic radiation and smooth change of frequency (sweep signal 20 – 100 Hz) at various power levels of the vibrator for different experimental conditions (Fig.8). Quasi-steady radiative stress (QSRS) is determined as displacement of the average line of the vibroseismic signal relative to its position with preliminary stationary loading (Fig.9).

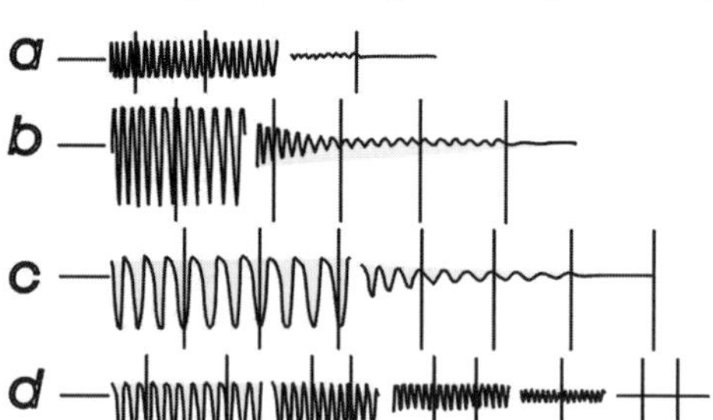

Fig. 8. Recordings of the contact stresses σ*(t)* under the vibrator plate for different operating conditions: a) full range of the recording *A* =

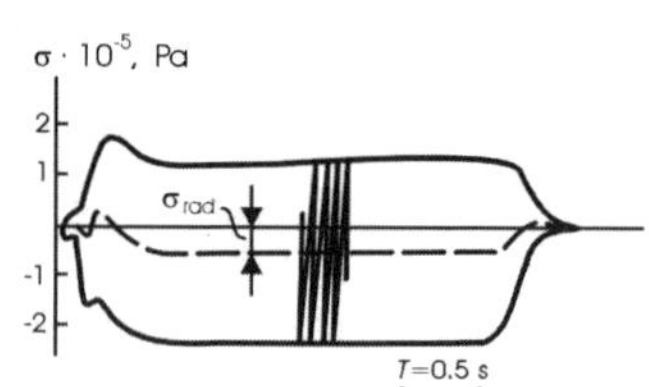

Fig.9. Schematic diagram of a recording of the contact stress $\sigma(t)$ plate for an intense vibrosignal.

60 kPa and frequency f = 60 Hz; b) A = 200 kPa and f = 30 Hz; c) A = 400 kPa and f = 30 Hz; d) sweep signal, A = 320 -80 kPa and f = 30 -70 Hz. For convenience certain parts of the recording have been condensed. The last portion of each trace is the recording after stopping the vibrator. The time markers are 0.01sec.

The drop in the vibrosignal and a short segment of the signal itself is shown. The stress $\sigma_{rad}(t)$ *is* noted, which occurs as a result of the vibration. The dashed line is the average for the oscillation, f = 20 Hz.

During the intense excitation directly under the plate of vibrator the observed values of QSRS (100-200 kPa) correspond to unloading of preliminary loading of the soil. It drops to 10 to 15 kPa at a depth of 50 cm (Fig.10). Noticeable QSRS up to 5 kPa outside the plate are observed to distances up to 2.5 m and practically disappear at the distances of 7 to 10 m (sensitivity limit is 10^3 Pa) (Fig.11).

The experimental quadratic dependence of QSRS on Mach number has been obtained and estimate of nonlinearity coefficient has been made (Fig.12). In sandy soil n is about 10^2 for a longitudinal wave velocity of 200 m/s and a density of 1.8 g/cm^3 [5].

Quasi-static deformations of the Earth's surface induced by vibrator has been measured by strainmeter at distances of (10 – 500) m from the vibrator [6].

The initiation of quasi-steady deformations of day surface under intense excitation by vibrator has been recognised in experiments on measuring a inclination of day surface near working vibrator. This additional inclination disappears at lockout of vibrator [7].

ANALYTICAL TREATMENT

The nonlinear seismic effects are found by Aptikaev according to the equation of the Feynman describing nonlinear response of absorbing medium. In geological media irrespective of the mechanism of absorption the sufficiently strong seismic waves leads to production of the following effects:

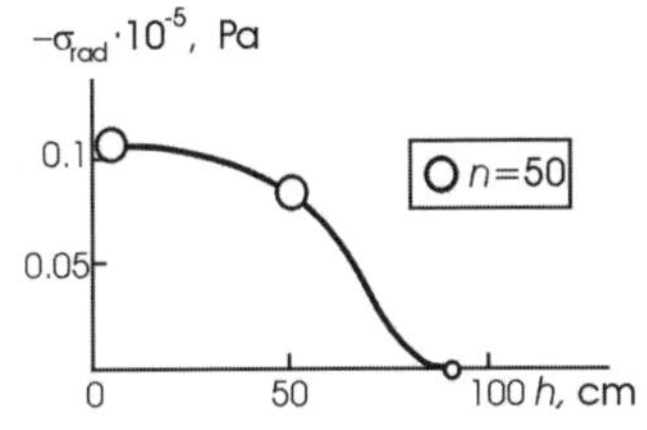

Fig.10. Seismic radiation stress $\sigma_{rad,}$ for the detector under the center of the plate at depths of 5, 50 and 90 cm. The area of the circles is proportional to the number of measurements, n.

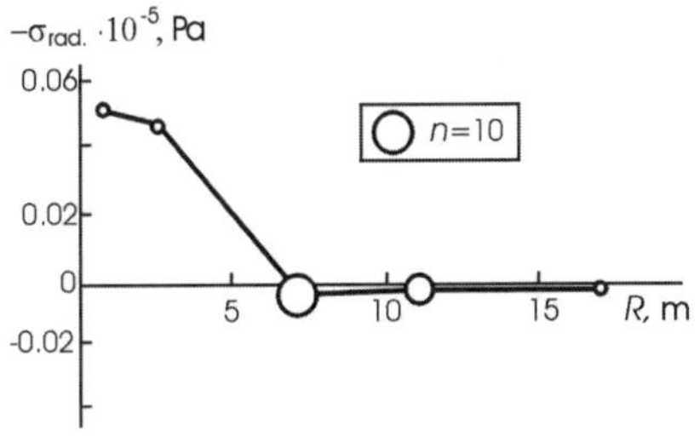

Fig.11. Relationship of the seismic radiation stress to the distance of the detector from the plate. The detectors are near the surface of the Earth, components are radial, f = (25 -35) Hz. The area of the circles is proportional to the number of measurements, n.

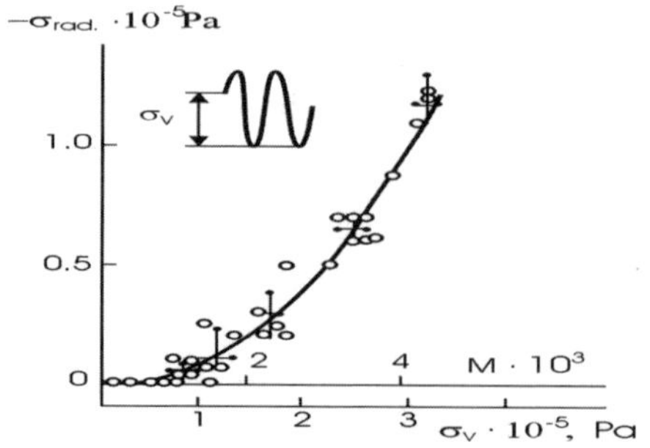

Fig.12. Relationship of the seismic radiation stress to the intensity of the vibroseismic signal. The Mach number is also given on the horizontal axis. The frequency interval is 20 to 40 Hz and Co = 200 m/sec. The results presented are for measurements directly under the plate of the vibrator at a depth of about 5 cm.

(1) steady component, equal to half of quadrate of the peak deformation; or, in other words, proportional to a quadrate of a vibration amplitude and inversely proportional quadrate of velocity of a propagation wave.
(2) second harmonic, which has dependence identical with those of zero harmonic and will give initiation of higher harmonics.
(3) combination frequencies, which level is proportional to product of amplitudes of initial tones and is inversely proportional to a quadrate of velocity of a propagation wave.

CONCLUSIONS

Wide search, analysis and understanding of significance of seismo-radiactive stresses and accompanied processes are necessary for fundamental and applied geophysical investigation.

Astroblems and multiple ring-structure as an object of nonlinear seismology
Preface of applied some direction

The unknown new factor is considered which forming astroblems and craters of the Earth and heavenly bodies. Seismic effect is given by strong nonlinearity of the powerful wave field generating seismic radiative forces in cratering process. Reflection of seismic wave effects in morphological peculiarities of impact structures is shown.

REFERENCES

1. Khavroshkin O.B. Some problems of nonlinear seismology. 1999. United Institute of Physics of the Earth Press., Moscow. P.286
2. Rykunov L.N., Khavroshkin O.B., Tsyplakov V.V. The effect of modulation of high frequency noise of the Earth // Discovery Diploma № 282 Goskomizobreteniy USSR. 1983 Moscow. P.1
3. Rykunov L.N., Khavroshkin O.B., Tsyplakov V.V. The modulation of high frequency microseism // J. Dokl. Science Section. 1978. V.238. Translated from Reports of the Academy of Sciences USSR. V.238, p.303-306
4. Diakonov B.P., Karryev B.S., Khavroshkin O.B., Nikolaev A.V., Rykunov L.N., Seroglasov R.R., Trojanov A.K., Tsyplakov V.V. Manifestation on earth deformation processes by high frequency seismic noise characteristics // Physics of the Earth and Planetary Interior. Amsterdam. 1990. 63. 151-162
5. Vasiliev Yu.I., Vidmont N.A., et al., "Experimental investigation of seismic radiation stress in soft soil" , Izvestya, Earth Physics. **22,** 38-41 (1986). Translated from Fizika Zemli, **1,** 52-56 (1986)
6. Nikolayev A.V., et al., "Quasi-static vibration-induced deformations of the Earth's surface and nonlinear properties of rocks", Physics of the Solid Earth, **30**, 1023-1031 (1995). Translated from Fizika Zemli, **12,** 3-11, (1994)
7. Nikolayev A.V., et al., "Static deformation of the Earth's surface in the vicinity of the harmonic source of seismic oscillations ", Physics of the Solid Earth, **32**, 622-624 (1996). Translated from Fizika Zemli, **7,** 72-74, (1996)

Dynamic principles of prognosis and control

Gennady A. Leonov

Saint-Petersburg State University, Russia

Abstract: The approaches to forecasting and control, based on general regularities of instability demonstration in dynamical systems are described. These approaches, developed in the frame of experimental mathematics, make it possible to avoid the attempts of construction, identification, and analysis of approximate models of highly complicated real dynamical objects. In place of that the efforts are made to collect certain experimental material for real model and apply it to the obtaining of forecasting and the construction of control. It is remarked that the appearance of instabilities is a result of general regularities, the account of which leads to some general principles of qualitative control theory.
Keywords: forecasting, control, the Klausewitz principle, master-slave principle, the Thermidor Law.

1 Introduction

Consider dynamical systems, generated by differential equations

$$\frac{dx}{dt} = f(x) \tag{1}$$

or the difference ones

$$x(t+1) - x(t) = f(x(t)), \quad x \in R^n. \tag{2}$$

For these equations for any constant $s > 0$ we have the following obvious property of solutions

$$x(t+s, s, x_0) = x(t, 0, x_0). \tag{3}$$

This property also holds true in the case of more general descriptions of dynamical systems (from the point of view of their phase spaces or nonlinear operators acting in these spaces).

Chaos Theory: Modeling, Simulation and Applications
C. H. Skiadas, I. Dimotikalis and C. Skiadas (Eds)

Relation (3) implies that the segment of trajectory $x(t, x_0)$, outgoing from the point x_0 at time $t = 0$, coincides with the segment of trajectory, outgoing from the point x_0 at time $t = \tau$,.

It follows that under the same conditions the physical experimental data are repeated and, therefore, we can theoretically forecast some processes and control them.

However by reason of arising the instabilities (which are an object of intensive consideration in recent years [1]), the coincidence, mentioned above, is often possible in sufficiently small time intervals $(0, \tau)$ and $(s, s + \tau)$ only.

We describe certain approaches to the prognosis and control, which are due to the general laws of instabilities arising in dynamical systems. These approaches, developed within the framework of "experimental mathematics", are based on that we do not try to construct, identify, and analyze the approximate models of rather complicated real dynamical objects but collect a certain experimental material connected with real models and then make use of it for the prognosis and the construction of control. The occurrence of instabilities obeys certain general regularities, the account of which results in certain general principles of the qualitative control theory. They are also discussed in the present paper.

2 Informative example. How to make a weather forecast for a week. The reason why it cannot be made for more than two weeks

In 50-60s of the 20th century the progress in the fields of continuum mechanics and computational mathematics makes it possible to produce more accurate mathematical models of atmospheric changes, to construct more effective algorithms for the solution of differential equations of these models, and to realize these algorithms with the help of more high-speed computers. Due to this breakthrough it occurred a widespread opinion that, having made some additional efforts in these directions, we can make the weatherforecast for many weeks, months, and even years.

However it turned out afterwards that a long weatherforecast is impossible in principle.

This fact was established theoretically in the works of E.Lorenz and his progeny, discovered the instability in the mathematical models of atmosphere. The latter means a strong sensitivity of solutions of differential equations, describing atmospheric processes, with respect to the initial data. The understanding of this fact gave rise to observational experiments in the framework of new direction, in "experimental mathematics", which will be considered below.

In Europe there was accumulated the large material of meteorological observations. Such observations were carried out regularly for many decades.

Consider the observations, for example, in May 9, 2004 in a certain region of Europe. We choose then a certain year (for example, 18: such that in May 9, 18: in this region there were observed approximately the same meteorological parameters (temperature, pressure, air moisture, wind strength and direction, cloudiness and so on). The mentioned-above parameters are used as initial (and boundary) conditions for the solutions of differential equations of atmospheric model.

These equations describe the laws of continuum mechanics, which hold always true, in any year. Therefore the equations hold also true and are the same in different years. The solutions of these equations are uniquely determined by initial data.

Since the equations and the initial data in May 9, 2004 and May 9, 18: are identical, we believe that the solutions, describing the change of meteorological data (such as a temperature, pressure, air moisture and so on), are also identical.

Hence for each of days from the chosen time interval, the parameters, which had been observed for a month (from May 9, 2004 to June 9, 2004 and from Nay 9, 18: to June 9, 18:), must coincide with adequate accuracy. It would seem, therefore, that the weather conditions of June 1, 2004, for example, and June 1, 18: must be very close to each other. However the experiments show that such a coincidence is possible only on the time intervals not exceeding two weeks. At the same time the coincidence of weather conditions for a week can be rather close and in the meteorology this fact is most often used for a short-range forecast. But the results of observations in the time intervals exceeding two weeks are highly diverged. Therefore as a rule the weather conditions in June 1, 2004 and June 1, 18: are different.

What's the matter? It turns out that a small divergence of initial data at the initial moments of observations results in the great divergence of observable parameters already after two weeks.

Thus, *even though the mathematical model of atmosphere is sufficiently correct, the computer technique is advanced, and the computer is high-speed the only result we obtain is that the correct weatherforecast for two weeks is impossible.*

For this reason Japanese refused to make a weatherforecast for more than ten days.

Here it should be remarked the following. While in conventional approach as the basic problems we regard the construction of more accuracy mathematical models, the development and realization of numerical algorithms for the solution of differential equations and for identification of parameters of these equations, in the considered approach the main problem is a generation of special databases.

Under the circumstances the principle formulated by famous Prussian general Klausewitz is of great importance.

3 Prognosis of market behavior as the analog of weather forecast

What analogy is it between the changes of weather and market? The answer to this question is as follows.

While the physical laws and the corresponding equations of convection are valid for any time interval, the similar laws of market depend on a policy, financial circumstances, and a drive of market participants. All of them are the same only on the short time interval $[t_0, T]$ (days or hours). It is clear that the change of the variable quantities of market $x_j(t)$ obey the laws of market. The number of such quantities (by using a comparison with a weatherforecost) must be about ten $(j = 1, \ldots, 10)$.

In the case of a market the varying of initial (and boundary) conditions like those, used for the weatherforecast, occurs by virtue of the change of variables on the certain "initial" time interval $[t_0, t_1]/$ (t_1 is much less than T.) This allows us to draw a certain analogue to an "initial function" for differential equations with delay.

Hypothesis. There exist classes of markets such that a "good" coincidence of all observable variable characteristics of market $x_j(t)\,(j = 1, \cdots, N)$ on the intervals $[t_0, t_1]$ and $[t_0+\tau, t_1+\tau](t_1+\tau < T)$ implies their "good" coincidence on the certain intervals $[t_1, t_1+\varepsilon]$ and $[t_1+\tau.t_1+\tau+\varepsilon]$, in which $t_1 + \tau + \varepsilon < T$.

Here the "good" means a certain preliminary "smoothing or averaging" of the quantities $x_j(t)$, what is similar, for example, to how we account a certain "average" velocity of wind, smoothing its counterblasts or weakening on small time intervals.

Thus, we can, apparently, forecast the behavior of certain markets on small time intervals like that of weather, making use of similar parameters (characteristic variables) of the previous observations.

Certainly, this hypothesis has need for the check on concrete multiparametric markets. Besides, in this case we need to choose happily (from the experiments, as before) the time scales (i.e. $t_1, T, \tau, \varepsilon$).

4 The Klausewitz principle

The Klausewitz principle is studied in any Academy of General Staff of any country, which looks to its safety. It is important for us that this principle permits a wide generalization and can be used not only in wartime.

We formulate now the Klausewitz principle for military operations.

Any military operation must be designed as the operation bounded in space and time. The next operation is designed with account for the resulting information on the previous operation.

So, any war, no matter what aims it sets, must be partitioned into separate operations such that they have their own tactical aims and follow in sequence.

The aims of operation and the forces and resources, used for its accomplishment, are corrected with account for the resulting information on the previous operations.

The choice of spatial and time restrictions of operation is an object of military art. They are often chosen as a result of a heavy and murderous preceding experience.

Note that in the considered case there exists the analogy with the well-known fact that the weatherforecast more than for two weeks is impossible!

Like the weatherforecast, in the complicated armed struggle it can also occur the instabilities and, therefore, the previously confirmed plans and models themselves can be the cause of destruction of military operation.

A shining example of the Klausewitz principle application is famous "Stalin's 10 drives" in 1944. Using the previous war experience, the General Staff of Red Army arrives at a conclusion that the optimal time period for operation is 1–2 months and its spatial framework is 200-300 kilometers. For these operations it was formed the corresponding organizations, named fronts (the Leningrad, Karelian, Byelorussian, Ukrainian, and Baltic ones). As a result of combination of war experience and the correct use of the Klausewitz principle all ten operations, followed in sequence, were brightly ended: it was run the blockade of Leningrad and was set free Crimea, the south-west Ukraina (the Korcun'-Shevchenko operation), Byelorussia (the "Bagration" operation), Moldavia (the Yassko-Kishinev operation), and the other regions of USSR.

At the same time the tendency: "to develop success" after the end of designed operation, i.e. to continue mechanically a motion, is often ended in disaster.

A shining example is a cruel rout of Red Army in 1920 about Warsaw and a collapse of German Army about Stalingrad.

These examples bring out clearly that the actions, which seem to be unnatural for arm laymen, such as the stopping of German Army about Dunkirk in 1940 (then English Army has been able to be evacuated over the Channel) and of Red Army about Warsaw in 1944 (at that time it has been begun Warsaw armed insurrection) were made just in accordance with the Klausewitz principle. In both cases the previous operations were finished and it was necessary to design the new operations for Dunkerk and Varshaw be taken and to make the corresponding preparations.

The Klausewitz principle must be taken into account also in the case of the putting in force of any global reforms (in a country, company, and public structure). In these cases it is necessary first to divide neatly the designed transformations into separate parts, to enforce, in series, separate transformations, to obtain results, and to design with their help the next transformations. Then it is necessary to obtain the results of second stage transformations and to account them in the planning of third stage reforms. Only after such careful design, the third stage can start and so on.

What devastating contrast between the reforms process control, following from the Klausewitz principle, and the undigested actions of the administrative authority of SSSR and Russia in 80-90s of the last century!

Certainly, the choice of depth and time of each stage of reforms is an object of administrative and economical art (as the similar parameters for the military operations are an object of military art). Also, it is necessary that the aims, which are set at each stage, were attainable (i.e. the tools and resources were sufficient to accomplish the posed aim).

In addition it is important for us that, keeping the above-mentioned scheme, we does not lose sight of the final global cause. (Similar to a victory is a final global aim of military operations).

5 The Klausewitz principle in the problems with bounded resources

One of modification of the Klausewitz principle is an segregation and sequential solution of priority problems in the case of bounded resources. Recall that the classical Klausewitz principle is a solution of basic problem via sequential solutions of specially segregated subproblems.

Below we give one of the most shining examples of applying the modified Klausewitz principle.

In the result of World War I, Germany suffered a cruel defeat. Its main co-belligerent, Austria-Hungary, was partitioned and ceased his existence as a powerful European State. On Germany it was imposed the enormous contributions and hard restrictions on creating the modern army. In the twenties of the last century Germany was poor and weak State. But all at once, in 1934-1936 years, it builds a powerful Army equipped by advanced armament. In 1936 the army of Germany exceeds in its power the arms of England and France together.

How came it?

After the defeat in World War I, Germany has kept the kernel of General Staff that consists of the well-educated officers with militant experience and with experience of organizing and mobilizing work. They helped to save old design collectives and create the new ones.

These design collectives developed and created test samples of new military technique. The test samples passed the tests only. The commercialization was omitted. Then, using the obtained results and recommendations of General Staff, the designers created, at once, the next generation of arms.

This made possible to develop secretly the best models of arms and, practically at a time in 1934, to commercialize these models, to mobilize the army, and to munition it by this arms within very short time. After that the recollected west allies arrived at the conclusion that the fight with Germany in 1937 was a forlorn hope.

The example considered shows once more the necessity of applying the the above-mentioned dynamical principles of control.

6 The "master-slave" principle

The “master-slave” principle is usual in modern technique. One has a set of similar devices ("slaves"), which are not related together, operate at a time, complying with the signals of one standard device ("master") only. For example, switching a television, you switch together the "sitting" in your television "slave", a horizontal generator, which governs the motion of beam in electron-beam tube. In the broadcasting station there is a master, a high-stable calibration oscillator, which transmits the information on its own frequency, using a television signal. Your television receives this information and a special device, a clock unit, tunes a generator- slave to the frequency of generator-master. The slave is not such high-stable as the master and all the time it is necessary to observe that it does not "sidetrack" its own frequency. Such an observation assumes the existence of feedback: as soon the slave begins to "sidetrack" as the clock unit compares master's and slave's frequencies and "enforces" the slave to operate again with master's frequency.

Thus, the "master-slave" principle assumes the existence of a monitoring and feedback for each of slaves in order to the slave executes its operation function.

An example of dynamical planning and control by the master-slave principle is a conveyor. The master is a high-stable conveyor velocity, the slaves are workers, performing independently of each other similar operations with the velocity, which is "enforced" by the master. The nonperformance of necessary operations with right velocity is detected at once (monitoring) and is replied to correct this nonperformance (feedback). Another shining example of the "master-slave" principle is a pirate team. Without the hardly synchronized fulfillment of captain commands the sailing ship control is impossible. Therefore the absolutely free men, rovers, organize a "master-slaves" system, in which they discard for a while their freedom.

This hard principle of control in the people collective one tries usually to soften by some illusions of "social copartnership" and "corporate responsibility". However, in fact, a master remains a master and a slave is a slave.

7 The principle of continuous successful process

The principle of continuous successful process is also a dynamical control principle for stabilizing a system and preventing the occurrence and development of instability, which can lead to the chaos and collapse of system (from within).

In the mathematical theory of dynamical systems it is well known (and we considered this before) that the fore-runners of the development of unstable processes are the oscillatory or "differently directed" motions. Therefore if in the dynamical system each of its subsystems evolves with positive derivative, then the instabilities, as a rule, are suppressed.

The above-mentioned principle is used, long ago, in a personnel policy of large west companies. In the case under consideration it is transformed in the following rule.

It is not recommended to demote a worker and to cut its salary. Only onward movement! To the point of the last day of the work in the company until he takes a notice about firing.

In other words: it is better to fire than to demote.

In this case the elementary subsystem, a concrete worker, moves upward continuously.

Also, it is well known that the nosedive of any activity ratio can lead to a partial or complete destruction of all system much like the Great American depression in 30s of twentieth century: at the time the stock prices in Wall Street have fallen down suddenly and roughly.

8 The principle of wide latitude

The principle of wide latitude (the principle of federalism) is, in some measure, opposite to the "master-slave" one.

It is clear that such a hard and single-channel control as the master-slave principle does not accept for all cases. A hierarchical federative control system is often more adaptive for providing preassigned requirements.

Here we also have a master (for example, president) but together with the master there are apprentices (for example, governors) to whose the master devolves some his credentials. These credentials must be rather wide since otherwise the system would be close to the operation by the master-slave principle. In this case the work is carried out on the boundaries of stability ranges. Therefore it is necessary to provide for a series of special administrative untidisaster measures in the structure of system.

Firstly, in the considered system the mechanism of a quick and resolute change of apprentice must be stipulated (and be applied automatically) when the apprentice does not deal with the assigned incident and requirements.

Secondly, it must be provided a system of matching, training, and education of apprentices, what is also a stabilizing factor.

Obviously, in the case of the well-developed and correctly operational federative system, all the players obey a certain system of the established and conventional rules. In this case there occur some traditions and a system of values, which all participants of process are oriented on.

The above implies that the frequent and regular removability of master and apprentices is by no means a stabilizing factor, rather it is contrariwise. In each of such alterations there are elements of destabilization. The more is complicated controllable dynamical system, the greater is probability that they can reveal themselves.

9 The Thermidor Law

After each revolution there arrives a dictatorship. This law is the generalization of many historical facts in different countries: England, France, Russia, and many others. How can this phenomenon be expressed in terms of the appearance and suppression of instabilities? During the revolution many restrictions turn out to be lost and many different "degrees of freedom" (in the mechanics this term has precise bearing) appear. A social system becomes more "multidimensional". Such multidimensionality may lead (and, as a rule, it does) to instabilities, which, in turn, lead to the chaotization of community. The community falls in chaos, which can be suppressed but only (unfortunately!) in unique manner: the sharp restriction of "degrees of freedom". The system becomes of "small dimensionality" and, sometimes, even one-dimensional, single-channel (recall the master-slave principle). Such a restriction of freedoms suppresses chaos and the community arrives at the dictatorship, which the many of the community welcomes during a certain (sometimes, very short) time.

References

1.Leonov G. A. Strange Attractors and Classical Stability Theory. Petersburg University Press, 2008.

On a problem of approximation of Markov chains by a solution of a stochastic differential equation

Gabriel V. Orman

Department of Mathematical Analysis and Probability,
"Transilvania" University of Braşov,
500091 Brasov, Romania
Email: ogabriel@unitbv.ro

Abstract: Much scientific works has been done on the applications of the Brownian motion in such diverse areas as molecular and atomic physics, chemical kinetics, solid-state theory, stability of structures, population genetics, communications, and many other branches of the natural and social sciences and engineering. We shall refer below to some aspects concerning the approximation of Markov chains by a solution of a stochastic differential equation to determine the probability of extinction of a genotype. Thus, the Markovian nature of the problem will be pointed out.
Keywords: Brownian motion, stochastic differential equations, Markov chains, transition probabilities, binomial distribution.

2000 MS Classification: 60H10, 60H30, 60J65, 60J20, 60J70

1 Introduction

It is known that a precise definition of the Brownian motion involves a measure on the path space, such that it is possible to put the Brownian motion on a firm mathematical foundation. Much scientific works has been done on its applications in such diverse areas as molecular and atomic physics, chemical kinetics, solid-state theory, stability of structures, population genetics, communications, and many other branches of the natural and social sciences and engineering. In this sense, many contributions have been done by P. Lévy, K. Itô, H.P. McKean, Jr., S. Kakutani, H.J. Kushner,

Chaos Theory: Modeling, Simulation and Applications
C. H. Skiadas, I. Dimotikalis and C. Skiadas (Eds)

A.T. Bharucha-Reid and other. Also some models based on Brownian motion are successfully applied to nucleotide strings analysis.

We shall refer here only to some aspects concerning the approximation of Markov chains by a solution of a stochastic differential equation to determine the probability of extinction of a genotype. Thus, the Markovian nature of the problem will be pointed out again, and we think that this is a very important aspect.

Obviously, the interaction of a population can have a great complexity, which lead to the enhancement of the interdisciplinary coordination in these studies.

When a differential equation is considered if it is allowed for some randomness in some of its coefficients, it will be often obtained a so-called *stochastic differential equation* which is a more realistic mathematical model of the considered situation.

For example, let us consider the simple population growth model (according to [8])

$$\begin{aligned} \frac{dN}{dt} &= a(t)N(t) \\ N(0) &= k \end{aligned} \tag{1}$$

where $N(t)$ is the size of the population at time t, and $a(t)$ is the relative rate of growth at time t. Obviously, it might happen that $a(t)$ is not completely known, but subject to some random environmental effects, so that one gets

$$a(t) = r(t) + \text{"}noise\text{"},$$

where the exact behaviour of the noise term is unknown, but only its probability distribution.

The function $r(t)$ is assumed to be nonrandom, and one put the problem to solve (1) in this case. Such a problem conduct us to the following notion: *the equation obtained by allowing randomness in the coefficients of a differential equation is called a* ***stochastic differential equation.*** Therefore, it is clear that any solution of a stochastic differential equation must involve some randomness. In other words one can hope to be able to say something about the probability distribution of the solutions.

2 Brownian motion

Regarding the Brownian motion we think that it is the most important stochastic process. As a practical tool, it has had profound impact on almost every branch of physical science, as well as several branches of social sciences. As a creation of pure mathematics, it is an entity of uncommon beauty. It reflects a perfection that seems closer to a law of nature than to a human invention.

In 1828 the English botanist Robert Brown observed that pollen grains suspended in water perform a continual swarming motion. The chaotic motion of such a particle is called ***Brownian motion*** and a particle performing such a motion is called a ***Brownian particle***.

The first important applications of Brownian motion were made by L. Bachelier and A. Einstein. L. Bachelier derived (1900) the law governing the position of a single grain performing a 1-dimensional Brownian motion starting at $a \in R^1$ at time $t = 0$

$$P_a[x(t) \in db] = g(t, a, b)db \tag{2}$$

where $(t, a, b) \in (0, +\infty) \times R^2$ and g is the Green (or the *source*) function

$$g(t, a, b) = \frac{1}{\sqrt{2\pi t}}\, e^{-\frac{(b-a)^2}{2t}}$$

of the problem of heat flow

$$\frac{\partial u}{\partial t} = \frac{1}{2}\frac{\partial^2 u}{\partial^2 a}, \qquad (t > 0).$$

A. Einstein (in 1905) also derived (2) from statistical mechanical considerations and applied it to the determination of molecular diameters.

Bachelier also pointed out the Markovian nature of the Brownian path expressed in

$$P_a[a_1 \leq x(t_1) < b_1, a_2 \leq x(t_2) < b_2, \cdots a_n \leq x(t_n) < b_n] =$$
$$\int_{a_1}^{b_1}\int_{a_2}^{b_2}\cdots\int_{a_n}^{b_n} g(t_1, a, \xi_1)\, g(t_2 - t_1, \xi_1, \xi_2) g(t_n - t_{n-1}, \xi_{n-1}, \xi_n)\, d\xi_1 d\xi_2 \cdots d\xi_n$$

where $0 < t_1 < t_2 < \cdots < t_n$, and used it to establish the law of maximum displacement

$$P_a[\max_{s \leq t} x(s) \leq b] = 2\int_0^b \frac{e^{-\frac{a^2}{2t}}}{\sqrt{2\pi t}}\, da, \quad t > 0,\ b \geq 0.$$

But he was unable to obtain a clear picture of the Brownian motion and his ideas were unappreciated at that time. This because a precise definition of the Brownian motion involves a measure on the path space, and it was not until 1908-1909 when the works of É. Borel and H. Lebesgue have been appeared. Beginning with this moment was possible to put the Brownian motion on a firm mathematical foundation and this was achived by N. Wiener in 1923.

A continuous-time stochastic process $\{B_t \,|\, 0 \leq t \leq T\}$ is called a ***standard Brownian motion*** on $[0, T)$ if it has the following four properties: *i.*

$B_0 = 0$; *ii.* The increments of B_t are independent; that is, for any finite set of times $0 \leq t_1 < t_2 < \cdots < t_n < T$, the random variables

$$B_{t_2} - B_{t_1},\, B_{t_3} - B_{t_2},\, \cdots ,B_{t_n} - B_{t_{n-1}}$$

are independent; *iii.* For any $0 \leq s \leq t < T$ the increment $B_t - B_s$ has the normal distribution with mean 0 and variance $t - s$; *iv.* For all ω in a set of probability one, $B_t(\omega)$ is a continuous function of t. The Brownian motion can be represented as a random sum of integrals of orthogonal functions. Such a representation satisfies the theoretician's need to prove the existence of a process with the four defining properties of Brownian motion, but it also serves more concrete demands. Especially, the series representation can be used to derive almost all of the most important analytical properties of Brownian motion. It can also give a powerful numerical method for generating the Brownian motion paths that are required in computer simulation.

3 Stochastic differential equations

To describe the motion of a particle driven by a *white noise* type of force (due to the collision with the smaller molecules of the fluid) the Langevin equation

$$\frac{d\nu(t)}{dt} = -\beta\nu(t) + \mathbf{f}(t) \tag{3}$$

is used, where $\mathbf{f}(t)$ is the white noise term. Its solution is the following

$$y(t) = y_0 e^{-\beta t} + e^{-\beta t} \int_0^t e^{-\beta t} \mathbf{f}(s) ds. \tag{4}$$

If we denote by $\mathbf{w}(t)$ the Brownian motion, then it is given by

$$\mathbf{w}(t) = \frac{1}{q} \int_0^t \mathbf{f}(s) ds, \tag{5}$$

so that $\mathbf{f}(s) = \dfrac{q d\mathbf{w}(s)}{ds}$. But $\mathbf{w}(t)$ is nowhere differentiable, such that $\mathbf{f}(s)$ is not a function. Therefore, the solution (4), of Langevin's equation, is not a well-defined function. This difficulty can be overcome, in the simple case, as follows. Integrating (4) by parts, and using (5), it results

$$y(t) = y_0 e^{-\beta t} + q\mathbf{w}(t) - \beta q \int_0^t e^{-\beta(t-s)} \mathbf{w}(s) ds. \tag{6}$$

But all functions in (6) are well defined and continuous, such that the solution (5) can be interpreted by giving it the meaning of (6). Now, such a procedure can be generalized in the following way. The functions $f(t)$ and $g(t)$ are considered to be defined for $a \leq t \leq b$. For any partition $P : a \leq t_0 < t_1 < \cdots < t_n$, we denote

$$S_P = \sum_{i=1}^{n} f(\xi_i)[g(t_i) - g(t_{i-1})],$$

where $t_{i-1} \leq \xi_i \leq t_i$. If a limit exists

$$\lim_{|P|\to 0} S_P = I$$

where $|P| = \max_{1\leq i\leq n}(t_i - t_{i-1})$, then it is said that I is the ***Stieltjes integral*** of $f(t)$ with respect to $g(t)$. It is denoted

$$I = \int_a^b f(t)dg(t).$$

Now the stochastic differential equation

$$\begin{aligned} &dx(t = a(x(t), t)dt + b(x(t), t)dw(t) \\ &x(0) = x_0 \end{aligned} \tag{7}$$

is defined by the Itô integral equation

$$x(t) = x_0 + \int_0^t a(x(s), s)ds + \int_0^t b(x(s), s)dw(s). \tag{8}$$

The simplest example of a stochastic differential equation is the following equation

$$\begin{aligned} &dx(t) = a(t)dt + b(t)dw(t) \\ &x(0) = x_0 \end{aligned} \tag{9}$$

which has the solution

$$x(t) = x_0 + \int_0^t a(s)ds + \int_0^t b(s)dw(s).$$

The *transition probability density* of $x(t)$ is a function $p(x, s; y, t)$ satisfying the condition

$$P(x(t) \in A \,|\, x(s) = x) = \int_A p(x, s; y, t)dy,$$

for $t > s$, where A is any set in R. It is supposed that $a(t)$ and $b(t)$ are deterministic functions.

4 Markov property

Some aspects regarding the Markov property, in a vision of Kiyosi Itô, are discussed below.

Let S be a *state space.* A Markov process is a system of stochastic processes

$$\{X_t(\omega), t \in T, \omega \in (\Omega, K, P_a)\}_{a \in S}.$$

For each $a \in S$, $\{X_t\}_{t \in S}$ is a stochastic process defined on the probability space (Ω, K, P_a). It is considered that $C = C(S)$ is the space of all continuous functions (it is a separable Banach space with a supremum norm).

The transition probabilities of a Markov process are $\{p(t, a, B)\}$.

Let us denote by $\{H_t\}$ the transition semigroup and let R_α be the resolvent operator of $\{H_t\}$. Now $p(t, a, B)$, H_t and R_α can be expressed in terms of the process as follows. Let f be a function in $C(S)$. Then

1°. $p(t, a, B) = P_a(X_t \in B)$.

2°. For $E_a(\cdot) = \int_\Omega \cdot P_a(d\omega)$ one has $H_t f(a) = E_a(f(X_t))$.

3°. $R_\alpha f(a) = E_a\left(\int_0^\infty e^{-\alpha t} f(X_t) dt\right)$.

One can observe that 1° and 2° are immediately. Now, for 3°, it is considered the following equality:

$$R_\alpha f(a) = \int_0^\infty e^{-\alpha t} H_t f(a) dt = \int_0^\infty e^{-\alpha t} E_a(f(H_t)) dt.$$

But $f(X_t(\omega))$ is right continuous in t for ω fixed and measurable in ω for t fixed and therefore it is measurable in the pair (t, ω). Thus, the Fubini's theorem can be used and one gets

$$R_\alpha f(a) = E_a\left(\int_0^\infty e^{-\alpha t} f(X_t) dt\right),$$

that is just 3°.

Notation. For $\mathcal{C}$ a σ-field on Ω, the space of all bounded $\mathcal{C}$-measurable functions will be denoted by $\mathbf{B}(\Omega, \mathcal{C})$, or simple $\mathbf{B}(\mathcal{C})$.

Now the *Markov property* is given in the following theorem. Let be given $\Gamma \in K$. The following is true

$$P_a(\theta_t \omega \in \Gamma | K_t) = P_{X_t(\omega)}(\Gamma) \quad \text{a.s.}(P_a);$$

that is to say

$$P_a(\theta_t^{-1} \Gamma | K_t) = P_{X_t(\omega)}(\Gamma).$$

Note 1. The following notation can be used

$$P_{X_t(\omega)}(\Gamma) = P_b(\Gamma)|_{b = X_t(\omega)}.$$

It will be suffice to show that

$$P_a(\theta_t^{-1}\Gamma \cap D) = E_a(P_{X_t}(\Gamma), D) \tag{10}$$

for $\Gamma \in K$ and $D \in K_t$.

One can distinguish three cases.

[**I.**] Let us consider Γ and D as follows:

$$\Gamma = \{X_{s_1} \in B_1\} \bigcap \{X_{s_2} \in B_2\} \bigcap \cdots \bigcap \{X_{s_n} \in B_n\},$$

and

$$D = \{X_{t_1} \in A_1\} \bigcap \{X_{t_2} \in A_2\} \bigcap \cdots \bigcap \{X_{t_m \in A_m}\}$$

with $0 \leq s_1 < s_2 < \cdots < s_n$, $0 \leq t_1 < t_2 < \cdots < t_m \leq t$ and $B_i, A_j \in K(S)$.

Now it will be observed that the both sides in (10) are expressed as integrals on S^{m+n} in terms of transition probabilities. Thus, one can see that they are equal.

[**II.**] Let now be Γ as in the case [**I.**] and let us denote by D a general member of K_t. For Γ fixed the family $\mathcal{D}$ of all D's satisfying (10) is a Dynkin class. If $\mathcal{M}$ is the family of all M's in the case [**I.**] then, this family is multiplicative and $\mathcal{M} \subset \mathcal{D}$. In this way it follows

$$\mathcal{D}(\mathcal{M}) \subset \mathcal{D} = K(\mathcal{M}) = K_t$$

and one can conclude that, for Γ in the case [**I.**] and for D general in K_t, the equality (10) holds.

[**III.**] (**General case.**) This case can be obtained in a same manner from [**II.**] by fixing an arbitrary $D \in K_t$.

It will be obtained that $P_a(\Gamma)$ is Borel measurable in a for any $\Gamma \in K$.

$$E_a(G \circ \theta_t, D) = E_a(E_{X_t}(G), D) \quad \text{for } G \in \mathbf{B}(K), D \in K_t,$$
$$E_a(F \cdot (G \circ \theta_t)) = Ea(F \cdot E_{X_t}(G)) \quad \text{for } G \in \mathbf{B}(K), F \in \mathbf{B}(K_t),$$
$$E_a(G \circ \theta_t | K_t) = E_{X_t}(G) \quad (a.s.)(P_a) \text{ for } G \in \mathbf{B}(K).$$

[For more details see [4], [5], [6], [1]].

5 The probability of extinction of a genotype

We now consider an application to genetics in the sense in which it was discussed by W. Feller. Results due to Z. Schuss will be considered too.

This problem was firstly discussed in detail by S. Wright and R. A. Fisher. Its Markovian nature was pointed out by G. Malécot in: *Sur un problèm de proba-*

As it is known the heritable characters depend on special carries, called *genes,* which appear in pairs. Each gene of a particular pair can suggest two forms A and a which determine a genetic type in a population. Therefore, three different pairs can be formed AA, Aa, aa, such that the organism belongs to one of these three *genotypes*. On the other hand, the reproductive celles, called *gametes*, are assumed to have one gene. In this way, the gametes of an organism of genotype AA or aa have the gene A or respective the gene a, whereas the gametes of an organism of genotype Aa may have the gene A or the gene a with equal probability. We can view such a problem in the context of the *binomial distribution*

$$P_n(k) = C_n^k p^k q^{n-k}.$$

We can imagine an urn with $2N$ elements (that is the genes of types A and a). Then, the genotype structure of N offsprings will be the result of $2N$ independent drawings from the urn. Furthermore, it is considered that a population consists of N individuals in each generation. Now, if i of the genes are of the type A $(0 \leq i \leq 2N)$ in some generation, then it is said that the generation is in the state E_i. In this way a Markov chain is connected to such a genetic process. It has $2N+1$ states $E_0, E_1, E_2, \cdots, E_{2N}$. Calculating the transition probability from the state E_i to E_j, in one generation, one gets

$$P_{ij}^N = C_{2N}^j \left(\frac{i}{2N}\right)^j \left(1 - \frac{i}{2N}\right)^{2N-j}. \tag{11}$$

Now, one considers a population of N individuals consisting of $X_N(n) = i$ individuals of type A in the nth generation. Then, the next generation consists of N individuals randomly selected from a practically infinite offspring of the previous generation. Obviously the selection process is binomial with probability $x = \frac{i}{N}$ for type A, the proportion $\frac{i}{N}$ of A types being equal to the probability in a large offspring population. Therefore, the transition probability is given by the following equality

$$P_{ij}^N = P(X_N(n+1) = j \,|\, X_N(n) = i) = C_N^j\, x^j (1-x)^{N-j}.$$

Let now denote by s the *fitness* of A relative to a when selection forces act on the population

$$s = \frac{\overline{x} - x}{x(1-\overline{x})}$$

where

$$\overline{x} = \frac{x(1+s)}{1+sx}.$$

---

bilités en chaine que pose la génétique, Comptes Rendus de l'Académie des Sciences, vol.219, 1944, pp.379-381. It is also presented at lenght in [10].

Therefore, one obtains

$$P_{ij}^{N} = C_N^j \, \overline{x}^j (1-\overline{x})^{N-j}.$$

Now, if $s = s_N(n)$ is a random variable then, the *probability of extinction* of a genotype, or the *time until extinction*, or the *total A population*, or other characteristics of interest, become very hard to calculate. For this reason the Markov chains $\{X_N(n)\}$ can be approximated by a diffusion process, or more exactly, by a solution of a stochastic differential equation.

To this end the following form is considered for the process

$$X_N([Nt]) = N\, x_N(t)$$

where $[Nt]$ is the greatest integer not exceeding Nt and t is any positive number. Thus, $x_N(t)$ represents the *proportion* of A types in population. Now it is supposed that

$$N\, E s_N(n) \to \sigma(t)$$

as $N \to \infty$, $n = [Nt]$, and $NEs_N^2(n) \to \nu(t)$ and $NEs_N^k(n) \to 0$ for all $k > 2$.

Let now be $\Delta t = \frac{1}{N}$. Then, one gets

$$\begin{aligned} a_N &\equiv \frac{1}{\Delta t} E_{x,t}\left[x_N(t+\Delta t) - x_N(t)\right] = \\ &= E[X_N([Nt]+1) - X_N([Nt]) \,|X_N([Nt]) = x_N]. \end{aligned}$$

But $X_N(n)$ is a binomial variable $B(N, \overline{x})$ such that it results

$$\begin{aligned} a_N = NE(\overline{x} - x) = NE\left(\frac{(1+s_N(n))x}{1+s_N(n)x} - x\right) \to \\ [\sigma(t) - \nu(t)x]x(1-x) \equiv a(x,t) \end{aligned} \tag{12}$$

and respective

$$\tfrac{1}{\Delta t} E_{x,t}[x_N(t+\Delta t) - x_N(t)]^2 \to x(1-x)[1+\nu(t)x(1-x)] \equiv b(x,t) \tag{13}$$

The moments of higher order tend to zero as $N \to \infty$. Now it can be shown that the convergence is sufficiently rapid as to satisfy the imposed conditions. Let now consider again the specified model. Thus, we have

$$E[x_N(t+\Delta t) - x_N(t) \,|\, x_N(t)] = 0$$

and

$$\begin{aligned} E[|x_N(t+\Delta t) - x_N(t)|^2 \,|\, x_N(t)] &= \frac{x_N(t)[1-x_N(t)]}{N} \\ E[|x_N(t+\Delta t) - x_N(t)\,|^4\, x_N(t)] &\leq \frac{K}{N^2} \end{aligned}$$

for a constant K.

Now, it can be seen that the conditions for convergence hold, provided that the stochastic differential equation

$$dx(t) = a(x,t)dt + \sqrt{b(x,t)}\, dw(t)$$
$$x(0) = x \tag{14}$$

has a unique solution (with absorption at $x = 0$ and $x = 1$). But the conditions of the existence and uniqueness theorem are not satisfied by the coefficient $\sqrt{b(x,t)}$ in (14). To show the existence and uniqueness it is necessary to consider (14) in the interval $I_\varepsilon[\varepsilon, 1-\varepsilon]$ with absorption at the boundary δI of I_ε. The conditions of the existence and uniqueness theorem are satisfied in I_ε. Thus, it is a unique solution $x_\varepsilon(t)$ in I_ε up to the time

$$\tau_\varepsilon = \inf\{t \,|\, x_\varepsilon(t) \in \delta I_\varepsilon\}.$$

If $\varepsilon_1 < \varepsilon_2$ then, $\tau_{\varepsilon_1} \geq \tau_{\varepsilon_2}$ and

$$x_{\varepsilon_1}(t) = x_{\varepsilon_2}(t) \quad \text{where} \quad 0 \leq t \leq \tau_{\varepsilon_2},$$

by the so-called the *localization principle.* Now, as $\varepsilon \to 0$, one gets that $x_{\varepsilon(t)}$ converges to a limit $x(t)$ and $\tau_\varepsilon \to \tau$ (here τ is the absorption of $x(t)$).

In this way it results that $x_N(t) \to x(t)$, with $x(t)$ the solution of the stochastic differential equation

$$dx(t) = a(x,t)dt + \sqrt{b(x,t)}\, dw(t)$$

with absorbing barriers at $x = 0$ and $x = 1$.

Thus, once a genotype is extinct, it will stay extinct for all future generations unless mutation occurs. Therefore, the pobability of extinction is the probability of exit of $x(t)$ from the interval (0, 1).

Conclusion. Obviously, verious situation may exist when the survival of a particular genotype can be very dynamic.

The *Localization principle.* Let us consider $a_1(x,t) = a_2(x,t)$ and $b_1(x,t) = b_2(x,t)$ for $c \leq x \leq d$ and $t \geq 0$. Suppose that $a_i(x,t)$ and $b_i(x,t)$, with $i = 1, 2$, satisfy the conditions of the existence and uniqueness theorem in R for the stochastic differential equations

$$dx_i(t) = a_i(x,t)dt + b_i(x,t)dw(t), \qquad i = 1, 2$$
$$x(0) = x_0 \in [c, d].$$

Let us denote $\tau_i = \inf\{t \,|\, x_i(t) \notin (c,d)\}$ with $i = 1, 2$. Then, $\tau_1 = \tau_2\, a.s.$ and $x_1(t) = x_2(t)$ for all $t \leq \tau_1\, a.s.$

But, as we have already emphasized, for a random variable $f = f_N(n)$ the *probability of extinction* of a genotype, or the *time until extinction*, or the *total A population*, or other characteristics of interest, become very hard to calculate. And this is the reason for which a Markov chain $\{X_N(n)\}$ is useful to be approximated by a solution of a stochastic differential equation.

References

1.A.T. Bharucha-Reid. Elements Of The Theory Of Markov Processes And Their Applications. Dover Publications, Inc., Mineola, New York, 1997.
2.W. Feller. An Introduction to Probability Theory and its Applications, vol.I, Wiley & Sons, Inc., New York, London, 1960.
3.I.I. Gihman and A.V. Skorohod. Stochastic Differential Equations, Springer-Verlag, Berlin, 1972.
4.K. Itô. Selected Papers, Springer-Verlag, Berlin, 1987.
5.K. Itô and H.P. McKean Jr. Diffusion Processes and Sample Paths, Springer-Verlag, Berlin Heidelberg, 1996.
6.K. Itô. Stochastic Processes, Eds. Ole E. Barndorff-Nielsen, Ken-iti Sato, Springer, 2004.
7.H.J. Kushner and G.G. Yin. Stochastic Approximation Algorithms and Applications, Springer-Verlag New York, Inc., 1997.
8.B. ksendal. Stochastic Differential Equations: An Introduction with Applications, Fifth Edition, Springer-Verlag, 1998.
9.G.V. Orman. Some general aspects of the Markov dependence in the process of generation of the words, Preprint & Communic., 20th *Conference on Stochastic Processes and their Application*, Nahariya, 1991.
10.G.V. Orman. Capitole de matematici aplicate, Ed. Albastră, Cluj-Napoca, 1999.
11.G.V. Orman. Lectures on *Some Problems Of Stochastic Approximation Algorithms and Applications*, Preprint, Gerhard Mercator University, Duisburg, 2001.
12.G.V. Orman. Handbook Of Limit Theorems and Stochastic Approximation, Transilvania University Press, Brasov, 2003.
13.G.V. Orman. On Markov Processes: A Survey Of The Transition Probabilities And Markov Property. In: *Chaotic Systems: Theory and Applications*, Eds. Christos H. Skiadas and Ioannis Dimotikalis, World Scientific Publishing Co Pte Ltd., 2010.
14.Z. Schuss. Theory and Applications of Stochastic Differential Equations, John Wiley & Sons, New York, 1980.
15.D.W. Stroock. Markov Processes from K. Itô Perspective, Princeton Univ. Press, Princeton, 2003.
16.M. T. Wasan. Stochastic Approximation. Cambridge University Press, 1969.

Scattering by many small inhomogeneities and applications

A. G. Ramm

Mathematics Department, Kansas State University,
Manhattan, KS 66506-2602, USA
Email: ramm@math.ksu.edu

Abstract: Many-body quantum-mechanical scattering problem is solved asymptotically when the size of the scatterers (inhomogeneities) tends to zero and their number tends to infinity.
A method is given for calculation of the number of small inhomogeneities per unit volume and their intensities such that embedding of these inhomogeneities in a bounded region results in creating a new system, described by a desired potential. The governing equation for this system is a non-relativistic Schrödinger's equation described by a desired potential.
Similar ideas were developed by the author for acoustic and electromagnetic (EM) wave scattering problems.
Keywords: wave scattering by small inhomogeneities, metamaterials, nanotechnology, refraction coefficient, negative refraction.

PACS 43.20.+g, 62.40.+d, 78.20.-e.
MSC 35J10; 45F05; 74J25, 81U10; 81U40; 82C22

1 Introduction

In a series of papers, cited in the references, and in the monograph [8] the author has developed wave scattering theory by many small bodies of arbitrary shapes and, on this basis, proposed a method for creating materials with some desired properties, in particular, with a desired refraction coefficient. The goal of this paper is to describe a new method for creating materials with a desired refraction coefficient and to compare it with the method proposed earlier. The new method is applicable to the problems which deal with vector fields and tensorial refraction coefficients. Our presentation deals with the simpler case of scalar fields and uses paper [27].

Chaos Theory: Modeling, Simulation and Applications
C. H. Skiadas, I. Dimotikalis and C. Skiadas (Eds)

Let us formulate the statement of the problem. Let $D \subset \mathbb{R}^3$ be a bounded domain filled in by a material with a known refraction coefficient. The scalar wave scattering problem consists of finding the solution to the Helmholtz equation:

$$L_0 u := [\nabla^2 + k^2 n_0^2(x)]u = 0 \quad \text{in} \quad \mathbb{R}^3, \quad \operatorname{Im} n_0^2(x) \geq 0, \tag{1}$$

$$u = u_0 + v, \quad u_0 := e^{ik\alpha\cdot x}, \tag{2}$$

$$v = A_0(\beta,\alpha)\frac{e^{ikr}}{r} + o(\frac{1}{r}), \quad r := |x| \to \infty, \quad \beta := \frac{x}{r}. \tag{3}$$

The function $n_0^2(x)$ is assumed bounded, Riemann-integrable,

$$n_0^2(x) = 1 \quad \text{in} \quad D' := \mathbb{R}^3 \backslash D, \quad \operatorname{Im} n_0^2(x) \geq 0. \tag{4}$$

The function $A(\beta,\alpha)$ is called the scattering amplitude. The wavenumber $k = \frac{2\pi}{\lambda}$, where λ is the wavelength in D', $\alpha \in S^2$ is the direction of the incident plane wave u_0, S^2 is the unit sphere in $\mathbb{R}^3$, $\beta \in S^2$ is the direction of the scattered wave. The solution to problem (1)-(3) is called the scattering solution. It is well-known ([5]) that this solution exists and is unique under our assumptions. These assumptions are: $n_0^2(x)$ is bounded: $\sup_{x\in\mathbb{R}^3} |n_0^2(x)| \leq n_0 = const$, $\operatorname{Im} n_0^2(x) \geq 0$, $n_0^2(x) = 1$ in D'. Since the solution u is unique, the corresponding scattering amplitude $A_0(\beta,\alpha)$ is determined uniquely if $n_0^2(x)$ is given. We assume $k > 0$ fixed and do not show the dependence of $A_0(\beta,\alpha)$ on k. The operator L_0 at a fixed $k > 0$ can be considered as a Schrödinger operator $L_0 = \nabla^2 + k^2 - q_0(x)$, where $q_0(x) := k^2 - k^2 n_0^2(x)$.

Problem RC:

We want to construct a material in D with a desired refraction coefficient, that is, with a desired function $n^2(x)$, $\operatorname{Im} n^2(x) \geq 0$.

Here RC stands for refraction coefficient.

Why is this problem of practical interest?

We give two reasons.

First, creating a refraction coefficient such that the corresponding material has *negative refraction* is of practical interest.

One says that a material has negative refraction if the group velocity in this material is directed opposite to the phase velocity.

Secondly, It is of practical interest to create a refraction coefficient such that the corresponding scattering amplitude $A(\beta) = A(\beta,\alpha)$ at a fixed $\alpha \in S^2$ (and a fixed $k > 0$) approximates an arbitrary given function $f(\beta) \in L^2(S^2)$ with any desired accuracy $\epsilon > 0$.

This problem we call *the problem of creating material with a desired wave-focusing property.*

In Section 2 we explain how to create a material with a desired refraction coefficient. In Section 3 we compare our recipe for creating material with a desired refraction coefficient with the recipe given in [24].

2 Creating material with a desired refraction coefficient by embedding small inhomogeneities into a given material

The basic idea of our method is to embed in D many small inhomogeneities $q_m(x)$ in such a way that the resulting medium can be described by a desired potential $q(x)$.

Let us assume that $p_M(x)$ is a real-valued compactly supported bounded function, which is a sum of small inhomogeneities:

$$p_M = \sum_{m=1}^{M} q_m(x),$$

where $q_m(x)$ vanishes outside the ball $B_m := \{x : |x - x_m| < a\}$ and $q_m = A_m$ inside B_m, $1 \leq m \leq M$, $M = M(a)$.

The question is:

Under what conditions the field u_M, which solves the Schrödinger equation with the potential $p_M(x)$, has a limit $u_e(x)$ as $a \to 0$, and this limit $u_e(x)$ solves the Schrödinger equation with a desired potential $q(x)$?

We give a complete answer to this question in Theorem 1 below.

The class of potentials q, that can be obtained by our method, consists of bounded, compactly supported, Riemann-integrable functions. It is known that the set of Riemann-integrable functions is precisely the set of almost everywhere continuous functions, that is, the set of bounded functions with the set of discontinuities of Lebesgue measure zero in $\mathbb{R}^3$. These assumptions on q are not repeated but are always valid when we write "an arbitrary potential".

In fact, a more general set of potentials can be constructed by our method. It is mentioned below that for some class of unbounded potentials, having local singularities, which are absolutely integrable, our theory remains valid.

Our result is as follows:

Assume that $q(x)$ is an arbitrary Riemann-integrable in D potential, vanishing outside D, where D is an arbitrary large but finite domain, and the functions $A(x)$ and $N(x)$, which we can choose as we wish, are chosen so that $A(x_m) = A_m$ and $A(x)N(x) = q(x)$, where $N(x) \geq 0$.

Then the limit $u_e(x)$ of $u_M(x)$ as $a \to 0$ does exist, and solves the scattering problem with the desired refraction coefficient $n^2(x)$:

$$\nabla^2 u_e + k^2 n^2(x) u_e := \nabla^2 u_e + k^2 u_e - q(x) u_e = 0, \tag{5}$$

$$u = u_0 + v, \tag{6}$$

where u_0 solves problem (5) and v satisfies the radiation condition.

The notation $u_e(x)$ stands for the effective field, which is the limiting field in the medium as $M \to \infty$, or, equivalently, $a \to 0$. Under our assumptions (see Lemma 1 below) one has $M = O(\frac{1}{a^3})$.

The field u_M is the unique solution to the integral equation:

$$u_M(x) = u_0(x) - \sum_{m=1}^{M} \int_D g(x,y,k) q_m(y) u_M(y) dy, \quad g(x,y,k) = \frac{e^{ik|x-y|}}{4\pi|x-y|}, \tag{7}$$

where $u_0(x)$ is the incident field, which one may take as the plane wave, for example, $u_0 = e^{ik\alpha\cdot x}$, where $\alpha \in S^2$ is the direction of the propagation of the incident wave.

We assume that the scatterers are small in the sense $ka << 1$. Parameter $k > 0$ is assumed fixed, so the limits below are designated as limits $a \to 0$, and condition $ka << 1$ is valid as $a \to 0$.

If $ka << 1$, then the following transformation of (7) is valid:

$$u_M(x) = u_0(x) - \sum_{m=1}^{M} \frac{e^{ik|x-x_m|}}{4\pi} A_m u_M(x_m) \int_{|y-x_m|<a} \frac{dy}{|x-y|} [1+o(1)]. \tag{8}$$

To get (8) we have used the folowing estimates:

$$|x - x_m| - a \le |x-y| \le |x-x_m| + a, \qquad |y - x_m| \le a.$$

These estimates imply

$$e^{ik|x-y|} = e^{ik|x-x_m|}[1 + o(1)],$$

provided that $|y - x_m| \le a$ and $a \to 0$.

We also have taken into account that, as $a \to 0$, one has:

$$\max_{x \in B_m} |u_M(x) - u_M(x_m)| = o(1).$$

A proof of this statement is given in [12].

Here is a sketch of another proof of the above statement. Equation (7) has a unique solution because it is an equation with a compactly supported bounded uniformly with respect to M potential p. Its solution satisfies a homogeneous Schrödinger equation and the radiation condition. Thus, this solution is unique. This solution is uniformly (with respect to M) continuous in D.

Yet another argument, different from the ones, given above, can be outlined as follows. The limiting funtion $u_e(x)$ is in $H^2(D)$ by the standard elliptic regularity results, so it is continuous in D. The function u_M converges to u_e uniformly. Therefore, it satisfies the above inequality as $M \to \infty$, or, equivalently, as $a \to 0$.

The fourth proof of the same statement can be based on a result from [26]. Namely, a justification of a collocation method for solving equation (17) (see below) for the limiting field u_e is given in [26]. From the arguments, given there, one obtains the uniform convergence of u_M to u_e:

$$\lim_{M\to\infty} \max_{x\in D} |u_M(x) - u_e(x)| = 0.$$

Let us return to equation (8).

We want to prove that the sum in (8) has a limit as $a \to 0$, and to calculate this limit assuming that the distribution of small inhomogeneities or, equivalently, the points x_m, is given by the following formula:

$$\mathcal{N}(\Delta) = |V(a)|^{-1} \int_\Delta N(x)dx[1 + o(1)] \qquad a \to 0, \tag{9}$$

where $V(a) = 4\pi a^3/3$, $N(x) \geq 0$ is an arbitrary given function, continuous in the closure of D, $0 \leq N(x) < \mathbf{p} < \mathbf{1}$, $\mathbf{p}$ is a constant arising in the problem of spheres packing, and $\mathcal{N}(\Delta)$ is the number of small inhomogeneities in an arbitrary open subset $\Delta \subset D$.

The maximal value of the constant constant $\mathbf{p}$ is the maximal ratio of the total volume of the packed spheres divided by $|\Delta|$, where $|\Delta|$ is the volume of Δ. The total volume of the balls, embedded in the domain Δ, is equal to

$$V_\Delta := V(a)\mathcal{N}(\Delta) = \int_\Delta N(x)dx[1 + o(1)].$$

If diam(Δ) is sufficiently small and $x \in \Delta$ is some point in Δ, then

$$V_\Delta = N(x)|\Delta|[1 + o(1)] < \mathbf{p}|\mathbf{\Delta}|.$$

Since the domain Δ is arbitrary, one lets $a \to 0$ and concludes that

$$N(x) \leq \mathbf{p} < \mathbf{1}.$$

It is conjectured that maximal value of $\mathbf{p}$ *is equal approximately* 0.74. (see, e.g., [32]).

There is a large literature on optimal packing of spheres (see, e.g., [3], [32]). For us the maximal value of $\mathbf{p}$ is not important. What is important is the following conclusion:

One can choose $N(x) \geq 0$ as small as one wishes, and still create any desired potential $q(x)$ by choosing suitable $A(x) > 0$.

The total number $M = \mathcal{N}(D)$ of the embedded small inhomogeneities is $M = O(\frac{1}{a^3})$, according to (9).

Our basic new tool is the following lemma.

Lemma 1. *If the points x_m are distributed in a bounded domain $D \subset \mathbb{R}^3$ according to (9), and $f(x)$ is an arbitrary Riemann-integrable in D function, then the following limit exists:*

$$\lim_{a\to 0} \sum_{m=1}^{M} f(x_m)V(a) = \int_D f(x)N(x)dx. \tag{10}$$

The class of Riemann-integrable functions is precisely the class of bounded almost everywhere continuous functions, i.e., bounded functions with the set of discontinuities of Lebesgue measure zero in $\mathbb{R}^3$. The class of such functions is precisely the class of functions for which the Riemannian sums converge to the integrals of the functions.

The result in (10) we generalize to the class of functions for which the integral in the right-hand side of (10) exists as an improper integral. *In this case $f(x)$ may be unbounded at some points y*, but the limit $\lim_{\delta\to 0}\int_{D_\delta} f(x)N(x)dx$ exists and

$$\lim_{\delta\to 0} \int_{D_\delta} f(x)N(x)dx := \int_D f(x)N(x)dx,$$

where $D_\delta := D \setminus B(y,\delta)$, and $B(y,\delta)$ is the ball centered at $y \in D$ and of radius δ. In this case the sum in (10) is defined as follows:

$$\lim_{a\to 0} \sum_{m=1}^{M} f(x_m)V(a) := \lim_{\delta\to 0}\lim_{a\to 0} \sum_{x_m\in D_\delta} f(x_m)V(a). \tag{11}$$

The same definition is valid for the conclusion of Theorem 1, which is our result.

Theorem 1. *If the small inhomogeneities are distributed so that (9) holds, and $q_m(x) = 0$ if $x \notin B_m$, $q_m(x) = A_m$ if $x \in B_m$, where $B_m = \{x : |x - x_m| < a\}$, $A_m := A(x_m)$, and $A(x)$ is a given in D function such that the function $q(x) := A(x)N(x)$ is Riemann-integrable, then the limit*

$$\lim_{a\to 0} u_M(x) = u_e(x) \tag{12}$$

does exist and solves problem (5)-(6) *with the potential*

$$q(x) = A(x)N(x). \tag{13}$$

There is a large literature on wave scattering by small inhomogeneities. A recent paper, which we use, is [27]. Some of the ideas of this approach were earlier applied by the author to scattering by small particles embedded in an inhomogeneous medium (see [3]-[23]).

We apply Lemma 1 to the sum in (8), in which we choose $A_m := A(x_m)$, where $A(x)$ is an arbitrary continuous in D function which we may choose as we wish. A simple calculation yields the following formulas:

$$\int_{|y-x_m|<a} |x-y|^{-1}dy = V(a)|x-x_m|^{-1}, \qquad |x-x_m| \geq a, \tag{14}$$

and

$$\int_{|y-x_m|<a} |x-y|^{-1}dy = 2\pi(a^2 - \frac{|x-x_m|^2}{3}), \qquad |x-x_m| \leq a. \tag{15}$$

Therefore, the sum in (8) is of the form (10) with

$$f(x_m) = \frac{e^{ik|x-x_m|}}{4\pi|x-x_m|}A(x_m)u_M(x_m)[1+o(1)]. \tag{16}$$

Applying Lemma 1, one concludes that the limit $u_e(x)$ in (12) does exist and solves the integral equation

$$u_e(x) = u_0(x) - \int_D \frac{e^{ik|x-y|}}{4\pi|x-y|}q(y)u_e(y)dy, \tag{17}$$

where $q(x)$ is defined by formula (13).

Applying the operator $\nabla^2 + k^2$ to (17), one verifies that the function $u_e(x)$ solves problem (5)-(6).

Thus, the conclusion of Theorem 1 is established.

In (16) $f(x_m)$ depends on a through $u_M(x_m)$. Nevertheless Lemma 1 is applicable because $u_M(x)$ converges to the solution $u_e(x)$ of eqution (17) as $a \to 0$. This follows from the results in [26] and [12]. These results give two independent ways to prove convergence of u_M to u_e as $a \to 0$. The results in [12] yield compactness of the set $\{u_M(x)\}$, while the results in [26] convergence of a version of the collocation method for solving equation (17) is proved. This version of the collocation method leads to solving the following linear algebraic system

$$u_j = u_{0j} - \sum_{m=1, m\neq j}^{n} g_{jm}q_m u_m V(a), \quad 1 \leq j \leq m, \tag{18}$$

where $u_j := u(x_j)$, $q_m := q(x_m)$, and $g_{jm} := \frac{e^{ik|x_j - x_m|}}{4\pi|x_j - x_m|}$. It is proved in [26] that the linear algebraic system (18) has a solution and this solution is unique for all sufficiently small a, and the function

$$u^{(n)}(x) := \sum_{m=1}^{n} u_j \chi_j(x)$$

converges to $u_e(x)$ uniformly:

$$\lim_{n\to\infty} ||u_e(x) - u^{(n)}(x)||_{C(D)} = 0,$$

where $\chi_j(x) = 1 \quad x \in D$, $\chi_j(x) = 0 \quad x \notin D$.

In particular, it is proved in [26] that a collocation method for solving equation (17) leads to sums, similar to the sum on the left-hand side of (10), and the collocation method, studied in [26], converges to the unique solution of the equation (17). This means that the function $u_M(x)$ converges to $u_e(x)$.

Remark 1. The quantities A_m in the definition of q_m can be it tensors. In this case *the refraction coefficient $n^2(x)$ of the created material is a tensor, and u is a vector field.*

Remark 2. The recipe for creating materials with the desired refraction coefficient by the method, based on Theorem 1, can be summarized as follows:

Recipe 1: Given a bounded domain filled with the material with the refraction coefficient $n_0^2(x)$, one embeds the small inhomogeneities q_m, $q_m = 0$ for $|x - x_m| > a$, $q_m = A_m := A(x_m)$, where the points x_m are distributed by the rule (9), and one chooses $A(x)$ and $N(x) \geq 0$, so that $A(x)N(x) = q(x)$, where $q(x)$ is the desired potential, or which is the same, the desired refraction coefficient $n^2(x) := 1 - k^{-2}q(x)$. The material, created by the embedding of these small inhomogeneities, in the limit $a \to 0$, has the refraction coefficient $n^2(x) = 1 - k^{-2}q(x)$.

One can fix $N(x) \geq 0$ and choose $A(x)$ such that the function $q(x) = A(x)N(x)$ is a desired function, so that $n^2(x) = 1 - k^{-2}q(x)$ is a desired function.

We call this recipe *Recipe i*, where i stands for inhomogeneities.

3 A discussion of two recipes

In [24] a different recipe was proposed for creating materials with a desired refraction coefficient. We call the recipe from [24] *Recipe p*, where p stands for particles. *Recipe p* consists of embedding small particles, for example, balls of radius a, $ka << 1$, into a given material according to the distribution law, similar to the one in (9), but with $a^{2-\kappa}$ in place of $V(a)$, where $\kappa \in (0, 1]$ is a parameter experimenter can choose at will. The physical properties of the embedded particles are described by the boundary impedance $\zeta_m = \frac{h(x_m)}{a^{\varkappa}}$. The function $h(x)$, determining the boundary impedances, can also be chosen as one wishes. We refer the reader to the paper [24] for details. Numerical results on the implementation of *Recipe p* are given in [2] and [4].

There are two technological problems that should be solved in order that *Recipe p* be implemented practically.

Technological Problem 1: *How does one embed many small particles in a given material so that the desired distribution law is satisfied?*

Technological Problem 2: *a) How does one prepare a small particle with the desired boundary impedance* $\zeta_m = \frac{h(x_m)}{a^{\varkappa}}$*?*

b) How can one prepare a small particle with the desired dispersion of the boundary impedance, that is, the desired ω*-dependence of* $h = h(x, \omega)$*?*

The motivation for the last question is the following: by creating material with the refraction coefficient depending on the frequency ω in a desired way, one can, for instance, prepare materials with negative refraction.

The first problem, possibly, can be solved by *stereolitography.*

The second problem one should be able to solve because the limiting cases $\zeta_m = 0$ (hard particles) and $\zeta_m = \infty$ (soft particles) can be prepared, so that any intermediate value of ζ_m one should be able to prepare as well.

The author formulates the above technological problems in the hope that engineers get interested and solve them practically.

In *Recipe p* the number M of the embedded particles is $M := M_2 = O(\frac{1}{a^{2-\kappa}})$, while in *Recipe i* the number M is $M := M_1 = O(\frac{1}{a^3})$. Thus, $M_1 >> M_2$ as $a \to 0$. So, in recipe 2 one has to embed much smaller number of small particles than in *Recipe i*. This is an advantage of *Recipe p* over *Recipe i*. The disadvantage of *Recipe p*, compared with *Recipe i*, is in the possible practical difficulties of creating particles with the prescribed very large boundary impedances $\zeta_m = O(\frac{1}{a^{\kappa}})$ when a is very small.

3.1 Negative refraction

Material with negative refraction is, by definition, a material in which group velocity is directed opposite to the phase velocity, (see [1] and references therein). Group velocity is defined by the formula $\mathbf{v}_g = \nabla_{\mathbf{k}}\omega(\mathbf{k})$. Phase velocity $\mathbf{v}_p$ is directed along the wave vector $\mathbf{k}^0 = \frac{\mathbf{k}}{|\mathbf{k}|}$. In an isotropic material $\omega = \omega(|\mathbf{k}|)$, and $\omega n(x, \omega) = c|\mathbf{k}|$. Differentiating this equation yields

$$\nabla_{\mathbf{k}}\omega\big[n(x,\omega)+\omega\frac{\partial n}{\partial\omega}\big]=c\mathbf{k}^0. \tag{19}$$

Thus,

$$\mathbf{v}_g\big[n+\omega\frac{\partial n}{\partial\omega}\big]=c\mathbf{k}^0. \tag{20}$$

Wave speed in the material is $|\mathbf{v}_p|=\frac{c}{n(x,\omega)}$, where c is the wave speed in vacuum (in D'), and $n(x,\omega)$ is a scalar, and $\mathbf{v}_p$ is directed along $\mathbf{k}^0$.

For $\mathbf{v}_g$ to be directed opposite to $\mathbf{v}_p$ it is necessary and sufficient that

$$n+\omega\frac{\partial n}{\partial\omega}<0. \tag{21}$$

If the new material has the function $n(x,\omega)$ satisfying (21), then the new material has negative refraction.

One can create material with the refraction coefficient $n^2(x,\omega)$ having a desired dispersion, i.e., a desired frequency dependence, by choosing function $h=h(x,\omega)$ properly.

Let us formulate the technological problem solving of which allows one to implement practically *Recipe p* for creating materials with negative refraction.

Technological Problem 3: *How does one prepare a small particle with the desired frequency dependence of the boundary impedance* $\zeta_m=\frac{h(x_m,\omega)}{a^\kappa}$, *where* $h(x,\omega)$ *is a given function?*

This problem is the same as *Technological Problem 2b)*, formulated above.

3.2 Wave-focusing property

Let us formulate the problem of preparing a material with a desired wave-focusing property. The refraction coefficient is related to the potential by the formula

$$q(x)=k^2-k^2n^2(x).$$

We assume in this Section that $k>0$ is fixed, and $\alpha\in S^2$, the incident direction (that is, the direction α of the incident plane wave $e^{ik\alpha\cdot x}$) is also fixed.

Problem: Given an arbitrary fixed $f(\beta)\in L^2(S^2)$, *an arbitrary small fixed* $\epsilon>0$, *an arbitrary fixed* $k>0$, *and an arbitrary fixed* $\alpha\in S^2$, *can one find* $q\in L^2(D)$, $q=0$ *in* $D'=R^3\backslash D$, *such that*

$$\|A_q(\beta)-f(\beta)\|_{L^2(S^2)}<\epsilon, \tag{22}$$

where $A_q(\beta) := A_q(\beta, \alpha, k)$ *is the scattering amplitude, corresponding to* q*, at fixed* α *and* k*, and* $q := k^2 - k^2n^2(x)$*?*

The answer is *yes*, and an algorithm for finding such a q, and, therefore, such an $n^2(x)$, is given in [17]. See also [18], [20]. The above problem is the inverse scattering problem with the scattering data given at a fixed k and a fixed direction α of the incident plane wave.

There are many potentials q *which solve the above problem.* It is possible to choose from the set of these potentials the one which satisfies some additional properties, for example, one can choose q to be arbitrarily smooth in D, one can try to choose q with non-negative imaginary part, etc.

The desired radiation pattern $f(\beta)$ can, for example, be equal to 1 in a given solid angle and equal to 0 outside this solid angle. In this case the scattered field is scattered predominantly into the desired solid angle.

This is why we call such a material a material with wave-focusing property.

References

1.V. M. Agranovich, Yu., Gartstein, Spatial dispersion and negative refraction of light, Physics-Uspekhi, 176, N10,(2006),1051-1068.

2.M.Andriychuk and A.G.Ramm, Scattering by many small particles and creating materials with a desired refraction coefficient, International Journ. Comp.Sci. and Math. (IJCSM), 3, N1/2, (2010), 102-121.

3.J. H. Conway and N. J. A. Sloane, Sphere packings, lattices and groups, Springer-Verlag, New York, 1998.

4.S. Indratno and A.G.Ramm, Creating materials with a desired refraction coefficient: numerical experiments, International Journ. Comp.Sci. and Math. (IJCSM), 3, N1/2, (2010), 76-101.

5.A. G. Ramm, Scattering by obstacles, D.Reidel, Dordrecht, 1986.

6.A. G. Ramm, Equations for the self-consistent field in random medium, Phys.Lett. A, 312, N3-4, (2003), 256-261.

7.A. G. Ramm, Inverse problems, Springer, New York, 2005.

8.A. G. Ramm, Wave scattering by small bodies of arbitrary shapes, World Sci. Publishers, Singapore, 2005.

9.A. G. Ramm, Electromagnetic wave scattering by many small particles, Phys. Lett. A, 360, N6, (2007), 735-741.

10.A. G. Ramm, Materials with the desired refraction coefficients can be made by embedding small particles, Phys. Lett. A, 370, 5-6, (2007), 522-527.

11.A. G. Ramm, Scattering by many small bodies and applications to condensed matter physics, Europ. Phys. Lett., 80, (2007), 44001.

12.A. G. Ramm, Many-body wave scattering by small bodies and applications, J. Math. Phys., 48, N10, (2007), 103511.

13.A. G. Ramm, Wave scattering by small particles in a medium, Phys. Lett. A 367, (2007), 156-161.

14.A. G. Ramm, Wave scattering by small impedance particles in a medium, Phys. Lett. A 368, N1-2,(2007), 164-172.
15.A. G. Ramm, Distribution of particles which produces a desired radiation pattern, Communic. in Nonlinear Sci. and Numer. Simulation, 12, N7, (2007), 1115-1119.
16.A. G. Ramm, Distribution of particles which produces a "smart" material, Jour. Stat. Phys., 127, N5, (2007), 915-934.
17.A. G. Ramm, Inverse scattering problem with data at fixed energy and fixed incident direction, Nonlinear Analysis: Theory, Methods and Applications, 69, N4, (2008), 1478-1484.
18.A. G. Ramm, Creating wave-focusing materials, LAJSS (Latin-American Journ. of Solids and Structures), 5, (2008), 119-127.
19.A. G. Ramm, Electromagnetic wave scattering by many conducting small particles, J. Phys A, 41, (2008), 212001.
20.A. G. Ramm, Some results on inverse scattering, Modern Phys. Lett. B, 22, N23, (2008), 2217-2240.
21.A. G. Ramm, A recipe for making materials with negative refraction in acoustics, Phys. Lett. A, 372/13, (2008), 2319-2321.
22.A. G. Ramm, Wave scattering by many small particles embedded in a medium, Phys. Lett. A, 372/17, (2008), 3064-3070.
23.A. G. Ramm, Electromagnetic wave scattering by small bodies, Phys. Lett. A, 372/23, (2008), 4298-4306.
24.A. G. Ramm, Preparing materials with a desired refraction coefficient and applications, In the book "Topics in Chaotic Systems: Selected Papers from Chaos 2008 International Conference", Editors C.Skiadas, I. Dimotikalis, Char. Skiadas, World Sci.Publishing, 2009, pp.265-273.
25.A. G. Ramm, Preparing materials with a desired refraction coefficient, Nonlinear Analysis: Theory, Methods and Applications, 70, N12, 92009), e186-e190.
26.A.G.Ramm, A collocation method for solving integral equations, Internat. Journ. Comp. Sci and Math., 3, N2, (2009), 222-228.
27.A. G. Ramm, Creating desired potentials by embedding small inhomogeneities, J. Math. Phys., 50, N12, (2009), 123525.
28.A. G. Ramm, Electromagnetic wave scattering by many small particles and creating materials with a desired permeability, Progress in Electromag. Research, M, 14, (2010), 193-206.
29.A. G. Ramm, Electromagnetic wave scattering by many small bodies and creating materials with a desired refraction coefficient, Progress in Electromagnetic Research M (PIER M), 13, (2010), 203-215.
30.A. G. Ramm, Materials with a desired refraction coefficient can be created by embedding small particles into the given material, International Journal of Structural Changes in Solids (IJSCS), 2, N2, (2010, 17-23.
31.A. G. Ramm, A method for creating materials with a desired refraction coefficient, Internat. Journ. Mod. Phys B, 24, (2010).
32.N.Sloane, The sphere packing problem, Documenta Mathematika, Vol. III (1998), 387-396.

Modeling recent economic debates

Christos H. Skiadas

Data Analysis and Forecasting Laboratory,
Technical University of Crete, Chania, Crete, Greece
Email: skiadas@asmda.net

Abstract: The previous years' disaster in the stock markets all over the world and the resulting economic crisis lead to serious criticisms of the various models used. It was evident that large fluctuations and sudden losses may occur even in the case of a well organized and supervised context as it looks to be the European Union. In order to explain the economic systems, we explore models of interacting and conflicting populations. The populations are conflicting into the same environment (a Stock Market or a Group of Countries as the EU). Three models where introduced 1) the Lotka-Volterra 2) the Lanchester or the Richardson model and 3) a new model for two conflicting populations. These models assume immediate interaction between the two conflicting populations. This is usually not the case in a stock market or between countries as delays in the information process arise. The main rules present include mutual interaction between adopters, potential adopters, word-of-mouth communication and of course by taking into consideration the innovation diffusion process. In a previous paper (Skiadas, 2010 [9]) we had proposed and analyzed a model including mutual interaction with delays due to the innovation diffusion process. The model characteristics where expressed by third order terms providing four characteristic symmetric stationary points. In this paper we summarize the previous results and we analyze the case of a non-symmetric case where the leading part receives the information immediately while the second part receives the information following a delay mechanism due to the innovation diffusion process (the spread of information) which can be expressed by a third order term. In the later case the non-symmetric process leads to gains of the leading part while the second part oscillates between gains and losses during time.
Keywords: stock market problem, stock market, the EU financial market, innovation diffusion modeling, Lotka-Volterra, Lanchester, Richardson, simulation, chaotic simulation.

1. Introduction

In the present paper we explore a two conflicting population model into the same environment (a Stock Market or a Group of Countries as the EU). There are several modeling approaches of the subject. The simpler but yet quite realistic are 1) the two populations' predator-prey Lotka-Volterra

Chaos Theory: Modeling, Simulation and Applications
C. H. Skiadas, I. Dimotikalis and C. Skiadas (Eds)

model expressed by two coupled differential equations, 2) the Lanchester [1] or the Richardson [2] model of two conflicting populations and 3) the model that we had proposed in a previous publication [9] of two conflicting populations. The later model is expressed by a general system of two coupled interacting non-linear equations of the third order. This model even in its simpler form provides a characteristic behavior over time. The simpler form of the three models and their solutions appear in Table I. To simplify things we assume the same parameter c for the coupling cxy in both populations. The three models have three fixed points at $(x, y)=(0, 0)$ and $(x, y)=(a/c, b/c)$. That is different is the stability of these points.

Table I

Proposed Model	Lotka-Volterra	Lanchester Richardson
$\dot{x} = -ay + cxy$	$\dot{x} = -bx + cxy$	$\dot{x} = -ay + cxy$
$\dot{y} = bx - cxy$	$\dot{y} = ay - cxy$	$\dot{y} = -bx - cxy$
$x + y + \frac{a}{c}\ln(x - a/c) + \frac{b}{c}\ln(y - b/c) = h$	$a \ln x + b \ln y - c(x + y) = h$	$x + y + \frac{a}{c}\ln(x - a/c) - \frac{b}{c}\ln(y + b/c) = h$

The Lanchester-Richardson model of conflict is the classical one used to express a battle where both parties have losses thus assuming negative values for the parameters of x and y. The same model used in an economic conflict in a market or between countries as is the case in the continuing economic crisis in the EU countries needs to be reformulated by assuming a positive parameter b for the leading country or group of countries. This is presented in Table I, in our proposed model. The Lotka-Volterra model is more difficult to apply in this form. It can be used in a more complicated form including both the conflict model and the predator-prey model. This general model was proposed in a previous publication and is presented in the following.

2. The Model and Simulations

To model the specific situation we take into account that two populations x and y are present into the stock market and interact each other. Even more to simplify the case we suppose that the variables x_t and y_t stand for the number of players or for the number of transactions or for the values of the stocks belonging to each part of the players at a specific time period t. The aim is to explore the behavior of the two populations during time and especially in the limits [9].

A model including the main characteristics of two interrelating or even conflicting populations into a stock market can be expressed by the following set of differential equations [9]:

$$\dot{x} = a_1 y + a_2 xy + a_3 x(l - y) + a_4 yx(k - x)$$
$$\dot{y} = b_1 x + b_2 xy + b_3 y(k - x) + b_4 xy(l - y) \qquad (1)$$

where a_1, a_2, a_3, a_4 and b_1, b_2, b_3, b_4 are parameters expressing the mutual interaction of the populations x and y. The parameters k and l express the upper limit of the populations x and y respectively.

The first term to the right stands for the flows from the one part to the other whereas the second term expresses the mutual interaction between the active parts of the populations x and y.

The non active parts of the populations are $(k - x)$ for x and $(l - y)$ for the population y. The interaction of these parts with the active populations y and x is expressed with the third terms to the right of the above differential equations (see [3, 4, 5, 6, 7, 8, 9]).

The fourth terms to the right include the terms $x(k - x)$ and $y(l - y)$ multiplied by y and x respectively. These terms account to the rates of adoption-diffusion that is extremely important in order to express the word-of-mouth communication between adopters (the active part of the players) and potential adopters (the non-active part of the players).

The Simplest Second Order Model

The simplest model, as presented in Table I, is expressed by the set of two coupling differential equations of the form:

$$\dot{x} = -ay + cyx$$
$$\dot{y} = bx - cxy \qquad (2)$$

Where a, b and c are parameters of interaction.

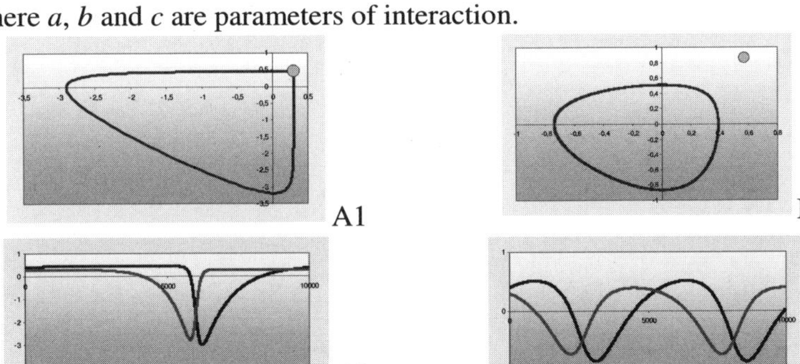

A1 B1

A2 B2

A1-2. High values of c=1.33 B1-2. Medium values for c=0.7

Figure 1. The simplest second order model

Especially the parameter c is selected to be the same in both equations expressing the coupling of both populations x and y. This is a second order differential equations system with two fixed points at $(x, y)=(0, 0)$ and $(x, y)=(a/c, b/c)$. The point (0, 0) is stable whereas the other one is unstable. The behavior of the system is highly influenced by the coupling parameter *c* as it is illustrated in Figure 1. In the case A1-A2 of this Figure a high value for *c* leads to sudden losses for both players with a small delay between each other. The recovery period is characterized by an almost linear part leading to a new sharp period of losses. Instead in the case B1-B2 a relatively medium value for parameter c leads to oscillating behavior with a small delay between each other.

The Simplest Third Order Model

This is a model of the form:

$$\dot{x} = -ay + cyx$$
$$\dot{y} = bx - cxy^2 \tag{3}$$

This is a third order differential equations system with three fixed points at $(x, y)=(0, 0)$, $(x, y)=(a/c, \sqrt{b/c})$ and $(x, y)=(a/c, -\sqrt{b/c})$. The point (0, 0) is stable whereas the other two are unstable.

$$\frac{dy}{dx} = \frac{bx - cxy^2}{-ay + cyx}. \tag{4}$$

The solution is: $c\ln(y^2 - b/c) - 2a\ln(x - a/c) + 2cx = h$ (5)

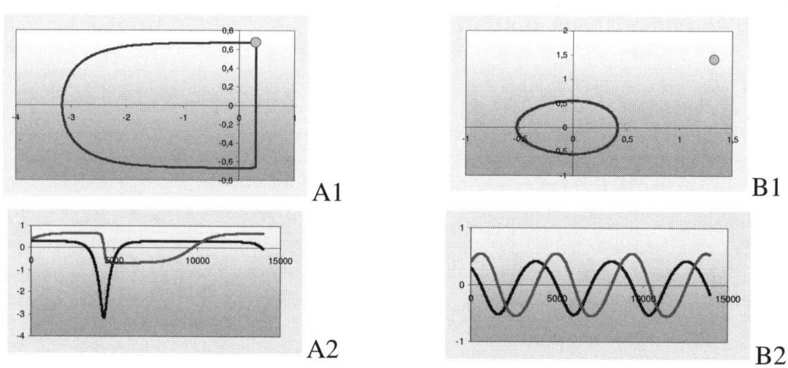

A1-2. High values of C=1.33 B1-2. Small values for C=0.3

Figure 2. A third order model

Figure 2 illustrates two simulation cases of this model. In the case A1-A2 of this figure a high value for *c* leads to strong losses for one player with a delay in the recovery period between each other and medium or small losses for the other player but with a delay. *This delay is the key point for the strong played. He has the time to benefit from the other and exit the system or change the rules and start again.* Instead in the case B1-B2 a relatively medium value for parameter *c* leads to oscillating behavior with a small delay between each other.

Looking back to the set of the two general differential equations (1) we see that we have two third degree equations for x and y. In the following we check the influence of the third order term to the stock-market model. To this end the above model is simplified to the following form [9]:

$$\dot{x} = -ay + cyx^2$$
$$\dot{y} = bx - cxy^2 , \tag{6}$$

The last equations include in the right part side two basic parts of the interaction and the diffusion process where $x = \sqrt{a/c}$ and $y = \sqrt{b/c}$ are the upper limit or the maximum for x and y respectively.
There is a fixed point at (0, 0) and four other characteristic points at

$$(x = \sqrt{a/c}, y = \sqrt{b/c}),\ (x = \sqrt{a/c}, y = -\sqrt{b/c})$$
$$(x = -\sqrt{a/c}, y = \sqrt{b/c}), (x = -\sqrt{a/c}, y = -\sqrt{b/c})$$

The last differential equations give the following differential equation for x and y [9]:

$$\frac{dy}{dx} = \frac{bx - cxy^2}{-ay + cyx^2} \tag{7}$$

The solution is: $(y^2 - b/c)(x^2 - a/c) = h$
Where h is the integration constant.
Figure 3 illustrates characteristic graphs of the last equation. The four characteristic points define a rectangle. Into this space are drawn few of the trajectories of the process. The trajectories outside of the rectangle diverge to infinity [9].

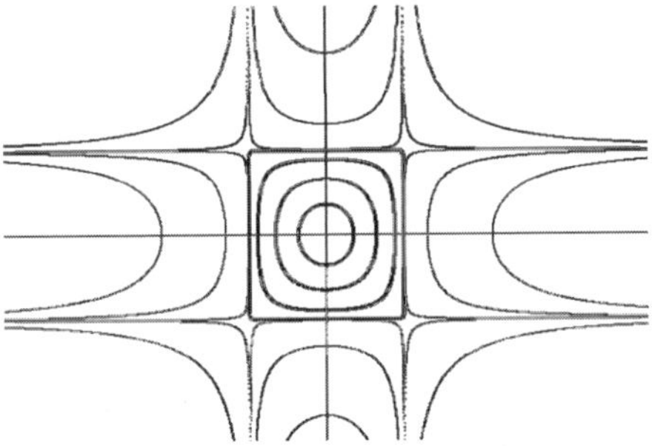

Figure 3. Characteristic Trajectories for various *h*.

As it is presented in Figure 3 there appear to be positive and negative values for the stock-market process. As it was expected from the symmetry of the process both players have equal chances for gains and losses. This could change radically if the one player succeeds to destroy symmetry and move to the second case expressed by the model expressed by eq. (3). To this end the players, the Countries in the case of EU, try to break the symmetry by using their local or global advantages as the position, size, economic situation, production level, information sources, strategic position, communication, and any other systematic or no-systematic movements tactical or strategic in order to take advantage leading to economic or other benefits.

3. Conclusions

Three models expressing two conflicting populations in the stock-market are developed. These models result from a more general one. The results support the well known process of fluctuations, oscillations and further sudden growth and decrease of gains-losses of the two conflicting populations. It was derived that depending on the type of model followed it could arise that both would be stabilized to gains or losses or to stabilize the one in gains and the other to losses. *However, in the long run, both will face gains or losses. The strategy of the leading or strong player is to benefit from the other and quit the game or change the rules and start again. This is the case of the recent maneuvers in the EU economic system.*

References

[1] F. W. Lanchester, F.W. "*Mathematics in Warfare*" The World of Mathematics, J. Newman ed., Simon and Shuster (New York), Vol. 4, 1956.

[2] L. F. Richardson, *The Collected Papers of Lewis Fry Richardson, Volume 2: Quantitative psychology and studies of conflict, in* Ashford, Oliver M; Charnock H; Drazin, P G; Hunt, J C R; Smoker, P, Sutherland, Ian. Eds, Cambridge: Cambridge University Press, 1993.

[3] C. H. Skiadas. Two generalized rational models for forecasting innovation diffusion. *Technol Forecast Soc Change* 27:39–61, 1985.

[4] C. H. Skiadas. Innovation diffusion models expressing asymmetry and/or positively or negatively influencing forces. *Technol Forecast Soc Change* 30:313–330, 1986.

[5] C. H. Skiadas. Two simple models for the early and middle stage prediction of innovation diffusion. *IEEE Trans Eng Manage* 34:79–84, 1987.

[6] C. H. Skiadas and A. N. Giovanis. A stochastic bass innovation diffusion model for studying the growth of electricity consumption in Greece. *Appl Stoch Models Data Anal* 13:85–101, 1997

[7] C. H. Skiadas, A. N. Giovanis and J. Dimoticalis. A sigmoid stochastic growth model derived from the revised exponential. In: Janssen J, Skiadas CH (eds) *Applied Stochastic Models and Data Analysis*. World Scientific, Singapore, pp 864–870, 1993.

[8] C. H. Skiadas and C. Skiadas. *Chaotic Modeling and Simulation: Analysis of Chaotic Models, Attractors and Forms*, Taylor and Francis/CRC, London, 2009.

[9] C. H. Skiadas, A two population model for the Stock Market problem. In: Skiadas CH and I. Dimotikalis (eds) *Chaotic Systems: Theory and Applications*. World Scientific, Singapore, pp 302–308, 2010.

Classical versus quantum dynamical chaos: Sensitivity to external perturbations, stability and reversibility

Valentin V. Sokolov[1,2], Oleg V. Zhirov[1,3] and Yaroslav A. Kharkov[3,1]

[1]Budker Institute of Nuclear Physics,
Novosibirsk, Russia
Email: V.V.Sokolov@inp.nsk.su
[2]Novosibirsk Technical University, Russia
[3]Novosibirsk State University, Russia

Abstract: The extraordinary complexity of classical trajectories of typical nonlinear systems that manifest stochastic behavior is intimately connected with exponential sensitivity to small variations of initial conditions and/or weak external perturbations. In rigorous terms, such classical systems are characterized by positive *algorithmic complexity* described by the Lyapunov exponent or, alternatively, by the Kolmogorov-Sinai entropy. The said implies that, in spite of the fact that, formally, any however complex trajectory of a perfectly isolated (closed) system is unique and differentiable for any certain initial conditions and the motion is perfectly reversible, it is impractical to treat that sort of classical systems as closed ones. Inevitably, arbitrary weak influence of an environment crucially impacts the dynamics. This influence, that can be considered as a noise, rapidly effaces the memory of initial conditions and turns the motion into an irreversible random process.

In striking contrast, the quantum mechanics of the classically chaotic systems exhibit much weaker sensitivity and strong memory of the initial state. Qualitatively, this crucial difference could be expected in view of a much simpler structure of quantum states as compared to the extraordinary complexity of random and unpredictable classical trajectories. However the very notion of trajectories is absent in quantum mechanics so that the concept of exponential instability seems to be irrelevant in this case. The problem of a quantitative measure of complexity of a quantum state of motion, that is a very important and nontrivial issue of the theory of quantum dynamical chaos, is the one of our concern. With such a measure in hand, we quantitatively analyze the stability and reversibility of quantum dynamics in the presence of external noise.

To solve this problem we point out that individual classical trajectories are of minor interest if the motion is chaotic. Properties of all of them are alike in this

Chaos Theory: Modeling, Simulation and Applications
C. H. Skiadas, I. Dimotikalis and C. Skiadas (Eds)

case and rather the behavior of their manifolds carries really valuable information. Therefore the phase-space methods and, correspondingly, the Liouville form of the classical mechanics become the most adequate. It is very important that, opposite to the classical trajectories, the classical phase space distribution and the Liouville equation have direct quantum analogs. Hence, the analogy and difference of classical and quantum dynamics can be traced by comparing the classical ($W^{(c)}(I,\theta;t)$) and quantum (Wigner function $W(I,\theta;t)$) phase space distributions both expressed in identical phase-space variables but ruled by different(!) linear equations.

The paramount property of the classical dynamical chaos is the exponentially fast structuring of the system's phase space on finer and finer scales. On the contrary, degree of structuring of the corresponding Wigner function is restricted by the quantization of the phase space. This makes Wigner function more coarse and relatively "simple" as compared to its classical counterpart. Fourier analysis affords quite suitable ground for analyzing complexity of a phase space distribution, that is equally valid in classical and quantum cases. We demonstrate that the typical number of Fourier harmonics is indeed a relevant measure of complexity of states of motion in both classical as well as quantum cases. This allowed us to investigate in detail and introduce a quantitative measure of sensitivity to an external noisy environment and formulate the conditions under which the quantum motion remains reversible. It turns out that while the mean number of harmonics of the classical phase-space distribution of a non-integrable system grows with time exponentially during the whole time of the motion, the time of exponential upgrowth of this number in the case of the corresponding quantum Wigner function is restricted only to the Ehrenfest interval $0 < t < t_E$ - just the interval within which the Wigner function still satisfies the classical Liouville equation. We showed that the number of harmonics increases beyond this interval *algebraically*. This fact gains a crucial importance when the Ehrenfest time is so short that the exponential regime *has no time to show up*. Under this condition the quantum motion turns out to be quite stable and reversible.
Keywords: quantum chaos, Wigner function, noise, fidelity, stability, reversibility, Markov's chain, entropy, purity.

1 Introduction

The exponential sensitivity of the dynamics of non-integrable classical systems to small variations of the initial conditions and/or to weak external perturbations constitutes the very essence of the classical dynamical chaos. Such an instability of the motion rapidly effaces the memory of initial conditions and makes impractical the treatment of such systems as closed ones. Inevitably, arbitrary weak interaction with an environment crucially influences the dynamics of such systems. Ordinary, this influence can be considered as a noise, that turns the motion into an irreversible random process.

In striking contrast, strong numerical evidence has been obtained long ago [1] that the quantum mechanics of the classically chaotic systems appears to exhibit much weaker sensitivity and strong memory of the initial state. Qualitatively, this crucial difference could be expected in view of a much simpler structure of quantum states as compared to the extraordinary complexity of random and unpredictable classical trajectories. The problem of a quantitative measure of complexity of a state of motion is a very important and nontrivial issue of the theory of dynamical chaos. In rigorous terms, classical chaotic systems are characterized by positive *algorithmic complexity* described by the Lyapunov exponent or, alternatively, by the Kolmogorov-Sinai entropy. But, being formulated in such a way, the concept of complexity cannot be immediately transferred to the quantum mechanics, where the very notion of trajectory is irrelevant and there is no quantum analogue to the Lyapunov exponent. However, just due to the very this fact, individual classical trajectories are, in essence, of minor interest if the motion is chaotic. All of them are alike in this case and rather the behavior of their manifolds carries really valuable information. Under such conditions the phase-space methods and, correspondingly, the Liouville form of the classical mechanics become the most adequate. It is very important that, opposite to the classical trajectories, the classical phase space distribution and the Liouville equation have direct quantum analogs. Hence, a comparison between classical and quantum dynamics can be made by studying the time evolutions of the classical ($W^{(c)}(I,\theta;t)$) and quantum (Wigner function $W(I,\theta;t)$) phase space distributions both ruled by linear equations. The great advantage of the phase space approach is that one operates with distributions that can be presented in the both cases in identical action - angle (I,θ) variables (introduced on the basis of the substantial unperturbed integrable part of the total system's Hamiltonian). Notice that there always exists one to one correspondence between the Wigner function and the system's quantum density matrix.

The paramount property of the classical dynamical chaos is the exponentially fast structuring of the system's phase space on finer and finer scales. On the contrary, degree of structuring of the corresponding quantum "distribution" is restricted by the quantization of the phase space. This makes Wigner function more coarse and relatively "simple" as compared to its classical counterpart. Fourier analysis affords quite suitable ground for analyzing complexity of a phase space distribution, that is equally valid in classical and quantum cases. It turns out that while the mean number of harmonics of the classical phase-space distribution of a non-integrable system grows with time exponentially during the whole time of the motion, the time of exponential upgrowth of this number in the case of the corresponding quantum Wigner function is restricted only to the Ehrenfest interval $0 < t < t_E$ - just the interval within which the Wigner function satisfies the classical Liouville equation. We will show below that the number of harmonics increases beyond this interval *algebraically*. This fact gains a crucial importance when the Ehrenfest time is so short that the exponential regime *has no time to show up*.

The *Peres fidelity* [2,3] i.e. the weighted-mean distance between two different distributions proved to be a suitable tool that allows for quantitative description of the degree of stability or reversibility of quantum motion [4,5]. It is of primary importance that this quantity, being expressed in terms of the Wigner function, has direct classical limit. On the other hand it can be directly *related to the number of harmonics* of the corresponding phase-space distributions. Thereby, the problems of stability and reversibility of dynamics are linked to that of the time behavior of the Fourier harmonics of the corresponding phase-space distributions.

Having in mind the said above we consider below the evolution during an arbitrary time interval t of an initially simple ($\equiv$ with small number of harmonics of the Wigner function) and pure quantum state $\hat{\rho}(0) = |\psi(0)\rangle\langle\psi(0)|$ against a persistent noisy background. This evolution is described by the unitary transformation

$$\hat{\rho}(\xi(t);t) = \hat{U}(\xi(t);t)\hat{\rho}(0)\hat{U}^{\dagger}(\xi(t);t) \qquad (1)$$

where the evolution operator $\hat{U}(\xi(t);t)$ depends on the realization $\xi(t)$ (or history) of the noise. Sensitivity of the motion to the noise can be characterized by the overlap (Peres fidelity) of the states developed by the time t during the evolutions *with*, $\hat{U}(\xi(t);t)$, and *without*, $\hat{U}(t)$, influence of the noisy environment,

$$F_{(sen)}(\xi(t);t) = \mathrm{Tr}\left[\hat{\rho}(t)\,\hat{\rho}(\xi(t);t)\right] . \qquad (2)$$

This quantity is bounded in the interval $[0,1]$ and the sensitivity is the weaker the closer the fidelity remains to one by the time t.

Let us, on the other hand, reverse the motion at the moment t and evolve the state (1) backward in time with no noise till the initial moment $t = 0$

$$\hat{\tilde{\rho}}(0|\xi(t);t) = \hat{U}^{\dagger}(t)\,\hat{\rho}(\xi(t);t)\,\hat{U}(t) = \hat{f}(\xi(t);t)\hat{\rho}(0)\hat{f}^{\dagger}(\xi(t);t), \qquad (3)$$

and compare after that the obtained reversed state with the initial one $\hat{\rho}(0)$. The fidelity defined as

$$F_{(rev)}(\xi(t);t) = \mathrm{Tr}\left[\hat{\rho}(0)\,\hat{\tilde{\rho}}(0|\xi(t);t)\right] = \left|\langle\psi(0)|\hat{f}(\xi(t);t)|\psi(0)\rangle\right|^2 \qquad (4)$$

describes the degree of reversibility of the motion. We have defined in (3) the unitary *fidelity operator*

$$\hat{f}(\xi(t);t) = \hat{U}^{\dagger}(t)\, U(\xi(t);t)\,. \tag{5}$$

In spite of the different interpretations the both introduced fidelities are numerically identical because of unitarity of the evolution operators:

$$F_{(sen)}(\xi(t);t) = F_{(rev)}(\xi(t);t) \equiv F(\xi(t);t)\,. \tag{6}$$

This relation directly *connects the reversibility of the motion with sensitivity to external perturbations.*

It has to be noted that the fidelity defined in such a way strongly depends on the noise history and therefore is not suitable yet for immediate use. The appropriate reduced measure is obtained by averaging over all possible noise realizations

$$F(\sigma;t) = \overline{F(\xi(t);t)} = \mathrm{Tr}\left[\hat{\rho}(t)\,\hat{\rho}^{(av)}(\sigma;t)\right] = \left|\langle\psi(0)|\overline{\hat{f}(\xi(t);t)}|\psi(0)\rangle\right|^2. \tag{7}$$

The fidelity (7) depends already on the only parameter σ - the noise level. The new definition brings into consideration the averaged density matrix $\hat{\rho}^{(av)}(\sigma;t) = \overline{\hat{\rho}(\{\xi\};t)}$ [6].

Guided by the notions and relations mentioned in this section we establish below a quantitative description and criteria of stability and reversibility of motion of classically chaotic quantum systems.

2 The model

As a typical example we use [4,5] the periodically kicked quartic oscillator whose one-step evolution is described by the Floquet operator $\hat{\mathcal{F}} = e^{-\frac{i}{\hbar}\hat{H}^{(0)}}\,\hat{D}\left(i\frac{g_0}{\sqrt{\hbar}}\right) = e^{-i\left(\omega_0\hat{n}+\hbar\hat{n}^2\right)}\,e^{i\frac{g_0}{\sqrt{\hbar}}\left(\hat{a}+\hat{a}^{\dagger}\right)}$, where $\hat{a},\hat{a}^{\dagger}$ are the bosonic ($[\hat{a},\hat{a}^{\dagger}]=1$) creation-annihilation operators and $\hat{n}=\hat{a}^{\dagger}\hat{a}$ is the excitation number operator. The driving force is given by $g(t) = g_0\sum_{\tau}\delta(t-\tau)$. The classical motion of this oscillator becomes chaotic when the kick strength g_0 exceeds unity. We suppose further that each kick is followed by an instant perturbation $\xi_{\tau}\,\hbar\,\hat{n}\delta(t-\tau)$ with Gaussian random intensity ξ_{τ}, ($\langle\xi_{\tau}\rangle = 0$, $\langle\xi_{\tau}\xi_{\tau'}\rangle = \sigma^2\delta_{\tau\tau'}$), which models a persistent external noise. Such a perturbation gives rise to the phase plane rotation by a random angle ξ_{τ} at the time moment τ. For any given realization $\{\xi\} \equiv \{\xi_1,\xi_2,...,\xi_t\}$ of the noise, the evolution is described by the unitary operator $\hat{U}(\{\xi\};t) = \prod_{\tau=1}^{\tau=t}\left[e^{-i\xi_{\tau}\hat{n}}\,\hat{\mathcal{F}}\right]$.

In the α-phase space representation [7,8], the quantum Wigner function W is obtained from the density matrix $\hat{\rho}(t)$ by the following Fourier transformation

$$W(\alpha^*,\alpha;t) = \frac{1}{\pi^2\hbar}\int d^2\eta\, e^{(\eta^*\frac{\alpha}{\sqrt{\hbar}}-\eta\frac{\alpha^*}{\sqrt{\hbar}})}\mathrm{Tr}\left[\hat{\rho}(t)\,\hat{D}(\eta)\right], \tag{8}$$

where the integration runs over the complex η-plane and the unitary operator $\hat{D}(\eta) = \exp(\eta\,\hat{a}^\dagger - \eta^*\hat{a})$ is the displacement operator of the coherent states $|\alpha\rangle = \hat{D}\left(\frac{\alpha}{\sqrt{\hbar}}\right)|0\rangle$ built upon the ground state $|0\rangle$ of the unperturbed Hamiltonian $\hat{H}^{(0)}$. The variables α are complex and independent of the effective Planck's constant $\hbar$. The Wigner function is normalized to unity, $\int d^2\alpha W(\alpha^*,\alpha;t) = 1$ and is real though, in general, not positive definite. It satisfies the evolution equation

$$i\frac{\partial}{\partial t}W(\alpha^*,\alpha;t) = \hat{\mathcal{L}}_q\,W(\alpha^*,\alpha;t), \tag{9}$$

with the Hermitian "quantum Liouville operator" $\hat{\mathcal{L}}_q$. This equation reduces in the case $\hbar = 0$ to the classical Liouville equation with respect to the canonical pair $\alpha, i\alpha^*$ with the classical Hamiltonian function being given by the diagonal matrix elements $H_c(\alpha^*,\alpha;t) = \langle\alpha|\hat{H}^{(N)}(\hat{a}^\dagger,\hat{a})|\alpha\rangle$ of the normal form $\hat{H}^{(N)}$ of the quantum Hamiltonian operator. In other words, this function is obtained from the quantum Hamiltonian by substituting $\hat{a} \to \alpha/\sqrt{\hbar}\,,\ \hat{a}^\dagger \to \alpha^*/\sqrt{\hbar}$.

The (classically canonical) transformation $\alpha = \sqrt{I}\,e^{-i\theta}$ converts the function (8) to $W(I,\theta;t)$. The Fourier amplitudes of the latter function are expressed in terms of the matrix elements $\langle n+m|\hat{\rho}|n\rangle$ of the density operator along the mth collateral diagonal as

$$\begin{aligned} W_m(I;t) &= \tfrac{2}{\hbar}\,e^{-\frac{2}{\hbar}I}\textstyle\sum_{n=0}^{\infty}(-1)^n\sqrt{\tfrac{n!}{(n+m)!}}\times \\ &(4I/\hbar)^{\frac{m}{2}}\,L_n^m\,(4I/\hbar)\,\langle n+m|\hat{\rho}(t)|n\rangle,\quad m\geq 0, \end{aligned} \tag{10}$$

with L_n^m the Laguerre polynomials and $W_{-m}(I;t) = W_m^*(I;t)$. The inverted relation reads

$$\begin{aligned} \langle n+m\big|\hat{\rho}(t)\big|n\rangle &= (-1)^n\,2\sqrt{\tfrac{n!}{(n+m)!}}\times \\ \textstyle\int_0^\infty dI\,e^{-2\frac{I}{\hbar}}&\,(4I/\hbar)^{\frac{m}{2}}\,L_n^m\,(4I/\hbar)\;W_m\,(I;t)\;. \end{aligned} \tag{11}$$

The normalization condition reduces to $\int_0^\infty dI\,W_0(I;t) = 1$.

It is convenient to start the evolution with the ground initial state $\hat{\rho}(0) = |0\rangle\langle 0|$. The corresponding Wigner function $W(\alpha^*,\alpha;0) = \frac{2}{\hbar}e^{-\frac{2}{\hbar}|\alpha|^2}$ is isotropic and occupies the cell $\hbar/2$ in the phase plane. Few (one in the case of parameters considered below) first kicks produce a state of a practically general form. At a running moment of time $t > 0$, the excitation of the oscillator and the degree of anisotropy of the Wigner function are characterized by the probability distributions [4,5],

$$w_n(\{\xi\};t) = \langle n|\hat{\rho}(\{\xi\};t)|n\rangle \,, \tag{12}$$

and

$$\begin{aligned} \mathcal{W}_m(\{\xi\};t) &= (2-\delta_{m0})\sum_{n=0}^{\infty}\left|\langle n+m|\hat{\rho}(\{\xi\};t)|n\rangle\right|^2 \\ &= (2-\delta_{m0})\sum_{n=0}^{\infty} w_n(\{\xi\};t)w_{n+m}(\{\xi\};t), \end{aligned} \tag{13}$$

respectively (both normalized to unity). Here $\hat{\rho}(\{\xi\};t) = \hat{U}(\{\xi\};t)\hat{\rho}(0)\hat{U}^{\dagger}(\{\xi\};t)$. With the noise history being *fixed*, the evolution is unitary so that the state remains *pure* during the whole time of the motion. The second equality in Eq. (13) establishes an *important connection* between degree of anisotropy of the Wigner function and the system's excitation.

3 Coarse-grained features of quantum evolution

3.1 Coarse-graining and mean values

The mean values $\langle n\rangle_{\{\xi\};t}$ and $\langle|m|\rangle_{\{\xi\};t}$ calculated with the help of the distributions (12,13) characterize respectively the degree of excitation and the number of θ-harmonics, i.e. complexity [4,5] of the quantum state developed by the time t. Our numerical simulations showed that these values do not depend on the noise history at a given noise level σ (i.e. they are self-averaging quantities). Moreover, the analysis of our numerical data (see the lower panels in Fig. 2) reveals that the low moments depend very weakly on the noise level σ as well. *The chosen noise does not influence appreciably not only the number of harmonics but also the degree of excitation.*

In accordance with what have been said in the Introduction, the number of θ-harmonics of the quantum Wigner function grows with time exponentially only during the Ehrenfest interval and then slows down to algebraical increase. A numerical illustration of this statement is given in Fig. 1 where the time behavior of the root-mean-square $\langle m^2\rangle_t$ for three different values of the effective Plank's constant $\hbar$. Just the quantum dynamics under the conditions when *the exponential regime does not show up* is the case of our concern below.

As to the distributions (12,13) themselves, our detailed numerical data indicate (see upper panels in Fig.2; we fix system parameters as $\omega_0 = 1, \hbar = 1, g_0 = 2$ throughout the paper; *the Ehrenfest time* $t_E < 1$ *in this case*) that at a given time t they undergo (as functions of n or m, respectively)

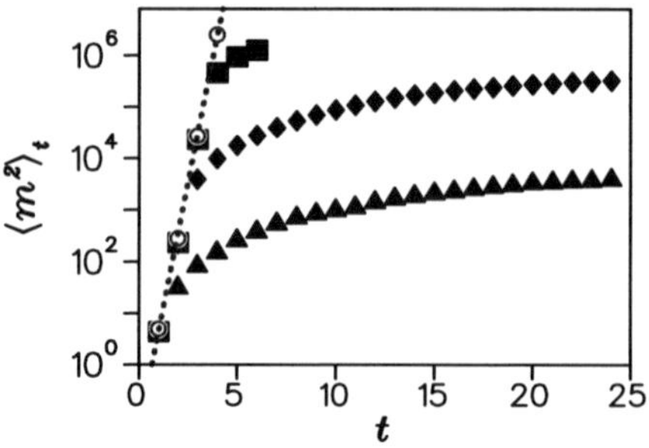

Fig. 1. (color online) Root-mean-square $\langle m^2\rangle_t$ versus time t at $g_0 = 1.5$. Squares, diamonds and triangles correspond to $\hbar = 0.01, 0.1$ and 1. Empty circles refer to classical dynamics and the dashed line fits these data.

fluctuations, rather strong in the first case and much weaker in the second one, around a *regular exponential decay* with identical slopes. These slopes, contrary to the fluctuations, are not sensitive to the noise histories. Such universal exponential laws represent the *coarse-grained* distributions [4,5]

$$w_n^{(c.g.)}(\sigma;t) = \frac{1}{\langle n\rangle_{\sigma;t}+1}\left[\frac{\langle n\rangle_{\sigma;t}}{\langle n\rangle_{\sigma;t}+1}\right]^n, \tag{14}$$

and

$$\mathcal{W}_m^{(c.g.)}(\sigma;t) = \frac{2-\delta_{m0}}{2\langle|m|\rangle_{\sigma;t}+1}\left[\frac{\langle|m|\rangle_{\sigma;t}}{\langle|m|\rangle_{\sigma;t}+1}\right]^m \tag{15}$$

which entirely ignore the fluctuations. The first moments $\langle n\rangle_{\sigma;t}$ and $\langle|m|\rangle_{\sigma;t} = \langle n\rangle_{\sigma;t}$ of these distributions are the free parameters to be used for fitting the slopes of the actual distributions calculated numerically and plotted in Fig. 2. Being defined in such a way, they approximate quite well the corresponding history-insensitive mean values numerically found with the help of the *exact* distributions (14,15). Simulations with truncated excitation bases in the Hilbert space of different dimensions, $N \gg 1$ up to N=6000, convinced us that the rate of growth of the quantum mean excitation is practically indistinguishable from the classical diffusion law $\langle n\rangle_t = \frac{g_0^2}{\hbar}t$ (straight lines in lower panels in Fig. 2). The increase of the number of angular harmonics $\langle m\rangle_t = \langle n\rangle_t$ is also linear as distinct from the classical exponential upgrowth. Furthermore, the amplitudes of fluctuations of the exact distributions (12,13) are appreciably reduced when the noise level σ is growing.

A natural way of the coarse graining consists in averaging over realizations $\{\xi\}$ of the noise [5]. Indeed, as it follows from the data presented in Fig. 2, such an averaging leaves the slopes practically unchanged but suppresses the fluctuations. The latter are entirely obliterated at the strong noise limit $\sigma \to \infty$. Hence we *define* finally the coarse-grained distributions as

$$w_n^{(c.g.)}(t) = \overline{w_n(\{\xi\};t)}\Big|_{\sigma=\infty} = w_n^{(c.g.)}(\sigma=\infty;t);$$
$$\mathcal{W}_m^{(c.g.)}(t) = \overline{\mathcal{W}_m(\{\xi\};t)}\Big|_{\sigma=\infty} = \mathcal{W}_m^{(c.g.)}(\sigma=\infty;t). \tag{16}$$

The bar indicates averaging over the noise. Now, the connection $\langle |m| \rangle_{\infty;t} = \langle n \rangle_{\infty;t}$ directly follows from the exact relation (13) thus leading to the identity of the slopes. Being plotted, the coarse-grained distributions (14,15) are indistinguishable from the distributions (16) shown by the red/darkgray lines in Fig. 2.

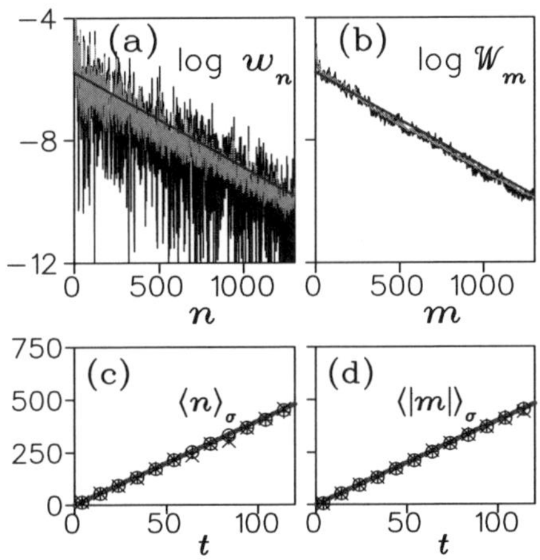

Fig. 2. (Color online). *Top*: (a),(b) - Probability distributions (12,13) at the moment $t = 80$ with no noise ($\sigma = 0$, black), weak noise ($\sigma = 0.001$, green/gray) and in the strong noise limit, (red/darkgray) In the two latter cases the data have been averaged over 10^3 noise realizations. *Bottom*: (c),(d) - Time evolution of the moments $\langle n \rangle_\sigma$, $\langle |m| \rangle_\sigma$ (crosses, pluses and circles for $\sigma = 0, 0.001$ and $\sigma = \infty$, respectively) as compared to the classical diffusion law (straight magenta/gray lines).

3.2 The information entropy

As has been mentioned above, the characteristic number $\langle |m| \rangle_{\sigma;t}$ of Fourier harmonics of the Wigner function at a given moment of time t can serve [4] as a measure of complexity of the current quantum state. Another possibility is to use [9] for the same purpose the information entropy. The latter, by virtue of the probabilistic interpretation, can be defined as

$$\mathcal{I}(t) = -\sum_{m=0}^{\infty} \mathcal{W}_m^{(c.g.)}(t) \ln \mathcal{W}_m^{(c.g.)}(t)$$
$$\approx \ln\langle|m|\rangle_{\infty;t} + 1 + \frac{1}{2\langle|m|\rangle_{\infty;t}} + \ldots, \quad (t \gg 1). \tag{17}$$

This entropy grows monotonically with time starting from $\mathcal{I}(0) = 0$. Generally speaking, during the Ehrenfest time $0 < t \lesssim t_E$ the growth is linear with a slope determined by the classical Lyapunov exponent. Afterwards it slows down to the logarithmic behavior.

4 Sensitivity of quantum evolution to the noise

The one step evolution of the averaged density matrix is described by the relation

$$\langle n'|\hat{\rho}^{(av)}(\sigma;\tau)|n\rangle = e^{-\frac{1}{2}\sigma^2(n'-n)^2} \langle n'|\hat{\mathcal{F}}\hat{\rho}^{(av)}(\sigma;\tau-1)\hat{\mathcal{F}}^{\dagger}|n\rangle \,. \tag{18}$$

The averaging over the noise suppresses the off-diagonal matrix elements and cuts down the number of harmonics of the corresponding averaged Wigner function. Notice, however, that the normalization condition $\mathrm{Tr}\hat{\rho}^{(av)}(\sigma;\tau) = 1$ holds independently of the noise level during the whole evolution.

The fidelity operator (5) reads in our case

$$\hat{f}(\{\xi\};t) = \hat{\mathcal{F}}^{\dagger\, t} \prod_{\tau=1}^{\tau=t} \left[e^{-i\xi_\tau \hat{n}}\, \hat{\mathcal{F}}\right] = \prod_{\tau=1}^{\tau=t} e^{-i\xi_\tau \hat{n}(\tau)}, \tag{19}$$

where $\hat{n}(\tau) = \hat{\mathcal{F}}^{\dagger\,\tau}\hat{n}\,\hat{\mathcal{F}}^{\tau}$ is the Heisenberg evolution of the operator $\hat{n}$. It is easy to calculate explicitly the fidelity (7) for the two limiting cases of weak and extremely strong noise.

4.1 Weak noise limit

Keeping in the last product the noise intensity $\xi_\tau \neq 0$ only in one certain exponential, averaging then over ξ_τ, expanding finally up to the term $\sim \sigma^2$ and summing finally over all the moments $1 \leq \tau \leq t$ we obtain

$$F_{(sen)}(\sigma;t) = 1 - \frac{1}{2}\sigma^2 \sum_{\tau=1}^{\tau=t}\sum_{m=1}^{\infty} m^2\, \mathcal{W}_m(0;\tau) + O(\sigma^4) =$$
$$1 - \frac{1}{2}\sigma^2 \sum_{\tau=1}^{\tau=t} \langle m^2\rangle_{0;\tau} + O(\sigma^4)\,. \tag{20}$$

Here $\langle m^2\rangle_{0;\tau}$ is the square mean number of harmonics developed by the time τ during evolution without noise. Thus the problem reduces in this approximation to the simpler one similar to that investigated in [4]: a single instant perturbation probes the state at the moment τ formed by this moment during preceding free evolution.

Single instant perturbation The fidelity equals in this case (see [4])

$$F^{(1)}(\sigma;t) = \overline{\left|f^{(1)}(\xi;t)\right|^2}, \quad f^{(1)}(\xi;t) = \langle 0|e^{-i\xi\hat{n}(t)}|0\rangle = \sum_{n=0}^{\infty} w_n(t)\, e^{-i\xi n} \tag{21}$$

Subsequent averaging over the strength parameter ξ results in

$$F^{(1)}(\sigma;t) = \sum_{m=0}^{\infty} e^{-\frac{1}{2}\sigma^2 m^2}\, \mathcal{W}_m(t) = 1 - \frac{1}{2}\sigma^2\, \langle m^2\rangle_{0;t} + \ldots, \tag{22}$$

where the harmonics probability distribution reads now (compare with (13))

$$\mathcal{W}_m(t) = (2-\delta_{m0}) \sum_{n=0}^{\infty} w_n(t) w_{n+m}(t)\,. \tag{23}$$

On the other hand, taking into account the expansion $f^{(1)}(\xi;t) \approx 1 - i\xi\langle 0|\hat{n}(t)|0\rangle + \frac{1}{2}\xi^2\langle 0|\hat{n}^2(t)|0\rangle + \ldots$ we find with the same accuracy

$$F^{(1)}(\sigma;t) \approx 1 - \sigma^2\, \left[\langle 0|\hat{n}^2(t)|0\rangle - \langle 0|\hat{n}(t)|0\rangle^2\right] + \ldots. \tag{24}$$

Comparing this with Eq. (22) we arrive at the following important *exact relation* between the number of harmonics and the root-mean-square deviation $\chi_2(t)$ of the excitation number.

$$\langle m^2\rangle_{0;t} = 2\chi_2(t), \quad \chi_2(t) = \langle n^2\rangle_{0;t} - \langle n\rangle^2_{0;t}\,. \tag{25}$$

Using the exponential coarse-grained ansatz (14,15) we find that after the Ehrenfest time

$$\langle m^2\rangle_{0;t} = 2\langle|m|\rangle_{0;t}(\langle|m|\rangle_{0;t}+1) = 2\langle n\rangle_{0;t}(\langle n\rangle_{0;t}+1) \tag{26}$$

that is $\langle|m|\rangle_{0;t} = \langle n\rangle_{0;t}$. *After the Ehrenfest time the mean number of the angular harmonics increases in the same manner as the excitation number, i.e. diffusively.*

The expansion (22) shows that fidelity $F^{(1)}(\sigma;t)$ stays close to unity as long as the noise level σ is noticeably smaller than the *critical value*

$\sigma_c(t) = \sqrt{2/\langle m^2\rangle_{0;t}}$ (compare with [4]). This critical value falls with the time t exponentially during the Ehrenfest time. The fidelity collapses very fast unless the Ehrenfest time is so small that the exponential regime does not show up. But in the latter (quantum) case the critical value decreases from the very beginning only linearly and fidelity retains a good value for much longer time.

The exponential ansatz allows us to advance beyond the lowest non-trivial approximation in σ^2. As long as the number of terms in the sum in Eq. (22) is large enough we can approximate this sum by the integral

$$\begin{gathered} F^{(1)}(\sigma;t) \approx \int_0^\infty d\mu\, e^{-\frac{1}{2}\sigma^2\mu^2}\, \mathcal{W}(\mu;t) = \lambda(t)\int_0^\infty d\mu\, e^{-\frac{1}{2}\sigma^2\mu^2}\, e^{-\lambda(t)\mu} \\ = \int_0^\infty d\eta\, e^{-\eta^2/4\gamma^2}\, e^{-\eta} = \sqrt{\pi}\gamma\, e^{\gamma^2}\left[1-\Phi(\gamma)\right], \\ \lambda(t) \approx \sqrt{\frac{2}{\langle m^2\rangle_{0;t}}}\,, \qquad \gamma = \frac{\lambda}{\sqrt{2}\sigma} \end{gathered} \tag{27}$$

where $\Phi(\gamma)$ is the probability integral. Though fidelity depends, generally, on the time t and the noise level σ separately, these variables combine in the only scaling variable $\gamma = \frac{\lambda(t)}{\sqrt{2}\sigma} = \left(\sqrt{2}\,\sigma/\sigma_c(t)\right)^{-1}$ in the integral approximation (27). This result is in good agreement with our numerical data in a wide though finite interval of the noise level values σ.

It is easy to estimate the η-integral in (27) in the limiting cases of small and large σ. In the first case $\gamma \gg 1$ the range of integration is determined by the second exponential factor. Expanding the Gaussian exponent in the power series we find

$$F^{(1)}(\sigma;t) = 1 - \frac{1}{2\gamma^2} \approx 1 - \frac{\sigma^2}{\sigma_c^2(t)} = 1 - \frac{1}{2}\sigma^2\langle m^2\rangle_{0;t}\,. \tag{28}$$

in agreement with Eq. (22). Quite opposite, when $\gamma \ll 1$ the range of integration is fixed by the Gaussian factor so that the integral approximation (27) yields the expression

$$F^{(1)}(\sigma;t) \approx \sqrt{\pi}\,\gamma = \frac{\sqrt{\pi}}{\sigma\sqrt{\langle m^2\rangle_{0;t}}} \tag{29}$$

that vanishes at any finite time t when $\sigma \to \infty$. This is in obvious contradiction with the finite correct result

$$F_\infty^{(1)}(t) = \mathcal{W}_0(t) = \sum_{n=0}^\infty w_n^2(t) = \frac{1}{2\langle n\rangle_{0;t}+1} = \frac{1}{\sqrt{2\langle m^2\rangle_{0;t}+1}} \tag{30}$$

that follows from the exact eq. (22). Indeed, the integral approximation (27) fails when the noise level σ is so large that the number of noticeable terms of the sum in (22) becomes small. A crossover from Eq. (27) to the asymptotic behavior (30) takes place near some new critical value $\tilde{\sigma}_c(t)$.

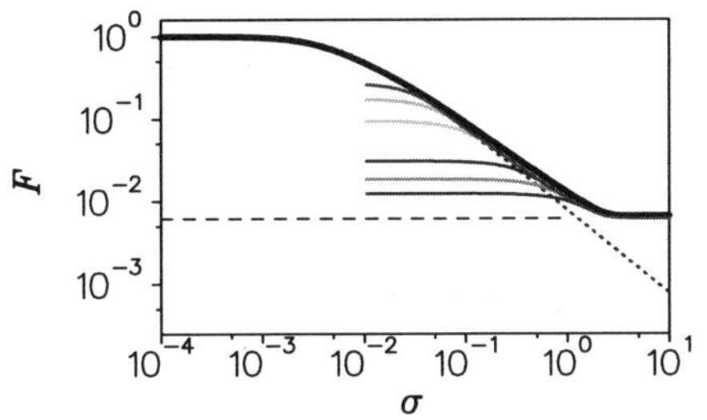

Fig. 3. (Color online). Fidelity (21) as a function of the noise level σ. The black full line shows our numerical data. The red dotted line corresponds to the integral approximation (27); the dashed horizontal line represents the asymptotic behavior (30); the colored full lines represent finite sums with (from bottom to top) more and more terms taken into account.

Persistent noise The critical value $\sigma_c(t) = \sqrt{2/\sum_{\tau=1}^{\tau=t}\langle m^2\rangle_{0;\tau}} \propto t^{-3/2}$ cumulates in this case single-pulse contributions at all moments $1 < \tau \leqslant t$. Though the critical value decreases with time faster than in the case of a single instant perturbation [4], the decrease is power-like as before. This fact results in *a crucial difference* in responses of classical and quantum systems to a weak external perturbation. The insert in Fig.4 elucidates our statement: we apply a weak with a level $\sigma \ll 1$ and compare the times t_c and t_q during which fidelity falls up the same value in the cases $\hbar \ll 1$ (classical exponential regime) and $\hbar \sim 1$ (quantum power-like regime). We have found that $t_q \gg t_c$; more than that, this inequality becomes the stronger the smaller is the noise level σ. The ratio t_q/t_c can reach hundreds times or even more.

4.2 Strong noise limit

In the opposite case of extremely strong noise $\sigma \to \infty$ we insert t times the completeness condition $\sum_{n=0}^{\infty}|n\rangle\langle n| = 1$ while calculating the matrix element $f(\xi;t) \equiv \langle 0|\hat{f}(\xi;t)|0\rangle$, then calculate the average value $\overline{|f(\xi;t)|^2}$ and take at last into account that $\lim_{\sigma\to\infty} e^{-\frac{1}{2}\sigma^2 m^2} = \delta_{m0}$. Finally we arrive at

$$F_\infty(t) = \sum_{n_t}\left|\langle 0|\hat{\mathcal{F}}^{\dagger\ t}|n\rangle\right|^2 \sum_{n_{(t-1)}}\left|\langle n_t|\hat{\mathcal{F}}|n_{(t-1)}\rangle\right|^2 \\ \ldots\sum_{n_2}\left|\langle n_3|\hat{\mathcal{F}}|n_2\rangle\right|^2 \sum_{n_1}\left|\langle n_2|\hat{\mathcal{F}}|n_1\rangle\right|^2\left|\langle n_1|\hat{\mathcal{F}}|0\rangle\right|^2. \tag{31}$$

Only the transition probabilities in the absence of noise are present, the forward evolution being a chain of non-interfering successive transitions. Quite opposite, the backward transition contains the entire set of possible quantum-mechanical paths.

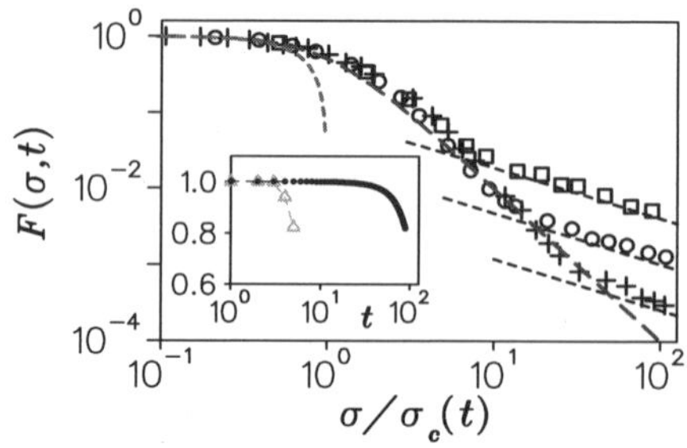

Fig. 4. (Color online). The scaling properties of the fidelity. Crosses, circles and squares correspond respectively to: $\sigma = 0.004, 0.032$ and 0.256. The magenta/grey and black short-dashed lines show the weak and strong noise limits (20) and (34); the blue/grey long-dashed line corresponds to the fit (35). Insert: fidelity decay in the classical Lyapunov ($\hbar = 0.01$, green/gray triangles) and quantum ($\hbar = 1$, black circles) regimes; in the both cases $\sigma = 0,0005$. Notice logarithmic scale along the t axis.

The evolution of the averaged density matrix is of special interest. Eq. (18) shows that in the strong noise limit the density matrix remains diagonal during the whole time of the motion: $\langle n'|\hat{\rho}^{(av)}(\infty;\tau)|n\rangle \equiv \langle n'|\hat{\rho}^{(d)}(\tau)|n\rangle = \delta_{n'n}\, w_n^{(d)}(\tau)$. The evolution equation (18) reduces in this case to

$$w_n^{(d)}(\tau) = \sum_{n'=0}^{\infty} \mathbb{Q}_{nn'}\, w_{n'}^{(d)}(\tau - 1), \tag{32}$$

$$\mathbb{Q}_{nn'} = \left|\langle n|\hat{D}\left(i\frac{g_0^2}{\sqrt{\hbar}}\right)|n'\rangle\right|^2 > 0\,. \tag{33}$$

The matrix $\hat{\mathbb{Q}}$ is symmetric, positively definite, and obeys the condition $\sum_{n'=0}^{\infty} \mathbb{Q}_{nn'} = \langle n|n\rangle = 1$. The equation (32) describes a homogeneous Markov's chain. Notice that the motion does not depend in this limit on the properties of the unperturbed Hamiltonian $\hat{H}^{(0)}$.

It immediately follows now that

$$F(\infty;t) = \sum_{n=1}^{\infty} w_n(0;t)\, w_n^{(d)}(t) \approx \frac{1}{2\langle n\rangle_{\infty;t} + 1}\,. \tag{34}$$

We have used here the exponential ansatz (14) for both w - distributions as well as the practical independence of the mean excitation number $\langle n\rangle_{\sigma;t}$ of the noise. This formula is in a good agreement with our numerical data.

4.3 Moderate noise: scaling properties

Analytical consideration is not possible in the general case of the moderate noise level $\sigma \gtrsim \sigma_c(t)$. Generically, the evolution of the averaged density matrix $\hat{\rho}^{(av)}(\sigma;t)$ is not unitary. This entails state mixing and suppression of the quantum interference, i.e. decoherence. Nevertheless, even if this ratio exceeds unity, the fidelity (7) continues, as Fig. 4 clearly demonstrates, to depend only on the ratio $\sigma/\sigma_c(t)$ up to the time $t_{(dec)}(\sigma)$, when the full decoherence takes place and the evolution becomes Markovian.

A simple fit (compare with [4])

$$F_f(\sigma;t) = \frac{1}{1+\sigma^2/\sigma_c^2(t)}, \quad t < t_{(dec)}(\sigma) \tag{35}$$

describes our numerical data rather well (Fig. 4). With the help of this fit, the decoherence time is estimated as $t_{(dec)}(\sigma) = \sqrt{\frac{6\hbar}{D\sigma^2}}$, where $D = g_0^2$ is the classical diffusion coefficient.

5 Reversibility: von Neumann entropy and purity

The state mixing induced by the averaging over the noise realizations leads to loss of memory about the initial state. Thereupon the notion of the invariant (independent of the basis) von Neumann entropy

$$\mathcal{S}(\sigma;t) = -\mathrm{Tr}\left[\hat{\rho}^{(av)}(\sigma;t)\ln\hat{\rho}^{(av)}(\sigma;t)\right] \tag{36}$$

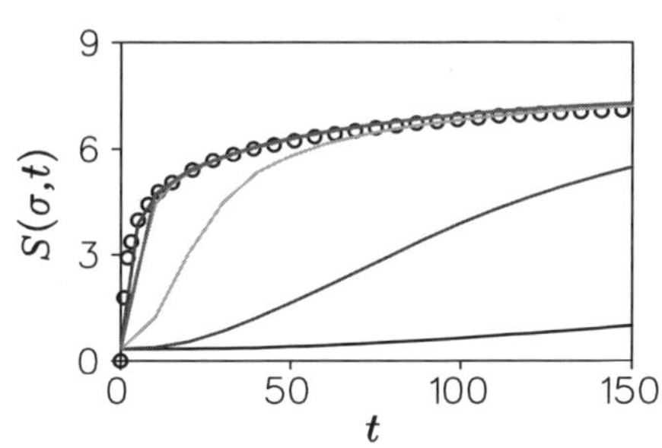

Fig. 5. (Color online). Von Neumann entropy versus the time. From bottom to top (black, blue, green, magenta and red): $\sigma = (0.125, 1, 8, 64, 512)\cdot 10^{-3}$. Circles show the information Shennon entropy $\mathcal{I}(t)$, see eq. (17).

This entropy increases with time monotonically, Fig. 5, approaching the function $\mathcal{I}(t)$ (17) from below when $t \to t_{(dec)}$. After the full decoherence takes place, the system occupies the whole phase volume accessible at the running degree of excitation $\langle n\rangle_{\sigma;t} \approx \langle n\rangle_{\infty;t} = \langle|m|\rangle_{\infty;t}$ thus reaching a sort of equilibrium. Henceforth the phase volume expands "adiabatically": the entropy $\mathcal{S}(\sigma; t > t_{(dec)}) \approx \mathcal{I}(t)$ remains practically constant when $\langle n\rangle_{\infty;t} \gg 1$. In particular case of the strong noise limit, $(\sigma \to \infty,\ t_{(dec)} \to 0)$, the averaged density matrix is diagonal and the von Neumann entropy equals

$$\begin{aligned}\mathcal{S}_\infty(t) &= -\sum_{n=0}^{\infty} w_n^{(d)}(t) \ln w_n^{(d)}(t) \\ &\approx \ln\langle n\rangle_{\infty;t} + 1 + \frac{1}{2\langle n\rangle_{\infty;t}} + \dots, \quad (t \gg 1)\,.\end{aligned} \tag{37}$$

We have arrived at a remarkable connection $\mathcal{S}_\infty(t) = \mathcal{I}(t)$. Similar connection between the information and invariant von Neumann entropies has been discovered also in the theory of random band matrices [10].

Another interesting aspect of the quantum dynamics is the degree of reversibility of motion influenced by a permanent noise acting during the backward as well as the forward evolution. This degree can naturally be measured by the mean overlap (Peres fidelity) of the initial state $\hat{\rho}(0)$ and the state $\hat{\rho}(0|\{\xi\},\{\xi'\};t)$ formed during the forward evolution for some time t under the influence of a stationary noise with the level σ and a history $\{\xi\}$ and then backward evolution for the same time under the same noise level but with an independent history $\{\xi'\}$:

$$\begin{aligned}\mathbb{F}(\sigma;t) = &\overline{\mathrm{Tr}\left[\hat{\rho}(0)\,\hat{\rho}(0|\{\xi\},\{\xi'\};t)\right]} = \\ &\overline{\mathrm{Tr}\left[\hat{\rho}(\{\xi\};t)\,\hat{\rho}(\{\xi'\};t)\right]} = \\ &\mathrm{Tr}\left[\hat{\rho}^{(av)}(\sigma;t)\,\hat{\rho}^{(av)}(\sigma;t)\right] \equiv \mathbb{P}(\sigma;t)\,.\end{aligned} \tag{38}$$

The quantity $\mathbb{P}(\sigma;t)$ is referred for as *purity* [6]. It can be expressed in terms of the mean number of harmonics $\langle|\mathfrak{m}|\rangle_{\sigma;t}$ of the *averaged* Wigner function. The corresponding probability distribution is related to the averaged density matrix $\hat{\rho}^{(av)}(\sigma;t)$ as

$$\mathbb{W}_m(\sigma;t) = (2-\delta_{m0})\frac{\sum_{n=0}^{\infty}\left|\langle n+m|\hat{\rho}^{(av)}(\sigma;t)|n\rangle\right|^2}{\mathbb{P}(\sigma;t)}\,. \tag{39}$$

(to be confronted with eq.(13)). This means in particular that, in the exponential approximation for mixed states [4],

$$\begin{aligned}\mathbb{P}(\sigma;t) = &\frac{1}{\mathbb{W}_0(\sigma;t)}\sum_{n=0}^{\infty}\left[w_n^{(av)}(\sigma;t)\right]^2 \approx \\ &\frac{2\langle|\mathfrak{m}|\rangle_{\sigma;t}+1}{2\langle n\rangle_{\sigma;t}+1} \xrightarrow{\sigma\to\infty} \frac{1}{2\langle n\rangle_{\infty;t}+1}\,.\end{aligned} \tag{40}$$

Contrary to $\langle|m|\rangle_{\sigma;t}$, the mean value $\langle|\mathfrak{m}|\rangle_{\sigma;t}$ strongly depends on the noise level and vanishes rapidly when σ grows. Comparison with Eq. (34) shows that, as in [4], the degrees of reversibility and sensitivity to external perturbations are directly connected, $\mathbb{F}(\infty;t) = F(\infty;t)$. Notice also the relation $\mathcal{S}_\infty(t) \approx -\ln \mathbb{P}_\infty(t)$, $(t \gg 1)$ that follows from Eqs. (37,40).

6 Summary

The main goal of this Letter was to investigate in detail the the dynamics of a classically chaotic quantum system with few (one in our illustrative model) degrees of freedom affected by a persistent external noise under the condition that the Ehrenfest time interval is so short that the classical-like exponential instability does not show up. We have shown first that the noise weakly influences the complexity of the quantum state, which we characterize by the mean number $\langle|m|\rangle_{\sigma;t}$ of θ-harmonics of the Wigner function. This number almost does not depend on the noise realization (self-averaging property) as well as on the level of the noise. At the same time the noise efficiently washes off fluctuations of the corresponding probability distribution thereby displaying the universal regular exponential decay of the coarse-grained distribution that describes the features of the motion independent of the realization of the noise.

The Peres fidelity that specifies a quantitative measure of sensitivity of the motion to the noise utilizes the density matrix $\hat{\rho}^{(av)}(\sigma;t)$ *averaged over the noise.* The sensitivity remains weak until the noise level σ exceeds some critical value. We have proved that with the assumptions indicated above the decrease of this critical value is power-like, $\sigma_c(t) \approx 1/\sqrt{\sum_{\tau=1}^{\tau=t}\langle|m|\rangle^2_{\infty;\tau}} \propto t^{-3/2}$.

A scaling behavior has been discovered: the Peres fidelity depends only on the ratio $\sigma^2/\sigma_c^2(t)$ in a wide interval of the noise level up to some value $\sigma_d(t)$ that is considerably larger than the critical value $\sigma_c(t)$. The scaling is destroyed only under influence of even a stronger noise $\sigma > \sigma_d(t)$. The evolution becomes Markovian in this case. This implies that decoherence takes place at the time $t_{(dec)}(\sigma) \sim \sqrt{\frac{\hbar}{D\sigma^2}}$. The information entropy and the invariant von Neumann entropy coincide, $\mathcal{S}_\sigma(t) \Rightarrow \mathcal{I}(t)$, when $t \gtrsim t_{(dec)}(\sigma)$. They are identical in the limit $\sigma \to \infty$. We have also noticed that the reversibility of the motion influenced by a persistent noise is measured by the purity of the state at the moment of time reversal, $\mathbb{F}(\sigma;t) = \mathbb{P}(\sigma;t)$.

References

1. D.L. Shepelyansky, Physica D **8**, 208 (1983); G. Casati, B.V. Chirikov, I. Guarneri, and D.L. Shepelyansky, Phys. Rev. Lett. **56**, 2437 (1986).
2. A. Peres, Phys. Rev. A **30**, 1610 (1984).
3. T. Gorin, T. Prosen, T.H. Seligman, and M. Žnidarič, Phys. Rep. **435**, 33 (2006).
4. V.V. Sokolov and O.V. Zhirov, Europhys. Lett.**84**, 30001 (2008); V.V. Sokolov, O.V. Zhirov, G. Benenti, and G. Casati, Phys. Rev. E **78**, 046212 (2008); see also in: *Topics on Chaotic Systems: Selected Papers from CHAOS2008 International Conference*, World Scientific, Page 314 (2009); G. Benenti and G. Casati, Phys. Rev. E **79**, 025201(R) (2009).
5. V.V. Sokolov and O.V. Zhirov, arXiv:0909.4179 [quant-ph] 28 Oct 2009, to appear in Europhys. Lett.
6. F.M. Cucchietti, D.A.R. Dalvit, J.P. Paz and W.H. Zurek, Phys. Rev. Lett.**91**, 210403 (2003); T. Gorin, T.H. Seligman, Phys. Lett. A **309**, 61 (2003).
7. R.J. Glauber, Phys. Rev. **131** 2766 (1963).
8. G.S. Agarwal and E. Wolf, Phys. Rev. D **2**, 2161 (1970); *ibid.* 2187.
9. A. Peres and D. Terno, Phys. Rev. E **53**, 284 (1996).
10. V.V. Sokolov, B.A.Brown and V. Zelevinsky, Phys. Rev. E **58**, 65 (1998).

On logistic-like iterative maps

Dimitrios A. Sotiropoulos

Department of Sciences, Technical University of Crete, Chania, 73100 Greece
Email: dimsotirop@yahoo.com

Abstract: Logistic-like first order iterative maps, defined here as $x_{n+1} = r\, x_n^{\lambda}(1-x_n)^{\mu}$, are examined. The parameters r, λ and μ are positive real numbers, while the variable x and its map range from 0 to 1, the latter yielding the upper value of r for which full chaos is obtained. Depending on the values of λ and μ, the resulting x's can have a totally different behavior from those of the logistic map, given by λ = μ =1. The focus here is on fixed points since their existence, for given values of λ and μ, is necessary for obtaining chaotic x's. The purpose of the paper is four-fold: first, to define regions of existence for the fixed point(s) in terms of the parameters r, λ and μ; second, to determine the nature of the fixed points, whether they are attractors (stable), repellors (unstable) or super-stable, according to the values of the parameters; third, to define those maps for which the fixed points can be written in explicit algebraic form; and fourth, for iterative nearby maps, to obtain their fixed points in an approximate algebraic form in terms of the exact fixed points. The approximation is based on Newton's method, one step from the nearest iterative map whose fixed points can be obtained exactly, in explicit form. The validity of the fixed point approximation depends on the stability of the fixed points and is, subsequently, established in respect of well defined surfaces of the parameters r, λ and μ.
Keywords: chaotic maps, logistic maps, iterative maps, recurrence equations, fixed points, attractors, stability.

1. Introduction

The present paper deals with first order nonlinear iterative maps. They are important in a variety of scientific fields such as computational mathematics, computational physics, nonlinear mechanics, electronics and chemical engineering. Applications are found, among others, in materials mixing (e.g. Wiggins & Ottino [1]), in digital watermarking (e.g. Mooney, [2]), in population growth modeling (e.g. Marotto [3]) in economic macro-models (e.g. Stutzer [4]), and in music composition (e.g. A. Sotiropoulos [5]).

Logistic-like are the iterative maps under examination here. These maps are defined here by the following recurrence equation:

$$x_{n+1} = r\, x_n^{\lambda}(1-x_n)^{\mu} \tag{1}$$

in which the parameters r, λ and μ are positive real numbers, while the variable x and its map range from 0 to 1. A review and analysis of the map

Chaos Theory: Modeling, Simulation and Applications
C. H. Skiadas, I. Dimotikalis and C. Skiadas (Eds)

for specific values of the exponent parameters λ and μ, such as $\lambda=1$ and $\mu<1$, may be found in a recent monograph by Skiadas & Skiadas [6]. A detailed study of the map's features for $\lambda = 2$ and $\mu = 1$ is given by Marotto [3] and Gottlieb [7]. The chaotic behavior of the case $\lambda=\mu=1/2$ was visited recently by V. Sotiropoulos [8], while that of the double logistic map by A. Sotiropoulos [5]. On maps of related functional dependence, Stutzer [4] studied the iterative map $x_{n+1} = r\, x_n\, (1-x_n^{1/2})$ as a macro-economic dynamic model, while Gottlieb [9] analyzed a similar iterative map of the form $x_{n+1} = r\, x_n^{3/2}\, (1-x_n^{1/2})$. Further, Skiadas & Skiadas [6] looked at the chaotic behavior of these and other maps with or without delay, including the map of the generalized rational iteration model $x_{n+1} = x_n + r\, x_n\, (1-x_n)/[1-(1-\sigma)x_n]$ in which σ is positive. This innovation diffusion modeling finds a number of applications including the social sciences as demonstrated recently by Skiadas and Skiadas [10].

In the present study, there are no limitations on the values of λ and μ other than they need to be positive real numbers. These values include the special cases, $\lambda = \mu =1$ which define the classical logistic map, $\lambda=\mu=1/2$ which yield the elliptic map, and $\lambda =2$, $\mu =1$ which give the map describing flame development in combustion. The behavior of the resulting x's for each map depends exclusively on the values of λ and μ and it can be totally different from that of the logistic map.

The focus in the present paper is on fixed points since for given values of λ and μ their existence is a necessary condition for obtaining chaotic x's. First, regions of existence of the fixed point(s) are defined in terms of the parameters r, λ and μ; second, the nature of the fixed points is determined, whether they are attractors (stable), repellors (unstable) or super-stable, according to the values of the parameters; third, maps are defined for which the fixed points can be written in explicit algebraic form; and fourth, for iterative nearby maps, the fixed points are obtained in an approximate algebraic form in terms of the exact fixed points. The approximation is based on Newton's method, one step from the nearest iterative map with fixed points that can be obtained exactly, in explicit form. Tests of the approximation are for the difference between the two pairs of λ and μ values, for each value of the free parameter r. For maps that are further away, the fixed points are obtained within an interval, with its upper and lower bounds given exactly by the fixed points of the nearest neighboring map on each side, respectively, of the map in question.

2. The upper limit of the parameter r

The mapping function of the logistic-like first order recurrence equation given in (1) is:

$$f(x) = r\, x^{\lambda}\, (1-x)^{\mu} \;;\; 0 \le x \le 1, \quad r, \lambda, \mu > 0 \qquad (2)$$

However, the parameter r is not free since its upper value is limited by the requirement that the maximum value of f be equal to one. If the values of λ and μ are such that the iteration x's exhibit chaotic behavior, then it is the maximum value of r that yields full chaos. As mentioned, the r maximum value is obtained from the maximum value of f(x) being equal to one. This maximum value of f(x) is at $x=x_0$ for which the derivative of f(x) of (2) is equal to zero, other than x=0 and/or x=1 which may yield a minimum, giving

$$x_0 = \lambda/(\lambda+\mu) \tag{3}$$

Then, $f(x_0) = 1$, results in the following upper limit of r:

$$r_{max} = (1 + \lambda/\mu)^{\mu} (1 + \mu/\lambda)^{\lambda} \tag{4}$$

which is symmetric in λ and μ and increasing with increasing λ and μ. Note that when $\lambda=\mu$, $r_{max}=2^{2\lambda}$, meaning that an increase in λ by ½ results in an increase in r_{max} by a power of 2. Of particular interest may be the maximum r's for the following (λ,μ) values: (1/2,1/2), $r_{max}=2$; (3/4,3/4), $r_{max}=2^{3/2}$; (1,1), $r_{max}=4$ which is well known corresponding to the logistic map; (2,1) or (1,2), $r_{max}=27/4$; (3/2,3/2), $r_{max}=8$; and (2,2), $r_{max}=16$.

3. The existence of fixed points

Fixed points of a first order iterative map are defined as solutions of the equation x=f(x) where f(x) is the mapping function. Since the mapping function defined in (2) is convex, this equation will have one unique solution, its fixed point, if the slope of f(x) at x=0 is greater than or equal to one, the slope of x. When $f'(0)=1$, the fixed point will be given by x=0. However, if $f'(0)=0$ then there will either be no fixed points or two fixed points depending on whether the line y=x crosses the curve y=f(x). This means that, for this case, there will be a minimum value of r above which the iterative map will have two fixed points. The mapping function in (2) is dependent on the parameters λ and μ, therefore, the values of λ and μ define regions of existence of fixed point(s). These are obtained following the above discussion. From (2), the slope of f (x) at x=0 becomes

$$\begin{aligned} f'(0) &= \infty\,, \quad \lambda<1 \\ &= r\,, \quad \lambda=1 \\ &= 0\,, \quad \lambda>1 \end{aligned} \tag{5}$$

independent of the parameter μ. From (5) it is concluded that:

a) when $\lambda<1$ there is *one* fixed point no matter what the value of the parameter r is, but with its location dependent on the value of r,

b) when $\lambda=1$ there is *one* fixed point only if r is greater than or equal to one. This case includes the logistic map where $\lambda=1$, $\mu=1$. If r=1 then the fixed point is located at x=0, and

c) when $\lambda>1$ there are either *no* fixed points or *two* fixed points depending on the value of r. The minimum value of r above which, for this case, there will be two fixed points is obtained as follows. The fixed point equation, x=f(x), for f(x) as given by (2), may be written as

$$r = x^{1-\lambda}(1-x)^{-\mu} \tag{6}$$

Then, the condition $r'(x)=0$ yields the following location (fixed point) of the minimum r

$$x^* = [1 + \mu/(\lambda-1)]^{-1} \tag{7}$$

It is noted that x^* lies in the interval [0,1] only if λ is greater than or equal to one, with $\lambda=1$ giving $x^*=0$ and $r_{min}=1$. Substitution of (7) into (6) gives the following value of r above which there are two fixed points when $\lambda>1$

$$r_{min} = [1+\mu/(\lambda-1)]^{\lambda-1}\,[1+(\lambda-1)/\mu]^{\mu} \tag{8}$$

It is remarked that when $\lambda<1$, the slope of r(x) is nonzero and the minimum r is zero.

Of particular interest may be the minimum r's in order to have fixed points, for the following (λ,μ) values: (1/2,1/2), $r_{min}=0$; (3/4,3/4), $r_{min}=0$; (1,1), $r_{min}=1$ which is well known corresponding to the logistic map; (3/2,3/2), $r_{min}=16/3^{3/2}$; (2,1), $r_{min}=4$; and (2,2), $r_{min}=27/4$.

In summary, for the iterative map at hand defined by (1), the regions of existence of fixed points are given as follows:

$\lambda<1$, $r = [0, (1 + \lambda/\mu)^{\mu}(1 + \mu/\lambda)^{\lambda}]$: *one* fixed point
$\lambda=1$, $r = [1, (1 + 1/\mu)^{\mu}(1 + \mu)]$: *one* fixed point (9)
$\lambda>1$, $r = [(\lambda+\mu-1)^{\lambda+\mu-1}(\lambda-1)^{1-\lambda}\mu^{-\mu}, (1 + \lambda/\mu)^{\mu}(1 + \mu/\lambda)^{\lambda}]$: *two* fixed points

with all other regions of the parameters λ, μ and r having *no* fixed points. It is remarked that chaos will be produced in regions contained within the parameter regions defined by (9).

4. The stability of fixed points

Fixed points are classified as superstable if $f'(x_f)=0$ with x_f the solution of x=f(x), stable (attractors) if $|f'(x_f)|<1$, and unstable (repellors) if $|f'(x_f)|>1$. If $f'(x_f)=1$ the fixed points are non-hyperbolic and stability must be examined in each case. For the iterative map at hand $f'(x_f)$ becomes

$$f'(x_f) = \lambda - \mu x_f / (1-x_f) \tag{10}$$

which, for $f'(x_f)=1$, yields $x_f=x^*$ where x^* is given by (7). Thus, as expected all the fixed points are hyperbolic. The stability of the hyperbolic fixed points is now examined in detail.

a) *Superstable* fixed points. The condition $f'(x_f)=0$ together with (10) gives

$$x_f = \lambda/(\lambda+\mu) \tag{11}$$

which substituted in x=f(x) yields

$$r = (1 + \lambda/\mu)^{\mu} (1 + \mu/\lambda)^{\lambda-1} \tag{12}$$

It is noted that the superstable fixed point is identical with the location of the peak of f(x) given by (3).

b) *Stable* fixed points – *Attractors*: The condition $|f'(x_f)| < 1$ together with (10), after some algebraic manipulations having re-substituted the original map for x_f, yields:

$$x_f < \lambda/(\lambda+\mu):\ \lambda \leq 1 \text{ or } (\lambda>1 \text{ and } x_f>[1 + \mu/(\lambda-1)]^{-1}) \tag{13}$$

$$x_f > \lambda/(\lambda+\mu):\ x_f < [1 + \mu/(\lambda+1)]^{-1} \tag{14}$$

as the conditions for stable fixed points. From the discussion on section 3 above, it is seen that r is monotonically increasing for both cases in (13) since for $\lambda \leq 1$ the minimum r is at x =0 and for $\lambda>1$ the minimum r is at the x given by (7) which is the minimum x appearing in (13). Therefore, using (12) and (8) the two conditions in (13) may equivalently be written as

$$r < (1 + \lambda/\mu)^{\mu} (1 + \mu/\lambda)^{\lambda-1},\quad \lambda \leq 1 \tag{15a}$$

$$[1+(\lambda-1)/\mu]^{\mu} [1+\mu/(\lambda-1)]^{\lambda-1} < r < (1 + \lambda/\mu)^{\mu} (1 + \mu/\lambda)^{\lambda-1},\ \lambda>1 \tag{15b}$$

defining regions of the parameters λ,μ and r inside which there are attracting fixed points. Furthermore, since r is also monotonically increasing in the interval of x defined by (14), condition (14) can be equivalently written as

$$(1 + \lambda/\mu)^{\mu} (1 + \mu/\lambda)^{\lambda-1} < r < [1+(\lambda+1)/\mu]^{\mu} [1+\mu/(\lambda+1)]^{\lambda-1},\ \lambda>0 \tag{16}$$

defining the only other parameter region besides (15a) and (15b) that yields attracting fixed points. The three regions of fixed points stability given by (15a), (15b) and (16) can be combined as:

$$r < [1+(\lambda+1)/\mu]^{\mu} [1+\mu/(\lambda+1)]^{\lambda-1},\quad \lambda \leq 1 \tag{17a}$$

$$[1+(\lambda-1)/\mu]^{\mu} [1+\mu/(\lambda-1)]^{\lambda-1} < r < [1+(\lambda+1)/\mu]^{\mu} [1+\mu/(\lambda+1)]^{\lambda-1},\ \lambda>1 \tag{17b}$$

c) *Unstable* fixed points – *Repellors*: Having determined the regions of stability for the fixed points as given by (17a) and (17b), their regions of instability are complementary to them and, therefore, are given by inspection of (17a) and (17b), as

$$r > [1+(\lambda+1)/\mu]^{\mu} [1+\mu/(\lambda+1)]^{\lambda-1},\quad \lambda>0 \tag{18a}$$

$$r < [1+(\lambda-1)/\mu]^{\mu} [1+\mu/(\lambda-1)]^{\lambda-1},\quad \lambda>1 \tag{18b}$$

inside which the fixed points are repelling.

5. The solution of the fixed points equation

The fixed points equation defined by x=f(x), using (2) yields

$$x^{(1-\lambda)/\mu} + r^{1/\mu} x - r^{1/\mu} = 0 \quad (19)$$

There are three cases for the values of (1-λ)/μ for which the left hand side of equation (19) becomes a polynomial. Namely, they are: 0, integer, or the reciprocal of an integer, resulting in an algebraic equation of certain degree. The solution of the fixed points equation (19) will now be examined for each of these three cases:

a) (1-λ)/μ = 0. This is equivalent to λ=1, a case which includes the logistic map (μ=1). With this condition, equation (19) becomes linear, with the following fixed point solution

$$x = 1 - r^{-1/\mu}, \qquad \lambda=1 \quad (20)$$

which monotonically decreases from 1 to zero as μ varies from 0 to ∞.

b) (1-λ)/μ = n, where n is an integer. With this condition, the degree of the resulting algebraic fixed points equation (19) depends on whether n is positive (λ<1) or negative (λ>1), as

$$x^{n} + r^{1/\mu} x - r^{1/\mu} = 0, \qquad n>0 \ \ (\lambda<1) \quad (21)$$

$$x^{(1-n)} - x^{-n} + r^{-1/\mu} = 0, \quad n<0 \ \ (\lambda>1) \quad (22)$$

Equations (21) and (22) because of their different forms, for the same absolute value of n, will have a different number of fixed points as solutions, with different parameter regions of existence. Indeed, from (9) in section 3 above it is seen that equation (21) will yield one solution in the interval [0,1] provided that the mapping function falls within the same interval, while (22) will yield either no solutions or two solutions under the same provisions. If n is even, the degree of the algebraic equations (21) and (22) becomes n/2 with the substitution y=x+1/x.

Fixed points solutions of equations (21) and (22) are obtained in *explicit* algebraic form for several values of n. Simple explicit solutions include:

$$x = (1 + r^{-1/\mu})^{-1}, \qquad n=1 \ \ (\lambda=1-\mu) \quad (23)$$

which itself includes the elliptic case, λ=μ=1/2,

$$x = \tfrac{1}{2}\,[1 \pm (1 - r^{-1/\mu})^{1/2}], \qquad n=-1 \ \ (\lambda=1+\mu) \quad (24)$$

with the maps defined by μ=1/2 and μ=1 included in this, and

$$x = \tfrac{1}{2}\, r^{1/\mu}\, [\,(1 + 4r^{-1/\mu})^{1/2} - 1], \qquad n=2 \ \ (\lambda=1-2\mu) \quad (25)$$

the other root of the quadratic equation for this case being larger than one, and, thus, unacceptable.

c) $(1-\lambda)/\mu = 1/n$, a reciprocal integer. Again, in this case the degree of the resulting algebraic fixed points equation (19) depends on whether n is positive ($\lambda<1$) or negative ($\lambda>1$), as

$$(1-x)^n - r^{-n/\mu} x = 0, \qquad n>0 \ (\lambda<1) \tag{26}$$

$$(1-x)^{-n} x - r^{n/\mu} = 0, \qquad n<0 \ (\lambda>1) \tag{27}$$

Equations (26) and (27) because of their different forms, for the same absolute value of n, will have a different number of fixed points as solutions, with different parameter regions of existence. Indeed, from (9) in section 3 above it is seen that equation (26) will yield one solution in the interval [0,1] provided that the mapping function falls within the same interval, while (27) will yield either no solutions or two solutions under the same provisions. If n is even, the degree of the algebraic equations (21) and (22) becomes n/2 with the substitution $y=x+1/x$.

Fixed points solutions of equations (26) and (27) are obtained in *explicit* algebraic form for several values of n. It is noted that the two fixed points for the map $\lambda=\mu=2$, are given as the two (of the three) real solutions of the cubic (27) that fall in the interval [0, 1]. As equations (26) and (27) become identical to equations (21) and (22) respectively for n=1 and n=-1, simple explicit solutions of (26) and (27) besides the ones given by (23) and (24) for these two cases, include:

$$x = (1 + 1/2\, r^{-2/\mu}) \{1 - [1-(1 + 1/2\, r^{-2/\mu})^{-2}]^{1/2}\}, \qquad n=2 \ (\lambda=1-\mu/2) \tag{28}$$

the other root of the quadratic equation for this case being larger than one, and, thus, unacceptable.

6. Approximate fixed points

In section 5 above there were examined cases of the parameters λ and μ for which fixed points are obtained exactly in explicit algebraic form. However, there are (λ, μ) parameter regions inside which the fixed points, as solutions of equation (19), can only be obtained numerically. One of the goals of the present study is to obtain analytical solutions of the fixed points equation. To this end, besides the exact fixed points obtained in section 3 above, fixed points are obtained approximately for maps that are nearby maps of which the fixed points are known exactly, in explicit algebraic form. The approximation is based on Newton's method one step away from a nearby map whose fixed points are known exactly. This gives the fixed point of a map, f(x), as

$$x \approx x_f - g(x_f) / g'(x_f), \qquad g(x)=x - f(x) \tag{29}$$

where x_f is the exact fixed point of a nearby map, while the accurate fixed point of the map in question is solution of the equation g(x)=0. In section 5 it was discussed that the maps whose fixed points are obtained exactly in explicit algebraic form are a subset of those in the parameter regions defined by (λ-1)/μ=m, where m is zero, integer or reciprocal integer. The nearby map will, therefore, be defined by (λ-1)/μ=m-ε, in which ε<<1. With this parameter definition, equations (31) together with (2) yield

$$x = x_f \{1 - r^{-1/\mu} x_f^{m-1}(x_f^{-\varepsilon} - 1)/[1 + x_f^{-1-\varepsilon}(m-\varepsilon)(1 - x_f)]\} \qquad (30)$$

It is noted that no explicit use of the fact that ε is small was made in deriving (30). It was only used implicitly in writing down the approximate equation in (29).

The fixed point approximation as given by (30) is valid on the assumption that $|g'(x_f)| < 1$. That is, the approximation is valid if the map f(x) has a stable (attracting) point at the fixed point, x_f, of the nearby map. For the functional dependence of the maps under examination here as given by (2), the stability will occur if the map f(x) is stable at a nearby point, which in our case is the map's fixed point. In other words, the approximation (30) is valid if $|f'(x)| < 1$ with x solution of the second equation in (29). The stability of maps given by f(x) as defined in (2) at their fixed points, was discussed extensively in section 4 above. For any positive λ and μ, parameter regions were given by (17a) and (17b) inside which stable fixed points exist. Therefore, with (λ-1)/μ=m-ε , the fixed point approximation given by (30) is valid under the conditions

$$r < [1+m-\varepsilon+2/\mu]^{\mu}\,[1+(m-\varepsilon+2/\mu)^{-1}]^{\mu(m-\varepsilon)}\,, \quad m \le \varepsilon \qquad (31a)$$

$$[1+m-\varepsilon]^{\mu}[1+(m-\varepsilon)^{-1}]^{\mu(m-\varepsilon)} < r < [1+m-\varepsilon+2/\mu]^{\mu}[1+(m-\varepsilon+2/\mu)^{-1}]^{\mu(m-\varepsilon)}, \quad m > \varepsilon \qquad (31b)$$

These conditions define three dimensional regions of the parameters r, m (or λ) and μ bounded by surfaces given by the limiting values of the inequalities in (31a) and (31b), inside which the fixed point approximation (32) is valid.

A demonstration of the approximation scheme for fixed points is now given for the map defined by (2) with λ=μ=0.9, that is, a map near the logistic map (λ=μ=1). In this case, m=0, ε=0.1 and (31a) gives the r's for which the approximation is valid as r < 2.78. Two values of r are picked, r=2 and r=2.5. For these r's the fixed points of the logistic map are given by (20) with μ=1 as 0.5 and 0.6 respectively. Furthermore, the fixed points of the nearby map (λ=μ=0.9) can be obtained by the numerical solution of the fixed point equation as 0.565 and 0.655 for the two values of r=2, 2.5 respectively. In turn, (30) gives the approximate fixed points as 0.574 and 0.661 respectively. It is seen that the approximate fixed points are off by 1.5% and 0.8% from the numerically calculated ones.

For maps that are not near maps whose fixed points can be obtained exactly in explicit algebraic form, the fixed points can either be numerically calculated or be given within an interval, with its upper and lower limits given exactly by the fixed points of the nearest neighboring map in explicit algebraic form on each side, respectively, of the map in question.

References

[1] S. Wiggins and J.M. Ottino. Foundations of chaotic mixing. *Phil Trans. Royal Soc. A*, 362 (1818): 937-970, 2004.

[2] A. Mooney, J. G. Keating, and I. Pitas. A comparative study of chaotic and white noise signals in digital watermarking. *Chaos, Solitons and Fractals,* 35: 913–921, 2008.

[3] F. R. Marotto. The dynamics of a discrete population model with threshold. Math. Biosci. 58:123-128, 1982.

[4] M. J. Stutzer. Chaotic dynamics and bifurcation in a macro-model. Journal of Economic Dynamics and Control 2: 353-376, 1980.

[5] A. D. Sotiropoulos. Composing chaotic music from the letter m. In *Chaos Theory: Modeling, Simulation And Applications*, C. H. Skiadas, I. Dimoticalis and C. Skiadas, eds. World Scientific, 2011.

[6] C. H. Skiadas and C. Skiadas. *Chaotic Modelling and Simulation.* CRC Press, N. York, 2009. [7] H. P. W. Gottlieb. Thresholds in a class of iterative maps. *International Journal of MathematicalEducation in Science and Technology* 30 (1): 93-101, 1999.

[8] V. D. Sotiropoulos. The rainbow effect on composing chaotic algorithmic music. In *Chaos Theory: Modeling, Simulation and Applications*, C. H. Skiadas, I. Dimoticalis and C. Skiadas, eds. World Scientific, 2011.

[9] H. P. W. Gottlieb. Properties of some generalized logistic maps. In *Complex Systems: Mechanism of Adaptation*, R. J. Stonier and X. H. Yu, eds. Amsterdam: IOS Press, pp. 307-312, 1994.

[10] C. H. Skiadas and C. Skiadas. Innovation Diffusion Modeling: The Deterministic, Stochastic and Chaotic Case, *Nonlinear Dynamics,Psychology, & Life Sciences* (in press).

PART II

Invited and Contributed Papers

Improved ε expansion in theory of turbulence: Calculation of Kolmogorov constant and skewness factor

L. Ts. Adzhemyan[1], **M. Hnatich**[2,3] **and J. Honkonen**[4]

[1]Department of Theoretical Physics, St. Petersburg University,
Ulyanovskaya 1, St.Petersburg, 198504 Petrodvorets, Russia
Email: Loran.Adzhemyan@pobox.spbu.ru
[2]P.J. Šafarik University, Faculty of Sciences, Moyzesova 16,
04001 Košice, Slovakia
Email: hnatic@saske.sk
[3]Institute of Experimental Physics Slovak Academy of Sciences,
Watsonova 47, 04001 Košice, Slovakia
[4]Department of Military Technology, National Defence University,
P.O. BOX 7, 00861 Helsinki, Finland
Email: juha.honkonen@helsinki.fi

Abstract: The ε expansion in the theory of developed d-dimensional turbulence is amended by renormalization of random forcing in the stochastic Navier-Stokes equation, which takes into account additional divergences in two-dimensional space. The Kolmogorov constant and skewness factor calculated in the one-loop approximation of the improved perturbation theory are in reasonable agreement with their recommended experimental values.
Keywords: developed turbulence, stochastic Navier-Stokes equation, renormalization group.

1 Introduction

Description of developed turbulence based on the stochastic Navier-Stokes equation in the form of an effective field-theoretic model opens new possibilities for understanding of both the Kolmogorov scaling and the deviation therefrom as well as for calculation of principal physical quantities. The renormalization-group (RG) approach with its various perturbation schemes (e.g. the famous ε expansion, see [1], with effective analytical and numerical algorithms for evaluation of quantities involved forms a robust method, which allows to put into practice this scenario.

A specific feature of the renormalization-group approach and the ε expansion in the theory of developed turbulence is that the formal small parameter ε is not connected with the space dimension and it is determined only by the noise correlator of random forcing in the stochastic Navier-Stokes equation

Chaos Theory: Modeling, Simulation and Applications
C. H. Skiadas, I. Dimotikalis and C. Skiadas (Eds)
 (pp. 89 - 96)

[2,3]. Its physical value $\varepsilon = 2$ is not small [4,5], hence reasonable doubts arise about the effectiveness of such an expansion. For some paramount physical quantities like the critical dimensions of the velocity field and the coefficient of viscosity the ε expansion terminates at the first term due to the Galilei invariance of the theory [2,3]. Therefore, exact values are predicted for these quantities. However, there are other physically important quantities, viz. the skewness factor, the Kolmogorov constant and critical dimensions of various composite operators, for which the ε series do not terminate [6-8], therefore the question about the effectiveness of the expansion remains open.

Consider a quantity A calculated at the fixed point of the RG in the renormalized field theory of developed turbulence. In d dimensions it is a function of the parameters ε and d: $A = A(\varepsilon, d)$. In practice, calculations are often carried out in the ε expansion, whose coefficients for the quantity $A(\varepsilon, d)$ depend on the space dimension d

$$A(\varepsilon, d) = \sum_{k=0}^{\infty} A_k(d)\varepsilon^k. \tag{1}$$

Analysis shows that these coefficients $A_k(d)$ have singularities at dimension $d = 2$, which are not eliminated by the renormalization of the d-dimensional theory [8,9]. These divergences manifest themselves in the form of poles in the parameter $(d-2) \equiv 2\Delta$ in the coefficients of the ε expansion $A_k(d)$, which therefore may be expressed as Laurent series of the form

$$A_k(d) = \sum_{l=0}^{\infty} a_{kl}\, \Delta^{l-k}. \tag{2}$$

Divergences in Δ may be absorbed to suitable additional counterterms, which give rise to a different renormalized field theory. Henceforth, we therefore consider the theory with two formal small parameters ε and Δ, which satisfy the relation $\zeta \equiv \varepsilon/\Delta = const$. This gives rise to the double series

$$A(\varepsilon, d) = \sum_{k=0}^{\infty}\sum_{l=0}^{\infty} \zeta^k a_{kl}\, \Delta^l. \tag{3}$$

Adzhemyan et al [7,8] have recently proposed to improve the ε expansion of the quantity A by a special summation procedure of the double series (3). This improved ε expansion has produced values of the Kolmogorov constant and skewness factor in a notably better agreement with their experimental values than in the usual ε expansion.

In this paper we strive for the same goal by the aid of a suitably modified renormalization procedure. Instead of the minimal subtraction scheme used in [7.8], we use the scheme with a normalization point, in which the renormalization constants are chosen by prescribing to suitable renormalized correlation functions finite values at given wave numbers. Such a calculation (*without* expansion in ε) guarantees true reproduction of terms of the minimal subtraction approach and leads to good agreement with experiment already in the one-loop approximation. From the point of view of renormalization of the d-dimensional theory this approach corresponds to the inclusion in the action (see below) of an infrared irrelevant at $d > 2$ operator, which becomes relevant, when $d \to 2$.

2 Stochastic model of developed turbulence

The statistical model of developed homogeneous and isotropic turbulence of incompressible fluid is based on the stochastic Navier-Stokes equation

$$\nabla_t \varphi_i = \nu_0 \partial^2 \varphi_i - \partial_i \mathcal{P} + f_i, \qquad \nabla_t \equiv \partial_t + (\varphi\partial). \tag{4}$$

Here, $\varphi_i(t, \mathbf{x})$ is transverse vector field of the fluid velocity, $\mathcal{P}(t, \mathbf{x})$ the pressure, $f_i(t, \mathbf{x})$ is the transverse external random force per unit mass and ν_0 stands for the coefficient of kinematic viscosity. The random force f mimics instabilities generated by initial and boundary conditions and it is assumed to have a Gaussian distribution with zero mean and the pair correlation function

$$\langle f_i(t, \mathbf{x}) f_j(0, 0) \rangle \equiv D_{ij}(t, \mathbf{x}; 0, 0) = \frac{\delta(t)}{(2\pi)^d} \int d\mathbf{k}\, P_{ij}(\mathbf{k})\, d_f(k)\, e^{\mathrm{i}\mathbf{k}\mathbf{x}}, \tag{5}$$

where $P_{ij}(\mathbf{k}) = \delta_{ij} - k_i k_j / k^2$ is the transverse projection operator and d the dimension of space. In the framework of the RG approach, the function $d_f(k)$ is assumed to have the powerlike form

$$d_f(k) = D_{10} k^{2-2\Delta-2\varepsilon} + D_{20} k^2, \quad D_{i0} = u_{i0} \nu_0^3 / \bar{S}_d, \;\; i = 1, 2\,, \tag{6}$$

where $\Delta \equiv (d-2)/2$, $\bar{S}_d \equiv S_d/(2\pi)^d$, $S_d \equiv 2\pi^{d/2}/\Gamma(d/2)$ is the surface area of the unit sphere in d-dimensional space and Γ is Euler's Gamma function. The quantities Δ, $\varepsilon > 0$ in (6) play the rôle of two formal small expansion parameters. The physical value $\Delta = 1/2$ corresponds to three-dimensional space. The physical value of the parameter ε is equal to 2 which corresponds to the energy injection by infinitely large eddies. The second quadratic term with the amplitude D_{20} was introduced to absorb the singularities, which appear in the model as $d \to 2$ (see below). The parameters u_{10} and u_{20} in (6) play the rôle of bare charges. The stochastic problem (4), (5) is equivalent to the field-theoretic model with two transverse vector fields $\{\varphi, \varphi'\}$ and the de Dominicis-Janssen action [2]

$$S = \varphi' D \varphi'/2 + \varphi'[-\partial_t \varphi + \nu_0 \partial^2 \varphi - (\varphi\partial)\varphi], \tag{7}$$

where D is the correlator of random force (5). In (7) all necessary integrals over time and space as well as sums over vector indices are implied. Model (7) is logarithmic (i.e. the coupling constants u_{i0} are dimensionless) at $\varepsilon = 0$,

$\Delta = 0$. Ultraviolet (UV) divergences appear in the form of poles in ε, Δ and their linear combinations in the correlation functions of the fields $\{\varphi, \varphi'\}$. The standard analysis of canonical dimensions supplemented by arguments of Galilei invariance shows that for $d \geq 2$ superficial UV divergences in model (7) appear only in the one-irreducible functions $\langle\varphi'\varphi\rangle_{1-\mathrm{ir}}$, $\langle\varphi'\varphi'\rangle_{1-\mathrm{ir}}$ and the counterterms are of the form $\varphi'\partial^2\varphi$ and $\varphi'\partial^2\varphi'$.

These counterterms give rise to multiplicative renormalization of the parameters ν_0 and u_{10}, u_{20} in action (7) with two independent renormalization constants for the coefficient of viscosity ν_0 and for the amplitude D_{20}. Relations between renormalized and bare parameters are

$$\bar{S}_d D_{10} = u_{10}\nu_0^3 = u_1\mu^{2\varepsilon}\nu^3,\ \bar{S}_d D_{20} = u_{20}\nu_0^3 = u_2\mu^{-2\Delta}\nu^3 Z_{D_2},\ \nu_0 = \nu Z_\nu$$

$$u_{10} = u_1\mu^{2\varepsilon} Z_{g_1},\ u_{20} = g_2\mu^{-2\Delta} Z_{u_2},\ Z_{u_1} Z_\nu^3 = 1,\ Z_{u_2} Z_\nu^3 = Z_{D_2}\,.$$

Here, μ is the scale setting parameter, ν is the renormalized viscosity and u_1, u_2 are the dimensionless renormalized coupling constants. The amplitude D_{10} of the non-local term of the correlator of the random force is not renormalized. The renormalization constants Z_ν and Z_{D_2} are found from the condition that the one-irreducible functions $\langle\varphi'\varphi\rangle_{1-\mathrm{ir}}\big|_{\omega=0}$ and $\langle\varphi'\varphi'\rangle_{1-\mathrm{ir}}\big|_{\omega=0}$ are UV finite (are free from poles in ε at $\Delta/\varepsilon = const$).

An RG analysis of the model (7) shows the presence of an IR-stable fixed point $u_1*, u_2 \sim \varepsilon$ for small $\varepsilon > 0$ at $\zeta = const.$, which governs the infrared scaling. The parameters of this scaling – critical dimensions and universal scaling functions – are calculated in the form of series in ε. Due to the Galilei invariance of the theory the series for critical dimensions of fields Δ_φ, $\Delta_{\varphi'}$ and the frequency Δ_ω are terminated at the first order:

$$\Delta_\varphi = 1 - 2\varepsilon/3, \qquad \Delta_{\varphi'} = d - 1 + 2\varepsilon/3, \qquad \Delta_\omega = 2 - 2\varepsilon/3. \tag{8}$$

These formulae are exact without higher order corrections in ε and at the realistic value $\varepsilon = 2$ assume Kolmogorov values $\Delta_\varphi = -1/3$, $\Delta_\omega = 2/3$. Critical dimensions (8) are free from singularities at $d \to 2$. However, other physical quantities like the skewness factor, the Kolmogorov constant and critical dimensions of various composite operators strongly depend on the space dimension. Contrary to (8), they are expressed in the form of infinite series of types (1), (3).

In order to determine the dependence of these quantities on space dimension we have calculated the renormalization constants, infrared fixed points and scaling exponents in the one-loop approximation. In [9] the renormalization constants were calculated in the minimal subtraction (MS) scheme [only poles with respect to ε, $\Delta = O(\varepsilon)$ and their linear combinations are subtracted]. To improve the numerical performance we have used a renormalization scheme with a normalization point (NP). It is convenient to introduce normalized scalar one-irreducible functions

$$\Gamma_{\varphi'\varphi} = \frac{\langle\varphi_i'\varphi_i\rangle_{1-\mathrm{ir}}\big|_{\omega=0}}{\nu p^2(1-d)}\,, \tag{9}$$

$$\Gamma_{\varphi'\varphi'} = \frac{\langle \varphi'_i \varphi'_i \rangle_{1-\mathrm{ir}}\big|_{\omega=0}}{\nu^3 p^2 (d-1)} - \frac{u_1 (\mu/p)^{2\Delta+2\varepsilon}}{\bar{S}_d} - \frac{u_2}{\bar{S}_d} \tag{10}$$

and determine the renormalization constants from the conditions

$$\Gamma_{\varphi'\varphi}\Big|_{\substack{p=0,\\ \mu=m}} = 1\,, \qquad \Gamma_{\varphi'\varphi'}\Big|_{\substack{p=0,\\ \mu=m}} = 0\,, \tag{11}$$

where m is an infrared cut-off. Since physical quantities – critical exponents and universal scaling functions – are independent of the renormalization scheme, the application of the RG approach to the renormalized model specified by the condition (11) leads to the same ε, Δ expansions as in [9]. Instead, we propose to use the renormalization scheme (11) *without* the ε, Δ expansion. Such a scheme reproduces correctly the leading terms of expansion in both regimes $\varepsilon \to 0\,, \Delta = const$ and $\varepsilon {\sim} \Delta \to 0$ simultaneously. In the former, the additional term $\varphi' \partial^2 \varphi'$ is an infrared irrelevant operator in the action and the renormalization constant Z_{D_2} describes the renormalization of this additional operator.

Crucial quantities for the RG analysis are $\beta-$functions and anomalous dimensions γ defined as follows

$$\gamma_i = (\beta_1 \partial_{u_1} + \beta_2 \partial_{u_2}) \ln Z_{u_i}\,, \qquad i = 1,\, 2\,, \tag{12}$$

$$\beta_1 = -u_1(2\varepsilon + \gamma_1)\,, \qquad \beta_2 = -u_2(-2\Delta + \gamma_2)\,. \tag{13}$$

One-loop calculation of the renormalization constants Z_{u_i} yields

$$\gamma_1 = -\frac{3(d-1)(u_1+u_2)}{4(d+2)}\,, \gamma_2 = \frac{(d^2-2)(u_1+u_2)^2}{4d(d+2)\,u_2} + \gamma_1. \tag{14}$$

With the use of (13)and (14) the coordinates of the nontrivial fixed point $u_{1*} > 0$, $u_{2*} > 0$ are found as the solution of the equations $\beta_1(u_*) = 0$, $\beta_2(u_*) = 0$ in the form

$$u_{1*} + u_{2*} = \frac{8\varepsilon(d+2)}{3(d-1)}, u_{2*} = \frac{\varepsilon^2}{\varepsilon+\Delta}\,\frac{8(d^2-2)(d+2)}{9d(d-1)^2}\,. \tag{15}$$

Stability of the fixed point is determined by the sign of the real part of the eigenvalues $\omega_\pm$ of the matrix $\partial_j \beta_i\big|_{u=u_*}$:

$$\omega_\pm = \Delta + \frac{2\varepsilon(2d^2-3d+2)}{3d(d-1)} \pm \tag{16}$$

$$\pm \sqrt{\Delta^2 - \frac{4(d^2-2)}{3d(d-1)}\,\varepsilon\,\Delta - \frac{4(d^2-2)(2d^2-3d+2)}{9d^2(d-1)^2}\,\varepsilon^2}\,.$$

We are interested in the region $\varepsilon > 0$, $\Delta > 0$, in which the eigenvalues $\omega_\pm$ are positive. Therefore, the fixed point (15) is infrared stable. In the regime $\varepsilon {\sim} \Delta \to 0$ with the use of (16) we obtain

$$\omega_\pm = \Delta + \frac{4\varepsilon}{3} \pm \sqrt{\Delta^2 - \frac{4}{3}\varepsilon\,\Delta - \frac{8}{9}\,\varepsilon^2} + O(\varepsilon^2) \tag{17}$$

which is consistent with the result of [9]. At $\varepsilon \to 0$, $\Delta = const$ it follows that

$$\omega_- = 2\varepsilon + \frac{2(d^2-2)}{3d(d-1)} \cdot \frac{\varepsilon^2}{\Delta} + O(\varepsilon^3)\,, \tag{18}$$

$$\omega_+ = 2\Delta + \frac{2(d^2-3d+4)}{3d(d-1)}\varepsilon - \frac{2(d^2-2)}{3d(d-1)} \cdot \frac{\varepsilon^2}{\Delta} + O(\varepsilon^3)\,. \tag{19}$$

The quantity ω_- plays the rôle of the correction index ω in the usual ε expansion in the theory of developed turbulence, whereas ω_+ determines the critical exponent of the infrared-irrelevant composite operator $\varphi'\partial^2\varphi'$.

The terms $\sim\varepsilon$ in (18) and (19) are reliable: for ω_- relation (18) reproduces the known one-loop expression [2] for the exponent ω; for ω_+ we have checked the result by a direct calculation of the critical dimension of the composite operator $\varphi'\partial^2\varphi'$ in the usual ε expansion. Moreover, the terms $\sim\varepsilon^2/\Delta$ yield the true singular part with respect to Δ in the coefficients of ε^2: for ω_- it was confirmed by the pioneering two-loop calculation [6], for ω_+ we have checked it by calculation of the critical dimension of the composite operator $\varphi'\partial^2\varphi'$ in two-loop approximation. Expression (16), in fact, correctly reproduces main singular terms of the form $\varepsilon(\varepsilon/\Delta)^k$ and all leading terms of the ε expansion, i.e. the first terms of the corresponding Laurent series (2).

Calculation of graphs with increasing number of loops in our renormalization scheme guarantees that results become more precise step by step in the sense that the number of true terms of the ε expansion and the number of singular in Δ contributions is increased: an n-loop calculation correctly reproduces first n of all coefficients $A_k(d)$ in (1), while simultaneously of the coefficients $A_k(d)$ with $k > n$ the first n terms of their Laurent series (2) will be reproduced correctly.

3 Calculation of the Kolmogorov constant and skewness factor

In the framework of the proposed renormalization scheme we have calculated the Kolmogorov constant and skewness factor.The skewness factor is defined as follows

$$\mathcal{S} \equiv S_3/S_2^{3/2}\,. \tag{20}$$

In the Kolmogorov theory the third-order structure function $S_3(r)$ (determined exactly) and the second-order structure function $S_2(r)$ (in the inertial range) have the forms [10,11]:

$$S_3(r) = -\frac{12}{d(d+2)}\,\bar{\mathcal{E}}\,r, \quad S_2(r) = C_K\bar{\mathcal{E}}^{2/3}r^{2/3} \tag{21}$$

where $\bar{\mathcal{E}}$ is the average energy dissipation rate per unit mass and C_K is the Kolmogorov constant, the value of which is not determined in the framework of the phenomenological approach. All these expressions allow to connect the Kolmogorov constant with the skewness factor:

$$C_K = \left[-\frac{12}{d(d+2)\mathcal{S}} \right]^{2/3}. \tag{22}$$

Among the three quantities $S_2(r)$, $S_3(r)$ and $\mathcal{S}$ only the last one has a unique well-defined ε expansion. Thus, relation (22) may be used to determine C_K by means of the calculated value $\mathcal{S}$.

To find the RG representation of the skewness factor (20) the RG representations of the functions $S_2(r)$ and $S_3(r)$ have to be determined. Using results of calculations of the previous section we obtain the following values for the Kolmogorov constant and the skewness factor at fixed point (15)

$$C_K = [3Q(2)/2]\,[12/d(d+2)]^{2/3}\,, \quad \mathcal{S} = -\,[3Q(2)/2]^{-3/2}\,, \tag{23}$$

where in the one-loop approximation

$$Q(\varepsilon) = \left[4(d-1)/9/u_{1*}^2\right]^{1/3} \cdot \frac{u_{1*} + u_{2*}}{2}. \tag{24}$$

Using the last relation for $\varepsilon = 2$ we arrive at the values $C_K \approx 1.889$ and $\mathcal{S} \approx -0.308$. The values $C_K \approx 2.01$ and $\mathcal{S} \approx -0.28$ are considered the most reliable experimental values of these quantities [12]. Therefore, the suggested scheme for calculations yields quite reasonable agreement with the experiment.

4 Conclusion

The version of the RG approach used in the present paper bears certain resemblance with the well known RG method in the real space, which is widely used in the theory of critical phenomena (it is also called the "g expansion"). In the theory of phase transitions the parameter ε has the meaning of deviation from the critical space dimension (e.g, $\varepsilon = 4 - d$ for the φ^4-model). In the framework of the g expansion renormalization constants are calculated in the form of power series in the coupling constant g directly at $\varepsilon = 1$, i.e. at real value of the space dimension $d = 3$. In the framework of the g expansion calculations are notably simplified because it is much easier to calculate finite integrals in three dimensions than to calculate integrals with singularities at $d \to 4$. This is the reason why the use of the g expansion in the theory of critical phenomena has allowed to achieve better accuracy in perturbative calculations than in the usual ε expansion.

In the present paper we have used an approach akin to the g expansion to achieve a different aim. The choice of the renormalization constants from natural normalization conditions for the response and correlation functions together with the additional renormalization of random forcing allowed to include singular in $d-2$ contributions to the coefficients in all orders of the ε expansion. This summation led to a remarkable improvement of agreement of the theoretical prediction with the experimental value of the Kolmogorov constant already in the one-loop approximation.

Acknowledgments

The authors thank N. V. Antonov and M. V. Kompaniets for discussions. The work was supported by the Russian Foundation for Fundamental Research (grant No 08-02-00125a), Academy of Finland (grant No 127436), VEGA grant 0173 of Slovak Academy of Sciences, and by Centre of Excellency for Nanofluid of IEP SAS.

References

1.K. G. Wilson, and J. Kogut. The renormalization group and the ε expansion. *Physics Reports*, 12:75199, 1974.
2.C. De Dominicis, and P. C. Martin. Energy spectra of certain randomly-stirred fluids. *Physical Review A*, 19:419422, 1979.
3.L. Ts. Adzhemyan, A. N. Vasil'ev, and Yu. M. Pis'mak. Renormalization-group approach in the theory of turbulence: The dimensions of composite operators (in Russian). *Teoreticheskya i Matematicheskaya Fizika*, 57:268281, 1983.
4.L. Ts. Adzhemyan,N. V. Antonov, and A. N. Vasil'ev. Quantum field renormalisation group in the theory of developed turbulence (in Russian). *Uspekhy Fizicheskikh Nauk*, 166:12571284, 1996.
5.L. Ts. Adzhemyan,N. V. Antonov, and A. N. Vasil'ev. *The Field Theoretic Renormalization Group in Fully Developed Turbulence.* Gordon & Breach, London, 1999.
6.L. Ts. Adzhemyan,N. V. Antonov,M. V. Kompaniets, and A. N. Vasil'ev. Renormalization-group approach to the stochastic Navier-Stokes equation: Two-loop approximation. *International Journal of Modern Physics B*, 17:2137 2170, 2003.
7.L. Ts. Adzhemyan, J. Honkonen, M. V. Kompaniets, and A. N. Vasil'ev. Improved ε expansion for three-dimensional turbulence: Summation of nearest dimensional singularities. *Physical Review E*, 68:055302(R) (4 pages), 2003.
8.L. Ts. Adzhemyan, J. Honkonen, M. V. Kompaniets, and A. N. Vasil'ev. Improved ε expansion for three-dimensional turbulence: Two-loop renormalization near two dimensions *Physical Review E*, 71:036305 (19 pages), 2005.
9.J. Honkonen, and M. Yu. Nalimov. Two-parameter expansion in the renormalization-group analysis of turbulence. *Zeitshrift fur Physik B*, 99:297 303,1996.
10.A. S. Monin, and A. M. Yaglom. Statistical Fluid Mechanics: *Mechanics of Turbulence*, Vol. 2. MIT Press, Cambridge, 1975.
11.U. Frisch. Turbulence: *The Legacy of A. N. Kolmogorov.* Cambridge University Press, Cambridge, 1975.
12.K. R. Sreenivasan. On the universality of the Kolmogorov constant. *Physics of Fluids*, 7:27782784, 1995.

Dynamics of a rubbing Jeffcott rotor with three blades

Jan-Olov Aidanpää and Göran Lindkvist

Luleå University of Technology, Luleå, Sweden
Emails: joa@ltu.se, Goran.Lindkvist@ltu.se

Abstract: The non-linear behaviour of rubbing cylindrical rotors have been studied in several papers. In such systems rich dynamics have been found for frequencies above the natural frequency. Below natural frequency the solution was found to be stationary. In this paper the influence of blades is studied. A Jeffcott rotor with three blades is used and the contacts are described by large displacement beam theory. The model shows that no stationary point will exist and complex behaviour will occur below the natural frequency. For the studied rotor, failure due to high stresses will occur at driving frequencies below 50% of the natural frequency and instability at 80% of the natural frequency. The paper shows that the dynamics of bladed rotors differs from the dynamics of rubbing circular rotors. If a bladed rotor is used it is essential to study a model with blades. Otherwise the general conclusions on the dynamics can be wrong.
Keywords: rotor, dynamic, impact, rubbing, beam, blade.

1. Introduction

In rotor dynamics there are several situations when non-linear problems can occur. One such example is rub-impact which is a highly non-linear phenomenon. The problem is of industrial interest since there are several applications where rub-impact is the main cause for unwanted vibrations e. g. gas turbines, centrifuges, compressors and generators. It has been reported that 10.2% of 275 reported jet engine failures during 1962 to 1975 were caused was rubbing between rotating and stationary parts [1]. Several studies have been performed on the Jeffcott rotor with this kind of rubbing impacts. Some of them are described below, with focus on findings and development of methods.

The jump phenomenon and the influence of radial clearance were studied analytically in [2]. A modified Harmonic Balance method has been used to predict the occurrence and analyse the stability of quasi-periodic motion [3]. In [4] Fourier series and Floquet theory was used for analysis of global bifurcation and stability. They also reported three routes to chaos; from stable periodic through period doubling bifurcations, grazing bifurcation and a sudden transition from periodic motion to chaos. The stability for the case of full annular rub and cross coupling stiffness was analysed in [5]. Chaos has been reported to exist over large parameter ranges and different solutions can coexist [6]. In [7] approximate analytical solutions was

Chaos Theory: Modeling, Simulation and Applications
C. H. Skiadas, I. Dimotikalis and C. Skiadas (Eds)
 (pp. 97 - 104)

developed for non-linear dynamical responses and in [8] the harmonic balance method was used to calculate periodic responses of the non-linear system. In most cases the contact was modelled with an increased stiffness and Coulomb friction. However, other frictional models has also been considered [9]. In all models with annular rub, the system reaches a stationary point below the natural frequency of the rotor. It is only above the natural frequency where complex dynamics and multiple solutions are found [10]. In these models the rotor and stator was assumed to be circular. Bladed rotors in aero engines have also been studied with complex FEM models [12]. Due to the complexity of these models only short time sequences has been analysed and therefore the dynamics on parameter ranges is still unknown.

In this paper an attempt is made to model a rubbing Jeffcott rotor with three blades and Coulomb friction. The blades are uniform and the displacements are described by large displacement beam theory. A specific rotor is selected and the dynamics analysed in bifurcation diagrams and Poincaré sections for different driving frequencies. In several industrial applications the rotor consists of blades which can get in contact with a circular stator. The target is to evaluate the difference between a bladed turbine and a perfect circular one.

2. The Model

The model of the Jeffcott rotor is shown in Figure 1. The mass of the rotor, m, is supported by the shaft with stiffness k and damping c. The rotor is amplitude limited by the stator which has a radius R. The rotor is described by a point mass m in the centre and three mass less beams of length L, Young's modulus E and area moment of inertia I. The rotor is rotating with the angular velocity ω. In Figure (C) the geometry of the contact is shown. When a blade is in contact the beam will be deformed transversally Δ and axially δ. Both deformations are necessary in order to keep the beam inside the limit radius R. The contact force is described by a radial force P normal to the circle pointing from the contact point towards o and a tangential force μP, where μ is a coefficient of friction. To simplify the analysis it is assumed that ωt-ϕ is small which imply that P is an axial force and μP a tangential force on the beam.

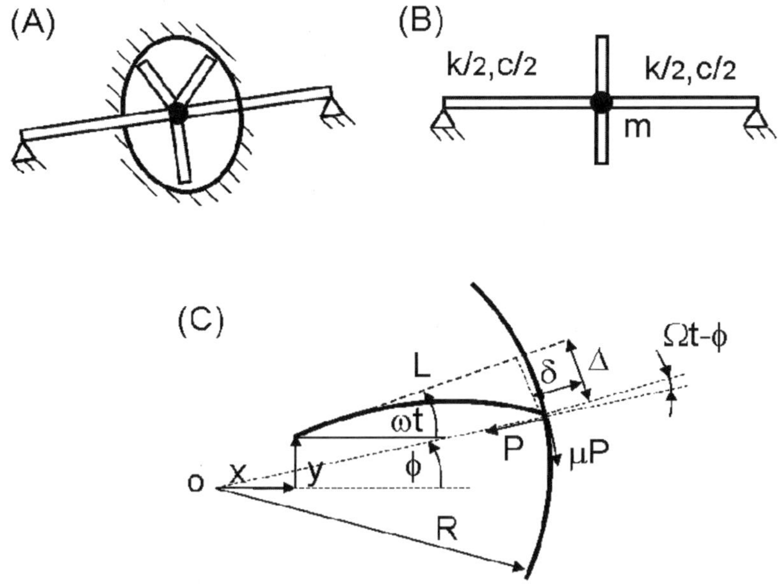

Figure 1. Rub impact model of the Jeffcott rotor. View of the whole system (A), side view (B) and geometry of the contact (C).

If o is the centre of the stator, the vector to the contact point of the beam can be described as

$$\bar{r} = \left(x + (L-\delta)\cos(\omega t) + \Delta\sin(\omega t)\right)\bar{i} + \left(y + y_0 + (L-\delta)\sin(\omega t) - \Delta\cos(\omega t)\right)\bar{j}$$

Where $\bar{i}$ and $\bar{j}$ are unit vectors in x and y direction respectively. The displacement y_0 is an initial misalignment of the rotor. When the rotor is in contact, the rotor is limited by the stator so that

$$|\bar{r}| = R$$

The contact force is described by the radial force P pointing toward o and a tangential component given by a coefficient of friction μP. As noted above, it is assumed that the contact angle $\phi = \omega t$ and therefore the forces on the beam are given by an axial compression force P and a transversal force μP. From beam theory the deformation of cantilevered beam is given by the equation

$$w''(\varepsilon) = \frac{P}{EI}\left(1 + (w')^2\right)^{3/2}\left((\Delta - w(\varepsilon)) + \mu P(L - \varepsilon - \delta)\right)$$

where " ′ " denotes derivation with respect to ε. The beam is assumed clamped at $\varepsilon = 0$ and subjected to the forces P and μP on the free end ($\varepsilon = L - \delta$). At the free end the beam will be displaced by the forces axially δ and transversally Δ. By numerical integration the values of P, Δ and δ can be found which satisfies $|\bar{r}| = R$. When a blade is in contact the forces in x and y directions are;

$$Fx = P\frac{\bar{r} \bullet \bar{i}}{R} - \mu P\frac{\bar{r} \bullet \bar{j}}{R} \quad ,$$

$$Fy = P\frac{\bar{r} \bullet \bar{j}}{R} + \mu P\frac{\bar{r} \bullet \bar{i}}{R} \quad .$$

For n blades there will be one set of such forces (Fx_i, Fy_i) for each blade i, hence the equation of motion for the Jeffcott rotor then becomes

$$\ddot{x} + 2\varsigma\omega_n\dot{x} + \omega_n^2 x = -\sum_{i=1}^{n} Fx_i / m$$

$$\ddot{y} + 2\varsigma\omega_n\dot{y} + \omega_n^2 y = -\sum_{i=1}^{n} Fy_i / m$$

The model is now ready for simulation.

3. Method

In this paper different simulation are carried out in order to evaluate the system. In these simulations a fourth order Runge-Kutta integration with adjustable time-step is implemented in an in-house code written in Fortran. In the bifurcation diagrams 100 Poincaré sections was collected after 500 periods of the excitation frequency ω. For the suggested model there are 5 dimensions in the state space: displacements x, y, velocities $\dot{x}$, $\dot{y}$ and the phase $\phi = \omega t$. Since the phase can be restricted to the interval $[0, 2\pi]$ it can be described as a circle S^1 with the period $2\pi / \omega$. Thus points in the 5 dimensional state space are given by $(x, \dot{x}, y, \dot{y}, \theta) \in R^4 \times S^1$. The Poincaré section $\sum$ is chosen as the cross section of this state space for a constant value of the phase θ_p. In this paper this constant value is chosen to 2π. Therefore the Poincaré section is defined as

$$\sum = \{x, \dot{x}, y, \dot{y}, \theta_p \mid \theta_p = 2\pi\}.$$

Points in this Poincaré sections are the intersections of a trajectory with the plane $\sum$ positioned at the constant phase $\theta_p = 2\pi$. If a root finding routine was applied at each contact, the time for finding P, Δ and δ would make the simulation time consuming. Analysis of bifurcation diagrams and any global dynamics would be difficult. With the assumption $\phi = \omega t$, there will be an unique relation between the forces acting on the beam and the forces in the contact point. In Figure 2 the displacements Δ and δ as function of P are shown one beam. In the figure the analytical models (dotted line for large deformation model and dashed for small deformations $(1 + (w')^2) \approx 1$) are shown together with nonlinear FEM analysis (solid).

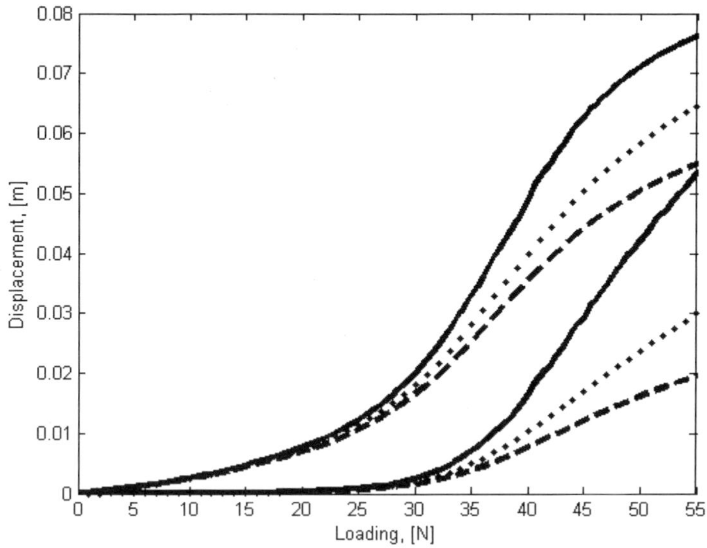

Figure 2. Displacements Δ and δ as function of P for a steel beam with $E=2.06\times10^{11}$, $I=0.01\times0.001^3/12$, $L=0.3$, $\mu=0.1$

The FEM program was LS-DYNA R4.2.1 with 10 default beam elements type 1 (Hughes-Liu with cross-section integration). The solution method was updated Lagrange solved with implicite solver. Hundred elements was also evaluated with a similar result. It is a clear difference between the FEM model and the analytical model which has not been understood yet. However, for a 50N load the studied beam will have a maximum stress of 2000 MPa which is far beyond most steel materials. Therefore a failure will occur before this load level. By curve fitting to the dotted functions of Figure 2, a simple model can be made of the displacements and forces in contact. An effective simulation program can thereby be made to analyse the system. Results for forces higher than 35 N must therefore be considered with caution due to this large deviation.

4. Results

A rubbing system is selected with $R=0.11$, $L=0.1$, $\zeta=0.1$, $y_0=0.010001$, $E=2.06\times10^{11}$, $I=0.01\times0.001^3/12$, $m=1$ and $\omega_n=10$ *[SI units]*. The values for EI corresponds to a blade in the shape of a rectangular steel beam with height *0.001[m]* and width *0.01[m]*. The dimensional form of the equation of motion is used to get some physical insight of the problem. In Figure 3 a bifurcation diagram (a) is shown for the system together with the maximum load (b).

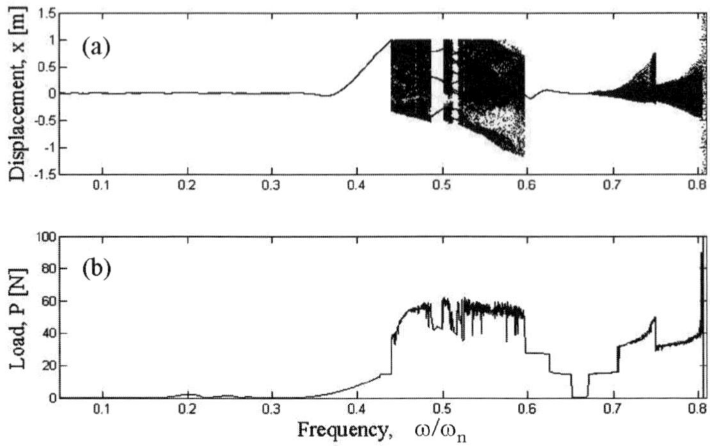

Figure 3. Bifurcation diagram for the displacement x (a) and maximum load P (b) for increasing driving frequency ration.

Figure 3 shows low vibrations and forces below 0.33. Above this frequency the amplitude and force increases. At ω/ω_n=0.45 the force will be about 60N. Above ω/ω_n=0.81 the forces grows beyond the validity of the model. In Figure 4 a blow up is shown for the frequency range [0.05-0.34].

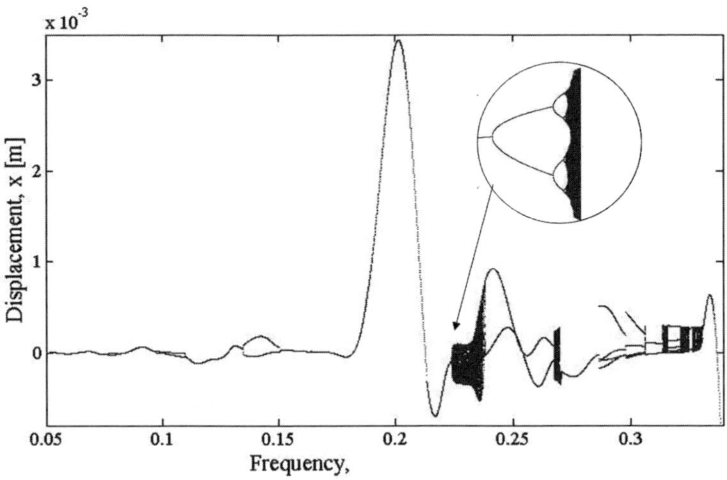

Figure 4 Bifurcation diagram for ω/ω_n=[0.05-0.34].

At low frequencies the system will move in period one. At ω/ω_n=0.078 the first region of period two is found and period two regions are also found in narrow frequency ranges at ω/ω_n=0.98, 0.136 and 0.164. The figure shows that the first region of chaotic motions is found at ω/ω_n=0.23. A blow up shows that the route to chaos is by period doublings. Note that regions of high amplitude period one are initiated at ω/ω_n=1/3, 1/6, 1/9, 1/12... and between those regions the amplitude is significantly lower. Poincaré sections are shown in Figure 5.

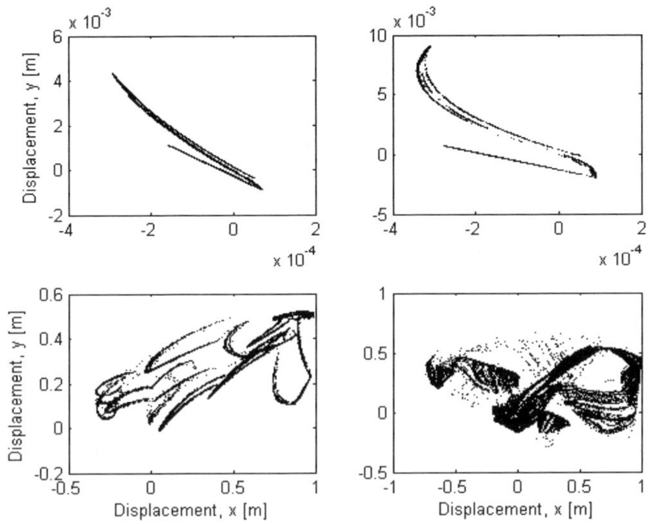

Figure 5. Projected Poincaré sections in x-y plane for the frequency ratios ω/ω_n=; 0.2245 (a), 0.228 (b), 0.442 (c), 0.54 (d).

In Figure 5 the development of the chaotic attractor is shown for different chaotic regions.

5. Discussion and Conclusion

Rubbing rotors have been studied extensively but mainly with models describing the rotor and the stator as cylinders. In several industrial applications the rotor consists of blades which conditions at contact significantly differs from the perfect circle. The target of this paper is to evaluate the difference between a bladed turbine and a perfect circular one. A turbine with three blades is selected for the study. In the case of perfect cylinder it has been shown that a smooth rubbing at a stationary position will remain beyond the natural frequency before impacts and complex dynamics will occur [6, 10]. In the case of bladed turbine there will not be any stationary position of the rotor. Each contact will lift the rotor and cause vibrations. For the studied rotor, period two motions occur already at driving frequency of 8% of the natural frequency of the rotor (ω/ω_n=0.08). At ω/ω_n=0.2 high amplitudes occurs due to periodic contact of each blade. Contact with each blade causes increased amplitude and occurs at integer fractions of $\omega_n/3$. In Figure 4 such regions of increased amplitudes are initiated at ω/ω_n=1/3, 1/6, 1/9, 1/12. Chaotic motions are first found in the range ω/ω_n=0.224-0.237. At ω/ω_n=0.33 a period one motion is initiated which grows as the driving frequency increases. Contact of the three blades excites the natural frequency with increased amplitude as a consequence. At ω/ω_n=0.44 the motion changes rapidly to chaotic with high amplitudes. This is the frequency where the blades of the studied system will fail due to high stresses. Finally, at ω/ω_n=0.81 the forces grows beyond the validity of the model and the program is stopped. The difference between the nonlinear FEM model and the analytical results was surprising. Different element types and number of elements was evaluated with a similar result. Experiment will therefore be performed to analyse the reason for this deviation. However, most of the dynamics presented in this paper was for low forces where the deviation between the models is small.

The paper shows that bladed turbines show a significantly different behaviour than the earlier studied circular ones. Complex dynamics will occur at low excitation frequencies and high forces will result in a failure of the turbine at frequencies below the natural frequency of the turbine.

6. Acknowledgement

The research presented in this paper has been carried out with funding by "Swedish Hydropower Centre - SVC". SVC has been established by the Swedish Energy Agency, Elforsk and Svenska Kraftnät together with Luleå University of Technology, The Royal Institute of Technology, Chalmers University of Technology and Uppsala University. www.svc.nu.

References

[1] A. F. Kascak, A. B. Palazzolo and G.Montaque. Transient Rotor Dynamic Rub Phenomena, Theory and Test. *Rotating Machinery Dynamics*. 2: 485-494, 1987.

[2] J. L. Isaksson. Dynamics of a rotor with annular rub. *Proceedings of IFTOMM 4th International Conference on Rotor Dynamics, Chicago*, 85-90, 1994.

[3] Y.-B. Kim and S. T. Noah. Quasiperiodic response and stability analysis for a nonlinear Jeffcott rotor. *J. of Sound and Vibration*. 190: 239-253, 1996.

[4] F. Chu and Z. Zhang. Bifurcation and chaos in a rub-impact Jeffcott rotor system. *J. of Sound and Vibration*. 210: 1-18, 1998.

[5] J. Jiang and H.Ulbrich. Stability analysis of sliding whirl in a nonlinear Jeffcott rotor with cross-coupling stiffness coefficients. *Nonlinear Dynamics*. 24: 269-283, 2001.

[6] E. V. Karpenko, M.Wiercigroch and M. P. Cartmell. Regular and chaotic dynamics of a discontinuously nonlinear rotor system. *Chaos, Solitons & Fractals*. 13: 1231-1242, 2002.

[7] E. V. Karpenko, M. Wiercigroch, E. E. Pavlovskaia and M. P. Cartmell. Piecewise approximate analytical solutions for a Jeffcott rotor with a snubber ring. *International Journal of Mechanical Sciences*, 44: 475-488, 2002.

[8] G. Von Groll and D. J. Ewins. The harmonic balance method with arc-length continuation in rotor/stator contact problems. *J. of Sound and Vibration*, **241**(2): 223-233, 2001.

[9] A. F. Kascak and J. J.Tomko. Effects of Different Rub Models on Simulated Rotor Dynamics. *NASA Technical Paper 2220*. 1984

[10] J. O. Aidanpää. Multiple solutions in a rub-impact Jeffcott rotor. *The 10th of International Symposium on Transport Phenomena and Dynamics of Rotating Machinery, Honolulu, Hawaii, March 07-11*. ISROMAC10-2004-120, 2004.

[11] M. Legrand, C. Pierre, P. Cartraud and J.-P. Lombard. Two-dimensional modeling of an aircraft engine structural bladed disc-casing modal interaction. *J. of Sound and Vibration,* 319: 366-391, 2009.

Exploring process of fibre breaking in tube samples of composite during quasi-static process of fracture

Dorota Aniszewska and Marek Rybaczuk

Institute of Materials Science and Applied Mechanics,
Wroclaw University of Technology, Smoluchowskiego 25,
50-370 Wroclaw Poland
Emails: dorota.aniszewska@pwr.wroc.pl, marek.rybaczuk@pwr.wroc.pl

Abstract: This paper presents numerical methods of modelling composites, which allow observation of fibre breaking during static loads without using invasive experimental methods. Our simulations are based on cellular automata, which is an alternative method to study behaviour of dynamical systems. We assume defects evolution in composite as a dynamical system depending on external and internal forces and properties of fibres.
Keywords: cellular automata (CA), fractal measure, fractal dimension, chaos.

1. Introduction

Using cellular automata (CA) discovered by Wolfram, 2002 and described by Ilachinski, 2002, to modelling composite allow us observation of defects propagation in meso- and macroscopic length scale. We examine fibre breaking in tube samples of fibre reinforced composite with constant matrix-fibres interaction.

2. Modelling of composite with cellular automata

Model of composite sample is presented in Fig. 1a and it corresponds with real samples of composite with carbon fibres and epoxy resin matrix presented in Fig. 1b. Experiments were described by Gasior, Sankowska, 2006.
Each long fibre is build with hundreds of cellular automata. Each CA has two possible states: 0- broken or 1-not broken. Under force *F* CA can break if force exceeds minimal value. Probability of breaking randomized for every cellular automata is compared with function of fatigue properties of single fibre described with Weibull distribution $p(F_i)=1-\exp(-(F_i/0.5)^a)$ presented in Fig. 2. Power *a* decides how many cellular automata breaks in one step of simulation and we used a=5.

a)

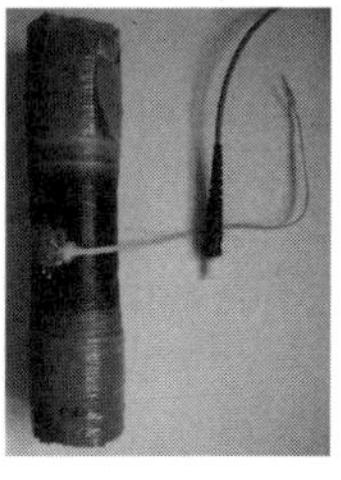

b)

Fig. 1. Tube samples of composite: a) modelled with cellular automata and b) real sample.

Chaos Theory: Modeling, Simulation and Applications
C. H. Skiadas, I. Dimotikalis and C. Skiadas (Eds)

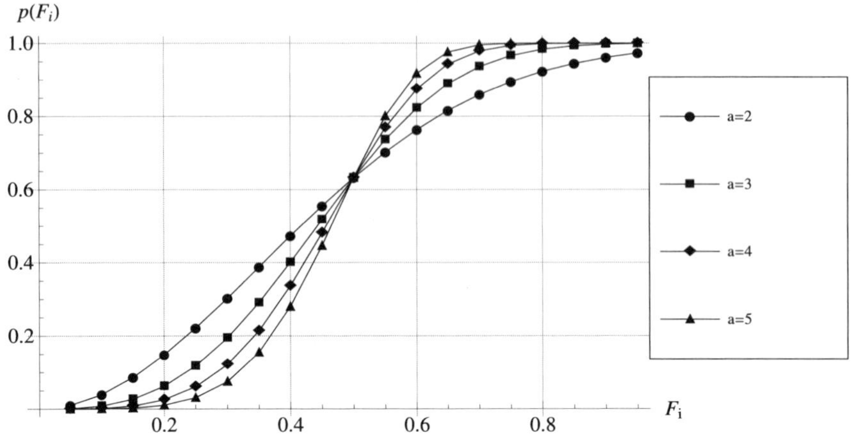

Fig. 2. Weibull distribution $p(F_i)=1-\exp(-(F_i/0.5)^a)$ for a = 2, 3, 4, 5, describing relationship between force F_i affecting single cellular automata and its breaking probability $p(F_i)$.

The behaviour of single CA building fibre is slightly different when it is a part of composite. Because of presence of matrix, given cellular automata in fibre can interact with other CA building adjacent fibres. Any broken fibre entails additional stress concentration in next fibres expressed by higher probability of breaking. If one cellular automata inside long fibre is broken, this fibre is weakened along and probability of breaking of other CA is reduced. Due to limited value of adhesive bond between fibre and matrix, the breaking probability for automata in close certain distance becomes 0. Probability of breaking depends also on interaction with fibres in neighbourhood with certain (correlation) radius. It increases when there are other broken CA in close range. Details are described in Aniszewska, Rybaczuk, 2009.

3. Numerical simulations

Numerical modelling of composite allows controlling material structure, properties of fibres and matrix. The simulations were performed for the same size of specimen with various correlation radius and various velocity of force growth.

At the beginning evolution of defects is absolutely random, what is presented in Fig. 3a. Finally defects group together, which is showed in Fig. 3b.

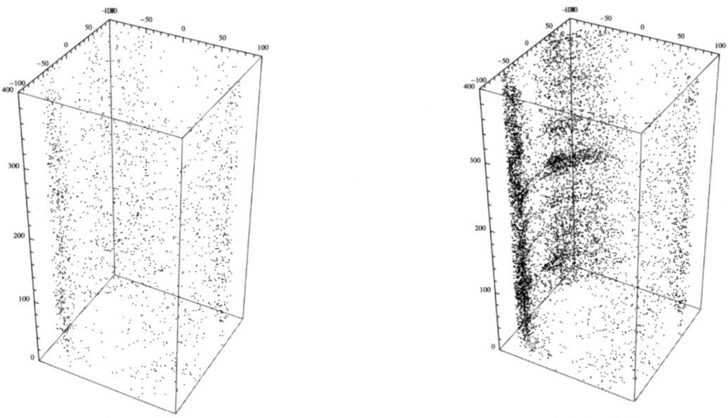

a) t=30 b) t=41

Fig. 3. Broken CA for the following time steps.

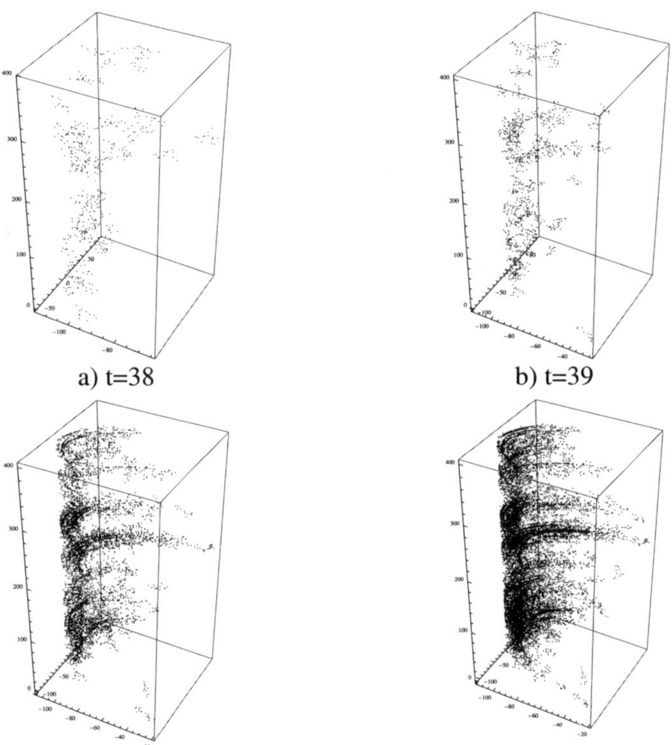

a) t=38 b) t=39

c) t=41 d) t=42

Fig. 4. The evolution of the correlated defects including at least 10 broken cellular automata.

Developing localization and finite volume of specimen are responsible for the material failure. The final failure occurs very rapidly in volume small comparing to the volume of entire sample. Fig. 4 represents correlated defects containing at least 10 broken CA in neighbourhood, which are responsible for sample breaking.

4. Fractal characteristics of defects growth

Our final goal is formulating defects evolution in terms of fractals. Fractal geometry is able to describe the complexity of defects in composite, as in any other material, using fractal dimension and their size with fractal measure according to Mandelbrot, 1983.

In mathematical considerations the most important fractal dimension has been defined by Hausdorff and described by Falconer, 1990. It can be applied to any bounded set and it is accompanied with fractal measure. At present time there isn't known algorithm allowing evaluation of Hausdorff measure and dimension, but there are alternative definitions of fractal dimensions. Box-counting dimension of set F is defined as:

$$D = \dim_B F = \lim_{\delta \to 0} \frac{\log N_\delta(F)}{-\log \delta},$$

where $N_\delta(F)$ is the smallest number of sets of diameter at most δ, which cover *F*. These sets can be also closed balls of radius δ or cubes of side δ.

For our practical application we have defined box-counting pseudo-measure of *F*, called $V_D(F)$, as follows:

$$\nu_D(F) = \lim_{\delta \to 0} \frac{N_\delta(F)}{\delta^{-D}}.$$

Calculation of box-counting dimension is implemented in FD3 program described by Myers, Sarraille, 1994. Fractal characteristics calculated for a few samples are presented in Fig. 5.

Fractal dimension and pseudo-measure were computed for defects including at least 10 broken cellular automata. Defects evolution corresponds with fractal dimension and measure.

The moment, when growing defect approaches macroscopic size is clearly visible as high increase of measure. which corresponds to the simultaneous change of state of very many fibres. The increasing radius causes higher volume of final failure, because defected area influences further situated fibres.

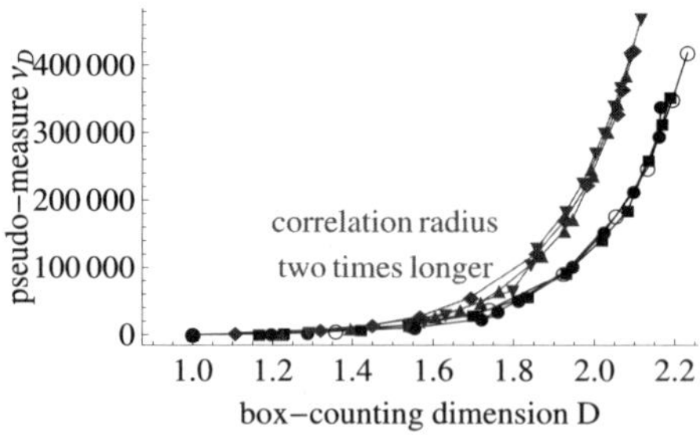

Fig. 5. Fractal characteristics of defects evolutions in tube samples of composite.

5. Discussion

This paper presents partially results of our continuous work of examination defects evolution with dynamical systems methods. This approach lets us answer the question if defects growth can be chaotic.

6. Acknowledgements

Calculations have been carried out in Wroclaw Centre for Networking and Supercomputing (http://www.wcss.wroc.pl), grant No. 112.

References

1. D. Aniszewska. M. Rybaczuk. Fractal characteristics of defects evolution in parallel fibre reinforced composite in quasi-static process of fracture. *Theoretical and Applied Fracture Mechanics*, 52:91-95,2009.
2. K. Falconer Fractal Geometry. Mathematical Foundations and Applications. *John Willey & Sons*, Chichester, 1990.
3. P. Gasior, A. Sankowska. Monitoring of the defects in epoxy-carbon composite tube with use of the optical fiber sensors. *23rd Danubia-Adria Symposium on Experimental Methods in Solid Mechanics*, Podbanske-Zilina, Slovak Republik: EDIS Editing Centre University of Zilina, 113-114, 2006.
4. A. Ilachinski. Cellular Automata. A Discrete Universe. Word Scientific, New Jersey, 2002.
5. B. B. Mandelbrot. The Fractal Geometry of Nature. *W.H. Freeman and Company*, New York, 1983.
6. S. Myers, J. J. Sarraille. Fd3: A program for measuring fractal dimension. *Educational and Psychological Measurement*, 54:94-97, 1994.
7. S. Wolfram. A New Kind of Science. *Wolfram Media*, 2002.

Non Hamiltonian chaos from Nambu dynamics of surfaces

Minos Axenides

Institute of Nuclear Physics, NCSR Demokritos
15310 Agia Paraskevi, Attiki, Greece
Email: axenides@inp.demokritos.gr

Abstract: We discuss recent work with E.Floratos (JHEP 1004:036,2010) on Nambu Dynamics of Intersecting Surfaces underlying Dissipative Chaos in R^3. We present our argument for the well studied Lorenz and Rössler strange attractors. We implement a flow decomposition to their equations of motion. Their volume preserving part preserves in time a family of two intersecting surfaces, the so called *Nambu Hamiltonians*. For dynamical systems with linear dissipative sector such as the Lorenz system, they are specified in terms of Intersecting Quadratic Surfaces. For the case of the Rössler system, with nonlinear dissipative part, they are given in terms of a Helicoid intersected by a Cylinder. In each case they foliate the entire phase space and get deformed by Dissipation, the irrotational component to their flow. It is given by the gradient of a surface in R^3 specified in terms of a scalar function. All three intersecting surfaces reproduce completely the dynamics of each strange attractor.

1 Introduction

Dissipative dynamical systems, with a low dimensional phase space, present an important class of simple non-linear physical systems with intrinsic complex behavior (homoclinic bifurcations, period doubling, onset of chaos, turbulence), which generated intense experimental, theoretical and numerical work in the last few decades [1,2].

Recently [3] we have reexamined dissipative dynamical systems with a 3-dimensional phase space from the perspective of Nambu-Hamiltonian Mechanics(NHM) [4,5]. The latter represents a generalization of Classical Hamiltonian Mechanics, mostly appropriate for the study of odd-dimensional phase-space volume preserving flows(Liouville's theorem). As such and in order to make it directly applicable to the dynamics of dissipative systems in R^3 we must associate it to a volume preserving dynamics sector. We have done so by introducing a flow decomposition to their equations of motion and therefore isolate in their flow vector field its rotational(solenoidal) part. It is manifestly

Chaos Theory: Modeling, Simulation and Applications
C. H. Skiadas, I. Dimotikalis and C. Skiadas (Eds)

volume preserving, non-dissipative and hence directly describable in terms of the intersecting surfaces of NHM. Along with the remaing irrotational component to the flow the decomposition fully recovers the dissipative dynamics of the system in question, presenting itself as an equivalent formulation of the odd-dimensional dynamical system.

We have applied the above recipe to the famous examples of Lorenz[6] and Rössler [7] chaotic attractors which represent the prototype models for the onset of turbulence.[2].

In sect. 2 we start off with a discussion of flow decompostion for the most elementary and familiar of all dissipative systems in R^2 without chaotic behavior : the dissipative harmonic oscillator (DHO). In this case the existence of a Hamiltonian formalism identifies for its "closed" nondissipative sector the harmonic oscilator (HO) with a well defined integrable classical and quantum evolution.

The presence of chaotic flows for dissipative systems in R^3 is argued, by analogy, to nescessitate intersecting Nambu Surfaces which define rerspectively "closed" physical systems with integrable R^3 periodic orbits and simultaneously a well defined classical and quantum behaviour.

In sect. 3 we apply the framework for the cases of the Lorenz and Rössler strange attractors. We isolate their non-dissipative sector parametrized by two intersecting surfaces: a cylinder and a paraboid for the Lorenz attractor as well as a helicoid with a cylinder for the Rössler system. They account, amazingly, for the double scroll topology of the full " butterfly" Lorenz attractor. and the single scroll topology for the Rössler case. We end our presentation with Conclusions and open problems.

2 Flow Decomposition in Dissipative Systems: R^2 versus R^3

Dissipation is a necessary condition for dynamical systems to exhibit chaos. Yet it is not suficient. This is enunciated through a powerful No-go theorem. Indeed the Poincare-Bendixon alows for only fixed points and limits cycles in two dimensions. Chaotic Flows need space to emerge. At the minimal level they emerge with dissipative systems in R^3. They are typically associated with Strange Attractors such as the famous ones of Lorenz and Rössler. They belong to a large class of dynamical systems whose dynamics is governed by continuous set of 1st order ordinary differental flow equations

$$\dot{\boldsymbol{x}} \;=\; \boldsymbol{v}(\boldsymbol{x}(t), t, \lambda) \tag{1}$$

where $\boldsymbol{v}$ is a velocity field flow with λ some control external parameter. Their phase space dynamics, depending on whether they exchange energy with their environment or not, can be either *open-dissipative* or *closed (conservative-Hamiltonian)*. This is reflected on their velocity flow field being divergenceless or not. Let us see all these issues for the simple case of the Dissipative

Harmonic Oscillator (DHO) whose phase space equations of motion are well known to be for $x_1 = q$ and $x_2 = p$:

$$\begin{aligned} \dot{q} &= p + \alpha q \\ \dot{p} &= -q + \beta p \end{aligned} \tag{2}$$

The velocity flow is given by $\nabla \dot{v}_{tot} = (\alpha + \beta) \neq 0$ for appropriate values $\alpha \neq -\beta \neq 0$ with a net outflow or inflow of energy depending on the sign of $\nabla \dot{v}_{tot}$ (inflow < 0 or outflow > 0). Nevertheless we split it into a nondissipative component $v_{ND} = (p, -q)$ with zero flow $\nabla \dot{v}_{ND} = 0$. Its dissipative part is given by $v_D = \alpha q + \beta p$ with its flow being simply the total flow of the HO $\nabla v_D = \alpha + \beta$. The dynamics of the HO in one dimension can take the form of a 2nd order differential equation

$$\ddot{q} + (\alpha + \beta)\dot{q} + (1 + \alpha\beta)q = 0 \tag{3}$$

The damping factor is $\gamma = \alpha + \beta$ and the effective natural frequancy which is dissipation induced is given by $\omega_{eff} = 1 + \alpha\beta$.

The existence of the Harmonic Oscillator Hamiltonian $H = \frac{1}{2}(p^2 + q^2)$ along with a Dissipation function $D = \frac{1}{2}(\alpha q^2 + \beta p^2)$ reproduce the eqs. of motion of (2.2) in a more compact form

$$\dot{x}^i = \epsilon^{ij}(\partial_j H + \partial_j D) \tag{4}$$

where $i, j = 1, 2$ and $\epsilon^{12} = -\epsilon^{21} = 1$ as usual. The non-dissipative sector of the HO is given by

$$\begin{aligned} \dot{q} &= \frac{\partial H}{\partial p} = p \\ \dot{p} &= -\frac{\partial H}{\partial q} = -q \end{aligned} \tag{5}$$

Firstly it identifies a circular *periodic* orbit, the famous HO as the *integrable progenitor* of the transient DHO. This is a well defined closed physical system with a most familiar quantum behavior, the Quantum Harmonic Oscillator.

Secondly the linear dissipation of the DHO a transient spirals it inwards to a fixed point of zero energy. The harmonic oscillator *localizes* in effect such a dissipative evolution.

Lastly the rate at which the energy of the harmonic oscillator loses its energy as it spirals in depends on its damping strength. It is the *Quality factor* which is easily computed to be

$$Q = \frac{\omega_{eff}}{\gamma} = \frac{1 + \alpha\beta}{\alpha + \beta} \tag{6}$$

All in all Flow decomposition for the transient DHO has three immediate implications for the DHO: 1. Existence of Integrable Harmonic oscillator associated with a periodic orbit 2. Localization of the DHO evolution and asymptotic fixed point state 3. Quality factor for DHO as a measure of its energy loss rate and in effect of the damping strenth.

We will proceed now to examine whether it is feasible to implement this methodology for the case of Strange attractors which possess chaotic flows in R^3.

Nambu-Hamiltonian mechanics is a specific generalization of classical Hamiltonian mechanics, where the invariance group of canonical symplectic transformations of the Hamiltonian evolution equations in 2n dimensional phase space is extended to the more general volume preserving transformation group $SDiff(\mathcal{M})$ with an arbitrary phase space manifold M of any dimension $d = \dim(\mathcal{M})$.

In [3] we work with the case of a three dimensional flat phase space manifold. Nevertheless our results could be generalized to curved manifolds of any dimension[5,8].

Nambu-Hamiltonian mechanics of a particular dynamical system in R^3 is defined once two scalar functions $H_i \in C^\infty(R^3), i = 1, 2$, the generalized Hamiltonians [4,5] are provided. The evolution equations are:

$$\dot{x}^i \ = \ \{x^i, H_1, H_2\} \qquad i = 1, 2, 3 \tag{7}$$

where the Nambu 3-bracket, a generalization of Poisson bracket in Hamiltonian mechanics, is defined as

$$\{f, g, h\} \ = \ \epsilon^{ijk} \partial^i f \partial^j g \partial^k h \qquad i, j, k = 1, 2, 3 \qquad \forall f, g, h \in C^\infty(R^3) \tag{8}$$

Any local coordinate transformation

$$x^i \ \rightarrow \ y^i \ = \ y^i(x) \qquad i = 1, 2, 3 \tag{9}$$

which preserves the volume of phase space

$$\det\left(\frac{\partial y^i}{\partial x^j}\right) \ = \ 1 \qquad \forall\ x = (x^1, x^2, x^3) \in R^3 \tag{10}$$

leaves invariant the 3-bracket and therefor it is a symmetry of Nambu-Mechanics. Except for the linearity and antisymmetry of the bracket with respect to all of its arguments it also satisfies an important identity, the so called "Fundamental identity"[FI] [3,8].

The evolution eq.(2.7) has a flow vector field:

$$v^i(x) \ = \ \epsilon^{ijk} \partial^j H_1 \partial^k H_2 \qquad i, j, k = 1, 2, 3 \tag{11}$$

which is volume preserving

$$\partial^i v^i \ = \ 0 \tag{12}$$

The reverse is also true. We will name the phase-space volume preserving flows "Non-dissipative" while the non-conserving ones "Dissipative" ($\partial^i v^i(x)$ $> 0(< 0)$).

The flow equations of a general dissipative system may take the general vector form

$$\dot{\boldsymbol{x}} = \boldsymbol{\nabla} H_1 \times \boldsymbol{\nabla} H_2 + \boldsymbol{\nabla} D \tag{13}$$

We notice that given a pair of functions H_1, H_2 such that

$$\boldsymbol{v}_{ND} = \boldsymbol{\nabla} H_1 \times \boldsymbol{\nabla} H_2 \tag{14}$$

any transformation of H_1, H_2

$$H_i \to H_i'(H_1, H_2) \qquad i = 1, 2 \tag{15}$$

with unit Jacobian

$$\det\left(\frac{\partial H_i'}{\partial H_j}\right) = 1 \tag{16}$$

gives also

$$\boldsymbol{v}_{ND} = \boldsymbol{\nabla} H_1' \times \boldsymbol{\nabla} H_2' \tag{17}$$

In ref.[8] we reduced the evolution equation of the form (2.32) Nambu Mechanics in Hamiltonian-Poisson form as follows:

$$\dot{x}^i = \{x^i, H_1,\}_{H_2} \tag{18}$$

where the induced Poisson bracket

$$\{f, g\}_{H_2} = \epsilon^{ijk} \partial^i f \partial^j g \partial^k H_2 \tag{19}$$

satisfies all the required properties like linearity, antisymmetry and the Jacobi identity.

In the following sections we are going to present a detailed investigation of the Lorenz and Rössler attractors from the point of view of Dissipative Nambu-Hamiltonian Dynamics.

3 The Lorenz Attractor from Dissipative Dynamics of Intersecting Quadratic Surfaces

The Lorenz model was invented as a three Fourier mode truncation of the basic eqs. for heat convection in fluids in Reyleigh-Benard type of experiments [12] The time evolution eqns. in the space of three Fourier modes x, y, z which we identify as phase-space are :

$$\begin{aligned} \dot{x} &= \sigma(y - x) \\ \dot{y} &= x(r - z) - y \\ \dot{z} &= xy - bz \end{aligned} \tag{20}$$

where σ is the Prandtl number, r is the relative Reynolds number and b the geometric aspect ratio.

The standard values for σ, b are $\sigma = 10, b = \frac{8}{3}$ with r taking values in $1 \leq r < \infty$. There are dramatic changes of the system as r passes through various critical values which follow the change of stability character of the three critical points of the system $P_1 : x = y = z = 0, P_{\pm} : x = y = \pm\sqrt{b(r-1)}, z = r - 1$.

Lorenz discovered the non-periodic deterministic chaotic orbit for the value $r = 28$, which is today identified as a Strange Attractor with a Hausdorff dimension of $(d = 2.06)$[13]. Standard reference for an exhaustive numerical investigation of the Lorenz system is the book by Sparrow [14]∗.

There have been various attempts made to localize the Lorenz attractor, by convex surfaces, in order to get information about Hausdorff dimensions [13] and other characteristics [15].

We will proceed to exhibit Localization of the full Lorenz attractor from the Nambu surfaces which we will determine shortly. At the Quantum level the existence of an attracting ellipsoid in a matrix model formulation of the Lorenz system is manifest [3] as one gets attracting ellipsoids in higher dimensional phase spaces.

We now proceed to describe the Lorenz system in the framework of section 2. The flow vector field $\boldsymbol{v}$ is analyzed into its dissipative and non-dissipative parts as follows:

$$\boldsymbol{v}_D = (-\sigma x, -y, -bz) = \boldsymbol{\nabla} D \tag{21}$$

with the "Dissipation" function

$$D = -\frac{1}{2}\left(\sigma x^2 + y^2 + bz^2\right) \tag{22}$$

and

$$\boldsymbol{v}_{ND} = (\sigma y, x(r-z), xy) = (0, y, z-r) \times (-x, 0, \sigma) \tag{23}$$

The two *Hamiltonians* or Clebsch-Monge potentials H_1, H_2 are determined by

$$\boldsymbol{\nabla} H_1 \times \boldsymbol{\nabla} H_2 = \boldsymbol{v}_{ND} \tag{24}$$

or equivalently

$$H_1 = \frac{1}{2}[y^2 + (z-r)^2] \tag{25}$$

and

$$H_2 = \sigma z - \frac{x^2}{2} \tag{26}$$

The Lorenz system (20) can thus be written in the equivalent form in terms of ∘($\boldsymbol{r} = (x, y, z)$):

$$\dot{\boldsymbol{r}} = \boldsymbol{\nabla} H_1 \times \boldsymbol{\nabla} H_2 + \boldsymbol{\nabla} D \tag{27}$$

In the Non-Dissipative part(ND) of the dynamical system

$$\dot{\boldsymbol{r}} = \boldsymbol{\nabla} H_1 \times \boldsymbol{\nabla} H_2 \tag{28}$$

the Hamiltonians H_1, H_2 are conserved and their intersection defines the ND orbit. Moreover if we get the reduced Poisson structure (sect.2) from H_2 we obtain the 2-dim phase space Σ_2 to be the family of parabolic cylinders with symmetry axis the y-axis:

$$H_2 = \text{constant} = H_2(\boldsymbol{r_o}) \tag{29}$$

Σ_2 is thus given by

$$z = z_o + \frac{x^2 - x_o^2}{2\sigma} \tag{30}$$

with x_o, z_o the initial condition for x, z. The induced Poisson algebra (rel.19) is given by

$$\begin{aligned} \{x, y\}_{H_2} &= \partial_z H_2 = \sigma \\ \{y, z\}_{H_2} &= \partial_x H_2 = -x \\ \{z, x\}_{H_2} &= 0 \end{aligned} \tag{31}$$

The dynamics on the 2d-phase space Σ_2 is given by H_1

$$\begin{aligned} \dot{x} &= \{x, H_1\}_{H_2} \\ \dot{y} &= \{y, H_1\}_{H_2} \\ \dot{z} &= \{z, H_1\}_{H_2} \end{aligned} \tag{32}$$

and H_1 is an anharmonic oscillator Hamiltonian with $(x/\sigma, y)$ conjugate canonical variables. Using rel.(25-26) we get on Σ_2:

$$H_1 = \frac{1}{2\sigma^2} \left[y^2 + \frac{1}{2} (x^2 - a^2)^2\right] \tag{33}$$

with

$$a^2 = x_o^2 - 2\sigma(z_o - r) = -2H_2 + 2\sigma r \tag{34}$$

where $\frac{1}{\sigma^2}$ plays the role of the mass. Depending on the initial conditions we may have a single well ($a^2 \leq 0, H_2 \geq \sigma r$) or a double well potential ($a^2 > 0, H_2 < \sigma r$) respectively. The trajectories , the intersections of the two cylinders , H_1 and H_2 with orthogonal symmetry axes (x, y) may either have one lobe left/right or may be running from the right to the left lobe. This is reminiscent of the topology structure of the orbits of the Lorenz chaotic attractor.

The full Lorenz system does not conserve H_1, H_2 and there is a random motion of the two surfaces against each other. Their intersection is time varying. In effect at every moment the system jumps from periodic to periodic orbit of the non-dissipative sector. Moreover the motion of the non-dissipative system around the two lobes, either left or right, can now jump from time to time from one lobe to the other.

4 The Rössler Attractor from Dissipative Dynamics of a Cylinder Intersecting with an Helicoid

Rössler introduced a simpler than Lorenz's nonlinear ODE system with a 3d-phase space, in order to study in more detail the characteristics of chaos, which is motivated by simple chemical reactions [7].

The Rössler system is given by the evolution eqns:

$$\begin{aligned} \dot{x} &= -y - z \\ \dot{y} &= x + ay \\ \dot{z} &= b + z(x - c) \end{aligned} \tag{35}$$

with parameters a, b, c usually taking standard values $a = b = 0.2, c = 5.1$ or $a = b = 0.1, c = 14$ for the appearance of the chaotic attractor.

We turn now to the study of the Rössler system as a Dissipative Nambu-Hamiltonian dynamical system.

The key difference with the Lorenz attractor is that the dynamics of the system is simpler. Chaos appears as random jumps outwards and inwards the single lob attractor.

In order to get the three scalars, the two generalized Hamiltonians H_1, H_2 which are conserved and characterize the non-dissipative part and D the dissipation term:

$$\dot{\boldsymbol{r}} = \boldsymbol{\nabla} H_1 \times \boldsymbol{\nabla} H_2 + \boldsymbol{\nabla} D \tag{36}$$

we checked after some guess work, that we must subtract and add a new term in the first equation. Indeed we find for the two parts

$$\boldsymbol{v}_{ND} = (-y - z - \frac{z^2}{2}, x, b) \tag{37}$$

$$\boldsymbol{v}_{D} = (\frac{z^2}{2}, ay, z(x - c)) \tag{38}$$

satisfying accordingly $\boldsymbol{\nabla} \cdot \boldsymbol{v}_{ND} = 0$ and $\boldsymbol{\nabla} \times \boldsymbol{v}_D = 0$

We must determine H_1, H_2 and D such that

$$\boldsymbol{\nabla} H_1 \times \boldsymbol{\nabla} H_2 = \boldsymbol{v}_{ND} \tag{39}$$

and

$$\boldsymbol{\nabla} D = \boldsymbol{v}_D \tag{40}$$

For D we find easily

$$D = \frac{1}{2}\left[ay^2 + (x - c)z^2\right] \tag{41}$$

To get H_1, H_2 we must integrate first the Non-dissipative system:

$$\begin{aligned} \dot{x} &= -y - z - \frac{z^2}{2} \\ \dot{y} &= x \\ \dot{z} &= b \end{aligned} \tag{42}$$

The general solution is :

$$x(t) = -b(1+z(t)) + (x_o+b(1+z_o)) \cos(\mathrm{t}) - (y_o+z_o+\frac{z_o^2}{2}-b^2) \sin(\mathrm{t}) \quad (43)$$

$$y(t) = b^2-z(t)-\frac{z^2(t)}{2}+(y_o+z_o+\frac{z_o^2}{2}-b^2) \cos(\mathrm{t}) + (x_o+b+bz_o) \sin(\mathrm{t}) \quad (44)$$

$$z(t) = bt + z_o \quad (45)$$

To uncover H_1, H_2 we introduce the complex variable

$$w(t) = w_1(t) + iw_2(t) \quad (46)$$

with

$$w_1(t) = x(t) + b(1+z(t))$$
$$w_2(t) = y(t) + z(t) + \frac{z^2(t)}{2} - b^2 \quad (47)$$

We obtain

$$w(t) = w_o \cdot e^{it} \quad (48)$$

with

$$w_o \equiv w(t=0) \quad (49)$$

We see that there are two constants of motion, the first one being :

$$| w(t) | = | w_o | \quad (50)$$

and we define correspondingly,

$$H_1 = \frac{1}{2} | w(t) |^2 = \frac{1}{2} (x + b(1+z))^2 + \frac{1}{2} (y + z + \frac{z^2}{2} - b^2)^2 \quad (51)$$

The second integral of motion is obtained through the phase

$$w(t) = | w_o | \cdot e^{i\varphi_o} \cdot e^{it} = | w_o | e^{i\varphi(t)} \quad (52)$$

or from (4.14)

$$\varphi(t) - \frac{z}{b} = \varphi_o - \frac{z_o}{b} \quad (53)$$

and we define appropriately the second constant surface H_2

$$H_2 = b \ \mathrm{arctg}\frac{y + z + \frac{z^2}{2} - b^2}{1 + b(1+z)} - z \quad (54)$$

We easily check that rel.(41) is satisfied:

$$\boldsymbol{\nabla} H_1 \times \boldsymbol{\nabla} H_2 = (-y - z - \frac{z^2}{2}, x, b) \quad (55)$$

The family of surfaces H_1 and H_2 are a quadratic deformation of a cylinder and respectively a quadratic deformation of a right helicoid. Their intersection is the trajectory (43-44).

5 Conclusions-Open Problems

The main result of our present work is the demonstration of Dissipative Nambu-Hamiltonian mechanics of intersecting surfaces as the conceptual framework that underlies strange chaotic attractors both in their classical as well as quantum-noncommutative incarnation. It reproduces the familiar and well studied attractor dynamics of Lorenz and Rössler in a very intuitive manner accounting of their gross topological aspects (double lobe Butterfly for the Lorenz system) or single lobe for the Rössler attractor.

Quantum Nambu Dynamics of Surfaces [3,8,9], raises also the issue of possible existence of Quantum Strange Attractors. Their Quantum behavior was built systematically through *fuzzifying* the classical intersecting surfaces of the ND sector. We demonstrated this for the simplest case of the Lorenz system with a linear dissipation.

References

1.P. Cvitanovic Ed., *Universality in Chaos* Adam Holger, Bristol 1984.

2.J.P. Eckmann, *Roads to Turbulence in Dissipative Dynamical Systems* Rev. Mod. Phys. **53** no.4 (1981) 643.

3.M. Axenides and E. Floratos, *Strange Attractors in Dissipative Nambu Mechanics: Classical and Quantum Aspects* JHEP **1004** (2010) 036 [arXiv:0910.3881 [nlin.CD]].

4.Y. Nambu, *Generalized Hamiltonian Dynamics* Phys. Rev. D **7**, (1973) 2403.

5.L. Takhtajan, *On Foundation Of The Generalized Nambu Mechanics (Second Version)* Commun. Math. Phys. **160** , (1994) 295.

6.E.N. Lorenz, *Deterministic Non-Periodic Flow* J.Atm.Sci. **20** (1963), 130.

7.O.E. Rössler, *An Equation for Continuous Chaos* Phys. Lett.**57A**, (1976) 397.

8.M. Axenides and E. Floratos, *Nambu-Lie 3-Algebras on Fuzzy 3-Manifolds* JHEP **0902** (2009) 039 [arXiv:0809.3493 [hep-th]].

9.M. Axenides, E. G. Floratos and S. Nicolis, *Nambu Quantum Mechanics on Discrete 3-Tori* J. Phys. A **42** (2009) 275201 [arXiv:0901.2638 [hep-th]].

10.U. Weiss, *Quantum Dissipative Systems* World Scientific- Series in Modern Condensed Matter Physics, vol.**13** 2008; C-I. Um, K-H. Yeon, T.F. George, *The Quantum Damped Harmonic Oscillator* Phys. Rep. **362** (2002) 63; M.C. Gutzwiller, *Chaos in Classical and Quantum Mechanics* Springer-Verlag (1990), New York.

11.A. Clebsch , J.Reine Angew. Math. **56** (1859) 1.

12.M. Tabor, *Chaos and Integrability in Nonlinear Dynamics: An Introduction* Wiley-Interscience (1989).

13.D. Farmer, E. Ott and J.A. Yorke, *Hausdorff Dimension computation* Physica **7D** (1983) 153.

14.C. Sparrow, *The Lorenz Equation, Bifurcations, Chaos and the Strange Attractors*, Springel-Verlag, New York 1987.

15.C. Doering and J. Gibbon, *On the shape and Dimension of the Lorenz Attractor* Dynamics and Stability of Systems , vol**10**, No.**3**, (1995), 255 ; ibid, vol**13**, No.**3**,(1998)299.

A methodology for classifying periodic orbits

Jayanta K. Bhattacharjee[1], **Sagar Chakraborty**[2] **and Amartya Sarkar**[1]

[1]S. N. B. N. C. B. S., JD-Block, Sector-III, Salt Lake, Kolkata-98
[2]NBIA, Niels Bohr Institute, Blegdamsvej 17, 2100 Copenhagen Ø, Denmark
Emails: jkb@bose.res.in, sagar@nbi.dk, amarta345@bose.res.in

Abstract: We propose a unified methodology, based on renormalization group theory, for finding out existence of periodic solutions in a plethora of nonlinear dynamical systems appearing across disciplines. The technique will be shown to have the non-trivial ability of classifying the solutions into limit cycles and periodic orbits surrounding a center.
Keywords: limit cycle, center, renormalization group, two-dimensional dynamical systems.

1. Introduction

Inspite of the tremendous progress made in the area of dynamical systems over the last few decades, it remains a problem to find the nature of and locate the periodic orbits of a two dimensional dynamical system. Periodic orbits come in two varieties – centre type oscillations and limit cycles. The centre type trajectories are a continuous family of closed curves in phase space – each determined by the prescribed initial condition while the limit cycle is an isolated periodic trajectory. A little more than a decade ago a different way of looking at problem of nonlinear dynamics of oscillators was proposed by Chen et al [1,2] and has been explored by several groups. This method involves a direct use of perturbation theory and renormalization group (RG). The RG naturally leads to flow equations. In this respect it is akin to the Bogoliubov-Krylov method. The advantage however lies in the fact that RG uses naïve perturbation theory. One does not need to anticipate scales (as in multiple scales method) or make an assumption about slowly varying amplitudes and phases (Bogoliubov-Krylov).

How does one apply the RG principle to a problem in dynamics? We begin by observing that a periodic solution can be expressed as a Fourier series with amplitude A and phase θ of the lowest harmonic determining the amplitude and phase of the higher order ones. The amplitude and phase are quantities that will *flow*. A naive perturbation expansion of the dynamical variable will lead to a divergent answer. If 't' is the time at which we want to know $x(t)$ and t_0 is the initial time, the $x(t)$ will diverge as $t - t_0 \to\infty$. This is completely similar to divergence in field theories where a physical quantity (two point correlation function e.g.) diverges as the cutoff $\Lambda \to \infty$. If we are discussing a physical variable, then the answer has to be finite and while this

Chaos Theory: Modeling, Simulation and Applications
C. H. Skiadas, I. Dimotikalis and C. Skiadas (Eds)

is achieved in field theory by constructing running coupling constants it is done for the differential equation by introducing an arbitrary time scale τ and letting the amplitude and phase depend on τ. This leads to the flow equations

$$\frac{dA}{d\tau} = f(A,\theta), \quad 1.1$$

$$\frac{d\theta}{d\tau} = g(A,\theta). \quad 1.2$$

For the autonomous system of f and g are generally function of A alone. We propose to use the flow equations (1.1) and (1.2) to differentiate between oscillators which are of the centre variety and limit cycles. The centre type oscillation consists of a continuous family of closed orbits in phase space, each orbit being determined by its own initial condition. This implies that the amplitude A is fixed, once the initial condition is set. This must lead to,

$$\frac{dA}{d\tau} = 0 \cdot \quad 1.3$$

This statement is exact and is not tied to any perturbation theory argument. For the limit cycle on the other hand,

$$\frac{dA}{d\tau} = f(A) \quad 1.4$$

And $f(A)$ must be such that the flow must have a fixed point. The fixed point has to be stable for the limit cycle to be stable. If $A=0$ is the only fixed point of $f(A)$ in (1.4), then we have an attractor: either a focus or a node.

The calculation of $f(A)$ requires the use of perturbation theory. Application of perturbation theory is possible only if one can locate a centre – this is the basic periodic state. Locating a centre can sometimes be straightforward e.g. $\dot{x}_1 = x_2$, $\dot{x}_2 = -\partial V/\partial x_1$ where V is a general anharmonic potential $V = \frac{1}{2}x_1^2 + \frac{\lambda}{3}x_1^3 + \frac{\mu}{4}x_1^4$. Here $(x_1,x_2) = (0,0)$ is a linear centre around which perturbation theory can be done. Similarly for the Van der Pol oscillator $\dot{x}_1 = x_2$, $\dot{x}_2 = -k\dot{x}_1\left(x_1^2 - 1\right) - \omega^2 x_1$, the origin is a centre, for $k=0$.

In the Lotka-Volterra model $\dot{x}_1 = x_1 - x_1x_2$, $\dot{x}_2 = -x_2 + x_1x_2$ the origin is a saddle and (1,1) is the centre. Shifting the origin to the centre is the first step of the process of determining the function $f(A)$. In this case of course $f(A) = 0$ since the periodic state in the predator-prey model is a centre like state.

A more complicated situation is the Belushov-Zabotinsky reaction [3]. In that case, a transfer of origin to the fixed point will have to be followed by a proper setting of parameters in the problem to make the origin a centre which is the starting point of all perturbation theory. This raises the problem that the given dynamical system may not have a relevant parameter, e.g. the well known paradigm for the limit cycle,

$$\dot{z} = (1+i)z - \beta|z|^2 z, \tag{1.5}$$

where $z = x + iy$ is the complex variable and $\beta > 0$. The only fixed point is the origin and it is an unstable focus for all β. We can overcome this dificulty by considering the more general system,

$$\dot{z} = (\alpha_1 + i\alpha_2) z - \beta|z|^2 z. \tag{1.6}$$

The origin is now a stable focus for $\alpha_1 < 0$, unstable focus for $\alpha_1 > 0$ and a centre for $\alpha_1 = 0$. It is this centre about which one can set up a perturbation theory. The perturbative evaluation of $f(A)$ and $g(A)$ consequently involves the following initial steps:

i) Find the fixed points of the system and identify linear centres.

ii) If there are no linear systems, extend the parameter space and see if linear centre can be located as the parameters are changed.

iii) For every linear centre, thus located, we need to check the existence of a limit cycle by perturbatively constructing $f(A)$ and $g(A)$.

2. A Lienard System

We begin by studying a particularly interesting example of Lienard system $\ddot{x} + \dot{x}f(x) + g(x) = 0$. We consider a linear function for $f(x)$ and a quadratic one for $g(x)$, i.e. $f(x) = \alpha + \beta x$ and $g(x) = x + \lambda x^2$. Redefining α and β we can write the Lienard system in the form,

$$\ddot{x} + k\dot{x}(1+\mu+x) + x + \lambda x^2 = 0. \tag{2.1}$$

For $k = \lambda = 1$, this reduces to the special case studied numerically in ref. [4]. The above second order system can be written as,

$$\dot{x} = y, \tag{2.2a}$$

$$\dot{y} = -x - \lambda x^2 - ky(1+\mu+x). \tag{2.2b}$$

There are two fixed points (0, 0) and $(-1/\lambda, 0)$. The former is a centre for $k = 0$ and the latter is always a saddle. We need to study the possibility of a periodic orbit around the origin.

Accordingly we expand x in a double power series

$$x = x_0 + kx_{10} + \lambda x_{11} + k^2 x_{20} + \lambda^2 x_{22} + k\lambda x_{12} + \cdots \quad 2.3$$

and insert in Eq.(2.1) and equate the coefficient of $k^m \lambda^n$ to zero for all *m, n*. At different orders we have,

$$\ddot{x}_0 + x_0 = 0, \quad 2.4$$

$$\ddot{x}_{10} + x_{10} = (1+\mu)\dot{x}_0 + x_0\dot{x}_0, \quad 2.5$$

$$\ddot{x}_{11} + x_{11} = -x_0^2, \quad 2.6$$

$$\ddot{x}_{20} + x_{20} = (1+\mu)\dot{x}_{10} + x_{10}\dot{x}_0 + \dot{x}_{10}x_0, \quad 2.7$$

$$\ddot{x}_{21} + x_{21} = (1+\mu)\,x_{11} + x_{11}\dot{x}_0 + \dot{x}_{11}x_0 - 2x_0x_{10}. \quad 2.8$$

Using the initial condition $x = A_0$, $\dot{x} = 0$ at $t = 0$, we now write,

$$x_0 = A_0 \cos t \quad 2.9$$

$$x_{10} = \frac{A_0}{2}(1+\mu)(t\cos t - \sin t) + \frac{A_0^2}{6}(\sin 2t - \sin t) \quad 2.10$$

$$x_{11} = -\frac{A_0^2}{2} + \frac{A_0^2}{3}\cos t + \frac{A_0^2}{6}\cos 2t \quad 2.11$$

correct to $O(k)$ and $O(\lambda)$.

The divergence in the perturbation theory is obvious as to this order. The expression for $x(t)$ upto first order in perturbation parameters reads,

$$x(t) = A_0 \cos t + \frac{kA_0}{2}(1+\mu)(t\cos t - \sin t) + \frac{A_0^2}{6}(\sin 2t - 2\sin t) + \lambda\left[-\frac{A_0^2}{2} + \frac{A_0^2}{3}\cos t + \frac{A_0^2}{6}\cos 2t\right]. \quad 2.12$$

The divergence occurs as time difference from the initial time i.e. $t - 0$ becomes very big. We introduce an arbitrary time scale τ to write the divergent term as $t - \tau + \tau - 0$ and define two renormalization constants Z_1 and Z_2 as

$$A_0 = Z_1(t_0, \tau)A(\tau), \quad 2.13$$

$$\theta_0 = 0 = \theta(t) + Z_2(t, \tau). \quad 2.14$$

We have the expansions,

$$Z_1 = 1 + \alpha_{10}k + \alpha_{11}\lambda + \cdots \quad 2.15$$

$$Z_2 = \beta_{10}k + \beta_{11}\lambda + \cdots \quad 2.16$$

Inserting Eqs. (2.13) and (2.14) in the expression (2.12) for $x(t)$ we have

$$x(t) = A(1+\alpha_{10}k+\alpha_{11}\lambda+\cdots)\left[\cos(t+\theta)-(\beta_{10}k+\beta_{11}k)\sin(t+\theta)\right]$$
$$+\frac{kA}{2}(1+\mu)(t-\tau+\tau)\cos(t+\theta)-\frac{kA}{2}\sin(t+\theta)$$
$$+\frac{kA^2}{6}\left[\sin(t+\theta)-2\sin(t+\theta)\right]$$
$$+\lambda\left[-\frac{A^2}{2}+\frac{A^2}{3}\cos(t+\theta)+\frac{A^2}{6}\cos 2(t+\theta)\right]+\cdots \qquad 2.17$$

The series is correct to $O(k)$ and $O(\lambda)$. We choose $\alpha_{10} = -(1+\mu)\tau/2$, $\alpha_{11} = \beta_{10} = \beta_{11} = O$ and write

$$x(t) = A(\tau)\cos(t+\theta(\tau))+\frac{kA}{2}(1+\mu)(t-\tau)\cos(t+\theta)+\cdots \qquad 2.18$$

Clearly $x(t)$ cannot depend on the choice of τ and hence, from $dx/d\tau = 0$

$$\frac{dA}{d\tau} = \frac{kA}{2}(1+\mu), \qquad 2.19$$
$$\frac{d\theta}{d\tau} = 0. \qquad 2.20$$

The remaining τ-dependence is removed by setting $\tau = t$ and thus

$$x(t) = A(t)\cos(t+\theta)+\frac{kA^2}{2}\left[\sin 2(t+\theta)-2\sin(t+\theta)\right]$$
$$+\lambda\left[-\frac{A^2}{2}+\frac{A^2}{3}\cos(t+\theta)+\frac{A^2}{6}\cos 2(t+\theta)\right]+\cdots \qquad 2.21$$

We now go to the next order and after a straightforward calculation along the above lines obtain

$$\frac{dA}{d\tau} = \frac{kA}{2}(1+\mu)-k\lambda\frac{A^3}{8}, \qquad 2.22$$
$$\frac{d\theta}{d\tau} = \frac{kA}{6}(1+\mu)(k-\lambda)-\frac{k^2A^2}{24}-5\lambda^2\frac{A^2}{12}. \qquad 2.23$$

Now, if $k = 0$, the R.H.S. of equation (2.22) vanishes for any A, and thus in accordance with the prescription of our methodology we must have a centre type oscillation with an angular frequency of $\Omega = 1-5\lambda^2A^2/12$, which is the well-known correct result for a cubic oscillator when the amplitude is small. Similarly for $\lambda=0$ (and non-zero k), one has a centre type oscillation if $\mu = -1$ with frequency $\Omega = 1-k^2A^2/24$, a result which can be verified from the exact answer which exists in this case.

Of course, the RG flow equations also suggest that there is also a limit cycle solution in the system if $1 + \mu > 0$ (i.e. $\mu > -1$) and both k and λ have the same sign. It is a stable limit cycle if $k > 0$ and unstable if $k < 0$. The frequency of the limit cycle is:

$$\Omega = 1 + \frac{kA}{6}(1+\mu)(k-\lambda) - \left(\frac{k^2}{24} + \frac{5\lambda^2}{12}\right)A^2 \qquad 2.24$$

As one may note, for $k = \lambda$, the amplitude of the limit cycle is given by $A^2 = 4(1+\mu)/k$ and its frequency $\Omega = 1-11k^2A^2/24 = 1-11k(1+\mu)/6$. A positive definite quantity to this order would be $\Omega = 1-11k(1+\mu)/3$ and since we want $\Omega^2 > 0$, we restrict $\mu < 3/11k-1$. For $k = 1$, this leads to $\mu < -0.73$. The limit found numerically [4] by for the case k = λ =1 is –0.84.

3. Advantage Over Linear Stability Analysis

Before we conclude, let us witness how useful this RG technique is when one deals with the subtle cases of centers in nonlinear dynamical systems. It is very well known that linearized version of a nonlinear dynamical system may not reproduce qualitatively correct picture of the phase portrait near a fixed point. We now showcase the fact that while linearization of a certain nonlinear dynamical system *wrongly* establishes a fixed point as center (which originally is an attractor), our methodology gives correct result. Consider the following dynamical system:

$$\dot{x} = -y + \varepsilon a x\left(x^2 + y^2\right), \qquad 3.1a$$

$$\dot{y} = x + \varepsilon a x\left(x^2 + y^2\right). \qquad 3.1b$$

Here, ε is a small positive parameter that facilitates a trial perturbative solution of the form: $x(t) = x_0 + \varepsilon x_1 + \varepsilon^2 x_2 + \ldots$. Linear stability analysis would show that the fixed point (0,0) is a centre for all a. It can however be easily shown [4] by making use of polar coordinates, in system (3.1), the origin is a stable spiral when $a < 0$ and an unstable spiral for positive a. Now, applying the RG methodology prescribed in this paper, one arrives at the following flow equations, upto $O(\varepsilon^2)$:

$$\frac{dA}{d\tau} = \varepsilon a A^3, \qquad 3.2a$$

$$\frac{d\theta}{d\tau} = 0. \qquad 3.2b$$

One immediately notes that in accordance with our scheme of classification we can distinguish a centre from a focus. From the above flow equations, one can easily extract the *correct* information regarding the nature of the fixed point in system (3.1): if $a = 0$, $dA/d\tau = 0$ for all A, implying that the origin is a centre; whereas if $a \neq 0$, $dA/d\tau = 0$ *iff* $A = 0$, making the origin an attractor.

4. Conclusions

To conclude, we again emphasize that this paper introduces a simple yet powerful methodology --- based on perturbative renormalization group theory --- of identifying and classifying a periodic solution (limit-cycle or orbit around center) in various types of two-dimensional nonlinear dynamical system. This very technique can also distinguish between a focus and a center.

The different types of two-dimesional systems that can be handled using this methodology include not only simpler autonomous systems but also forced non-autonomous systems and time-delayed systems which are discussed in detail elsewhere [5].

Also, it has been shown that our technique yields the correct nature of the fixed point of a nonlinear dynamical system when the linearization about it gives a completely wrong idea regarding its true nature.

Acknowledgements

SC acknowledges academic and financial support from NBIA, Copenhagen and a post-doctoral research FNU grant no. 505100-50-30, 168 by Danish Research Council.

References

[1] L.Y. Chen, N. Goldenfeld and Y. Oono, Renormalization group theory for global asymptotic analysis. *Phys. Rev. Lett.*, **73:**1311, 1994.

[2] L .Y. Chen, N. Goldenfeld and Y. Oono, Renormalization group and singular perturbations: Multiple scales, boundary layers, and reductive perturbation theory. *Phys. Rev. E.*, **54:** 376, 1996.

[3] I. Lengyel, G Rabai and I Epstein, Experimental and modeling study of oscillations in the chlorine dioxide-iodine-malonic acid reaction. *J. Am. Chem. Soc.*, **112** (25):9104, 1990.

[4] Steven H. Strogatz, *Nonlinear Dynamics and Chaos: With Applications to Physics, Biology, Chemistry and Engineering* Westview Press, USA, 1994.

[5] A. Sarkar, J. K. Bhattacharjee, S. Chakraborty and D. Banerjee, arxiv., 1005.2858 v1 (2010)

Chaoticity in the time evolution of foreign currency exchange rates in Turkey

O. Cakar[†], O. O. Aybar[†*], A. S. Hacinliyan[†‡*] and I. Kusbeyzi[†*]

[‡]Yeditepe University, Department of Physics, Istanbul, Turkey,
[†]Yeditepe University, Department of Information Systems and Technologies, Istanbul, Turkey
[*]Gebze Institute of Technology, Department of Mathematics, Kocaeli, Turkey
Emails: oaybar@yeditepe.edu.tr, ahacinliyan@yeditepe.edu.tr, ikusbeyzi@yeditepe.edu.tr

Abstract: Tools from chaos theory that have found recent use in analysing financial markets have been applied to the US Dollar and Euro buying and selling rates against the Turkish currency. The reason for choosing the foreign exchange rate in this analysis is the fact that foreign currency is an indicator of not only the globalization of economy but also savings and investment. In order to test the globality assumption and to ascertain the degree of involvement of local conditions in Turkey, the Euro and US dollar exchange rates have been subjected to the same analysis.

Keywords: financial markets, time series analysis, detrended fluctuation analysis, r/s analysis on noisy data.

1. Introduction: The Fractal Market Hypothesis

Fractal Geometry is the geometry of real nature and intimately related to chaos. Fractal structures are suitable for applying to social sciences because of their self similarity and dynamic dimensions [1].

Time series analysis have been extensively used in market analysis. Financial time series are best analysed by using nonlinear time series analysis tools such as rescaled range (R/S) analysis, detrended fluctuation analysis (DFA) if heteroscedastity is dominant and nonlinear time series analysis including mutual information combined with False Nearset Neighbours (FNN) [2, 3, 4, 5]. These techniques reveal possible fractal structures. The Fractal Market Hypothesis of Edgar Peters lies at the root of the analysis of financial markets with the help of chaotic structures [1]. It emphasizes the effecct of liquidity and investment horizons on the behavior of investors. It aims to give a model of investor behavior and market price movements that is observed [1].

This hypothesis states that;

1. A market consists of many investors with different investment horizons.

Chaos Theory: Modeling, Simulation and Applications
C. H. Skiadas, I. Dimotikalis and C. Skiadas (Eds)
 (pp. 127 - 134)

2. The information set that is important to each investment horizon is different. As long as the market maintains this fractal structure, with no characteristic time scale, the market remains stable. When the market's investment horizon becomes uniform, the market becomes unstable because everyone is trading based upon the same information set.
3. Prices reflect a combination of short-term technical trading which is likely to contain noise and long term fundamental valuation. The market responds to the perception of expected earnings based on the changing economic environment. There is no reason to believe that the length of the short-term trends is related to the long-term economic trend. At any one time, prices may not reflect all available information, but only the information important to that investment horizon.

Stable foreign currencies have always been a source of investment in Turkey, hence this becomes a plausible parameter for analysis of saving and investment trends as well as the relation of the Turkish economy to global economies. For comparison purposes, the US Dollar-Euro rates have also been considered.

2. Time Series Analysis

In this part dollar buying, dollar selling, euro buying and euro selling data between 1.1.2005 and 6.2.2009 is analyzed. The buying and selling rates show closely similar behaviors. The linear trend analysis consistently gives a slightly perceptible increase in time with a very low coefficient of determination. The euro dollar exchange rate shows a comparable behavior with smaller fluctuations. This probably reflects the increased stability of the latter market [1, 5].

Figure 1 – Dollar & Euro-Buying & Selling Trend Lines

In all cases, looking at the fluctuations, the data seems to behave as a random walk and can not be explained with linear analysis. Dividing the data into years does not help us to understand the behavior and necessites a nonlinear analysis.

The US Dollar Buying Data is first divided into four one year periods and subjected to R/S analysis for each year. The R/S results of the US Dollar Buying Data shows similar characteristics in their behavior. In Figure 2, we see two regimes corresponding to slopes of 0.421818 and 0.204483 representing the regime of dollar buying among years. For the remaining data sets (Dollar Selling, Euro Selling and Euro Buying) we observe approximately the same characteristic regimes therefore there is no yearly change and the whole data can be used for R/S analysis.

For the overall data, again, two regimes are observed. The slope of regime one is 0.391818 (dimension=1.61) and the slope of regime two is 0.224483 (dimension=1.77). Both Hurst exponents are smaller than 0.50. So these series are not changing as a function of time. These series are not coincidental and prices disperse deterministically.

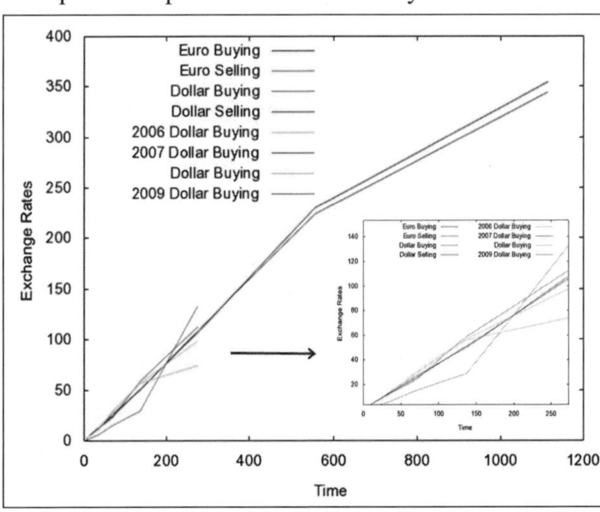

Figure 2 – R/S Analysis of Years

Division of the Dollar Buying Data into four yearly period for DFA shows a similar behavior for each year of the split data and combined data. The plot line with slope 1.465 represents a single regime of dollar buying among years. Dollar Selling, Euro Selling and Euro Buying rates also have the same characteristic behavior. We can therefore use the combined data for all years [6, 7, 8].

The Detrended Fluctuation Analysis results of Euro and Dollar Data show fluctuations that are similar to the divided sets. If we define a linear function with slope of 1.47 and passing through these fluctuations, approximately one satisfying region is observed. It is thus difficult to understand presence of multiple regimes in the exchange rates markets with detrended fluctuations analysis [2, 3, 5, 6, 7].

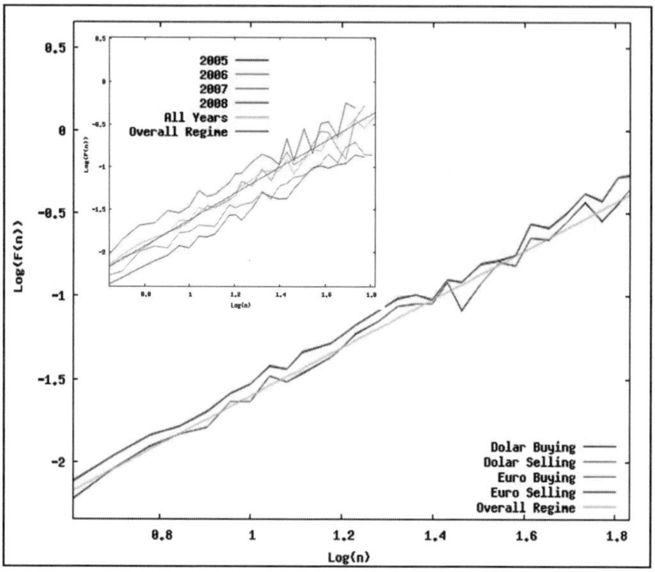

Figure 3 – DFA of overall data and split data into years

The behavior of the mutual information analysis shows that when one divides the data into years, all sets of Dollar buying data are very close to each other and also approximately have the same minimum representing the delay time of 20 days.

It is very interesting that dollar buying and dollar selling rates show exactly the same behavior. Red lines that represent the dollar buying rate have been masked because of this. The first minimum of all values are nearly at 20. US Dollar buying and selling , Euro buying and selling graphs all indicate a delay time of 20. We now proceed to the False Nearest Neighbors calculation. If data is divided into years, all sets of Dollar buying data are very close to each other and have the same minimum representing the embedding dimension as 5 or 6 approximately. Therefore data for all years can be used instead of splitting into years. We also notice that Euro buying and selling are intersecting each other but on the other hand Dollar buying and selling start intersecting and somewhere they diverge then again intersect each other. Since the embedding dimension of combined exchange rates is 5 or 6, the system can be modelled as a three dimensional one [2, 3, 4].

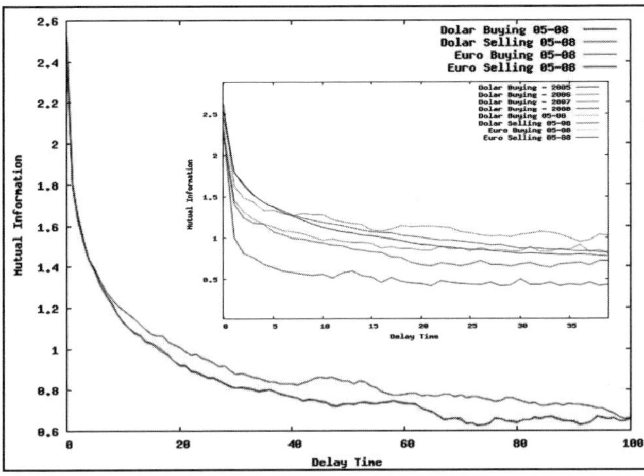

Figure 4 – Overall Mutual Information

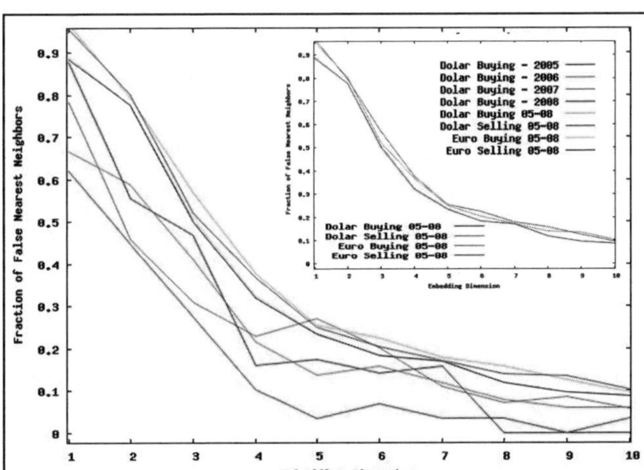

Figure 5 – FNN of Years

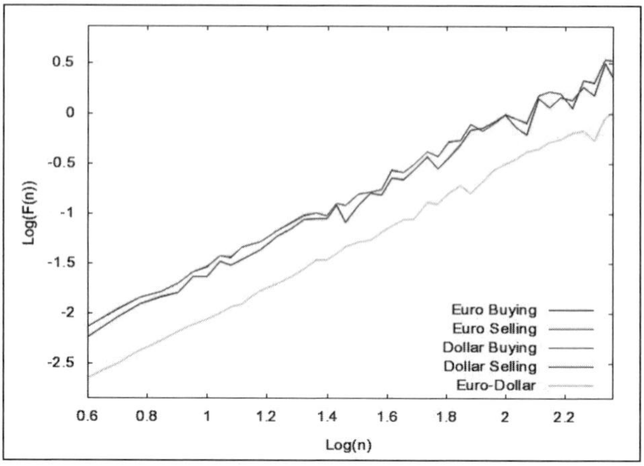

Figure 6 – Overall Euro Dollar Exchange Rates DFA

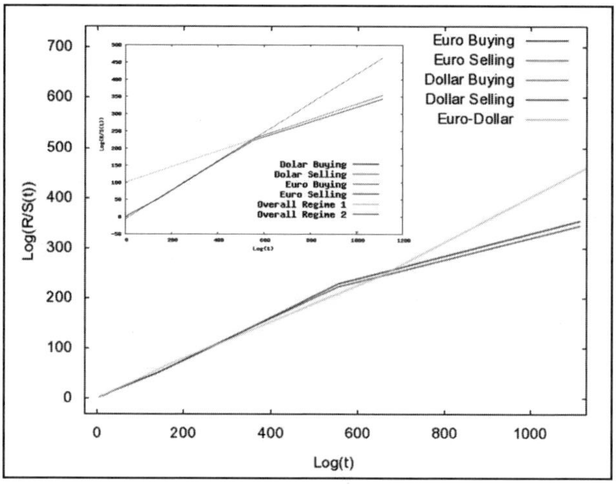

Figure 7 – Overall R/S Analysis

3. Conclusion

The R/S analysis reveals two regimes while only one regime is observed in DFA. The reason for this is that the R/S analysis is affected by extreme values that can occur in day to day trends. Both regimes show long range effects. These disappear in DFA which detrends the long range trend, so that the second regime probably shows the short-term fluctuations.

The results of the analyses show that fractal structures can explain the analysed data more closely than classical linear time series analyses. Moreover, in most cases, the behavior of the combined data is nearly the same as the behavior observed when the data is split into yearly bins. This indicates that relatively long term trade and investment trends have been uniform during the period under study. More specifically, if one looks at the analysis types one by one:

R/S analysis indicates the presence of two regimes in the data and is therefore incompatible with the results from the other analysis methods which do not give sufficient evidence for two regimes. Since the data contains fluctuations and noise by R/S analysis involves the effect of extreme values, we may be seeing two regimes. R/S analysis has been extensively used for financial data and the two different regimes that it suggests can come from short and long term trends in the yearly data and overall data involving the exchange rates, however one of the regimes can alternatively come from day to day random fluctuations [8, 9 ,10].

The DFA analysis usually shows one regime, so that it is mainly affected from short term fluctuations. Long range effects have been detrended. This indicates that the market behaves in a relatively consistent and stable way if fluctuations are ignored.

Nonlinear time series analyses reveals that the embedding dimension to be nearly 3. This would indicate a one dimensional trend according to Takens's theorem. On the other hand, if we look at the Hurst results, the dimension is nearly 1.7. We can say that FNN can better eliminate the fluctuations and noise in this case and give us more clear cut results. On the other hand, this can also be interpreted as a sign of fractal behavior [3, 4, 5, 10, 11, 12].

References

[1] Abhyankar, A., L. Copeland, and W. Wong. Uncovering Nonlinear Structure in Real-Time Stock-Market Indexes. *Jounal of Business & Economic Statistics*, 15N1: 1-14, 1997.

[2] Hacınlıyan, A., Sahin, G., Erenturk, M., Detrended fluctuation analysis in natural languages using non-corpus parametrization, *Chaos, Solitons and Fractal*, 41, 1, 198-205, 2009.

[3] Atak K., O. O. Aybar, G. Sahin, A. Hacınlıyan and Y. Skarlatos. Chaoticity analysis of the current through pure, hydrogenated and hydrophobically modified PEG-Si thin films under varying relative humidity. *Central European Journal of Physics.* 7, 3, 568-574, 2009.

[4] Hacınlıyan A., Y. Skarlatos, G. Sahin, K. Atak and O. O. Aybar. Possible Stretched Exponential Parametrization for Humidity Absorption in Polymers. *The European Physical Journal E.* 28, 4, 369-376, 2009.

[5] Hacınlıyan A., Y. Skarlatos, H. A. Yıldırım and G. Sahin. Characterization of chaoticity in the transient current through PMMA thin films. *Fractals.* 14, 125-131, 2006.

[6] Hacınlıyan A., Y. Skarlatos, H. A. Yıldırım and G. Sahin. Characterization of Chaocity in The Transient Current Through PMMA Thin Films, *Physical Review B*. 73, 13, 2006.

[7] Barkoulas, John, and Nickolaos Travolos. Chaos in an emerging capital market? The case of the Athens Stock Exchange. *Applied Financial Economics*, 8: 231-243, 1998.

[8] Crilly , A., R. Earnshaw, and H. Jones. *Applications of Fractals and Chaos*. 1. Berlin, Germany: Springer-Verlag, 1993.

[9] Hallegatte, Stephane, Michael Ghil, and Patrice Dumas. Business Cycles, Bifurcations and Chaos in a Neo-Classical Model with Investment Dynamics. *Elsevier Science*, 1-40, 2006.

[10] Liebovitch, Larry, and Daniela Scheurle. Two Lessons from Fractals and Chaos. *Complexity*, 5N4: 34-43, 2000.

[11] Richards, Gordon. A Fractal Forecasting Model for Financial Time Series. *Journal of Forecasting*, 23: 587-602, 2004.

[12] Peters E. E. *Fractal Market Analysis: Applying Chaos Theory to Investment and Economics*. New York: John Wiley & Sons, Inc., 1994.

Symbolic dynamics and chaotic synchronization

Acilina Caneco[1], Clara Grácio[2] and J. Leonel Rocha[3]

[1]Mathematics Unit, Instituto Superior de Engenharia de Lisboa, Lisboa and CIMA-UE Évora, Portugal
[2]Department of Mathematics, Universidade de Évora and CIMA-UE, Évora, Portugal
[3]Mathematics Unit, Instituto Superior de Engenharia de Lisboa and CEAUL, Lisboa, Portugal
Emails: acilina@deetc.isel.ipl.pt, mgracio@uevora.pt, jrocha@deq.isel.ipl.pt

Abstract: Chaotic communications schemes based on synchronization aim to provide security over the conventional communication schemes. Symbolic dynamics based on synchronization methods has provided high quality synchronization [5]. Symbolic dynamics is a rigorous way to investigate chaotic behavior with finite precision and can be used combined with information theory [13]. In previous works we have studied the kneading theory analysis of the Duffing equation [3] and the symbolic dynamics and chaotic synchronization in coupled Duffing oscillators [2] and [4]. In this work we consider the complete synchronization of two identical coupled unimodal and bimodal maps. We relate the synchronization with the symbolic dynamics, namely, defining a distance between the kneading sequences generated by the map iterates in its critical points and defining n-symbolic synchronization. We establish the synchronization in terms of the topological entropy of two unidirectional or bidirectional coupled piecewise linear unimodal and bimodal maps. We also give numerical simulations with coupled Duffing oscillators that exhibit numerical evidence of the n-symbolic synchronization.

Keywords: chaotic synchronization, symbolic dynamics, symbolic synchronization, kneading theory.

1. Introduction

Synchronization of two or more systems, is a process wherein the systems adjust a given property of their motion to a common behaviour, due to coupling or forcing. Chaos synchronization is the agreement or correlation of different chaotic processes in time. A great deal of researches has been undertaken on this subject, since the pioneering works of Pecora and Carroll ([9], [10]). Various types of synchronization have been studied. This includes complete synchronization, phase synchronization, lag synchronization, generalized synchronization, anticipated synchronization, and so on [1]. The coupled systems might be identical or different, the coupling might be unidirectional, (master-slave or drive-response), or bidirectional (mutual coupling) and the driving force might be deterministic or stochastic [11].

Chaos Theory: Modeling, Simulation and Applications
C. H. Skiadas, I. Dimotikalis and C. Skiadas (Eds)

In this paper we relate the synchronization with symbolic dynamics, namely, defining a distance between the kneading sequences generated by the map iterates in its critical points and defining n-symbolic synchronization. We will see that this distance is a good measure of synchronization, since it decreases to zero, as the coupling parameter increases and the oscillators begin to synchronize. Moreover we establish in two theorems the synchronization in terms of the topological entropy of two unidirectional or bidirectional coupled piecewise linear unimodal and bimodal maps. The n-symbolic synchronization is numerically showed for Duffing oscillators.

2. The Synchronization Problem

Consider two coupled systems $x_{n+1} = f(x_n)$ and $y_{n+1} = f(y_n)$. The two identical systems are in complete synchronization, if the difference $z_n = y_n - x_n$ converges to zero, as $n \to \infty$. The coupling may be unidirectional

$$\begin{cases} x_{n+1} = f(x_n) \\ y_{n+1} = f(y_n) + k[f(x_n) - (y_n)] \end{cases}$$

$$z_{n+1} = (1-k)[f(y_n) - f(x_n)], \tag{1}$$

or bidirectional

$$\begin{cases} x_{n+1} = f(x_n) - k[f(x_n) - (y_n)] \\ y_{n+1} = f(y_n) + k[f(x_n) - (y_n)] \end{cases}$$

$$z_{n+1} = (1-2k)[f(y_n) - f(x_n)], \tag{2}$$

where k is the coupling parameter. In [9] and [10] it is establish that this kind of synchronization can be achieved provided that all the Lyapunov exponents are negative.

Motivated by the fact that any transitive m-modal map with positive topological entropy and growth number $\pm s$, is topologically semi-conjugated to a piecewise linear map with slope $\pm s$ everywhere, see [8], we will study the synchronization of unimodal and bimodal piecewise linear maps.

2.1. *Coupling two unimodal piecewise linear maps*

Consider the tent map $f_s : [0,1] \to [0,1]$ defined by $f_s(x) = sx - s + 2$, if $0 \le x \le 1 - 1/s$ and $f_s(x) = s - sx$, if $1 - 1/s \le x \le 1$. This map f_s can be written as

$$f_s(x) = sx - s + 2 + 2\theta(x)(s - sx - 1) \tag{3}$$

with $\theta(x) = 1$, if $0 \le x \le 1 - 1/s$ and $\theta(x) = 0$, if $1 - 1/s \le x \le 1$.

Note that, there exist a one-to-one correspondence between the images of the map θ, $\{0,1\}$, and the usual alphabet $A=\{L, R\}$, associated to the unimodal map.

With the kneading theory we construct symbolic sequences associated with the critical points of the map, see [3], [7] and [8]. Noticing that, as the coupled systems became synchronized, the number of equal initial symbols in the kneading sequences grows (see Table 1), we are led to the following.

Definition 2.1. Let $S_z = S_{z_1} S_{z_2} \ldots S_{z_n} \ldots$, be a symbolic sequence, associated to the orbit of a point z of the map f, using in position i a symbol S_{z_i} belonging to some alphabet A. Define a distance between two sequences $S_x = S_{x_1} S_{x_2} \ldots S_{x_p} \ldots$ and $S_y = S_{y_1} S_{y_2} \ldots S_{y_q} \ldots$ by

$$d(S_x, S_y) = e^{-n}, \text{ where } n = \min\{n \geq 1 : S_{x_n} \neq S_{y_n}\}.$$

Definition 2.2. In the conditions of the previous definition, two sequences $S_x = S_{x_1} S_{x_2} \ldots S_{x_n} \ldots$ and $S_y = S_{y_1} S_{y_2} \ldots S_{y_q} \ldots$ are said n-symbolic synchronized iff $\exists n \in \mathbb{N}$: $d(S_x, S_y) \leq e^{-n}$.

Note that $d(S_x, S_y) = e^{-n}$, means that, $S_{x_i} = S_{y_i}$ $(1 \leq i < n)$ and $n = \min\{n \geq 1 : S_{x_n} \neq S_{y_n}\}$. For the unimodal map f_s, we used $\theta(x)$ to represent the symbolic sequence S_x. In this case, the symbol $S_{x_i} = \theta(x_i)$, in the position i of S_x, can only be 0 or 1. As we have $x_{n+1} = f(x_n)$, we must consider $\theta(x_n)$ and $\theta(x_{n+1})$.

Theorem 2.1. Let $x_{n+1} = f(x_n)$ and $y_{n+1} = f(y_n)$ be two identical coupled systems, with f given by (3) and $1 < s \leq 2$. Let h be the topological entropy of (3) ($h = \log s$) and $k \in [0, 1]$ the coupling parameter. If $\theta(y)$ and $\theta(x)$ are n-symbolic synchronized, then

1) The unidirectional coupled systems (1) are synchronized if $k > 1 - e^{-h}$.
2) The bidirectional coupled systems (2) are synchronized if

$$(1/2)(1 - e^{-h}) < k < (1/2)(1 + e^{-h}). \tag{4}$$

Proof. Attending to (1) and (3), we have for the unidirectional case $z_{n+1} = (1-k)[sy_n + 2\theta(y_n)(s - sy_n - 1) - sx_n - 2\theta(x_n)(s - sx_n - 1)]$. From the considered hypothesis, it follows that, for n sufficiently large, $\theta(y_n) = \theta(x_n) = \theta$, and $z_{n+1} = (1-k)(1-2\theta)sz_n$.

Then, $z_{n+m} = [(1-k)(1-2\theta)s]^m z_n$. We may writte $z_{n+m} = q^m z_n$, with $q = (1-k)(1-2\theta)s$. So, we have $\lim_{m \to \infty} q^m = 0$, iff $|q| < 1$. From the definition of θ, $|1 - 2\theta| = 1$, then, to have complete synchronization of the unidirectionally coupled maps it suffices $k > 1 - s^{-1}$, i.e., $k > 1 - e^{-h}$.

On the other hand, attending to (2) and (3), we have for the bidirectional case $z_{n+1} = (1-2k)[sy_n + 2\theta(y_n)(s - sy_n - 1) - sx_n - 2\theta(x_n)(s - sx_n - 1)]$. From the considered hypothesis, it follows that, for n large enough, $\theta(y_n) = \theta(x_n) = \theta$, and $z_{n+1} = (1-2k)(1-2\theta)sz_n$.

Then, $z_{n+m} = [(1-2k)(1-2\theta)s]^m z_n$. So, we have $\lim_{m\to\infty} z_{n+m} = 0$, iff $|1-2k||1-2\theta|s < 1$. From the definition of θ, $|1-2\theta| = 1$, then, to have complete synchronization of the bidirectionally coupled maps it suffices (4).

□

2.2. *Coupling two bimodal piecewise linear maps*

Consider the bimodal piecewise linear map $f_{s,r} : [0,1] \to [0,1]$, with slopes $\pm s$, defined by

$$f_{s,r} = \begin{cases} -sx+1 & if \quad 0 \le x < c_1 \\ sx+r-1 & if \quad c_1 \le x < c_2 \\ -sx+s & if \quad c_2 \le x \le 1 \end{cases}$$

with $r = (3+s)/2 - s(c_1+c_2)$ and critical points $c_1 = (2-r)/2s$ and $c_2 = (1+s-r)/2s$, see [7]. Recall that any transitive bimodal map is topologically semi-cojugated to such a map. The map $f_{s,r}$ can be written as

$$f_{s,r}(x) = -sx+1+\theta_{c_1}(x)(2sx+r-2)+\theta_{c_2}(y)(-2sx+s-r-1), \quad (5)$$

with
$$\theta_{c_i}(x) = \begin{cases} 0, & if \quad 0 \le x < c_i \\ 1, & if \quad c_i \le x \le 1 \end{cases} \qquad (i = 1,2).$$

In this case, we may define $\theta(x) = \theta_{c_1}(x) + \theta_{c_2}(x)$, i.e.,

$$\theta(x) = \begin{cases} 0 & if \quad 0 \le x < c_1 \\ 1 & if \quad c_1 \le x < c_2 \\ 2 & if \quad c_2 \le x \le 1 \end{cases}.$$

Note the existence of a one-to-one correspondence between the images of the map θ, {0, 1, 2}, and the usual alphabet associated to the bimodal map *A*={*L*, *M*, *R*}, see [3], [8] and [6]. For the bimodal map $f_{s,r}$, the symbol $S_{x_i} = \theta(x_i)$, in the position *i* of $S_x = \theta(x)$, can be 0, 1, or 2. Now, we can state a theorem for bimodal maps similar to the one to unimodal maps. The conclusions are the same, although for a different local map *f*.

Theorem 2.2. Let $x_{n+1} = f(x_n)$ and $y_{n+1} = f(y_n)$ be two identical coupled systems, with *f* given by (5) and $1 < s \le 2$. Let *h* be be the topological entropy of (5) ($h = \log s$) and $k \in [0, 1]$ the coupling parameter. If $\theta_{c_i}(x)$ and $\theta_{c_i}(y)$ for *i*=1,2, are *n*-symbolic synchronized, then
1) The unidirectional coupled systems (1) are synchronized if $k > 1 - e^{-h}$.
2) The bidirectional coupled systems (2) are synchronized if

$$(1/2)(1-e^{-h}) < k < (1/2)(1+e^{-h}). \quad (6)$$

Proof. Attending to (1) and (5), we have for the unidirectional case

$$z_{n+1} = (1-k)\{[-sy_n + 1 + \theta_{c_1}(y_n)(2sy_n + r - 2) + \theta_{c_2}(y_n)(-2sy_n + s - r + 1)] - [-sx_n + 1 + \theta_{c_1}(x_n)(2sx_n + r - 2) + \theta_{c_2}(x_n)(-2sx_n + s - r + 1)]\}.$$

From the considered hypothesis, it follows that, for n large enough,

$$\theta_{c_1}(y_n) = \theta_{c_1}(x_n) = \theta_{c_1} \text{ and } \theta_{c_2}(y_n) = \theta_{c_2}(x_n) = \theta_{c_2} \text{ and}$$

$$z_{n+1} = (1-k)[-1-2(\theta_{c_1} - \theta_{c_2})]sz_n.$$

Then, $z_{n+m} = [(1-k)[-1-2(\theta_{c_1} - \theta_{c_2})]s]^m z_n$. We may writte $z_{n+m} = q^m z_n$, with $q = (1-k)[-1-2(\theta_{c_1} - \theta_{c_2})]s$. So, we have $\lim_{m\to\infty} z_{n+m} = 0$, iff $|q| = |1-k||-1-2(\theta_{c_1} - \theta_{c_2})|s < 1$. Attending that,

$$\theta_{c_1} - \theta_{c_2} = \begin{cases} 0 & if \quad x < c_1 \quad or \quad c_2 \le x \\ 1 & if \quad c_1 \le x < c_2 \end{cases}$$

then,

$$q = \begin{cases} -(1-k)s & if \quad x < c_1 \quad or \quad c_2 \le x \\ (1-k)s & if \quad c_1 \le x < c_2 \end{cases}.$$

So, $|q| = (1-k)s$, and to have complete synchronization it suffices, $k > 1 - s^{-1}$, i.e., $k > 1 - e^{-h}$.

For the bidirectional case and attending to (2) and (5), we have

$$z_{n+1} = (1-2k)\{[-sy_n + 1 + \theta_{c_1}(y_n)(2sy_n + r - 2) + \theta_{c_2}(y_n)(-2sy_n + s - r + 1)] - [-sx_n + 1 + \theta_{c_1}(x_n)(2sx_n + r - 2) + \theta_{c_2}(x_n)(-2sx_n + s - r + 1)]\}.$$

From the hypothesis, $\theta_{c_1}(y_n) = \theta_{c_1}(x_n) = \theta_{c_1}$ and $\theta_{c_2}(y_n) = \theta_{c_2}(x_n) = \theta_{c_2}$, for n large enough. Then, $z_{n+m} = [(1-2k)[-1+2(\theta_{c_1} - \theta_{c_2})]s]^m z_n$. So, $\lim_{m\to\infty} z_{n+m} = 0$, iff $|1-2k||-1+2(\theta_{c_1} - \theta_{c_2})|s < 1$. As in the previous case, $|-1+2(\theta_{c_1} - \theta_{c_2})| = 1$, so the complete synchronization of the bidirectionally coupled maps happens if (6) is verified.

3. Duffing Application

Consider two identical bidirectionally coupled Duffing oscillators, [12].

$$\begin{cases} x''(t) = x(t) - x^3(t) - \alpha x'(t) + k[y(t) - x(t)] + \beta Cos(wt) \\ y''(t) = y(t) - y^3(t) - \alpha y'(t) - k[y(t) - x(t)] + \beta Cos(wt) \end{cases}. \tag{7}$$

We consider parameter values for which each uncoupled (k = 0) oscillator exhibits a chaotic behavior, so if they synchronize, that will be a chaotic synchronization. A basic tool is to do an appropriate Poincaré section. In our case, we did a section defined by $y = 0$, since it is transversal to the flow, it contains all fixed points and captures most of the interesting dynamics. In a previous work we have found in the parameter plane (α, β) regions U and $\mathfrak{B}$ where the first return Poincaré map behaves like a unimodal map and like a bimodal map respectively, see [4].

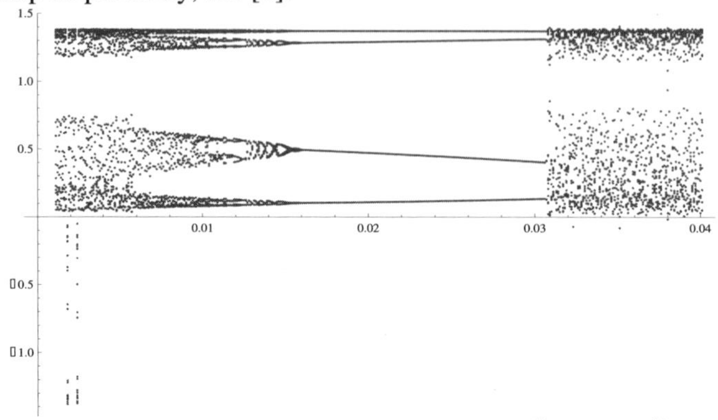

Figure 1. Bifurcation diagram for x_n as a function of $k \in [0.001, 0.04]$, and parameter values, in the unimodal region.

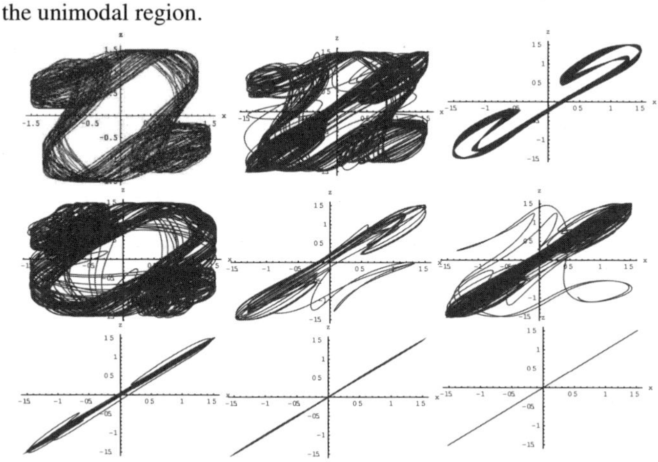

Figure 2. x versus y in the bimodal case for $(\alpha, \beta) = (0.5, 0.719)$ for increasing k.

Fig. 1 shows the bifurcation diagram for x_n as a function of $k \in [0.001, 0.04]$, for parameter values for which the Poincaré map behaves like a unimodal.

We computed the kneading sequences, the kneading determinant and the topological entropy for some values of the parameters (α, β), see [2] and [4]. The synchronization will occur when $x = y$. See in Fig.2 the evolution of x versus y for k=0.008, 0.014, 0.016, 0.018, 0.022, 0.023, 0.048, 0.061, 0.097.

4. Symbolic Dynamics

From Table 1 we verify that, as the coupling parameter k grows and the two coupled maps begin to synchronize, the number of initial equal symbols in the x and y symbolic sequences, grows also and the distance previously defined, decreases. We have verified the same situation for others values of the (α, β) parameters. This gives numerical evidence of the quality of the synchronization between the two oscillators.

Kneading sequences for x and y	k	n
S_x : RLRRRLRLRRRLRRRLRRRLRLRLRRRLRL	0.00601	
S_y : RLRRRLRRRLRRRLRRRLRRRLRRRLRRRL		7
S_x : RLRRRRRLRRRLRRRLRRRLRRRLRRRLRL	0.05	
S_y: RLRRRLRLRLRLRRRLRRRLRLRRRLRRRL		5
S_x : RLRRRLRLRLRLRRRLRRRLRLRLRLRLRL	0.06	
S_y : RLRRRLRLRRRLRRRLRRRLRLRLRLRLRL		9
S_x : RLRRRLRLRLRRRLRRRLRRRLRRRLRRRL	0.064	
S_y : RLRRRLRLRLRLRLRRRLRRRLRLRRRLRR		11
S_x : RLRRRLRLRRRLRRRLRLRLRLRRRLRRRL	0.065	
S_y : RLRRRLRLRRRLRRRLRRRLRLRLRLRLRL		17
S_x : RLRRRLRLRRRLRRRLRLRRRLRRRLRRRL	0.07	
S_y : RLRRRLRLRRRLRRRLRLRRRLRRRLRRRL		30
S_x : RLRRRLRLRRRLRRRLRLRRRLRRRLRRRL	0.08	
S_y : RLRRRLRLRRRLRRRLRLRRRLRRRLRRRL		30

Table 1: Symbolic sequences S_x and S_y, with n identical initial symbols, for the bidirectional coupled Duffing oscillators, for some values of the coupling parameter k, with $(\alpha, \beta) = (0.4, 0.3578)$ fixed.

The number of symbols used in information transmission depends on the limitation of the channel capacity [13]. Only finitely precise synchronization is achievable when oscillators are joined by channel with limited capacity. On this basis, we established in Definition 2.1 and 2.2. finitely precise synchronization with symbolic dynamics.

5. Conclusions

Doing Poincaré sections in the system of two coupled Duffing oscillators, we obtained regions U and B where the Poincaré map behaves like a unimodal and bimodal map respectively. By a result from Milnor and Thurston [8] it is known that every m-modal map $f : I = [a,b] \subset \mathrm{R} \rightarrow I$, with growth rate s and positive topological entropy $h_{top}(f)$ ($\log s = h_{top}(f)$) is topologically semi-conjugated to a $p+1$ piecewise linear map T, with $p \leq m$, defined on the interval $J = [0,1]$, with slope $\pm s$ everywhere and $h_{top}(f) = h_{top}(T) = \log s$. Motivated by this fact, we study the synchronization of piecewise linear unimodal and bimodal maps and obtained the results in theorems 2.1 and 2.2 expressing the synchronization interval in terms of the topological entropy of the coupled maps. From the previous theorems we may also verify that the bidirectional coupled maps begin to synchronize at half the coupling parameter for the unidirectional case. We relate the synchronization with symbolic dynamics, namely, defining a distance between the kneading sequences and defining n-symbolic synchronization.

References

[1] S. Boccaletti, J. Kurths, G. Osipov, D. L. Valladares, C. S. Zhou, The synchronization of chaotic systems, *Physics Reports* 366: 1-101, 2002.

[2] A. Caneco, C. Grácio and J. L. Rocha, Symbolic dynamics and chaotic synchronization in coupled Duffing oscillators, *J. Nonlinear Math. Phys.* 15, 3: 102-111, 2008.

[3] A. Caneco, C. Grácio and J. L. Rocha, Kneading theory analysis of the Duffing equation, *Chaos Solit. Fract.* 42, (3): 1529-1538, 2009.

[4] A. Caneco, C. Grácio and J. L. Rocha, Topological entropy for the synchronization of piecewise linear and monotone maps. Coupled Duffing oscillators, *Int. J. Bifurcation and Chaos*, 19 (11): 3855-3868, 2009.

[5] A. P. Kurian and S. Puthusserypady, Secure digital communication using chaotic symbolic dynamics, *Turk J. Engin* **14** (1): 195-207, 2006.

[6] J. P. Lampreia and J. Sousa Ramos, Symbolic Dynamics for Bimodal Maps, *Portugaliae Math.* 54(1): 1-18, 1997.

[7] M. M. Ramos, C. C. Ramos, R. Severino and J. S. Ramos, Topological invariants of a chaotic pendulum, *Int. J. Pure Appl. Math.* 10 (2): 209-226, 2004.

[8] J. Milnor and W. Thurston, On iterated maps of the interval I and II, *Lect. Notes inMath*, 1342, Springer-Verlag: 465-563, 1988.

[9] L. M. Pecora and T. L. Carroll, Synchronization in chaotic systems, *Phys. Rev. Lett.* 64, 821-824, 1990.

[10] L. M. Pecora and T. L. Carroll, Driving systems with chaotic signals, *Phys. Rev. A* 44: 2374-2383, 1991.

[11] A. Pikovsky, M. Rosenblum and J. Kurths, *Synchronization:a universal concept in nonlinear sciences*, Cambridge University Press, 2001.

[12] U. E. Vincent and A. Kenfack, Synchronization and bifurcation structures in coupled periodically forced non-identical Duffing oscillators, *Phys. Scr.* 77: 045005 (7pp), 2088.

[13] X. Wang, M. Wang and Z. Liu, Chaos synchronization basing on symbolic dynamics with nongenerating partition, *Chaos* 19: 023108, 2009.

New enciphering algorithm based on chaotic generalized Hénon map

Octaviana Datcu[1], Jean-Pierre Barbot[2] and Adriana Vlad[1,3]

[1]Faculty of Electronics, Telecommunications and Information Technology, Politehnica University of Bucharest, Romania
[2]Equipe Commande des Systèmes, Ecole Nationale Supérieure de l'Electronique et de ses Applications, Cergy-Pontoise, France
[3]Research Institute for Artificial Intelligence, Romanian Academy, Romania

Emails: octavianadatcu@yahoo.com, barbot@ensea.fr, avlad@racai.ro

Abstract: The paper belongs to the field of chaotic based cryptography. It relies on some ideas from classical fundamentals as mixing functions suggested by C.E. Shannon for secrecy systems, and on the known publication of M.S. Baptista, a paper essentially implying the ergodicity assumption of the chaotic signal in cipher design. In this paper, the first step was to choose a chaotic system of a higher dimension than Baptista used, aiming to obtain a more complex system having a hyperchaotic behavior. The proposed algorithm is based on Generalized Hénon Map (3D), stated in literature as hyperchaotic for a number of bifurcation parameters. The paper advances a new mixing enciphering scheme based on the Generalized Hénon Map (GHM), which may be used as an inner element in a cipher, providing a good practical diffusion and confusion. A random variable transform is applied on the state of the chaotic system at each iteration in order to obtain a new random variable of a quasi uniform law. This new random variable is further transformed, through a series of other functions containing elements of the secret key, into a discrete random variable. The discrete values – which are ASCII numbers - are combined by a simple relation with the plain message, also in ASCII format. It is obtained a first mask of the original message, involving the GHM. On this result (in its binary representation form) other simple transformations that depend on the state of the GHM are applied. That finally allows getting a transformed version of the message that can be included in one of the states of the GHM without disturbing its chaotic behavior. The results, including a perception of the diffusion and the confusion involved, are illustrated on natural text and *jpeg* image.
Keywords: Generalized Hénon Map, chaos based enciphering algorithm, mixing scheme, random variable transform.

1. Introduction

M.S. Baptista promotes chaotic systems as a possibility of enciphering and simultanously embedding secret information (Baptista, 1998). Thus, chaotic cryptography is based on the pseudo-random behavior of dynamical systems

---

Chaos Theory: Modeling, Simulation and Applications
C. H. Skiadas, I. Dimotikalis and C. Skiadas (Eds)

despite their deterministic nature. This behavior is controlled by particular values of parameters used to describe the system, the bifurcation parameters. Baptista uses the ergodicity feature of the chaotic logistic map to inspect and to reveal diffusion and confusion of his enciphering method.

In order to embed the message, the present paper uses the three dimensional discrete Generalized Hénon Map, described by equations (1):

$$\begin{aligned} x_{n+1} &= a - y_n^2 - b \cdot z_n \\ y_{n+1} &= x_n \\ z_{n+1} &= y_n \end{aligned} \tag{1}$$

where states $x_n, y_n, z_n \in (-2;2)$ and bifurcation parameters, $a \in [0;2)$ and $b \in [-0.3;0.3]$.

Suppose that the plain message is added by inclusion to the evolution of the GHM, to its first state, see equations (2). Knowing bifurcation parameters (a, b), the cryptanalyst could easily, **(3),** determine the plain text from the series $z_1, z_2, \ldots, z_{L+3}$, the output of the system. L is the length of the plain text and m_n is the $n-th$ character of the plain message.

$$\begin{aligned} x_{n+1} &= a - y_n^2 - b \cdot z_n + m_n \\ y_{n+1} &= x_n \\ z_{n+1} &= y_n \end{aligned} \tag{2}$$

$$\rightarrow m_n = z_{n+3} + z_{n+1}^2 + b \cdot z_n - a\,, \tag{3}$$

Anstett studies this drawback of polynomial nonlinear systems (Anstett *et al.*2006[1]).

Hence, the necessity to include the message into the evolution of the system not in its original form, but already enciphered.

We borrow from Baptista the idea of dividing the attractor of the chaotic system in subintervals. We take into account the requirement that good cipher has to provide the equiprobability of the elements of the key, (Arroyo *et al.*,2008[2] and Arroyo *et al.*,2009[3]).

We analyze the distribution of the states of the GHM. We take into consideration the shift register structure described by the three equations in (1), and we apply to the semi-difference of two of its states a trigonometric function. We further assign to this trigonometric function 10 distinct discrete values by dividing the interval $[-1;1]$ in 10 subintervals of equal length, as in Table 1. This finally enables to obtain the quasi-uniform distribution revealed in Figure 1. These 10 subintervals will be chosen by the evolution of the GHM in a deterministic-random manner.

Important note: The experimenter may use the algorithm proposed in literature (Wolf *et al*,1985[6]) to determine the pairs (a,b) of bifurcation parameters that induce hyperchaotic behavior of the GHM. The experimenter has also to pay attention to the fact that not all these (a,b) parameters allow the distribution seen in Figure 1.

TABLE 1. The ten subintervals of equal length.

Box No., ξ	Box (Interval)
1	[-1.0;-0.8)
2	[-0.8;-0.6)
3	[-0.6;-0.4)
4	[-0.4;-0.2)
5	[-0.2;0.0)
6	[0.0;-0.2)
7	[0.2;0.4)
8	[0.4;0.6)
9	[0.6;0.8)
10	[0.8;1.0)

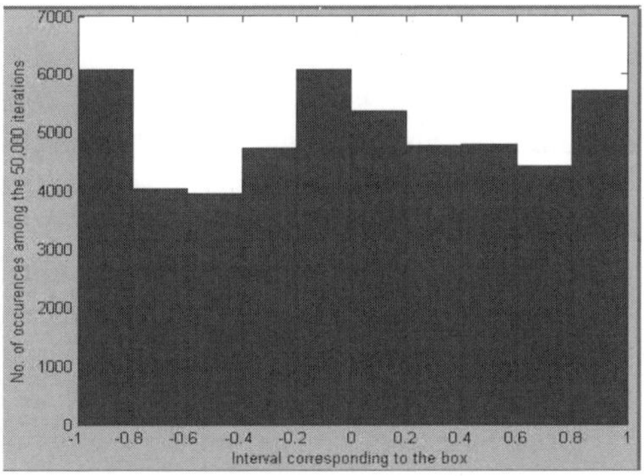

Figure 1. Histogram corresponding to $\sin((z-y)/2)$ values in the ten intervals from Table 1, for bifurcation parameters $a=1.76$ and $b=0.1$ and on a trajectory of 50,000 iterations.

2. The proposed enciphering method

2.1. *Scheme of the proposed enciphering method*

The scheme of the proposed enciphering method is illustrated in Figure 2.

- $[X \quad Y \quad Z]$ denotes the random vector of the GHM states. For fixed (a,b) and initial conditions (x_0, y_0, z_0) randomly chosen in the intervals mentioned in Section 1, there are 3 random processes corresponding to the 3 state variables in GHM. Thus, at each and every n iteration we can speak of a random vector (x_n, y_n, z_n).

- f is a random variables transform that produces a discrete random variable $\xi = f(\sin((z_n - y_n)/2)), \xi \in \{1,2,...,10\}$. The probabilities of occurrence of the values of ξ are almost equal, see Figure 1. ξ values are chosen by the states of the GHM, assigned as in Table 1.
- g_1 and g_2 are two new random variable transforms applied to ξ; they imply the external secret key; g_1 is a transform that assigns to the random variable ξ a discrete random variable with 10 values, bytes, in $0 \div 255$, ASCII interval, $g_1(\xi) = \zeta$. To choose ζ we have the set of possibilities presented in Figure 3, see also Remark 3. Applying g_2, a second set of 10 bytes is assigned to the variable ξ. They are 8 bits (1 byte) permutations, described in Figure 3, see also Remark 3, in correspondence with variable ξ, $g_2(\xi) = P$.
- m is the sequence of 8-bit characters, being the ASCII values of the characters in the plain message.
- R is the cryptogram obtained from the first enciphering step. It is a byte resulted as the addition modulo 2 without carry (bitxor) between the character that is enciphered and the byte ζ. The value ζ is the one assigned to the interval selected by the function $\sin((z_n - y_n)/2)$ depending on the state of the GHM at the moment of the enciphering of the $n-th$ character.
- m' is the byte obtained after the second enciphering step, meaning the permutation of the resulted byte, involving the rule established by the state of the GHM at the enciphering moment.
- Fixed reversible transformations are represented by a binary to decimal conversion and scaling by a factor. The scaling allows the inclusion of the masked character (after bitxor and permutation transformations) into the evolution of the GHM without disturbing its dynamics.
- $\tilde{m}$ is the masked and scaled message, that is added to the evolution of the hyperchaotic system, in system (2) instead m.
- In order to allow the recovery of the masked character from the series of the output z of the GHM, it is imperative to iterate the system three more times, after the last character of the message is enciphered, as it can be seen from the observer formula:

$$\tilde{m}_n = z_{n+3} + z_{n+1}^{\ 2} + b \cdot z_n - a \tag{4}$$

Remark 1: Depending of the application, the enciphering scheme from Figure 2 enables two variants for the cryptograms.

A. The user may transmit on the communication channel only the series $\{z_3, z_4, ..., z_{L+3}\}$ of the output of the system, where L is the length of the message to be transmitted. Knowing the initial conditions $\{z_2, z_1, z_0\} = \{x_0, y_0, z_0\}$ and bifurcation parameters (a,b), the receiver can easily synchronize the reception system with the emission one, and recover the set of masked characters $\{m'_0, m'_1, ..., m'_L\}$, having also the value of the scaling factor. In this case, we can consider $\{\tilde{m}_0, \tilde{m}_1, ..., \tilde{m}_L\}$ being the enciphered message.

B. On the channel the user may also transmit the document containing the masked message, m'. So, m' will be the cryptogram. In this case, the only role of the GHM is to embed the secret information and to provide the random non-preferential distribution of the bytes for the bitxor and the permutations which follow this operation.

Actual encoding of the message is delivered in two stages within the dotted lines in Figure 2. The two stages consist in multiple substitution and transposition, on 8 bit segments, following a random rule carried by GHM.

Remark 2: The security of the cipher can be increased by increasing the key dimension, selecting more than 10 bytes both for the bitxor operation and for the permutation as well. Another method is working on groups of two (digrams, 16b), three (trigrams, 24b) or four (tetragrams, 32b) characters. The transformations applied to the masked version may contain, for the same purpose, mixing transformations, as the ones suggested by Shannon (Shannon,1949[5]).

The proposed cipher is a closed secrecy system. Having finite key number, it will have unique solution. When the elements of the key involved in transformations g_1 and g_2 are not known, important delays in recovering the solution are caused, even knowing the masked version of the plain.

Remark 3: The key elements used in this cipher are presented in Figure 3. They are disposed in two matrices. Matrix K_ζ contains elements of the key which intervene in transformations of type g_1. So, on every column of matrix K_ζ there are 10 ASCII characters which are distinct each one from the others and randomly chosen, $\alpha_{1,1} \neq \alpha_{1,2} \neq ... \neq \alpha_{T,10}$. In this paper, the illustration is done for a simple case, where the matrix was reduced to one column, but the procedure may be extended to multiple columns.

Similarly, matrix K_P has columns formed with any 10 bytes randomly chosen, constraint to be distinct $\beta_{1,1} \neq \beta_{1,2} \neq ... \neq \beta_{T,10}$. They are the permutation P chosen by the transformation of type g_2.

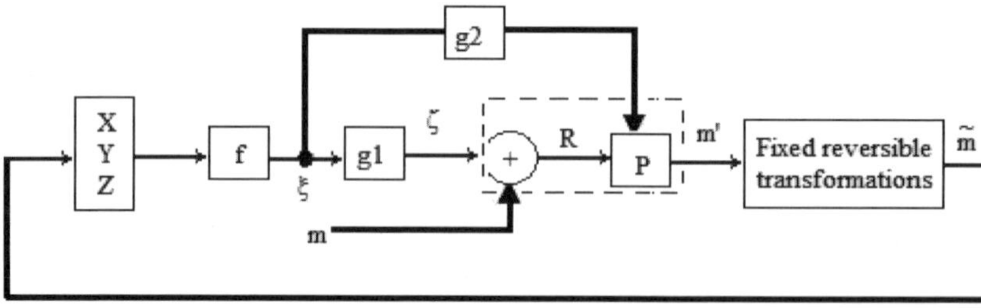

Figure 2. The enciphering scheme.

ξ	K_ζ				K_P			
1	$\alpha_{1,1}$	$\alpha_{2,1}$	...	$\alpha_{T,1}$	$\beta_{1,1}$	$\beta_{2,1}$	...	$\beta_{T',1}$
2	$\alpha_{1,2}$	$\alpha_{2,2}$	...	$\alpha_{T,2}$	$\beta_{1,2}$	$\beta_{2,2}$	...	$\beta_{T',2}$
3	...	...	...	...	...	...	...	...
4	...	...	...	...	...	...	...	...
5	...	...	...	...	...	...	...	...
6	...	...	...	...	...	...	...	...
7	...	...	...	...	...	...	...	...
8	...	...	...	...	...	...	...	...
9	...	...	...	...	...	...	...	...
10	$\alpha_{1,10}$	$\alpha_{2,10}$	...	α_{10}	$\beta_{1,10}$	$\beta_{2,10}$	...	$\beta_{T',10}$

Figure 3. Key elements disposed in two matrices.

2.2. *Deciphering*

Once the receiver gets the set of masked characters m', either directly, or deduced from the series $\{z_3, z_4, \ldots, z_{L+3}\}$, knowing the 10 bytes for bitxor and the 10 bytes for the permutation, in dependence with the evolution of the GHM he is led directly to the message. He ensures the synchronization between emission and reception systems and he benefits from the easily reversibility of the two enciphering transformations.

3. Analyze of results of the proposed enciphering algorithm

We illustrate the results obtained using the simple case when the key has only $2 \cdot 10 = 20$ bytes as key elements, a column in K_ζ and a column in K_P, Figure 3.

3.1. *Results of encryption of natural language text*

Figure 4 presents the frequency of occurrence of ASCII characters in natural language texts. It can be observed the preference for the interval corresponding to letters, numbers and spacing characters. In Figure 5, after encryption, one can observe the tendency of equalization of the frequencies of occurrence of ASCII characters.

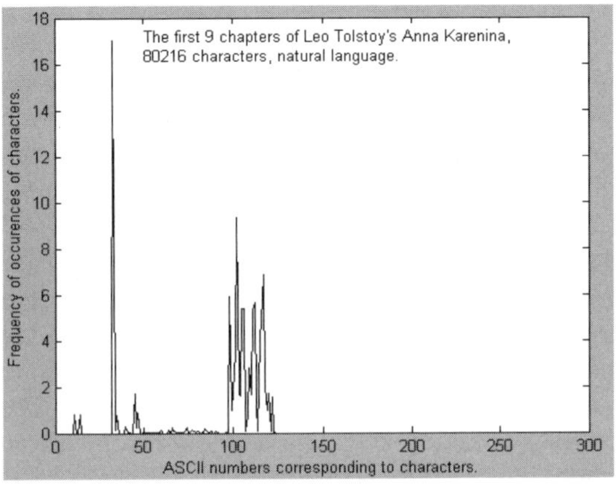

Figure 4. Frequency of occurrence of ASCII characters in a natural language text.

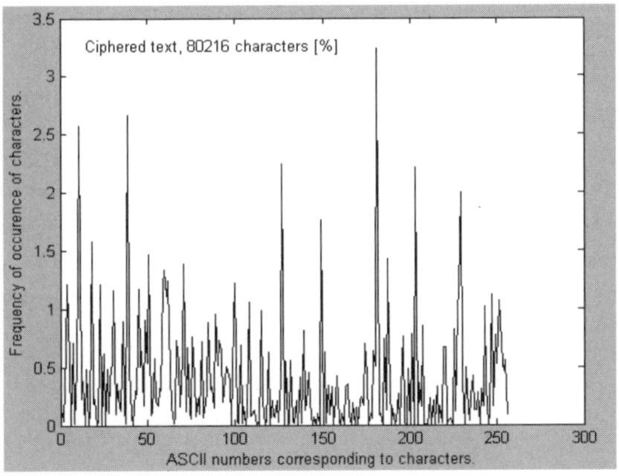

Figure 5. Frequency of occurrence of ASCII characters in ciphered version.

3.2. *Results when enciphering images*

The proposed enciphering algorithm has quite good results on images also, when every pixel of the image, ASCII number, is analogously enciphered. In Figure 6 it can be observed that pixels mix in irrecognizable manner.

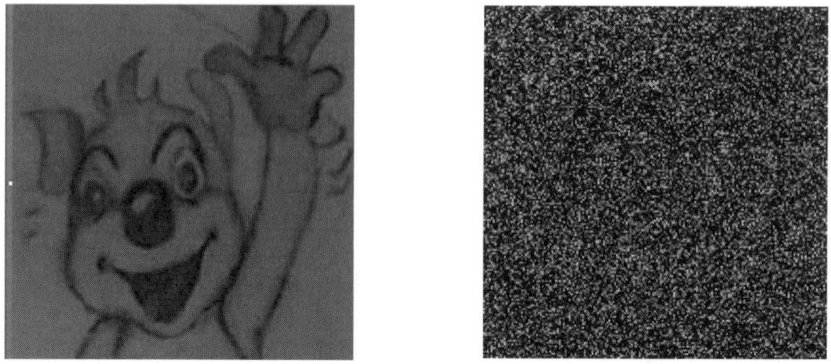

Original image. Enciphered image

Figure 6. JPEG image encryption.

4. Conclusions

We propose an algorithm that uses simple transformations, multiple substitutions and fixed transposition type. They are embedded by a hyperchaotic system having a simple form that allows easy synchronization between emission and reception, offering a random-deterministic rule to choose in a non-preferential way the enciphering key. The method has the advantage of multiple possibilities of improvement and extension.

5. Acknowledgement

The work has been funded by the Sectoral Operational Programme Human Resources Development 2007-2013 of the Romanian Ministry of Labour, Family and Social Protection through the Financial Agreement POSDRU/88/1.5/S/61178.

References

[1] F. Anstett, G. Millerieux and G. Bloch, Chaotic cryptosystems: cryptanalysis and identifiability. *IEEE Transactions on Circuits and Systems I.* v53 i12. 2673-2680, 2006.

[2] D. Arroyo, G. Alvarez, V. Fernandez. On the inadecuacy of the logistic map for cryptographic applications. ACTAS de la XRECSI, Salamanca, 2008

[3] D. Arroyo, G. Alvarez, Shujun Li. Some Hints for the Design of Digital Chaos-Based Cryptosystems: Lessons Learned from Cryptanalysis. *International Federation of Automatic Control, 2nd IFAC conference on analysis and control of chaotic systems*, London, 2009

[4] M.S. Baptista. Cryptography with chaos. *Physics Letters A* 240, 50-54, Elsevier, 23 March 1998.

[5] C.E. Shannon. Communication Theory of Secrecy Systems. *Bell Systems Technical Journal*, Vol. 28, 656:715, 1949.

[6] A. Wolf, J.B. Swift, H.L. Swinney , J.A. Vastano. Determining Lyapunov Exponents from a time series. Physica 16D (1985) 285-317, North-Holland, Amsterdam

Noise influence on the characteristic relations and reinjection probability densities of type-II and type-III intermittencies

Ezequiel Del Rio[1], Sergio Elaskar[2], J. M. Donoso[1] and L. Conde[1]

[1]Dept. Física Aplicada. ETS Ingenieros Aeronáuticos, Univ. Politécnica de Madrid. 28040 Madrid, Spain; Email: ezequiel.delrio@upm.es
[2]Dept. Aeronáutica. Facultad de Ciencias Exactas, Físicas y Naturales. Univ. Nacional de Córdoba. Avda. Vélez Sarfield 1611. Córdoba 5000. Argentina; Email: elaskar@efn.uncor.edu

Abstract: This paper explores the effect of the noise in the reinjection probability densities (RPD) for type-II and type-III intermittencies by using the temporal series of iterative maps. The RPD are calculated by means of a new method proposed in Refs. [1] and [2] and the results are compared with both, numerical simulations and analytical calculations. In addition, we provide an explanation for the gap observed in early experiments around the unstable point in the Poincaré map. We show that and added white noise approaches the RPD to the case of uniform reinjection for small distances of iterations to the unstable point. For large distances the RPD should be incremented with respect to the noiseless case. These numerical results suggest the existence of a noise induced reinjection mechanism.
Keywords: intermittency, characteristic relations, reinjection probability density.

Introduction

The intermittency is a particular form of deterministic chaos where transitions between different behaviours of the system take place. The system exhibits a regular response until a small change in a given parameter brings it into a different regime. This results into a time dependent response where chaotic bursts are separated at regular intervals by a *laminar* or regular responses [3]. The original idea was introduced in connection with the Lorentz system [4,5] but the intermittent behaviour is widespread in nature. It has been observed in the Bénard convection experiments [7], nonlinear electronic circuits [2,8,9,10] or in the human heartbeats time series [11].

In the type-II and III intermittencies, the reinjection mechanism from the chaotic motion into the laminar region depends on the dynamic of the chaotic phase. This constitutes global property and it was pointed out as a relevant factor in the scaling time duration of the laminar region [8, 10, 12]. This reinjection process from the chaotic response back into the regular or laminar response could be characterized by means of a reinjection probability density (RPD). However, only in few cases it is possible to derive an analytical

Chaos Theory: Modeling, Simulation and Applications
C. H. Skiadas, I. Dimotikalis and C. Skiadas (Eds)

expression for the RPD $\varphi(x)$ or to obtain it numerically or from experimental data or time series. The reason is the huge number of sampling data required to cover each small interval of initial conditions located within the reinjection zone.

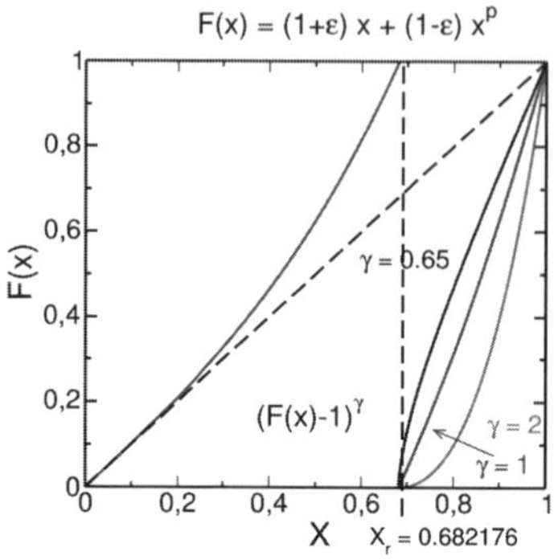

Figure 1. The map of Eq. (1) with $p = 3$ and $\varepsilon = 10^{-3}$.

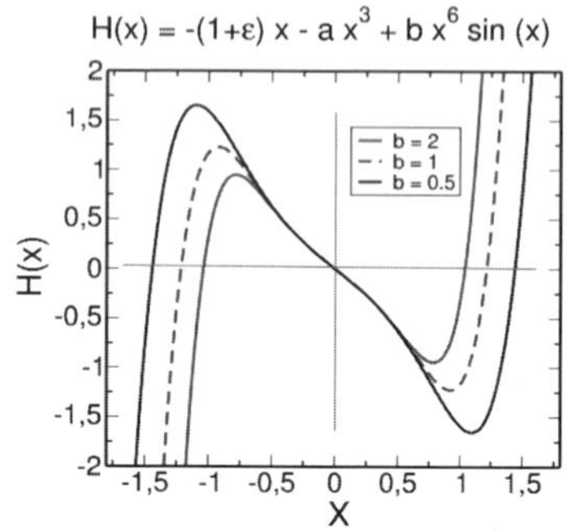

Figure 2. The map of Eq. (2) for $a = 1$ and $\varepsilon = 10^{-3}$.

In consequence, different simplifying assumptions have been proposed for $\varphi(x)$. The widespread approximation considers a uniform RPD independent of the reinjection point [3,4,7-10,13-16] or a rather artificial assumption considers that the reinjection occurs at a fixed point [10,15].

In this note we make use of a new approach to derive the RPD from numerical data of time series obtained from iterative maps. The RPD is calculated using iterative data series of one dimensional maps which exhibit type-II and type-II intermittencies.

Model maps

In order to model the type-II intermittency we analyze the following one dimensional map,

$$x_{n+1} = \begin{cases} F(x_n) & x_n \leq x_r \\ (F(x_n)-1)^{\gamma} & x_n \geq x_r \end{cases} \tag{1}$$

where $F(x_n) = (1+\varepsilon)\,x_n + (1-\varepsilon)\,x_n^p$ and x_r is the root of $F(x_r) = 1$ as it could be observed in Fig. 1. The type-III intermittency is modelled by using the map $x_{n+1} = H(x_n)$ where the function,

$$H(x_n) = -(1+\varepsilon)\,x_n - a\,x_n^3 + b\,x_n^6 \sin(x_n) \tag{2}$$

is represented in Fig. 2 and is usually approximated by,

$$H(x_n) \approx -(1+\varepsilon)\,x_n - a\,x_n^3$$

and this last expression corresponds to the local Poincaré map for type-III intermittency.

In Fig. 1 the origin $x = 0$is always an stable fixed point for $\varepsilon < 0$ that becomes unstable for $\varepsilon > 0$The successive iterations from an initial value x_o close to the origin, bring x_n away in a process driven by the parameters ε and p. When x_n becomes larger than x_r, a chaotic burst occurs, that will be interrupted when x_n is again mapped back into the laminar region of low values of x_n.

For $\gamma = 1$ for the map of Eq. (1) we have $x_{n+1} = F(x_n)$ (mod 1) while for $p = 2$ we recover the map used by Manneville in his pioneer paper [4]. The case $p = 3$ corresponds with the local Poincare map of type-II intermittency for points close to $x = 0$. Both parameters ε and p modify the duration of the so called *laminar phase*, where the system exhibits a quasiperiodic or regular response.

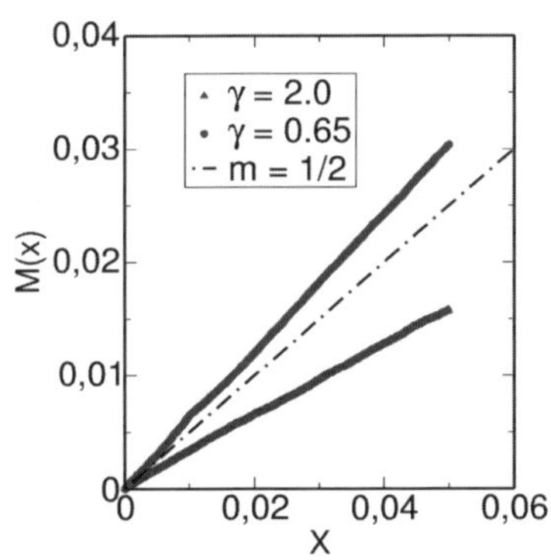

Figure 3. The values of $M(x)$ (solid points) for the map of Eq. (1) for two values of γ

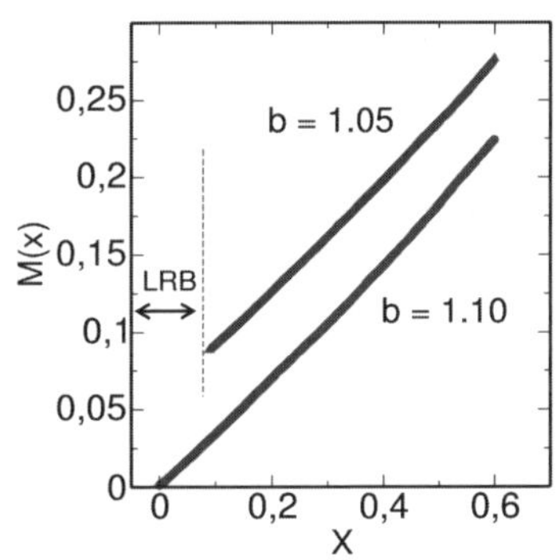

Figure 4. The function $M(x)$ (solid points) for the Eq. (2) for two values b showing $x_i > 0$.

The RPD is strongly dependent on the parameter γ which, as it could be observed in Fig. 1, determines the curvature of the map in region past x_r Then, only points $x > x_r$ are iterated back into of the laminar region around the unstable fixed point at the origin. As it could be observed in Fig. 1 larger values of γ increases the curvature of the function and, consequently, a larger fraction of iterated points are therefore mapped close to the unstable fixed point at $x = 0$.

Reinjection Probability Distribution

It is difficult to derive an analytical expression for $\varphi(x)$ because it relies on the chaotic response of the system. In order to obtain a semi-empirical expression for the RPD, we make use the function $M(x)$ defined in Refs. [1] and [2],

$$M(x) = \frac{\int s\varphi(s)\,ds}{\int \varphi(s)\,ds} \text{ if } \int \varphi(s)\,ds \neq 0 \tag{3}$$

and $M(x) = 0$ otherwise. This average allows a more reliable experimental and numerical access than $\varphi(x)$. The evaluation of $M(x)$ for the maps of Eqs. (1) and (2) are shown in Figs. 3 and 4 where the case of uniform reinjection corresponds to the dot dashed line with slope $\varphi(x)$.

The numerical values of $M(x)$ could be approximated by a linear function $M(x) = m(x - x_i) + d$, where $x_i > 0$ is the low boundary reinjection (LBR) point. This latter is defined as the lower value of x where a reinjection event occurs and is also indicated in Fig. (4). In addition, $x_c > x_i$ represents the maximum value where a laminar or regular response is observed. Assuming a linear expression for $M(x)$ we can solve the Eq. (3) and we have,

$$\varphi(x) = b\,(x - x_i)^{\alpha} \quad \text{with} \quad \alpha = \frac{1 - 2\,m}{m - 1} \tag{4}$$

The constant b is determined by the by the normalization condition,

$$N \int b\,(x - x_i)^{\alpha}\,dx = 1 \tag{5}$$

The RPD must be symmetric in the map of Eq. (2), and the factor $N = 2$ in Eq.(6) is introduced to account for the negative $(x < 0)$ reinjections, while for Eq. (1) we have $N = 1$. When $\alpha > -1$, or equivalently, for $0 < m < -1$, the above integral gives,

$$b = \frac{1}{N}\frac{\alpha + 1}{(x_c - x_i)^{\alpha+1}} = \frac{1}{N}\frac{m}{1 - m}(x_c - x_i)^{m/(m-1)} \tag{6}$$

which permits to evaluate the RPD. Therefore, the value of m obtained from the numerical calculations of Figs. 2 and 3 determines the RPD using the Eqs. (5) and (6). Then, the approximation $M(x) = m(x - x_i) + d$, determines $\varphi(x)$ which only rely on the density is determined only by the parameters m and d that are easier to calculate than $\varphi(x)$.

The RPD obtained using the linear approximation for Eq. (3) are in agreement with the numerical calculations of $\varphi(x)$ as evidenced in Figs. 2 and 3. The RPD obtained in the numerical calculations using a huge number of iterations of both maps are represented together with the functions $\varphi(x)$ given by Eq. (4) obtained by the linear fits of $M(x)$ in Figs. (2) and (3).

From Eq. (5), is also deduced that the properties of the RPD close to $x = x_i$ differ accordingly to the value of m. That is, we have $\lim_{x \to x_i} \varphi(x) = \infty$ when $0 < m < 1/2$ while for $1/2 < m < 1$ we obtain $\lim_{x \to x_i} \varphi(x) = 0$ and the value $m = 1/2$ corresponds to the uniform RPD $\varphi(x) = 1/x_c$. Therefore, from Eq. (3), we expect for $\gamma > 1$ values of m in the interval $0 < m < 1/2$ and for $\gamma < 1$ values in the interval $1/2 < m < 1$. These properties are also in agreement with the results of the numerical calculations as it could be observed in Figs. 5 and 6.

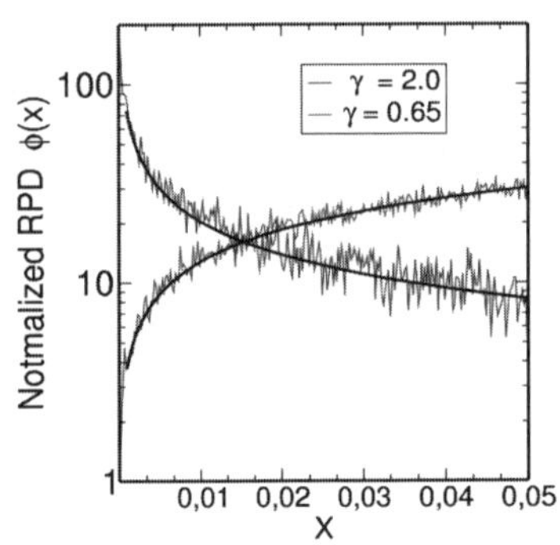

Figure 5. Comparison of $\varphi(x)$ for the map (1) compared with Eq. (4) (solid line) using the numerical data of Fig. 3.

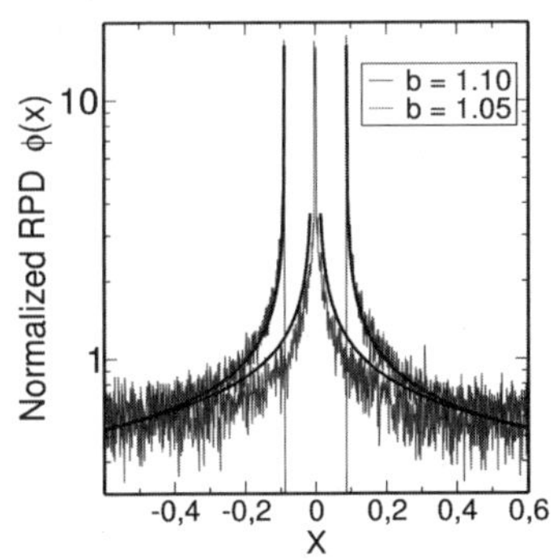

Figure 6. The same of Fig 5 for the map of Eq. (2) using the data of Fig. 4. The gap around the unstable point is evidenced.

The expression (3) filters the usual noise of the numerical data; and allows us to obtain a better description than the numerical reinjection density. It is important to emphasize that a LBR different from zero (see Figs 4 and 6) produces a gap around the unstable point in the Poincaré map. Such gap around the unstable point was observed in early experiments [7, 2]. Its existence is justified by the linear dependence of $M(x)$ according to the values of d. In first place, when $d \approx 0$ and simultaneously we have $0 < \alpha$, the function $\varphi(x)$ approaches to zero as x tends to the unstable point. In this case, the gap would be filled when we have in the Poincaré return map a large enough number of points. However, for a finite sampling $d > 0$ and the aforementioned gap develops as shown in Fig 5.

The effect of the Noise on the RPD

Most of investigations on the role of an external noise in the Pomeau intermittency introduce random perturbations in the local map around the unstable point, without considering its effect on the RPD. In this section we study the effect of a superposed random signal to the maps of Fig. 1 and 2 as,

$$x_{n+1} = G(x_n) + C\,\zeta(t) \tag{7}$$

where $-1 < \zeta(t) < +1$ is an uncorrelated white noise with amplitude C.

For non uniform RPD important changes would be expected in this distribution. On the contrary, for the particular case of the uniform reinjection the RPD still would remain independent of x. In Figures 7 and 8 are represented the numerical evaluation of $M(x)$ for different noise amplitudes C in Eq. (7). These results are calculated with the same parameters used for the lower and upper lines in Figs. 2 and 3. The solid bullets corresponding to the noiseless case $(C = 0)$ and its linear fit (solid line) are represented in Fig.7. These previous results are compared with those with an added $(C > 0)$ white noise.

For low values of x the function $M(x)$ approaches the line with slope $1/2$ (dashed lines) corresponding to the case of uniform RPD. This effect is caused by the noise induced spread of iterations for small of x. For different noise amplitudes, the results of Figs. 9 and 10 suggest that $\varphi(x)$ increases as the slope of $M(x)$ grows. This fact suggests, a new noise induced reinjection process that. For the map of Eq. (1) this means there points with $x < x_r$ are iterated downwards, back into the laminar region. This would explain the upwards legs of Figs. 9 and 10 where two reinjection mechanisms coexist.

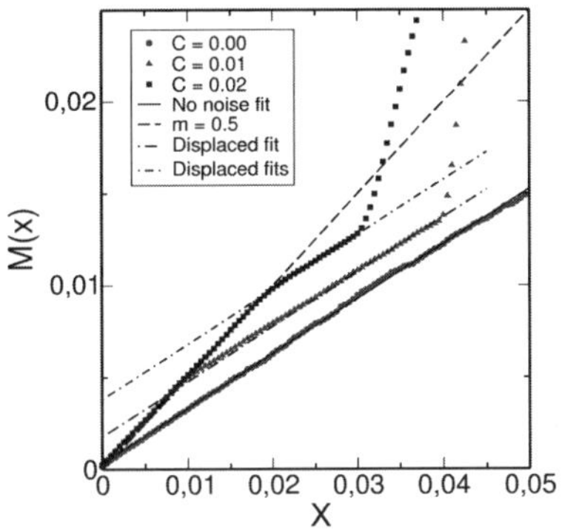

Figure 9. The numerical evaluation of $M(x)$ for the map (1) for different noise amplitudes in Eq. (8) for the same parameters as in Fig. 5.

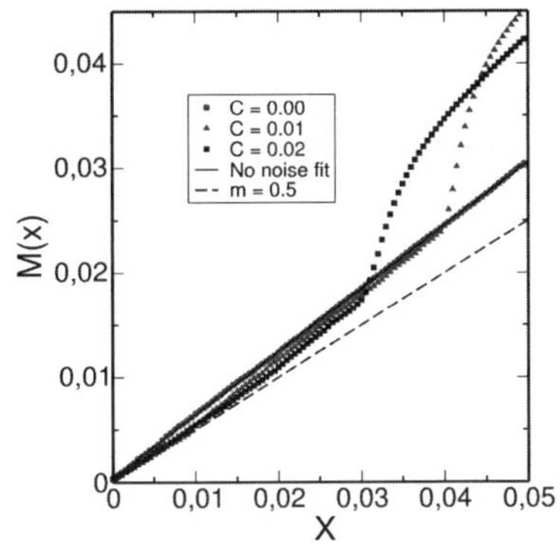

Figure 10. The numerical evaluation of $M(x)$ for the map (2) for different noise amplitudes in Eq. (8) for the same parameters as in Fig. 7.

Conclusions and Discussion

The function $M(x)$ constitutes a tool to evaluate the RPD and, in a number of cases, it is found that $M(x) = m(x - x_i) + d,$ leading to $\varphi(x) = b(x - x_i)^{\alpha}$. Then the usual approximation of uniform reinjection approximation is recovered when $\alpha = 0$ or equivalently $m = 1/2$. Similar power dependence for the RPD have been observed in different maps which exhibit type II and III intermittencies. The existence of a minimum initial value x_r for reinjection also provides an explanation for the gap around the unstable point experimentally reported [7]. It is currently assumed that the effect of the noise on the type-II and III intermittencies only affects the properties of the map locally around a stable point. On the contrary, our calculations using the function $M(x)$ indicate that the RPD is also strongly affected.

The departure from the linear response observed in $M(x)$ in Figs. 9 and 10 evidence an increment of the RPD for larger values of x_r and suggests the existence of a noise induced reinjection mechanism. The nonlinear expression for $M(x)$ and its relation between $\varphi(x)$ and the noise amplitudes will be the subject of further investigations.

Acknowledgements

This research was supported by the Technical University of Madrid (UPM) under grant AL09-PID-03 and by CONICET under grant PID-5269,

Universidad Nacional de Córdoba and Ministerio de Ciencia y Tecnología de Córdoba. Dr. J.M Donoso and Dr. L. Conde acknowledge the partial support of Spanish Ministry of Science and Innovation under Grant ENE2007-67406-C02-01.

References

[1] E. Del Rio and S. Elaskar. New characteristic relations in type-II intermittency. *Int. J. Bif. Chaos*, **20**. (4). pp. 1–7 (2010)
[2] E. Del Río, M.G. Velarde and A. Rodríguez-Lozano. Long time data series and difficulties with the characterization of chaotic attractors: a case study with intermittency III. *Chaos. Sol. Frac.*, **4**. (12). pp. 2169–2179 (1994)
[3] H.G. Schuster and W. Just. *Deterministic Chaos: An Introduction*. Wiley-VCH Verlag GmbH and Co. KGaA, Weinheim, Germany (2005)
[4] P. Manneville. Intermittency, self-similarity and 1/f spectrum in dissipative dynamical systems. *J. Physique*, **41**. pp. 1235–1243 (1980)
[5] Y. Pomeau and P. Manneville. Intermittent transitions to turbulence in dissipative dynamical systems. *Comm. Math. Phys.*, **74**. pp. 89–197 (1980)
[6] F. R. C. Grebogi, E. Ott and J. Yorke. Critical exponents for crisis induced intermittency. *Phys. Rev. A*, **36**. pp. 5365–5380 (1987)
[7] M.A. Rubio M. Dubois and P. Bergé. Experimental evidence of intermittencies associated with a subharmonic bifurcation. *Phys. Rev. Lett.* **51**. pp. 1446–1449 (1983)
[8] Lee Eok-Kyun Kim Chil-Min, O.J. Kwon and Lee Hoyun. New characteristic relations in Type-I intermittency. *Phys. Rev. Lett.* **73**, pp. 525–528 (1994)
[9] Kim Yeon-Soo Kim Jeong-Moog Kim Chil-Min, Yim Geo-Su and H.W. Lee. Experimental evidence of characteristic relations of type-I intermittency in an electronic circuit. *Phys. Rev. E*, **56**. pp. 2573–2577 (1997)
[10] Ryu Jung-Wan Kim Chil-Min, Yim Geo-Su and Park Young-Jai. Characteristic relations of type-III intermittency in an electronic circuit. *Phys. Rev. Lett.* **80**. pp. 5317–5320 (1998)
[11] J.J. Zebrowski and R. Baranowski. Type-I intermittency in nonstationary systems: models and human heart-rate variability. *Physica A*. **336**. pp. 74–83 (2004)
[12] J.P. Farmer D. Crutchfield and B.A. Huberman. Fluctuations and simple chaotic dynamics. *Phys. Rep.* **92**. pp. 45–82 (1982)
[13] A.S. Pikovsky. A new type of intermittent transition to chaos. *J. Phys. A*. **16**. L109–112 (1983)
[14] Park Young-Jai Cho Jin-Hang, Ko Myung-Suk and Kim Chil-Min. Experimental observation of the characteristic relations of type-I intermittency in the presence of noise. *Phys. Rev. E*. **65**. 036222 (2002)
[15] Kye Won-Ho and Kim Chil-Min. Characteristic relations of type-I intermittency in the presence of noise. *Phys.Rev. E*. **62**. pp. 6304–6307 (2000)
[16] Rim Sunghwan Kye Won-Ho and Kim Chil-Min. Experimental observation of characteristic relations of type-III intermittency in the presence of noise in a simple electronic circuit. *Phys.Rev. E*. **68**. 036203 (2003)
[17] S. Elaskar, E. Del Rio and J. Donoso. Reinjection probability density in type-III intermittency. *Int. J. Bif. Chaos*, Submitted.

Multifractal and wavelet analysis of epileptic seizures

Olga E. Dick[1] and Irina A. Mochovikova

Pavlov Institute of Physiology of Russian Academy of Science,
St. Petersburg, Russia
Email: [1]glazov.holo@mail.ioffe.ru

Abstract: The aim of the study is to develop quantitative parameters of human electroencephalographic (EEG) recordings with epileptic seizures. We used long-lasting recordings from subjects with epilepsy obtained as part of their clinical investigation. The continuous wavelet transform of the EEG segments and the wavelet-transform modulus maxima method enable us to evaluate the energy spectra of the segments, to find lines of local maximums, to gain the scaling exponents and to construct the singularity spectra. We have shown that the significant increase of the global energy with respect to background and the redistribution of the energy over the frequency range are observed in the patterns involving the epileptic activity. The singularity spectra expand so that the degree of inhomogenety and multifractality of the patterns enhances. Comparing the results gained for the patterns during different functional probes such as open and closed eyes or hyperventilation we demonstrate the high sensitivity of the analyzed parameters (the maximal global energy, the width and asymmetry of the singularity spectrum) for detecting the epileptic patterns.
Keywords: EEG, wavelet, multifractal, epileptic.

1. Introduction

Bioelectric activity of the human brain (EEG time series) is rather nonstationary, noisy and has chaotic and fractal dynamics [9]. Fractal signals have a self-similarity property at least stochastically, i.e., their patterns on small scales are not identical to the whole signal but the self-similarity remains after averaging by statistically independent samples of the signal. Monofractal signals such as white noise or synchronous chaos can be described by one value, fractal dimension, determining the maintenance of statistical characteristics with changing the scale [8]. Kannathal et. al. [6] and Indiradevi et. al. [5] discriminate normal and epileptic EEG signals by an increase of Hurst exponent showing the enchancement of long-range correlations in time series during epileptic seizure. On the contrary, Nurujjaman et. al. [9] describe a decrease of the Hurst exponent during epileptic discharge. Wink et al. [13] provide evidence that changes in the fractal dimensions of EEGs are influenced not only by neuropathology and but also by performance of a cognitively effortful task. They show that higher

Chaos Theory: Modeling, Simulation and Applications
C. H. Skiadas, I. Dimotikalis and C. Skiadas (Eds)

values of Hurst exponent correlate with shorter response latency in a facial encoding task. However, due to the paradoxical combination of short-term decorrelation in EEG patterns caused by chaos and noise and long-range correlation connected with the fractal structure, EEG time series demonstrate a different local scaling [3, 10]. Therefore, EEGs as the "stochastic chaos" can not be described by a single nonlinear parameter such as largest Lyapunov exponent, correlation dimension, or Hurst exponent [4]. The fluctuation dynamics in EEGs could be adequately described by a set of scales, and characterized by multifractals, in both healthy and pathologic conditions [12]. Multifractality of the healthy brain is revealed in EEG time series during complex imaginary and real visual-motor task performance [8], during awake and various sleep stages [11], during epileptic seizures [12].
There are several methods for determining the degree of multifractality by the distribution of the local Hölder exponents (the singularity spectrum) instead of one Hurst exponent describing the degree of monofractality. The most popular are the multifractal detrended fluctuation analysis (MFDFA) [7] and the wavelet transform modulus maxima method (WTTM) [2]. These methods applied to the EEGs give various results: an increase of the multifractality during the epileptic seizure [3, 12] (WTTM) or an absence of changes [3] (MFDFA). Thus, the question about feasibility of multifractal properties of the brain for the possibility to predict epileptic seizures remains open.
The aim of the work is to evaluate which the wavelet and multifractal parameters are most sensitive for estimating differences in the epileptic brain states free from epileptic discharges and containing epileptic seizures.

2. Experimental procedure

The scalp EEG data were collected from repeated recordings with Ag/AgCl electrodes placed at the F3, F4, Fz, C3, C4, Cz, P3, P4, Pz, O1, O2, Oz loci from 20 healthy subjects and 18 epileptic patients with complex partial seizures. The data were sampled at a rate 256 samples/sec with a resolution of 12 bits/sample during 60 minutes. Then the data were digitally filtered using 1–45 Hz band pass filter. The EEG recordings were obtained in three states: under resting condition with eyes closed and open and during hyperventilation. The each state included 307200 samples and it was divided into 120 segments of the duration 10 seconds. Among EEGs from epileptic patients the physician isolated the ictal period during seizures and interictal period between seizures. After repeated recordings 80 non- artifact segments of equal duration (10 seconds) were randomly chosen from the sets: "closed eyes", "open eyes", "hyperventilation", "before epileptic discharge", "during discharge" and "after discharge".

3. Multifractality and wavelet transform

To estimate multifractal scaling properties of EEG time series we use the wavelet transform modulus maxima (WTMM) method [2]. The method is based on the continuous wavelet transform of a time series describing the examined signal x(t):

$$W(a,t_0)=\frac{1}{a}\int_{-\infty}^{+\infty} x(t)\psi^*\left(\frac{t-t_0}{a}\right)dt,$$

where a is the scale parameter, t_0 is the space parameter, $\psi((t-t_0)/a)$ is the wavelet function obtained from the basic wavelet $\psi(t)$ by scaling and shifting along the time, symbol * means the complex conjugate. So, the wavelet transform of the signal consists in decomposing it into elementary space-scale contributions associated to wavelets which are constructed from one function by means of scaling and shifting. As the basic wavelet we use the complex Morlet wavelet:

$$\psi(t)=\pi^{-1/4}e^{i\omega_0 t}e^{-t^2/2}.$$

The value $\omega=2\pi$ gives the simple relation between the scale a and the frequency f: $f=1/a$.

The modulus of the wavelet spectrum $|W(a, t_0)|$ characterizes the presence and intensity of the frequency f at the moment t_0 in the signal. $|W(a, t_0)|^2$ determines the distribution of the signal energy and:

$$E(a)=\int_{-\infty}^{+\infty} |W(a,t_0|^2 dt$$

describes the global wavelet spectrum, i.e., the integral of distribution of energy over space parameters. The important information about the wavelet spectrum contains in the lines of local maxima of the surface for which the condition $\partial|W(a,t_0)|/\partial t=0$ is true at each scale a.

Information about possible fractal feature of the signal and its localization t_0 reflects in the asymptotic behavior of coefficients $|W(a, t_0)|$ at small a values. The faster the wavelet coefficients decrease at $a\to 0$, the more regular the signal is around that point. Abnormal small decrease of the wavelet coefficients at $a\to 0$ in a neighborhood of the point t_0 testifies about singularity of the signal at the point. Thus, the rate of the change of the modulus of the wavelet coefficients enables to analyze the structure of singularities of the signal.

The degree of singularity of the signal $x(t)$ at the point t_0 is described by the Hölder exponent, $h(t_0)$, the largest exponent such that the analyzed signal in a neighborhood of the point t_0 can be represented as the sum of the regular component (a polynomial $P_n(t)$ of order $n< h(t_0)$) and a member describing non-regular behavior [2]: $x(t)=P_n(t)+c|t-t_0|^{h(t_0)}$.

The value $h(t_0)$ is the measure of singularity of the signal at the point t_0 since the smaller $h(t_0)$ value, the more singular the signal.

In view of the simple dependence $W(a,t_0) \sim a^{h(t_0)}$ at $a \to 0$ [2], the Hölder exponent can be calculated by $h(t_0) \sim \log_{10} W(a,t_0)/\log_{10} a$.

However, with increasing the scale a the influence of neighbouring nonregularities can lead to inaccuracy and in practice the Hölder exponents are found on the basis of statistical description of local singularities by partition functions [1, 2].

4. Algorithm of the wavelet-transform modulus maxima (WTMM) method

The algorithm consists of the following procedures.

1) The continuous wavelet transform of the time series is used.

2) A set $L(a)$ of lines of local modulus maxima of the wavelet coefficients is found at each scale a.

3) The partition functions are calculated by the sum of q - powers of the modulus maxima of the wavelet coefficients along the each line at the scales smaller the given value a:

$$Z(q,a) = \sum_{l \in L(a)} \left(\sup_{a^* \le a} \left| W(a^*, t_l(a^*)) \right| \right)^q ,$$

$t_l(a^*)$ determines the position of the maximum corresponding to the line l at this scale.

4) By the fact that the partition function is $Z(q,a) \sim a^{\tau(q)}$ at $a \to 0$ [2], the scaling exponent can be extracted as $\tau(q) \sim \log_{10} Z(q,a)/\log_{10} a$.

5) Choosing different values of the power q one can obtain a linear dependence $\tau(q)$ with a constant value of the Hölder exponent $h(q) = d\tau(q)/dq = const$ for monofractal signals and nonlinear dependence $\tau(q) = qh(q) - D(h)$ with large number of the Hölder exponents for multifractal signals.

6) The singularity spectrum (distribution of the local Hölder exponents) is calculated from the Legendre transform [2]: $D(h) = qh(q) - \tau(q)$.

5. Estimated parameters and statistical analysis

Using the the global wavelet spectrum and the WWTM algorithm for the different EEG segments we obtain 1) the maximum of the global signal energy E_{max} and 2) two multifractal parameters: a) the width of the singularity spectrum $\Delta h = h_{max} - h_{min}$, where $h_{max} = h\ (q = -5)$ and $h_{min} = h\ (q = 5)$ are the maximal and minimal values of the Holder exponent corresponding to minimal and maximal fluctuation of the brain activity, respectively; b) the asymmetry of the singularity spectrum $\Delta = \Delta_2 - \Delta_1$, where

$\Delta_1 = h_{max} - h_0$ and $\Delta_2 = h_0 - h_{min}$, $h_0 = h\ (q = 0)$. Smaller Δh indicates that the time series tends to be monofractal and larger Δh testifies the enhancement of multifractality. The asymmetry parameter Δ characterizes where, in the region of q>0 (strong singularities) or in the region of q<0 (weak singularities), the singularity spectrum is more concentrated [11].

To examine the differences between the mean values of the parameters obtained for all the segments of different sets of one subject and to compare the mean parameters obtained for the same electrode loci for different subjects the non-parametric Mann-Whitney test ($p < 0.05$) was applied.

6. Results and discussion

Unlike the states "closed" or "open eyes" and "hyperventilation", the neighbouring events "before", "during" and "after epileptic seizure" are characterized by most inhomogenety in distribution of the signal energy. During the ictal period, i.e., time when the seizure takes place and develops, the energy of the signal significantly increases (Figure 1). This maximum corresponds to the well-known for epileptic patients 3 Hz frequency. Calculation of the global signal energy performed separately for the preictal, ictal and postictal periods enables us not only to isolate the seizure state but to distinguish states before and after seizure since the values of the global energy differ in 10 times in these states.

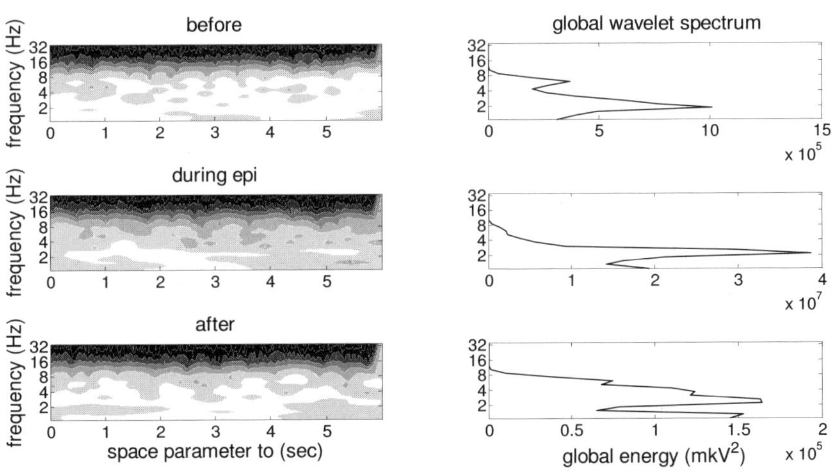

Fig. 1. The example of distributions of the EEG energy $|W(a, t_0)|^2$ and the global wavelet spectra for segments before, during and after epileptic seizure (site Oz).

The similar dynamics of the maximal global energy E_{max} during development the epileptic seizure is found for all the examined segments of the epileptic patients.

Figure 2 gives examples of the averaged (over the segments) dependences *h*(*q*) and singularity spectra *D*(*h*) for the various states of the epileptic patient. The shapes of *D*(*h*) indicate the multifractality of EEG time series in all the states. Neither opening eyes no hyperventilation effect essentially on the curves. Contrary, the Holder exponents differ before, during and after the epileptic discharge. The differences appear at weak fluctuations ($q<0$). In preictal period the singularity spectrum is expanded so that the dynamics of persistent sequences in the epileptic EEGs exhibit both anticorrelated (for $h<0.5$) and correlated (for $h>0.5$) behavior. In states between seizures the consequent EEG values are anticorrelated ($h<0.5$), i.e., persistent sequences are characterized by stochastically up-down patterns and large values are more likely to be followed by small values and vice versa. During epileptic seizure the degree of multifractality and inhomogeneity increases. These results are consistent with previous studies [12]. The largest width Δh and the asymmetry Δ of the singularity spectrum are observed during the ictal period. The enhancement of the both parameters in preictal period is due to small fluctuations so that the singularity spectrum is more concentrated in the region of weak singularities though the anticorrelated dynamics keeps. Opening eyes in the rest state leads to the increase of the multifractal parameters and opening eyes during hyperventilation nonsignificantly decreases the width and increases the asymmetry of the singularity spectrum. The similar dynamics of the multifractal parameters is observed for all the subjects having the seizures in occipital zone. It enables us to use the common practice of averaging the brain activity of all subjects for testing significant variations among the conditions [10].

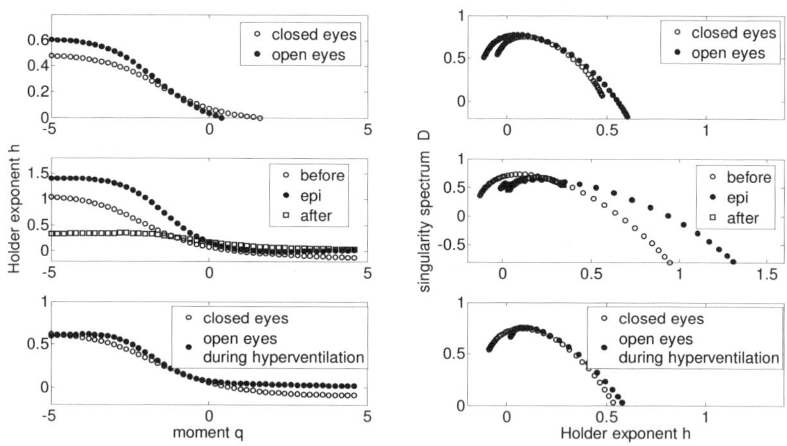

Fig. 2. The averaged over segments curves *h*(*q*) and sigularity spectra, *D*(*h*), for the patient 1, site Oz

There are no statistical differences between multifractal parameters for the healthy and epileptic subjects in the state with open eyes ($p = 0.92$). The distinctions between the states with open and closed eyes of the epileptic patients are identified by the maximal global energy E_{max} and the singularity spectrum asymmetry Δ (p = 0.01 and p = 0.04, respectively). The singularity spectrum width Δh is not distinguishable statistically for the states with open and closed eyes (p = 0.84). These findings show that the maximal global energy and the singularity spectrum asymmetry give better discrimination between the brain states than the Δh parameter.

	global energy E_{max}	width Δh	asymmetry Δ
healthy	(1.2±0.4)*10^5	0.59±0.09	0.54±0.08
epileptic closed eyes	(1.5±0.5)*10^5	0.48±0.07	0.25±0.06
open eyes	(2±0.6)*10^5	0.62±0.08	0.37±0.05
preictal period	(11±3)*10^5	1.11±0.12	0.62±0.08
ictal period	(410±35)*10^5	1.43±0.15	1.07±0.11
after ictal period	(1.8±0.4)*10^5	0.35±0.06	0.08±0.03

Table 1. Comparison of the mean parameters, averaging over 12 patients having the dominant seizures in the occipital zone. The data are given for the site Oz

The Mann-Whitney test founds the variations both between the ictal and preictal periods (p = 0.005) and between preictal and postictal periods (p = 0.009) by Δh, Δ and E_{max} values. Thus, these parameters can be used for predictability of epileptic seizures.

Table 2 illustrates the variability of the singularity spectrum width and the global energy E_{max} among electrode sites caused by domination of epileptic discharges in one or other site of the brain.

	left sites			right sites		
	C3	F3	O1	C4	F4	O2
singularity spectrum width Δh						
preictal	0.91±0.09	0.99±0.10	1,15±0.14	0.85±0.09	0.81±0.09	1.25±0.19
after ictal	0.29±0.02	0.41±0.03	0.65±0.09	0.45±0.07	0.39±0.05	0.56±0.07
global energy E_{max}*10^5						
preictal	12±3.6	15±4.4	21±5.3	11±3.1	10±2.7	23±4.1
after ictal	1.9±0.3	2.1±0.5	3.7±0.7	1.3±0.6	1.1±0.2	1.7±0.6

Table 2. Comparison of the parameters Δh and E_{max} in different electrode sites before and after epileptic seizure

In spite of the variety in the parameter values obtained for the sites the significant statistical difference (p <0.05) between preictal and after ictal periods is established in all the sites. Hence, the considered parameters allow to distinguish reliably preictal and postictal states. The rise in the singularity spectrum width and in the global energy correlates with the site of the seizure.

7. Conclusions

We suggest inhomogenety of EEG time series both for healthy and epileptic subjects. The enhancement of inhomogenety takes place not only during the epileptic seizure but in preictal period due to the rise of weak fluctuations of persistent sequences in the epileptic EEGs. The increase of the contribution of weak fluctuations and weak singularities, respectively, tends to the increase of the singularity spectrum asymmetry and the shift the spectrum into the region which is characterized by stochastically up-up or down-down patterns. Presence of the patterns can be the reason of significant rise of the maximal global energy before the epileptic seizure. It allows us to consider that the changes in the multifractal parameters and the maximal global energy give a good ability to estimate changes in the pathological brain activity and have predictive power for epileptic seizures.

References

1. A. Arneodo, E. Bacry and J. F Muzy. The thermodynamics of fractals revisited with wavelets. *Physica A* 213: 232-275, 1995.
2. E. Bacry, J. F. Muzy and A. Arneodo. Singularity spectrum of fractal signals: exact results. *J. Statist. Phys.* 70: 635-674, 1993.
3. R. Cardo, et al. Wavelet Leaders versus Multifractal Detrended Fluctuation Analysis in EEG time series. in *XII Reunión de Trabajo en Procesamiento de la Información y Control,* 78-83, 2007.
4. M. A. F. Harrison, et al. Correlation dimension and integral do not predict epileptic seizures. *Chaos* 15: 33106-33121, 2005.
5. K. P. Indiradev, E. Elias, P. S. Sathidevi. Complexity analysis of electroencephalogram records of epileptic patients using Hurst exponent International. *J. Medical Engineering and Informatics* 1: 368-380, 2009.
6. N. Kannathal, et al. Characterization of EEG—A comparative study. *Computer Methods and Programs in Biomedicine*, 80: 17-23, 2009.
7. J. W. Kantelhardt, et al. Multifractal detrended fluctuation analysis of nonstationary time series. *Physica A* 316: 87-114, 2002.
8. B. B. Mandelbrot. *The Fractal Geometry of Nature,* Freeman, San Francisco, 1983.
9. M. Nurujjaman, R. Narayanan and A. N. Sekar Iyengar. Comparative study of nonlinear properties of EEG signals of normal persons and epileptic patients. *Nonlinear Biomedical Physics* 3: 6-11, 2009.
10. D. Popivanov, et. al. Multifractality of decomposed EEG during imaginary and real visual-motor tracking. *Biological Cybernetics* 94: 149-156 , 2006.
11. M. A. Qianli, et al. A new measure to characterize multifractality of sleep electroencephalogram. *Chinese Science Bulletin* 51: 3059-3064, 2006.
12. I. H. Song and D. S. Lee. Fluctuation Dynamics in Electroencephalogram Time Series. In: Mira J, and Alvarez JR(eds) *Lecture Notes in Computer Sci.* Springer-Verlag, Berlin, pp.195-202, 2005.
13. A. M. Wink, et al. Monofractal and multifractal dynamics of low frequency endogenous brain oscillations in functional MRI. *Human Brain Mapping* 29: 791-801, 2008.

Fractal based curves in musical creativity: A critical annotation

Anastasia Georgaki and Christos Tsolakis

University of Athens, Music Department, Greece
Emails: georgaki@music.uoa.gr, tsolakis.chris@gmail.com

Abstract: In this article we examine fractal curves and synthesis algorithms in musical composition and research. First we trace the evolution of different approaches for the use of fractals in music since the 80's by a literature review. Furthermore, we review representative fractal algorithms and platforms that implement them. Properties such as self-similarity (pink noise), correlation, memory (related to the notion of Brownian motion) or non correlation at multiple levels (white noise), can be used to develop hierarchy of criteria for analyzing different layers of musical structure. L-systems can be applied in the modelling of melody in different musical cultures as well as in the investigation of musical perception principles. Finally, we propose a critical investigation approach for the use of artificial or natural fractal curves in systematic musicology.
Keywords: self-similarity, correlation, fractal motion, fractal curves, dimension, L system, power spectrum, random.

1. Introduction

The principles of symmetry and self-similarity were known in music and geometry well before the emergence of the notion of fractals. In the early 70'ies, Richard Voss introduced the concept of power spectrum, i.e. the autocorrelation of fluctuating density [1]. Mandelbrot gives examples of autocorrelation found in structures [2]: the perceived characteristics of most recorded natural sounds, as for example the sound of a flute, change if they are played back at different speeds. However, there is a category of sonances which sound exactly the same, irrespective of the playback speed. Mandelbrot calls these sounds scaling noises. The simplest of these is "white noise", i.e. a sound whose spectrum has equal power throughout all frequencies. The development of personal computers in the 80'ies, enabled a different approach to the synthesis of sound for composers and researchers. In 1985, Dodge proposed the creation of sound structures by a Koch curve in his project "Profile"[3] for creating sound structures. Nelson presented in 1988, the electronic composition "Fractal Mountains", which uses a different mapping strategy [4]. In 1990 Wallin used fractals in his percussion work "Stonewall", as a skeleton for structuring his musical intuition [5]. In the same year McNabb introduced interactive fractal algorithms in his music-performance work "The far and brilliant night". From 1989 to 1992 Evans

Chaos Theory: Modeling, Simulation and Applications
C. H. Skiadas, I. Dimotikalis and C. Skiadas (Eds)
 (pp. 167 - 174)

presented articles in which he studied how non-symmetry is related to the characteristics of virtual sound expression [6]. Gogins presented in 1991 his successful system of Iterated Function Systems (IFS) [7]. Besides Nelson, Kyburz used L-systems in his composition "Cells" to achieve self similarity [8]. From 1995 onwards, experiments with fractals are being conducted at an increasingly frequent rate on different aspects of sound and music ranging from audio signal processing to orchestral works. This has also led to the creation of music software that focus on the exploration of fractals in sound and music. Some of these are "FractMus", "QuasiFractalComposer", "A Musical Generator", "Bio2Midi", "Well Tempered Fractal", and "Lmuse" together with "Symbolic composer" are systems with additional capabilities in the domain of L-systems. General purpose systems for algorithmic composition and sound design such as Csound, MaxMSP and SuperCollider also have libraries for creating fractal structures, many of which are available online.

2. Fractal curve methods

2.1. *White tune*

Frequencies generated by Gaussian distribution have a uniform distribution in spectrum. In white noise, each impulse positive or negative returns to zero where the next begins. This process coils amplitudes around zero. The"whistling" tone whose spectral density ratio is $1/f^0$ remains constant, regardless sample's playback speed -a fractal property [9].

Figure 1. a)white noise b)pink noise c)Brown noise

This usually happens, when signal is self similar to its spectrum [10]. Melody of fig.2a, taken from Voss' article was created from white noise amplitudes [11]. These data create a scatter diagram (fig.3) which shows the relationship of each musical datum to its precedents.

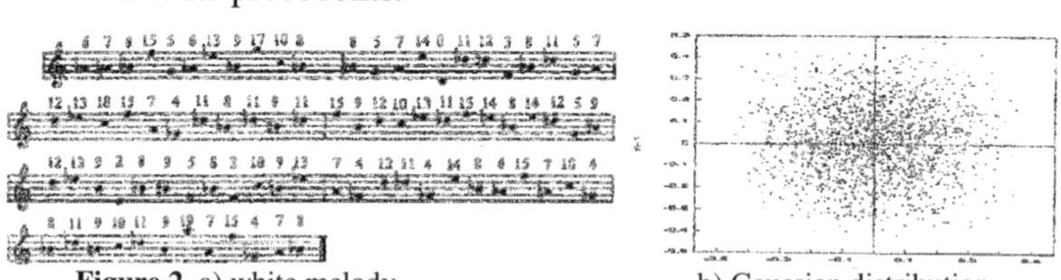

Figure 2. a) white melody b) Gaussian distribution

The result is quite interesting, as there is a similarity with Gaussian distribution in fig.2b. The relations of pitches are concentrated at the center (vertical columns 8 and 9 on both axes) and there are no data at the corners, especially in the upper left and lower right. The concentration of pitches' relation and the avoidance of large intervals (top left and bottom right corners, respectively) generate a degree of correlation which is contrary to what happens in amplitude distribution. Considering white noise as a form of fractal method,

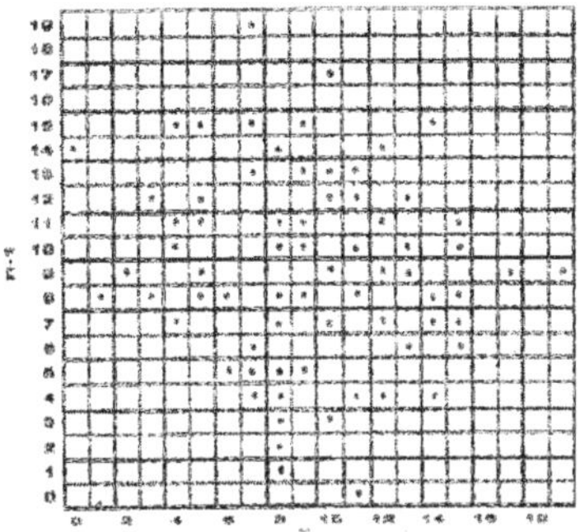

Figure 3. White melody scatter gram

$$D = \frac{\log(70/40)}{\log(400/100)} = \frac{0.55961}{1.38629} = 0.40368$$

we should investigate its dimension. Dimension in fractal melodies can inform us how colorful a melody is and how varied its motifs' repetitions. In this way, we can measure the "psychological" density of a musical work.

It is important to remember that dimension is the space in which the iterative process takes place [12]. For the purposes of this example we will use grids to measure dimension [13]. Given that the total length's logarithm divide by the number of boxes and as denominator the logarithm of one side's length (of the box) defines dimension, we modify the formula so that we can count the points from the previous chart: D=log (N/n)/log (B/b)[14]. Observing fig.3 we see a grid of four hundred boxes (20×20). Calculations give us the value of a dimension D=0.40368. It should be noted that this is not the dimension of white noise but that of a small melody sample, which is synthesized according to Gaussian distribution. Our first conclusions are that there is a wide variety of musical interval relations in the midrange. It is very likely that fractal dimension will rise if more notes are added to the example.

2.2. *Brown tune*

The term "Brownian motion" is named after R. Brown (1773-1858) who reintroduced the seemingly random motion of particles suspended in a liquid [15]. Mandelbrot notes that there are endless variations of Brownian motion [16]. In brown motion, the position of a point is given by $x_n = U_s + x_{n-1}$, where U_s is like a random sample.

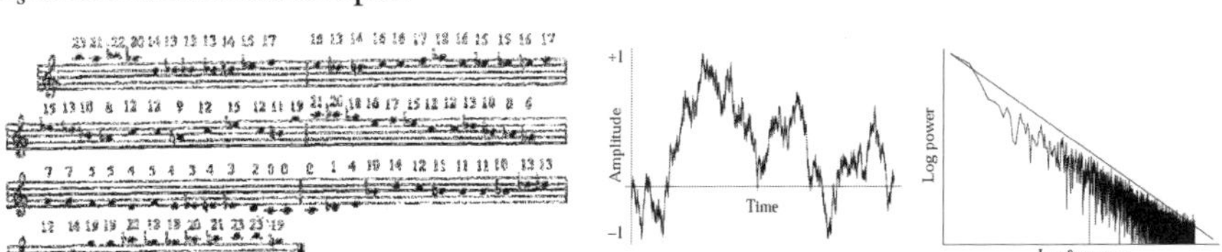

Figure 4. a) Brown melody b)Brown noise spectrum

Observing the equation seems that each step is based upon the former. The appearance of local similarity is obligated in consecutive points, which are placed relatively close to former, creating relationships not only with each previous point but as the sum of all previous movements. In fig 4b we see that the range of a non limiting-random walk derives from the square of the frequency $1/f^2$. Melody in fig4a is a sample of Voss' brown tune, displaying the characteristic of memory, (a brown motion feature), as it begins and ends with the same note.

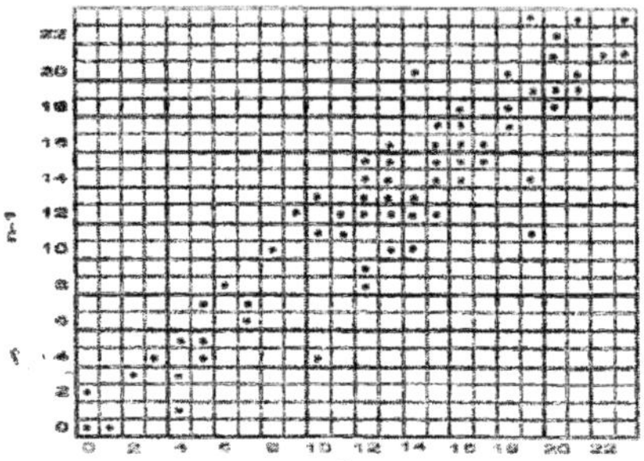

$$D = \frac{\log(66/39)}{\log(576/144)} = \frac{0.52609}{1.38629} = 0.37949$$

Figure 5. Brow melody dimension plot

Scatter diagram shows the concentration of musical intervals on a diagonal, which translates a high degree of correlation. Unlike the random-sounding, this melody has a statistical correlation in structure. Calculating the dimension as in the previous example we obtain D = 0.37949. As we have seen, each step is based on precedent but the increase of randomness is independent. Self-similarity is a characteristic of a particular region of the spectrum while fractal quality degenerates. The ridge of a mountain is a scalable fractal. The same statistical characteristics are observed at subsequently smaller scales of dimension. However, this process cannot be repeated indefinitely but ends after a certain minimum scale is reached [17].

2.3. *Pink tune*

Pink noise, is white noise with higher frequencies filtered out, 3dB per octave [18]. Including Gaussian distribution and denser frequencies at the bottom of the scale, has $1/f^1$ range (between that of white $1/f^0$ and brown noise $1/f^2$). Pink noise or flicker noise is one of the most representative model to generate intending melodies. This point of view should be kept under review, as fractals are patterns and their application depends on composer's mapping. It is true, that 1/f spectral density can be found in underwater currents, in variations of sunspots, even as noise inside the regions of our nervous system [19]. This aspect comprises an interesting sound philosophical basis. In any way, it doesn't mean that "good" music has 1/f ratio, non comprising the opposite. A clever way of creating 1/f scaling melody is given below. We use three dice [20]. Given sixteen notes chosen from a chromatic scale we range them in a predefined arithmetical order from number three (the smallest set of three dice 1+1+1=3) to eighteen. Mark the numbers from zero to seven in binary notation, like in fig.6 and assign one die in each column. To get the first note, cast three dice and write in another column the numbers that arise. The next note will not be produced in the same way. Beholding the sequence number column, it is obvious that only one digit changes, moving from zero to one. The procedure to be followed will be to throw dice that correspond to binary changes. In this particular case, for receiving the result in column 2, we choose only the cube corresponding to a binary change (the first from bottom) and reformulate the other two unchanged.

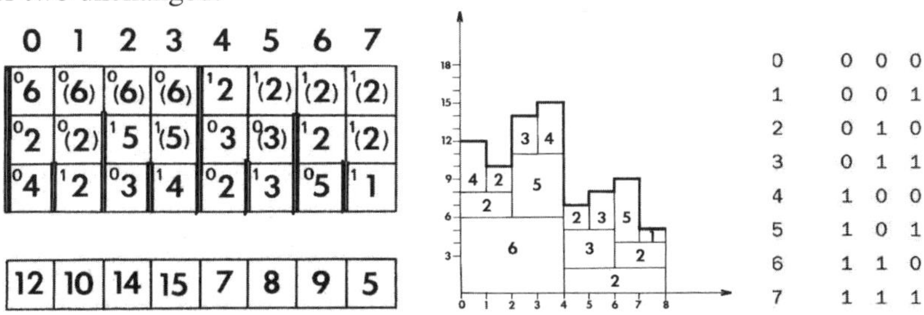

Figure 6. Die cast table and the melody obtained from the data

For resulting data from column 3, we'll choose the first two dice from the bottom up and just reformulate the third, etc. The space between two adjacent musical intervals is a random variable that depends on the number of dice thrown, as to formulate the next interval. We also note that melodies built as pairs of juxtaposed melodic lines, divided into sub-phrases. It is obvious that the hierarchical levels are as much as the number of dice!

2.4. L *systems in music*

An L-system is a regularly recurring process, which creates a large range of sequence symbols derived from a small initial sequence. This sequence is considered as an axiom accompanied by a set of production rules (one for each system's symbol) [21]. Thus, we create longer sequences by replacing each symbol with the corresponding rule and by repeating this process, we get the desired result. Types of L systems are in different categories such as stochastic, deterministic, context free, context sensitive, etc. The original goal on L-systems research was to develop patterns for biological organisms (shrubs, plants, etc.).

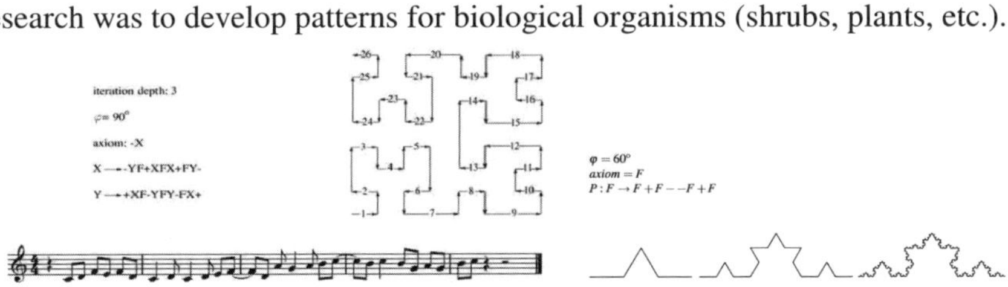

Figure 7. L system in music

An L-system consists of a 1) initial rule 2) additional rules to compensate the axiom when it is resumed, 3) a variable (α)that indicates the turtle's direction[22], 4) a variable that indicates the number of degrees the turtle turns.

For example, suppose the axiom X, d= 90, and rules: X→XFYFX+F+YFXFY-F-XFYFX, Y→YFXFY-F-XFYFX+F+YFXFY. The first iteration of the L-system axiom, substitute X, for rule X, giving the sequence XFYFX+F+XFXFY-F... As for the second iteration, the sequence generated by the first iteration, substitutes rule Y. Each iteration can be translated into geometric shape (using turtle). Prusinkiewicz[23] described a method for creating music, using L-systems curves. This method involves the reading of the horizontal lines as notes in time and vertical axis as pitch intervals between notes. Thus, each horizontal F is read as a common unit of duration. Each vertical axis is read as a common musical interval between notes. There is no distinction between left (for direction)and right (0.0 way)direction in interpreting curves and therefore melodies travel always forward in time (even if someone move the turtle left).

3. Curve aesthetics

By regarding curve as a fabricated argument we could remark how it permeates Western musical thought over the centuries. Curve is located in the rhetoric of baroque music [24], the linear sequences of classical composer's harmonic conversions until Wagner's endless melody which leads to non temporality [25]. The advent of twelve-tone technique signifies a cut off with "Newtonean" linearity since its aesthetics are based on the idea of successive cycles [26]. Thus serialism can be seen as based on aesthetics of "inner spinning". Several decades later (50'ies) Xenakis restored curve as a system of thought and meaning [27]. As well, many 20th century composers overloaded musical space with curves creating vast volumes of sound. This shift in the mode of curve's application (although successfully employed by a minority of composers), was a Trojan horse for musicians who did this kind of usage without critical thinking. The largest proportionate dominance of chart as an absolute value against the auto-generated musical code (which includes environment, social culture and a kind of social "fermentation") caused dislocation of the individual inside the imaginary human space. From this point of view, image, mathematical graph and static beauty appeared as a projection of the "proper" which can be certainly found outside of us! However, it would be unfair to say that this unfortunate event derives only from the "mean" knowing that composer's selecting and applying properties lie in the human creative process. The fact that fractal systems don't depend on a defined size of measurement, offers an open-work tool for a creative compositional process based on auto-similarity. Thus the organization of the inner world which consists of individual (musical) elements can receive stimuli from the outside world of human subjacent and translate these data as an organization and musical effect. We might even assume that the non-acoustic characteristics such as rhythm, musical forms and melodic movement (according to Ingarden [28]), present musical opportunities knowing that hierarchies as illustrated by the example of fig.6 are related to the number of dice which in turn depends on the composer. More generally, motion includes time and duration [29]. Temporality inside a musical work is the result of actions made by the human subjacent, through the musical space. Fractal dimension may not really exist in a musical work as a non-integer process, but it is fair to investigate the space where repetition and human subjective action are creating the musical temporality. So, all these processes emphasize the role of subjacent rather than obscure it. Moreover, researchers have noted that there is a low level of 1/f ratio in western music structure according to western tuning [30]. This does not mean that those composers created 1/f music nor that the 'good' music is the one with fractal proportions. On the other hand, recently scholars realized that plants for example are dominated by algorithmic beauty although beauty is presented by plants as an advantage for hundreds of years.

References

[1] Voss, R. & Clarke, J.(1975). 1/f Noise in Music and Speech. Nature 258, pp. 317-318

[2] Mandelbrot, B. (1983).The Fractal Geometry of Nature. Freeman, NewYork.pp.249

[3] Dodge, C. (1988). Profile: a Musical Fractal. Computer Music Journal, 12(3). pp.10-14.

[4] Nelson, G. L. "Fractal Mountains" in *Computer Music Currents*, vol 10 Wergo Schallplaten, Mainz, 1992

[5] Wallin,R.(1989)Fractal Music:Red Herring or Promised Land? Lecture given at the Nordic Symposium for Computer Assisted Composition Stockholm

[6] Evans, B.(1992) Number as Form and Content: A Composer's Path of Inquiry. Leonardo, Vol. 25, No. 3/4, Visual Mathematics: Special Double Issue (1992), pp. 303-311

[7] Gogins, M. Iterated Functions Systems Music. Computer Music Journal, 15, 1 (1991), pp. 40-48

[8] Supper, M. A Few Remarks on Algorithmic Composition. Computer Music Journal, Vol. 25, No. 1, Aesthetics in Computer Music , pp. 48-53

[9] Mandelbrot, B.(1983) pp.355

[10] Loy,G.(2006) Musimathics.The M.I.T. Press.pp.357

[11] Voss, R. & Clarke, J.(1978). 1/f Noise in Music. The Journal of Acoustical Society of America, 63, pp. 258-263

[12] For more information see Barnsley, M.(1993)Fractals Everywhere.2nd edit. Academic Press Professional

[13] Madden C.(1999) Fractals in Music:Introductory Mathematics for Musical Analysis.High Art Press Salt Lake City.pp.119-132

[14] b is the smallest number of boxes in the grid, N is the larger numbe of boxes that enclose one or more data points, B is the larger number of boxes in the grid(smaller box size) and n is the smaller number of boxes that enclose one or more data.

[15] Jan Ingenhousz described in 1785 the irregular motion of coal dust, upon the surface of alcochol.

[16] Mandelbrot, B.(1983)pp.350

[17] Loy,G.(2006)Musimathics.The M.I.T. Press.pp.357

[18] Everest, A.(2001)The Master Hanbook of Acoustics.4th edit.NewYork Tab Books.pp.111

[19] Musha T.(1981)1/f Fluctuation in Biological System. Sixth International Symposium on Noise in Physical Systems. National Bureau of Standards. pp.143-146

[20] Gardner, M.(1992). Fractal Music, Hypercards and More : Mathematical Recreations from Scientific American Magazine. W. H. Freeman and company,New York.pp.5

[21] Prusinkiewicz, P. & Lindenmayer, A.(1990). The Algoritmic Beauty of Plants. Springer-Verlag

[22] The well known turtle program of M.I.T. in logo language

[23] Prusinkiewicz, P.(1986). Score Generation with L systems. International Computer Music Conference 86'. Proceedings,pp. 455-457

[24] Bartel,D.(1997)Musica Poetica:musical-rhetorical figures in german Baroque music.Univ. of Nebraska Press

[25] Strawinsky,I(1945)Poetique musicale sous forme de six lecons.Flammarion edit 1997

[26] Hermand,J.&Richter,G.(2006)Sound Figures of Modernity.pp.181

[27] Solomos,M.(2004)Xenakis' Thought through his Writings.Journal of New Music Research,VOl(33)2pp.125-136

[28] Hans Rainer sepp&Lester Embree(2010)Handbook of Phenomenological Aesthetics.Springer.pp.167-170,223-229

[29] Lang, H.(1998)The Order of Nature in Aristotle's Physics.Cambridge University Press.pp.34

[30] Beran J.(2004)Statistics in Musicology.Chapman&Hall/crc

Time variant chaos encryption

Victor Grigoras[1] **and Carmen Grigoras**[2]

[1]Faculty of Electronics, Telecommunications and Information Technology, 'Gheorghe Asachi' Technical University of Iasi, Iasi, Romania
[2]Faculty of Medical Bioengineering, 'Gr.T. Popa' University of Medicine and Pharmacy of Iaşi, Iasi, Romania
Romanian Academy – Iaşi Branch, Institute of Computer Science, Romania
Email: grigoras@etc.tuiasi.ro

Abstract: Chaos synchronization was extensively studied the last two decades leading to several synchronization and modulation methods. The main aim of the research was to develop wide bandwidth, spread spectrum like and secure communication systems. Taking into account the high sensitivity of all synchronization and modulation methods to channel noise and parameter mismatch, digital implementations aiming at chaos encryption were preferred lately. Several authors developed, in their studies, different encryption approaches, most of them being successfully cryptanalized, due to the direct influence of the constant system parameters onto their nonlinear dynamics.

The present contribution proposes a time variant approach to chaotic encryption. The proposed method is based on chaos synchronization and plaintext modulation onto the chaotic dynamics of the emitter. In order to improve communication confidentiality, one or several parameters of the emitter system are modulated with pseudo-random digital sequences, thus drastically increasing the length of the encryption key. At the decryption end, the corresponding receiver parameters are also time variant. Exact knowledge of the emitter pseudo-random digital sequence shape and timing are necessary for correct decryption of the cipher-text. It is also worth noting the importance of the modulating sequence amplitude or dynamic range due to its influence on the instantaneous value of the emitter sensitivity and, by consequence, its capacity in hiding the transmitted plaintext. Thus a parametric analysis is presented in order to find a valid parameter range for possible modulation. For the digital implementation of the proposed encryption/decryption systems, although input and output are fixed point, the internal structure of both emitter and receiver must be implemented in floating point to obtain the closest behavior with the chaotic prototype that has analog valued state variables.

Both analog and discrete examples for the encryption/decryption system are analyzed. Presented simulations confirm the theoretical results and highlight the great improvement of the communication security by the proposed approach. The concluding remarks point towards some directions in further research.
Keywords: chaos synchronization, chaos encryption, parameter modulation.

Chaos Theory: Modeling, Simulation and Applications
C. H. Skiadas, I. Dimotikalis and C. Skiadas (Eds)
 (pp. 175 - 182)

1. Introduction

The process of chaos synchronization study is an ongoing field of research. Its aim is to identify the structure and parameters of many natural phenomena exhibiting this type of nonlinear dynamics and to find engineering applications for this complex dynamical behavior of nonlinear systems [1]. Secure communication systems based on chaos synchronization were proposed in both analogue and discrete versions, e.g. [2-3]. Due to synchronization sensitivity to channel perturbations, analog modulation is less attractive for chaos communication. Instead, it can be a promising prototype for discrete-time chaos encryption, if analog state equations are discretized and input and chaotic signals are quantized to digital values. From the cryptographic point of view, the key to decrypt a chaos modulated message consists of the emitter parameters. Although these parameters are real numbers, i.e. floating point in digital rpresentation, thus leading to a long and complex enough key, for some chaos encryption schemes, at least approximate emitter identification was possible, thus leading to brute force cryptanalysis succes. If constant parameters were replaced with time variant versions, breaking the chaos encryption would be much harder as both parameter variation law and time synchronization would be necessary to correctly decrypt the code.

Parameter modulation type chaos synchronization, allows using the same communication channel for the transmission of modulating signal and parameter time variation, as suggested in [4-5]. This approach suggests two alternatives for chaos encryption improvement:

- applying the plaintext as a modulating signal and the secret sequence as a parameter variation law, or
- applying the plaintext as a parameter variation law and the secret sequence input signal to the chaotic emitter.

The first version is efficient in chaos synchronization setups that do not benefit from an adaptive demodulation algorithm, thus efficiently hiding the time-variant encryption key. The latter needs an adaptive receiver scheme to decrypt the encrypted message. The secret sequence is necessary as a desired signal at the receiver. Modulating the encyption key onto the chaotic carrier seems less secure compared to the previous alternative, as the secret sequence might be discovered by frequency spectrum attack or adaptive receiver system attack.

In the present paper, we analyze both analog and discrete-time encryption systems, based on the plaintext modulating – time variant parameter (first version) synchronization setup. Examples were developed using the Lorenz system but any other prototype would behave in a quite similar way from the cryptographic point of view. The next section presents the principle of the analog chaos synchronization, developing the basic prototype for time variant private communication. In the third section, a discrete-time version of the proposed encryption system is developed.

Concluding remarks highlight the advantages and drabacks of the proposed approach, suggesting further development paths.

2. The Analog Modulation Case

As an analog prototype, we use a chaotic emitter and a synchronizing receiver, designed by the system partitioning (Pecorra and Carroll) method. The general state equations of the time variant chaotic emitter are of the form:

$$\begin{cases} \mathbf{x}'(t) = f\left(\mathbf{x}(t), \mathbf{p}(t), m(t)\right) \\ y(t) = x_1(t) \end{cases}$$

where $\mathbf{x}(t)$ is the state vector of dimension at least three, $\mathbf{p}(t)$ is the time variable parameter vector, $m(t)$ is the scalar modulating signal, $f(.)$ is the nonlinear state transition function and the output, $y(t)$ is chosen to be the first state variable (after some re-aranging, if necessary).

We adopt the system partitioning method for designing the sinchronizing receiver, leading to the general state equations for the receiver:

$$\mathbf{x}_r'(t) = f_r\left(\mathbf{x}_r(t), \mathbf{p}_r(t)\right)$$

Where $\mathbf{x}_\mathbf{r}(t)$ and $f_r(.)$ are the receiver, reduced state vector and state transition function respectivelly, i.e. without the first state variable and state equation of the emitter. It is implicitlly assumed that the modulating signal, $m(t)$, is included in the first state equation only. The receiver parameter vector, $\mathbf{p}_r(t)$, is time variable too and has the same dimension as the emitter one. The demodulator is built based on the first state equation of the emitter, replicated at the receiver end:

$$\begin{cases} x_{1r}'(t) = f_1\left(x_{1r}(t), \mathbf{p}_r(t)\right) \\ m(t) = y(t) - x_{1r}(t) \end{cases}$$

If the receiver is designed to synchronize with the chaotic emitter, for constant parameter vector, $\mathbf{p} = \mathbf{p}_r$, then it will synchronize also when coefficients are synchroneously time-variant, $\mathbf{p}(t) = \mathbf{p}_r(t)$, as the reciver and emitter parameters will be identical for each time moment. The same obsarvation is valid for the demodulation process, because the demodulator is identical to the coresponding state equation of the emitter, down to every parameter for each time moment. It is important to note that the synchronization process depends on the emitter and receiver state equations and not on the nonlinear dynamics the emitter exhibits. Thus both synchroniztion and demodulation perform well, even if the parameter vector exits the chaotic region of the emitter, for some moments of time. Although good functionality is preserved, the confidentiality of the transmission will suffer. That is why, a parametric analysis has to be performed before choosing the parameter variation range, in oder to keep equations coefficients in the chaotic region at all times, for security reasons only.

In the following, a Lorenz based chaotic communication system is simulated in order to check the previous theoretical assertions. The modulated emitter state equations are:

$$\begin{cases} x' = \sigma \cdot (y - x) \\ y' = \rho \cdot x - y - x \cdot z + m(t) \\ z' = x - \beta \cdot z + x \cdot y \end{cases}$$

Choosing to transmit the *y* state variable:

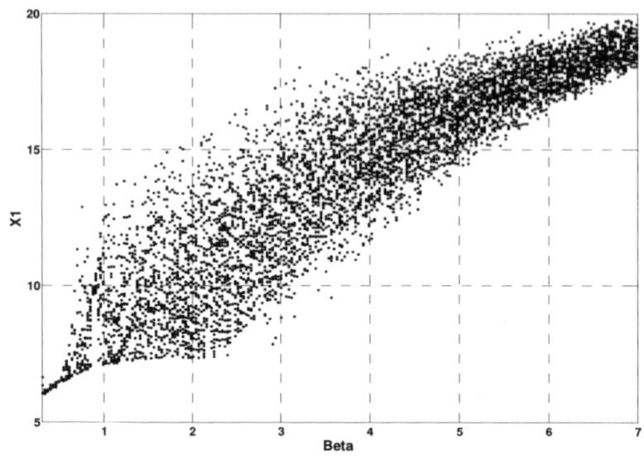

Figure 1. Bifurcation diagram for the first state variable of the Lorenz system at the variation of the β coefficient

$$u(t) = y(t)$$

the receiver state equation result in the form:

$$\begin{cases} \tilde{x}' = \sigma \cdot (y - \tilde{x}) \\ \tilde{z}' = \tilde{x} - \beta \cdot \tilde{z} + \tilde{x} \cdot y \end{cases}$$

with the coresponding demodulator equations:

$$\begin{cases} \tilde{y}' = \rho \cdot \tilde{x} - \tilde{y} - \tilde{x} \cdot \tilde{z} \\ \tilde{m}(t) = u(t) - \tilde{y}(t) = y(t) - \tilde{y}(t) \end{cases}$$

If we choose to make the coefficient β time-variant, we must perform a parametric analysis of the nonlinear dynamics upon the variation of this parameter. The bifurcation diagram in Figure 1, shows that a reliable range for this coefficient is [1, 3]. With this result in mind, we simulated the chaotic communication system, with the sequence in Figure 2 modulating both emitter and receiver b parameter. Figure 3 represents the results, namely the modulating signal and the synchronization error. Correct synchronization is obvious, after a short transient the demodulation error becoming negligible.

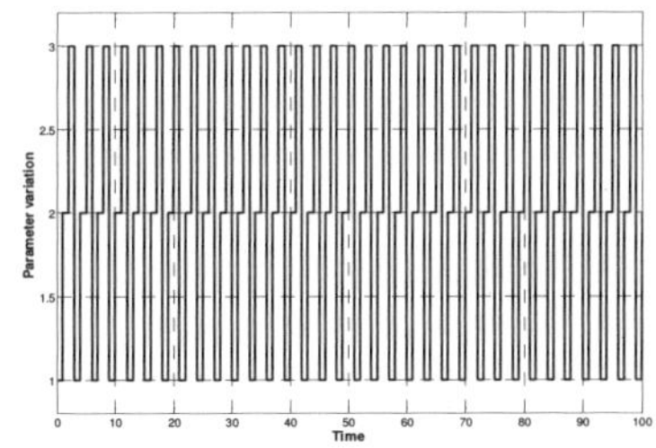

Figure 2. Parameter time evolution

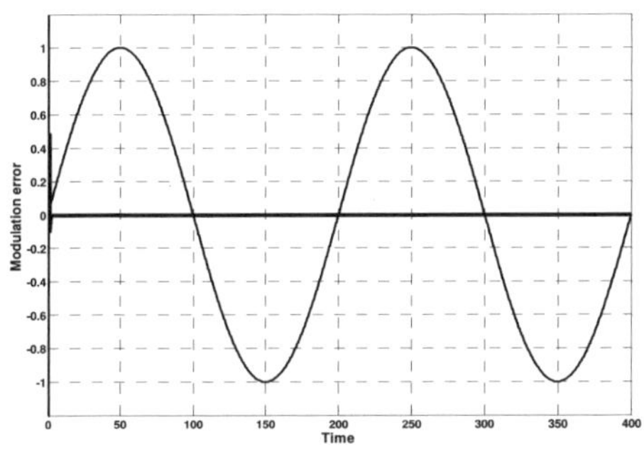

Figure 3. Modulating signal (thin line) and demodulation error (thick line)

Using the same parameter variation and modulating signal, we applied a small delay of $\Delta\tau = 0.03$ for the receiver parameter and obtained the results in Figures 4 and 5. The presented graphs show the high sensitivity of the system to the parameter sequence highlighting the higher security of the time-variant approach.

The presented examples use only one time-variant parameter, but any number of such coefficients (even ones introduced over the canonic minimum number) can be used for improved confidentiality.

3. The Discrete-Time Case

As a discrete example we chose and additive nonlinear system that allows the design of a synchronizing receiver by means of the inverse system approach. The difference equation for an additive nonlinear system of order N is given by:

$$y[k] = r\left(m[k] - \sum_{n=1}^{N} h[n] \cdot y[k-n] \right)$$

The algebraic nonlinear function $r(.)$ is the additive residue function:

$$r(x) = x - round(x)$$

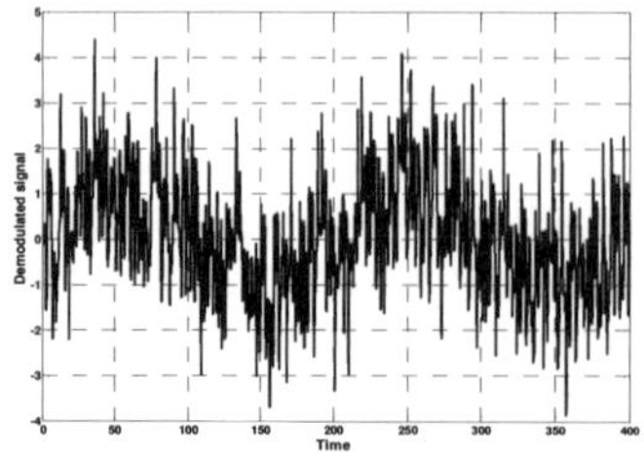

Figure 4. Demodulated signal for small error in the parameter variation

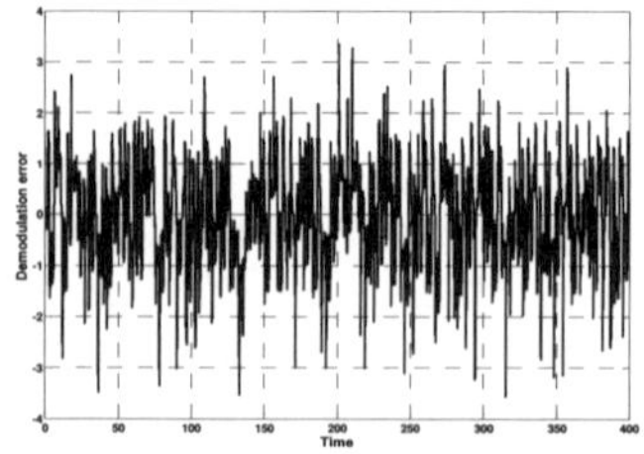

Figure 5. Demodulation error for small delay in the parameter variation

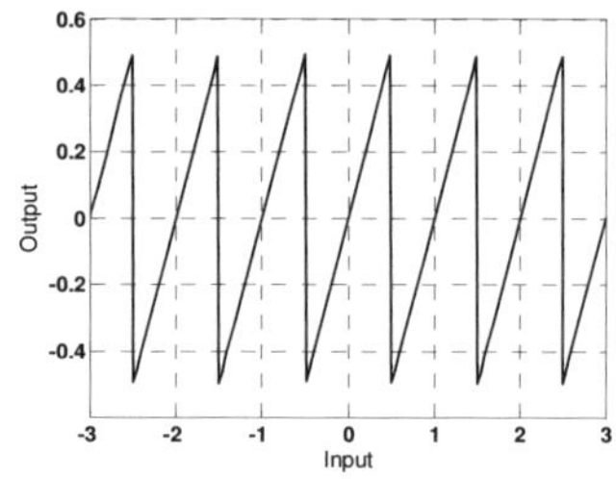

Figure 6. The additive residue nonlinear function

The graphical representation of the additive residue function in Figure 6 shows that the slope of this piecewise linear function is constant, 1.

In order to perform chaotically, the additive system must have an unstable linear prototype:

$$H(z) = \frac{1}{1 + \sum_{n=1}^{N} h[n] \cdot z^{-n}}$$

By using the inverse system approach, and taking into account the additivity property of the emitter, the receiver system results:

$$\tilde{m}[k] = r\left(\tilde{y}[k] + \sum_{n=1}^{N} h[n] \cdot \tilde{y}[k-n] \right)$$

The block diagrams of the resulting systems are depicted in Figures 7 and 8.

In the case of an N -th order communication system, we have naturally N coefficients available. In the following example, we used a second order system, with nominal values of the coefficients vector **h** = [1, -1.65, 1.4], ensuring chaotic behavior, for a variation range of 0.4, because of the linear prototype poles $z_{1,2}$ = 0.825+/-0.84816 i, outside the unit circle. The variation law used for the coefficents is presented in Figure 9. In the case of the authorised receiver, the modulating signal and demodulation error are presented in Figure 10, where zero demodulation error is visible after a transient of two samples. Unauthorised receivers will not be able to recover the modulating signal, as shown in Figures 11 and 12.

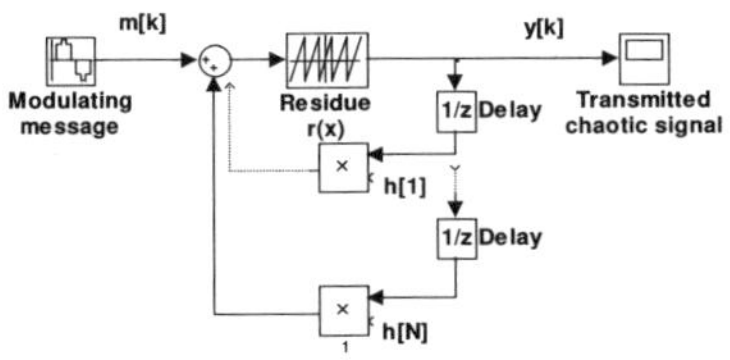

Figure 7. Block diagram of the additive nonlinear emitter

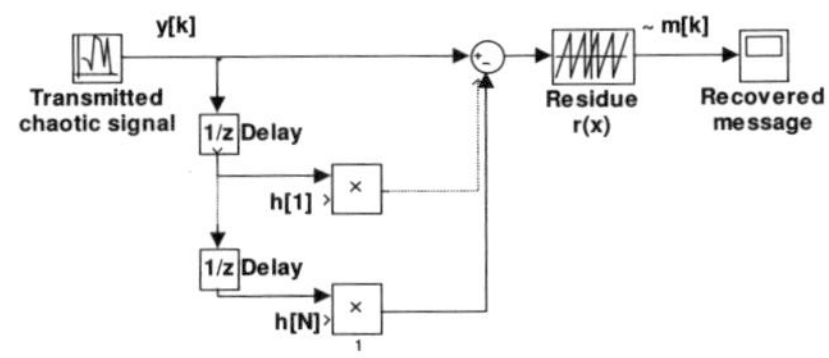

Figure 8. Block diagram of the additive nonlinear receiver

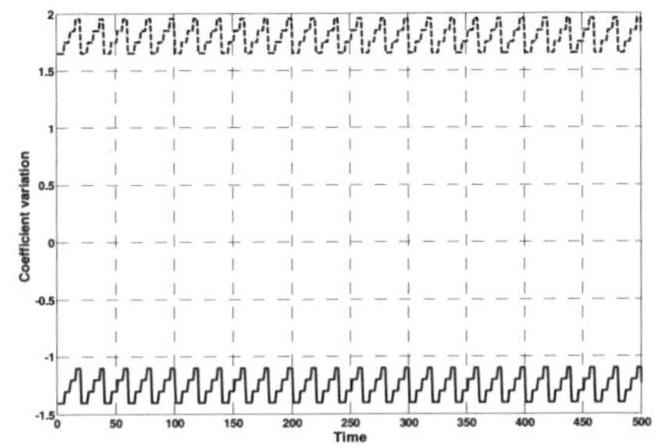

Figure 9. Parameter time evolution: second coefficient (continous line) and third one (intrerupted line)

4. Conclusions

The present contribution aims at improving the secrecy of a chaos encryption system by using time variant parameters in both emitter and receiver blocks. Both analog and discrete versions of the proposed approach are analized. Simulations confirm the theoretical demonstration and highlight that leghthening the encryption key with the variation sequence makes unauthorised decryption more dificult.

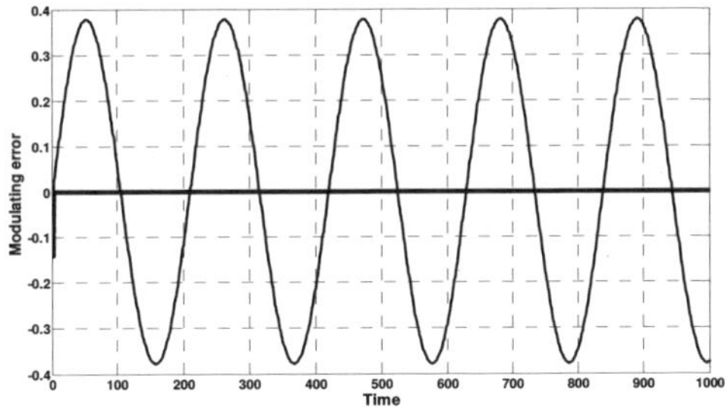

Figure 10. Modulating signal (thin line) and demodulation error (thick line)

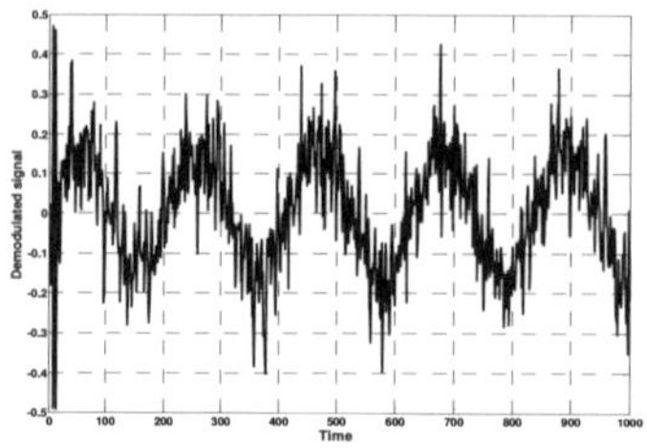

Figure 11. Demodulated signal for small error in the parameter variation

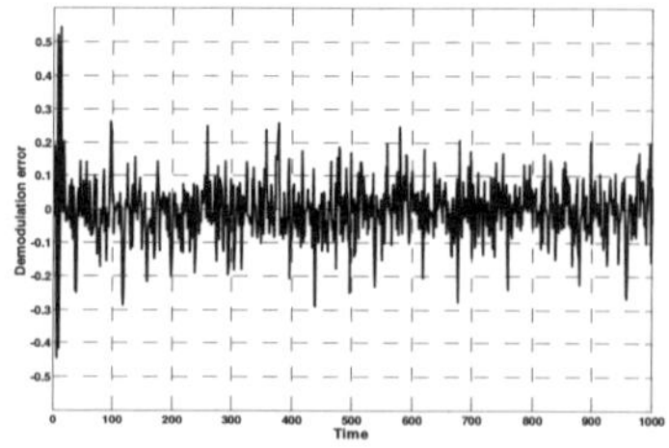

Figure 12. Demodulation error for small delay in the parameter variation

Further research is needed for designing a synchronization algorithm of the coefficent sequence for the authorised receiver without opening a back gate to unauthorised one.

References

[1] L. M. Pecorra, T. L. Carroll: ‘Synchronizing Chaotic Circuits’, IEEE Transactions on Circuits and Systems, Vol. CAS 38, No. 4, April 1991, pp. 453-456.

[2] V. Grigoras, V. Tataru, C. Grigoras: ‘Chaos Modulation Communication Channel: a Case Study’, Proceedings of the International Symposium on Signals, Circuits and Systems – ISSCS 2009, Iasi, Romania, July 9-10 2009, pp. 489-492.

[3] V. Grigoras, C. Grigoras: ‘Digital Data Encryption Based on Discrete Additive Systems’, in Topics on Chaotic Systems, Selected Papers from CHAOS 2008 International Conference, World Scientific Publishing Co., 2009, pp.149-157.

[4] Leuciuc, V. Grigoras: ‘Multi-parameter Chaos Modulation of Discrete Time Filters’, Proceedings of the 5th International Specialist Workshop on ‘Nonlinear Dynamics in Electronic Circuits’ NDES-97, Moscow, Russia, June 26-27, 1997, pag. 81-86.

[5] Leuciuc, A.; ‘Information transmission using chaotic discrete-time filter’ Circuits and Systems I: Fundamental Theory and Applications, IEEE Transactions on, Vol. CAS 47, No. 1, Jan. 2000, pp. 82 - 88

[6] L.O. Chua, T. Lin: "Chaos in Digital Filters", IEEE Trans. Circ. Sys., Vol. CAS-35, No. 6, pp. 648-658, June 1988

[7] U. Feldmann, M. Hasler, W. Schwarz, On the Design of a Synchronizing Inverse of a Chaotic System, Proceedings of the ‘European Conference on Circuit Theory and Design’ ECCTD'95, Istanbul, 1995, pp. 479-482.

[8] V. Grigoras, L. Goras, Additive Nonhomogeneous Discrete Systems, Proceedings of the 2nd International Specialist Workshop on ‘Nonlinear Dynamics in Electronic Circuits’ NDES'94, Krakov, Poland, 29-30 July 1994, pp. 185-189.

[9] V. Grigoras, L. Goras, A Z-Transform Approach to Additive Nonlinear Discrete Systems Analysis, Proceedings of the ‘European Conference on Circuit Theory and Design’ ECCTD'95, Istanbul, 1995, pp. 324-327.

[10] V. Grigoras, Digital Encryption Based on Exact Chaos Synchronization, International Journal on Chaos Theory and Applications, No. 4, Vol. 4, December 1999. pp. 15-20.

[11] Leuciuc, V. Grigoras, Finite Time Chaos Synchronization in Bijective Triangular Form Systems, International Journal of Chaos Theory and Applications, No. 1, Vol. 2, May 1997, pp. 3-16.

Aesthetic considerations in algorithmic and generative composition

Kerry L. Hagan

Digital Media and Arts Research Centre (DMARC), University of Limerick, Limerick, Ireland
Email: kerry.hagan@ul.ie

Abstract: Models of chance operations, random equations, stochastic processes, and chaos systems have inspired composers as historical as Wolfgang Amadeus Mozart. As these models advance and new processes are discovered or defined, composers continue to find new inspirations for musical composition. Yet, the relative artistic merits of some of these works are limited. This paper explores the application of extra-musical processes to the sonic arts and proposes aesthetic considerations from the point of view of the artist. Musical examples demonstrate possibilities for working successfully with algorithmic and generative processes in sound, from formal decisions to synthesis.
Keywords: algorithmic and generative composition, aesthetics, random and stochastic processes, chaos systems, sound synthesis.

1. Introduction

In many ways, the Western classical music tradition is steeped in numerical methods. Numbers define musical intervals and rules in the earliest examples of Western counterpoint. Numerical systems identify chords and inversions within the Western tonal system, as well. Classically trained musicians learn these systems and express the music theory of Western classical with these numbers.

Although there are the odd examples of musical numerical games emerging from history (e.g., Mozart's "Musical Games"), numerical systems came to the fore in the 20th century with the serializing of musical attributes such as pitch, duration, and timbre. The Second Viennese School freed music from the Western system of dissonance, consonance, and tonal center, allowing set operations to determine musical content through numeric representations of the twelve equal-tempered pitches in Western music.

Once the numbers of music were freed from the harmonic system, composers could look beyond set operations alone to other sources of numerical patterns, systems, rules, and models. With the advent of computers and ultimately computer music, composition with any kind of numerical system

Chaos Theory: Modeling, Simulation and Applications
C. H. Skiadas, I. Dimotikalis and C. Skiadas (Eds)

became more commonplace. Once audio itself was digitized, the numbers representing music - sound itself, in fact - expanded beyond the restrictions of Western music notation of pitch and duration.

Despite this, the application of extra-musical models and systems to the numbers of sound and music is relatively new to music-making. As a result, composers are still exploring the true potential of using scientific or mathematical models to provide numbers for music. The results can be surprising and occasionally dissatisfying. This paper seeks to address some arenas where pieces can fail to achieve the promise suggested by scientific or mathematical inspirations. Thus, this paper is a qualitative study of musical practices with extra-musical numbers.

In order to make useful formalizations of musical practice through qualitative assessment, the domain of practice must be defined and limited to a particular mode of artistic work. So, firstly, this paper defines the music by the historical context, intended reception, audience, and compositional methods.

Secondly, to draw conclusions of the effectiveness of a musical strategy pre-supposes a system of judgment, a basis on which music is to be compared. It is unethical to apply all systems of judgment to all systems of music. Therefore, this paper then defines the means by which a work is judged to be successful and, more importantly, how the musical experience is then critiqued.

Based on the constraints of compositional domain, the categories of human/model/computer interaction, and a critical system of comparisons, this paper then proposes some crucial questions and considerations a composer can make when creating algorithmic or generative music. Ultimately, this paper offers some conclusions regarding successful practice. However, it does not do so to set requirements or rules for "good" music, only to identify similarities between successful examples of algorithmic and generative composition.

2. What is Algorithmic and Generative Composition?

Musicians are not scientists, and the vocabulary they use shifts and crosses meanings and often implies methods of practice that may seem entirely independent of the name given. Therefore, it is not enough to simply say this paper addresses "algorithmic and generative composition."

Algorithmic and generative composition is used to include computer music, computer synthesis and computer-assisted composition, but also includes earlier musical examples created by hand calculation. *Generative* composition is also included in order to accommodate a means of creating music with processes that may not fall within a more carefully circumscribed definition of algorithm. Therefore, this paper addresses algorithmic and generative music, which includes any systems from the strictly algorithmic to more open-ended generative processes. More importantly, algorithmic and generative *composition* presupposes an artistic or aesthetic endeavor and not simply the act of demonstrating data for informational purposes.

Finally, this phrase refers to music that is meant to be experienced within the concert hall paradigm, which relies on a general type of audience within certain traditional conventions including concerts and/or sound recordings associated with Western classical music performance.

Additionally, the use of the phrase "algorithmic music" does not imply the occasional meaning of "automatic composition." Many composers who may rely on mathematical or numerical systems will deny their inclusion in the field of algorithmic and generative music on this basis alone. The use of the phrase in this paper supposes that this connotative definition can be overlooked for the greater issue at hand.

Some of the composers who have works falling into these definitions include Lejaren Hiller, Iannis Xenakis, Brian Ferneyhough, David Cope, Eduardo Reck Miranda, Julio d'Escrivan, and Hans Tutschku.

3. What is "Successful"?

Given that the domain represented in this discussion is algorithmic and generative composition within the concert hall paradigm, then the metric for the relative success or failure of a work can be determined by the genre itself. There are works throughout history whose success is measured by endurance, impact, innovation, or similar features. These works are accepted as part of a canon in a tradition, perhaps what Goehr [4] would call a musical museum, despite any individual's personal preference for these works.

This paper will review some of these canonic works and some newer examples that, nonetheless, offer examples of musical success to some degree. Newer works were decided by whether they stand on their own without recourse to scientific or extra-musical explanation and whether they withstand intense and attentive musical scrutiny. Additionally, this paper will comment on the published methodologies of self-identified algorithmic composers.

In the following examples, there appears to be three main aspects that determine the relative success of an algorithmic composition. Firstly, there is a correlation between successful works and the degree to which the composer intervenes, influences, adds to, or shapes the composition. Alternatively, successful works can also be the result of human performer intervention, e.g., algorithmic compositions performed on acoustic instruments by experienced performers.

Secondly, works that tend to stand without the aid of explanation or background have complex mapping paradigms of the data to musical parameters. There is a correspondence between the success of a work and the obfuscation of the underlying process.

The third characteristic of a successful work is whether or not care has been taken in orchestration or design of timbre. At the same time, timbral considerations are intricately linked to decisions made by the composer in instrumentation and mapping. For this reason, timbre may not be considered as a third aspect. However, its immediate effect on the reception of the work makes it necessary to address it separately.

3.1. *Human Intervention*

Perhaps the most telling evidence of the need for human intervention comes from the evident bias against the term algorithmic composition since it can mean automatic composition (without human intervention.) Many composers who may rely on mathematical or numerical systems will deny their inclusion in the field of algorithmic and generative music on this basis alone.

History tends to support this bias in musical examples. A very compelling example comes from a work that is often regarded to be musically weak but is nonetheless historically important. Namely, *Analogique A + B* (1959) by Iannis Xenakis demonstrates very clearly that composer intervention and further human performance and interpretation result in more successful outcomes. In two parts, *Analogique A* and *Analogique B*, Xenakis implements the same Markov process in different media. One utilizes short, electronic sine tones arranged and recorded on tape. The other is translated into quantized notes according to Western music notation and scored for strings. Although the process is the same, the electronic version allows Xenakis greater freedom in both pitch and duration, since sounds could be arbitrarily generated. However, it is the string version that was more successful as a musical work.

In many cases, limiting material to a musical scale can reduce the complexity and interest of a governing algorithm. Yet, the acoustic version (*Analogique A*) is a stronger work. The first impression may be that the more familiar and acoustic timbre of the strings makes the version more successful. Taken on its own, *Analogique B* does provide a rich timbral experience, as Xenakis' method of composition is a precursor to the common practice of granular synthesis today. Rather, the effect that human performance, hence human intervention, of the given musical material is far more salient than the sound source.

There is strong precedent for composer intervention or interference within algorithmic composition. Composers such as Di Scipio [3], Ariza [1], Gogins [5], and Xenakis [7] refer to integrating their feedback into systems, trial and error methods of fine-tuning systems, or "bending" by the composer in the process of writing algorithmic music. Essentially, the composer can make crucial decisions at various stages of the composition. For some interactive systems, composers make subjective judgment calls and guide algorithms. In other cases, composers will tweak systems several times before getting the musical material they need. And, whether the composer is "fudging" the output of a system or perturbing a steady state, composers will interfere and introduce error into the output of an algorithmic system. The resulting complexity of these composers' works, especially the musicality of them, makes the case for composer intervention.

3.2. *Complexity of Mapping*

The second, obvious attribute contributing to the success or failure of a work is the complexity of the mapping of the system or data output to the musical parameters. To accurately measure complexity is to pursue a study beyond the scope of this paper. Additionally, most composers do not publish their processes to the degree to which it would be necessary to quantify the complexity of their individual mapping. However, by looking at common student practices compared to those of established composers, the extreme ends of the spectrum can be identified.

Students learning the practice of algorithmic composition must learn a variety of skills relevant to the multi-disciplinary field. For example, they need to learn the mathematics behind a particular equation or iterated function. Then, they need to learn computer science skills that enable them to realize the mathematics in an environment that can lead to music or sound production. Finally, they must learn the art of composing and music-making, which is inherently a critical and subjective domain learned and advanced only by continual practice. To some degree, it is this last aspect that students cannot learn quickly, though some may have better natural instincts than others. Therefore, it is this final objective that students fail to realize in their short careers as students. As a result, student works in the algorithmic domain may show advanced understanding of scientific or mathematical systems and may exhibit sophisticated implementation of the systems in programming environments. Yet, students can fail dramatically to make music from these skills.

The fundamental difference between the student work and the established composer is the way the algorithms are translated into music. A student does not have the experience or depth of musical understanding to do much more than the most straightforward mapping of data to sound. For example, a two-dimensional system may be very simply translated into pitch and duration values, or pitch and instrument assignment.

The fundamental problem with this note-based approach to algorithmic composition is that it is inherently simple. Despite the complexity of the stochastic or chaotic system, the output is directly mapped to obvious musical qualities, making the reception unchallenging and less engaging. Though the systems exhibit compelling qualities, the results fall flat when translated to the temporal domain in the form of sound.

More seasoned composers searched for a way to complicate the relationship of the systems to output. For example, Bidlack [2] utilized a note-based approach to algorithmic composition realized in MIDI. In order to complicate the mapping while working within the same constraints, Bidlack implemented multi-dimensional systems (Lorenz and Hénon-Heiles) so that the third and fourth dimensions could be utilized for duration and loudness in addition to pitch. Additionally, it appears that Bidlack chose timbres independent of a deterministic system, an example of composer intervention.

Other successful works exhibited a complexity of mapping to the degree that any sign of the original algorithm is hidden beneath the immediacy of musical material. The listener hears music, not an algorithm. Complex patterns may emerge or dissipate, but they seemingly follow an internal musicality, not deterministic rules.

One example, *Olivine Trees* (1994) by Eduardo Miranda, demonstrates that complexity of the application of the algorithm can be manufactured not simply by the number of dimensions or parameters the algorithm controls, but by the time scale on which it is controlled. Miranda uses cellular automata to control grains of sound (reminiscent of the Xenakis example above). In his own description, Miranda likens the blending of these small grains to that of impressionistic painting [6]. The composer is inspired by the artistic potential of the system and reflects it in a sophisticated approach to implementation. In addition to CA, Miranda uses other processes and composers' tools to further influence the sound of the piece. In this sense, Miranda not only obfuscates his mapping, but he introduces a free reign of composerly intervention.

3.3. *The Importance of Timbre*

From the first works of electroacoustic music composition, timbre was the primary material of music. Pitch, rhythm, loudness, silence – these were the brushes, but timbre was the paint. The fascination with electroacoustic means of making music was the palette it provided.

As more composers began working with the new tools, new foci entered the domain of practice. As a result, later music relied less on timbre as its primary material. This was especially true for algorithmic composers whose primary material was the system or algorithm itself.

Yet, this attitude has been a detriment to the field of algorithmic composition. Timbre is an immediate attribute that anyone can perceive, regardless of his or her musical sophistication. It does not take a musical education to be aware of badly synthesized sounds, nor to grow fatigued with poor orchestration. Therefore, the most successful algorithmic compositions are examples of excellent instrumental or sound design choices.

It is difficult to address the timbre problem without addressing the human factor or the complexity question. Therefore, timbre can really be seen as a consequence of those aspects.

For example, if human intervention happens at the performance stage, i.e., live instrumentalists realize a work, then timbre is a consequence of the acoustic instruments. As a result, it has the richness, variety, and familiarity these instruments have in the culture. Likewise, if the mapping of a system is to the microsound level or to sound synthesis, then the complexity and interest of the system used is mapped into timbre, resulting in rich and interesting sounds.

However, since timbre is the immediate attribute first perceived in a work, it warrants separate consideration. Recognizable timbres can improve or ruin a work. In the case of works by Lejaren Hiller, notably the *Quartet No. 6 for Strings* (1973), the natural instrument timbres performed by live humans create music that stands on its own, regardless of its algorithmic roots. Hiller augments his timbral possibilities with the extended playing techniques developed in the 20th century. In comparison, much more algorithmically complex works that were realized with MIDI and commercial synthesizers now sound dated and awkward.

In the earlier examples, both Xenakis and Miranda synthesize complex timbres through algorithmic processes applied to granular synthesis. However, there are many different ways in which algorithmic processes can enhance timbre. Chaos and stochastic systems have enhanced spectral synthesis as well as physical modeling synthesis methods. These systems applied to acoustic science in this way are effective tools for sound design.

This aspect is an extension of the notion of complexity; an algorithm applied to the microsound of digital synthesis is rarely perceptible in its original form. Rather, algorithmic systems generate interesting and complex timbres in and of themselves, without immediate recognition.

4. Conclusion

Given a restricted domain of music that is made using algorithms and/or generative processes, it is possible to identify three aesthetic considerations common to the most successful works of the genre.

First, an algorithmic or generative system is rarely sufficient on its own. In the most enduring examples, the composers had significant input to the musical material, reserving some musical parameters for subjective judgment and development. Additionally, many influential composers intervened and modified their systems, either by perturbing them or by bending the outcomes.

Secondly, the systems used must be mapped complexly to musical parameters. Though it is not always the case, this usually means that the algorithms are not easily detected or identified.

Works that fail to reach these first two points, namely employ a lack of human intervention and simplistic mapping systems, create results similar to scientific, auditory display. Though informative in its own right and occasionally interesting sonically, this rarely withstands aesthetic scrutiny.

Thirdly, and intricately dependent on the first two points, the choice of timbre in an algorithmic work plays a significant and immediate role in the relative success of a work. Timbre is dependent on the first two points, because in one possible instantiation, acoustic instruments realize musical material generated by systems. In this case, human performative interpretation fundamentally changes the reception of a work. In some other cases, systems operated on the microsound level, the level of sound synthesis. This complex mapping of system to musical parameter often results in rich, varied, and interesting timbres. However, timbre can be the result of many other compositional decisions. And, its impact in the reception of a work is quite large. Therefore, it requires separate consideration.

Pieces that were used as examples include works by Iannis Xenakis, Lejaren Hiller, and Eduardo Reck Miranda, while the methodologies published by Rick Bidlack, Agostino Di Scipio, Christopher Ariza, and Michael Gogins supported these conclusions.

5. References

[1] C. Ariza, Automata Bending: Applications of Dynamic Mutation and Dynamic Rules in Modular One-Dimensional Cellular Automata. *Computer Music Journal*, 31: 29-49, 2007.

[2] R. Bidlack, Chaotic Systems as Simple (But Complex) Compositional Algorithms. *Computer Music Journal*, 16: 22-47, 1992.

[3] A. Di Scipio, On Different Approaches to Computer Music as Different Models of Compositional Design. *Perspectives of New Music*, 33: 360-402, 1995.

[4] L. Goehr, *The Imaginary Museum of Musical Works: An Essay in the Philosophy of Music*. Oxford: Clarendon Press, 1992.

[5] M. Gogins, Iterated Functions Systems Music. *Computer Music Journal*, 14: 40-48, 1991.

[6] E.R. Miranda and J.A. Biles, eds. *Evolutionary Computer Music*. 2007, Springer-Verlag: London.

[7] I. Xenakis, *Formalized Music: Thought and Mathematics in Composition*. Revised ed. Harmonologia Series No. 6. Stuyvesant, NY: Pendragon Press, 1992.

Optimizing nonlinear projective noise reduction for the detection of planets in mean-motion resonances in transit light curves

N. Jevtic[1,2], J. S. Schweitzer[2] and P. Stine[1]

[1]*Bloomsburg University, 400 second St, Bloomsburg, PA 17815, USA*
[2]*Physics Department, University of Connecticut, Storrs, CT 06269, USA*
Email: njevtic@bloomu.edu

Abstract: In the search for extrasolar planets in mean-motion resonance using the transit method, the delay in time-delay phase space reconstruction may be chosen to either lower noise at maximum depth or, to lower noise where one looks for planets in resonance, as the transits begin and end. The trade-offs in the selection of the delay will be discussed for the case of a simulated light curve.
Keywords: extrasolar planets, transit method, mean-motion resonance, nonlinear phase-space time series analysis, nonlinear projective noise reducation, optimal time delay.

1. Introduction

One of the most exciting recent developments has been the detection of extrasolar planet candidates. As of the writing of this paper, the tally stands at 463 planet candidates around other stars. The largest number has been identified using the radial velocity method. Using this method, over 40 multiple planet systems have been detected, with the star 55 Cancri boasting 5 planets. A conservative estimate is that at least 10% of these star systems are in mean-motion resonance [8]. Some estimates go as high as 30% with the majority of these systems in 2:1 resonances [7]. However, in the radial velocity method, if several planets are present, their signals will appear combined in the radial motion of the star, leading to potential misinterpretations of the data. Specifically, two planets in 2:1 resonant orbits can mimic the signal of a single planet in an eccentric orbit [1].

With the launch of the COROT space telescope and, more recently, the US Kepler space telescope that detect the presence of planets using the transit method, long-term uninterrupted light curves have become available. We discuss the potential of nonlinear time-delay time series analysis as a tool to detect planets in 2:1 mean-motion resonance through the optimization of choice of time delay. The choice of time delay in time-delay phase space reconstruction has been traditionally taken at the minimum of the average mutual information, the information theory analog of the autocorrelation function. We show that deviation from this prescription for the delay in time-

Chaos Theory: Modeling, Simulation and Applications
C. H. Skiadas, I. Dimotikalis and C. Skiadas (Eds)

delay phase space reconstruction allows us to either lower noise at maximum depth of the transit curve or, where one would actually look for planets in resonance, as the transits begin and end. The trade-offs in the selection of the delay will be discussed for the case of a simulated light curve modelled after the transit light curve of HD 209458b [2].

2. Transit Method for Detecting Mean-Motion Resonances

2.1. *The transit method for detecting extrasolar planets*

The transit method involves measuring the decrease in the brightness of a star as a planet passes in front of it. When observing stars the size of the Sun, the decrease in brightness for giant planets, such as Jupiter or Saturn is approximately 1%. For Earth-size planets it is about 0.01%. However, even from space, host-star light curves are contaminated with noise which makes such detections difficult.

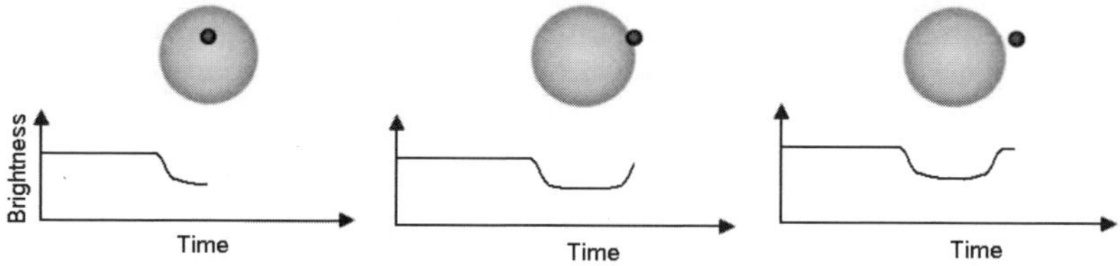

Figure 1. The transit method for detecting extrasolar planets: The star's brightness decreases during the planet's transit. (After ITA, University of Oslo, Norway.)

2.2. *Mean-motion resonance*

In an orbital resonance two orbiting bodies exert a regular, periodic gravitational influence on each other, which greatly enhances the net gravitational effect. A mean-motion orbital resonance (MMR) occurs when two bodies have periods of revolution that are a simple integer ratio of each other. In our solar system, Jupiter's moons, Io, Europa and Ganymede are an example of a Laplace resonance with orbital periods that are related as 4:2:1, with each successive neighbouring pair being in a mean-motion resonance. The extrasolar planets Gliese 876b and Gliese 876c are an example of an extrasolar 2:1 orbital resonance [8].

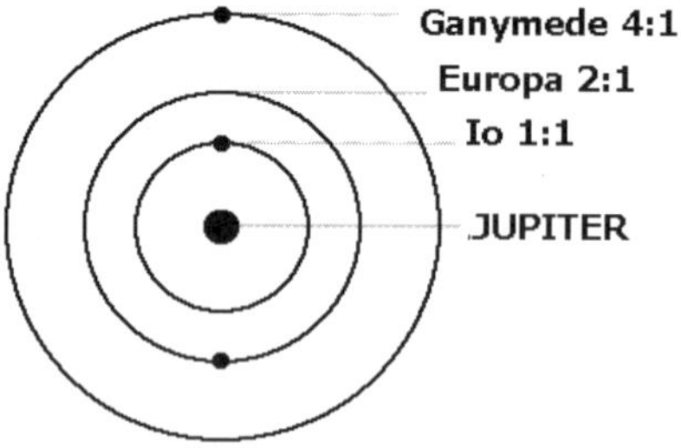

Figure 2. The Galilean moons of Jupiter: a) Io and Europa and b) Europa and Ganymede are in 2:1 resonances

3. The Methodology

3.1. *Phase-space portrait*

A simulated time series of two planets in MMR will be used to obtain a time -delay phase space portrait. According to the embedding theorems [9, 11, 12], the geometry of this time-delay phase-space portrait is diffeomorphic to the geometry of the phase-space representation that would be obtained if the equations governing the system were known. Thus this object is a surrogate and can be studied in lieu of the system.

3.2. *Parameters of the reconstruction*

Using the False Nearest Neighbour (FNN) method the uniformly sampled time series yields the dimension of the reconstructed phase space.

The optimal time delay is chosen at the minimum of the average mutual information (AMI) [3] which is an information theory analog to the autocorrelation function. It is expressed in terms of probabilities as:

$$\mathbf{AMI}(\tau) = \sum_{s(t),s(t+\tau)} \mathbf{P}(s(t),s(t+\tau))\log_2\left[\frac{\mathbf{P}(s(t),s(t+\tau))}{\mathbf{P}(s(t))\mathbf{P}(s(t+\tau))}\right]$$

Here **s** is the sampled scalar time series and τ is the time delay. However, whether the first, local minimum, or the global minimum determine the optimal delay is still a point open to discussion.

3.3. *Locally-projective nonlinear noise reduction*

The measured data are assumed to be the output of a low-dimensional dynamical system and random high-dimensional noise. This noise spreads the data off its manifold in phase space. For small amplitude noise, it is assumed the data are distributed closely around the manifold [4, 5].

It is therefore important to first identify this manifold and then to orthogonally project the data points onto it in order to lower the noise. For a trajectory whose delay vectors $\mathbf{a}_n$ lie inside a *m*-dimensional manifold ***M*** **"**, and the dynamical system with noise forms a *q*-dimensional manifold ***M,*** we over-embed and repeat the process until, due to iteration, the average correction ceases to decrease rapidly with each successive iteration.

The method is detailed in Jevtic et al. [6].

4. The Simulated Data

A simulated light curve modelled after the transit light curve of HD 209458b [2] was used, Figure 3. Transit light curves usually span hours while planet orbital periods are on a scale of days, weeks or months. Such intervals between transits are too long to be properly handled by nonlinear time series analysis so a string of transits are added together to generate a light curve.

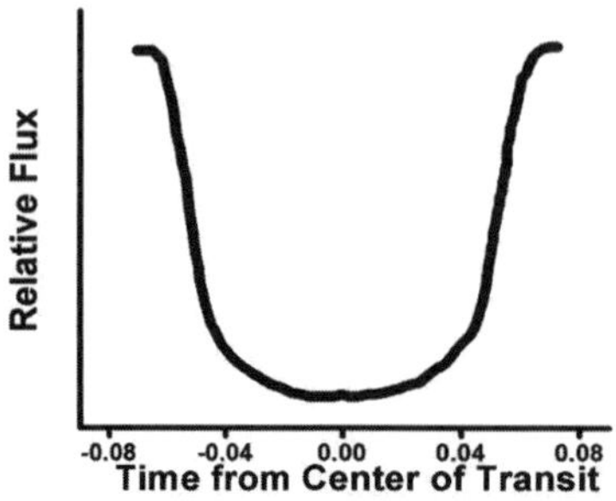

Figure 3. Light curve of a simulated transit of one planet

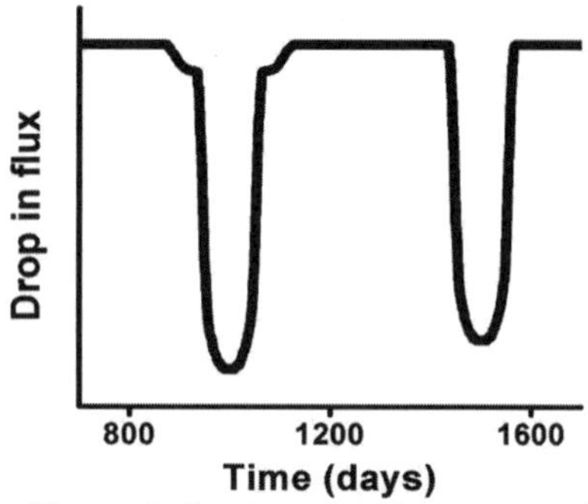

Figure 4. Section of a light curve with a mean-motion resonance (left) and "normal" single transit (right). The times between transits have been made shorter in respect to the duration of real transits and orbital periods.

We assume the existence of a second, slower moving companion in a 2:1 MMR, which results in a 10% increase in transit depth. The light curve consists of alternating "normal" single transits and transits due to both planets (Figure 4.).

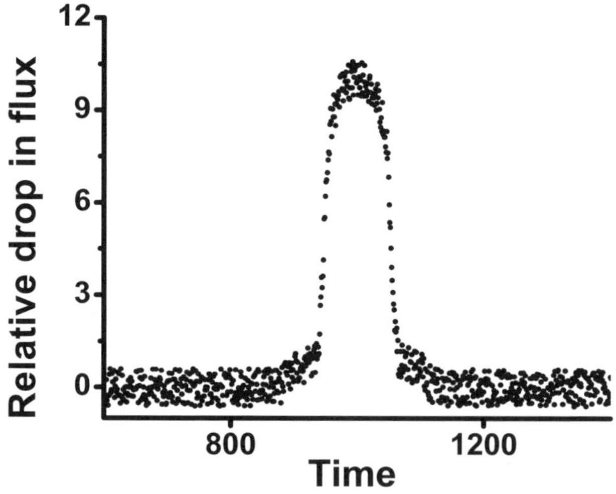

Figure 5. Transit of 2:1 MMR with 10% zero-mean random noise added. The contribution of the outer companion is now almost completely lost in the noise.
Note: This graph is inverted in respect to previous figures.

Ten percent zero-mean random white noise is added to this light curve to simulate real measurements (Figure 5.).

5. Results

Previous work [6] supplied the rationale for the nonlinear noise reduction. It was shown in [6] that in the case of a fundamental (inner planet) and a second harmonic (outer planet), a surface in phase space can easily be identified. The goal was to explore whether, for a simulated light curve with 10% noise, the small changes due to a planet in 2:1 MMR can be detected.

5.1. *Average mutual information*

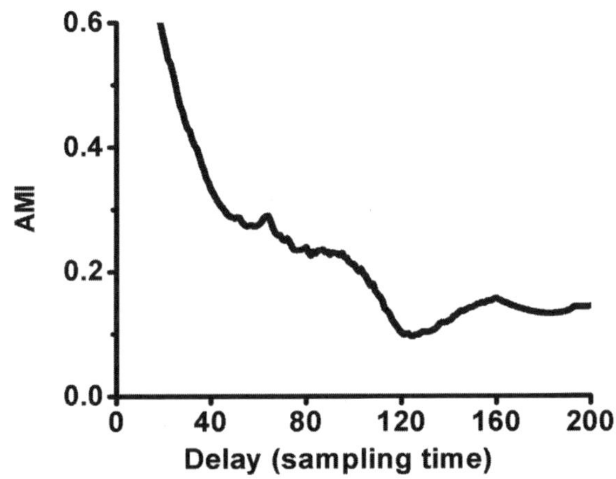

Figure 6. Average Mutual Information (AMI): First Local Minimum is at a delay of 56, the Global Minimum is at a delay of 125

The optimal time-delay for phase-space reconstruction is at the minimum of the average mutual information (AMI) which is an information-theory nonlinear analog to the autocorrelation function. For our simulated data (Figure 6.) the first local minimum is at a delay of 56, and the global minimum is at a delay of 125.

5.2. *Choice of global-minimum delay optimizes noise reduction at maximum transit depth*

The depth of the light curve transit is important for determining the ratio of the radii of the planet and the host star. The time delay at the AMI global minimum resulted in significantly lower noise at maximum transit depth (Figure 7.).

5.3. *Choice of local AMI minimum time delay optimizes baseline noise reduction*

In Figure 8 the shorter time delay at the first minimum of the AMI yields a cleaner baseline. The contribution of the second member of the system is more pronounced and symmetrical

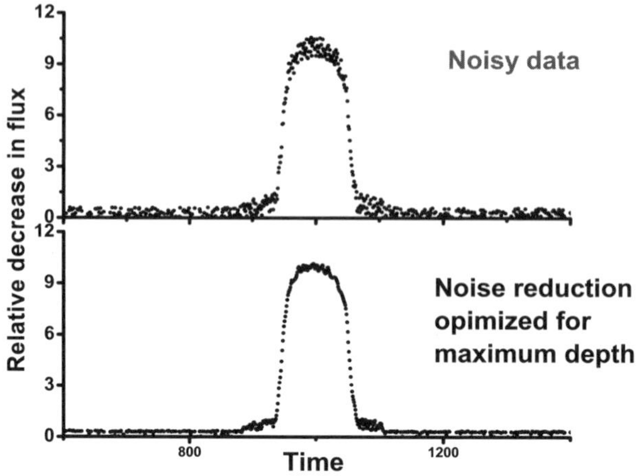

Figure 7. Delay taken at the global minimum of the AMI yields a clean depth of transit but it also yields a separation of the noise in the non-transit sections of the light curve into two levels due to the binary nature of depth values, resulting in a non-unique identification of the noisy and non-symmetric pedestal due to the second planet in MMR.

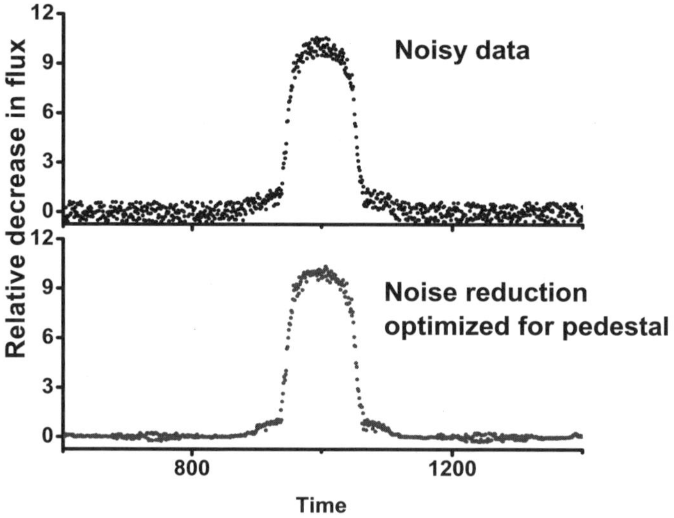

Figure 8. The shorter time delay at the first minimum of the AMI yields a cleaner baseline and easier identification of a pedestal. The contribution of the second member of the system has been efficiently recovered.

5.4. *Comparison of optimization*

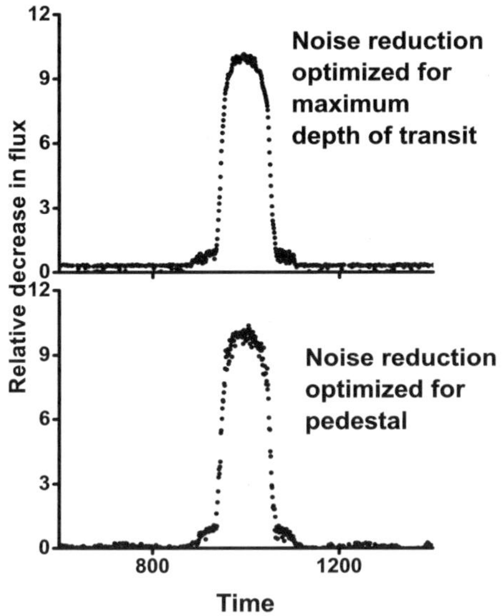

Figure 9. Delay taken at the global AMI minimum yields a clean depth of transit but the identification of a pedestal is difficult (top). Delay at first minimum of AMI yields a cleaner baseline and easier identification of a pedestal. Baseline is at 0. (bottom).

The comparison of the transit shapes in Figure 9 indicates that for efficient identification of planets in mean-motion resonance, the local minimum of AMI is the more efficient choice.

6. Conclusions

Nonlinear projective noise reduction preserves information on the dynamics of a system at the higher frequencies allowing for the search for smaller structures. When used to search for planets in mean-motion resonance on long, continuous simulated light curves of a system with a principle planet and a smaller, outer companion and added 10% random noise, nonlinear noise reduction is shown to be:

- Efficient in lowering noise and recovering the lower-level structure due to a second planet in MMR that was obscured by the noise for a delay at the first minimum of average mutual information.
- Adaptable to the search for structures in different regions of the light curve by a simple change of the time delay between that determined by the local to that at the global minimum of the average mutual information.

In the search for extrasolar planets in mean-motion resonance, the delay in time-delay phase space reconstruction may be chosen to either lower noise at maximum depth or where one looks for planets in resonance, as the transits begin and end. Future work will be focused on determining the limits of applicability of the method since transits occur at greater intervals than in our simulated data.

References

[1] G. Anglada-Escude, M. Lopez-Morales, J. E. Chambers, *How eccentric orbital solutions can hide planetary systems in 2:1 resonant orbits,* ApJ. 709:168-178, 2010

[2] T.M Brown et al.: *Hubble space telescope time-series photometry of the transiting planet of HD 209458*, ApJ, 552, 699-709.

[3] A.M. Fraser, H.L. Swinney, *Independent coordinates for strange attractors from mutual information,* PhysRevA,33, 2, 1986, pp.1134-1140

[4] P. Grassberger, , et al., *On noise reduction methods for chaotic data*, CHAOS 3, 127 (1993)

[5] R. Hegger, H. Kantz, and T. Schreiber, *Practical implementation of nonlinear time series methods: The TISEAN package*, CHAOS, 9, 413 (1999)

[6] N. Jevtic et al., *Nonlinear Time Series Analysis of PG1351+489 Light Intensity Curves* ApJ, 635:527-539, 2005

[7] W. Kley, and S. Zsolt, *The Formation of Resonant Planetary Systems*, Extrasolar Planets, Formation, Detection and Dynamics, Ed. R. Dvorak, 2008, Wiley-VCH GmBh&Co, KGaA, Weinheim

[8] Marcy *et al.*; P.R. Butler, D. Fischer, S.S. Vogt, J.J. Lissauer, E. J. Rivera, (2001). *"A Pair of Resonant Planets Orbiting GJ 876"*, *ApJ,* 556 (1): 296–301.

[9] H. Rein and J. C. B. Papaloizou, *Mean-Motion Resonances of High Order in Extrasolar Planetary Systems,* EAS Publications Series, Volume 42, 2010, .299-302

[10] J. Stark, *Delay Embeddings for Forced Systems I: Deterministic Forcing,* Journal of Nonlinear Science, 9, 255-332, 1999

[11] J. Stark et al. *Delay Embeddings for Forced Systems II: Stochastic Forcing*, Journal of Nonlinear Science, 13, 519, 2003.

[12] Takens-Mane, *Detecting Strange Attractors in Turbulence,* Dynamical Systems and Turbulence, 898, Lecture Notes in Mathematics (Springer-Verlag, 1981)

Acknowledgement

The Max Planck Institute for Complex Systems for the TISEAN software: R. Hegger, H. Kantz, and T. Schreiber, *Practical implementation of nonlinear time series methods: The TISEAN package*, CHAOS, 9, 413 (1999)

Dissipative solitons: Perturbations and chaos formation

Vladimir L. Kalashnikov

Instutute for Photonics, Technical University of Vienna, Vienna, Austria
Email: kalashnikov@tuwien.ac.at

Abstract: The perturbations of chirped dissipative solitons are analyzed in the spectral domain. It is shown, that the structure of the perturbed chirped dissipative soliton is highly nontrivial and has a tendency to an enhancement of the spectral perturbations especially at the spectrum edges, where the irregularities develop. Even spectrally localized perturbations spread over a whole soliton spectrum. As a result of spectral irregularity, the chaotic dynamics develops due to the spectral loss action. In particular, the dissipative soliton can become fragmented though remains localized.
Keywords: dissipative soliton, complex nonlinear Ginzburg-Landau equation, perturbation theory, chaotic soliton dynamics.

1. Introduction

The complex nonlinear Ginzburg-Landau equation (CGLE) has so broad horizon of application, that the concept of "the word of the Ginzburg-Landau equation" [1] has become broadly established. The CGLE demonstrates its effectiveness in quantum optics, condensate-matter physics, study of non-equilibrium phenomena and nonlinear dynamics, etc. In optics and laser physics, the CGLE provides an adequate description of ultra-short pulse formation and dynamics [2,3].

The CGLE is multiparameter and not integrable in a general form. Hence, a study of its solutions needs extensive numerical simulations. The exact analytical solutions are known for a few of cases, when they represent the solitary waves (dissipative solitons) [2]. As a rule, one presumes some class of functional expressions to construct such a restricted class of solutions. As a result, the solutions beyond a given class are missed. Moreover, the issues of stability and dynamical properties of the analyzed dissipative solitons remain open. These challenges stimulate interest in the approximate methods of integration of the CGLE [4] and the study of the dissipative soliton perturbations. In many cases, an excitation of such perturbations can result in the chaotic soliton dynamics [5,6].

A physically important sector, which permits an approximate analysis, is represented by the chirped dissipative solitons (CDSs) of the CGLE. The very important property of CDS is that such a soliton is energy scalable [7].

Chaos Theory: Modeling, Simulation and Applications
C. H. Skiadas, I. Dimotikalis and C. Skiadas (Eds)

This allows obtaining the femtosecond laser pulses with about of and over microjoule energies [8]. A large chirp causes the spectral extra-broadening of CDS and an appearance of spectral perturbations [5,6,9]. As a result, the chaotic dynamics of CDS (so-called, chaotic mode locking) can develop.

In this work, I present the extension of the approximate analytical technique [4,10,11,12] to analysis of CDS of the perturbed complex nonlinear CGLE. Three main sources of perturbations will be considered: i) higher-order (quintic) nonlinearity, ii) higher-order dispersions, and iii) narrowband absorption. As will be shown, the perturbations under consideration distort the CDS spectrum that affects the soliton dynamics and, in particular, leads to the chaos appearance.

2. Perturbative analysis in spectral domain

To be obeying the CGLE, the electromagnetic field with the amplitude $A(z,t)$ (t is the local time, and z is the propagation coordinate) has to be described on basis of the slowly varying amplitude approximation, that is provided by the relation $\omega_0 >> 1/T$ (ω_0 is the field carrier frequency, and T is the soliton width). When one can neglect the field variation along a cavity round-trip (for an oscillator) or the variation of material parameters along a fiber (for intra-fiber propagation), the field dynamics can be described on basis of the perturbed cubic CGLE [2]

$$\frac{\partial A(z,t)}{\partial z} = [-\sigma + (\kappa - i\gamma)P(z,t)]A(z,t) + \\ + (\alpha + i\beta)\frac{\partial^2}{\partial t^2}A(z,t) + \Gamma\{A(z,t)\}. \tag{1}$$

Here $P=|A|^2$ is the instant field power, α is the squared inverse gain bandwidth, β is the group delay dispersion (GDD) parameter (the case of $\beta>0$ corresponds to a normal dispersion and is under consideration in this work), and σ is the net-loss parameter. The nonlinear terms in Eq. (1) describe i) self-amplitude modulation with the parameter κ, and ii) self-phase modulation with the parameter γ. Γ is the perturbation, which functional form depends on the type of perturbation (see below).

The unperturbed Eq. (1) has the exact CDS solution in the form [2]

$$A(z,t) = \sqrt{P_0}\,\mathrm{sech}(t/T)^{1-i\psi}\exp(-iqz), \tag{2}$$

where T is the soliton width, P_0 is the soliton peak power, ψ is the chirp, and q is the soliton wavenumber. However, this exact solution is not promissory for a further perturbative analysis because it results from the assumed restriction on the CDS phase profile and, as a result, the constraint on β. Moreover, such a form of solution has the Fourier image, which is expressed in terms of the beta function, that complicates an anlysis in the spectral domain. Therefore it is convenient to use the technique developed in [4,10-13]. This technique allows obtaining the approximate expression for the Fourier image of (2)

$$a(\omega) \approx \sqrt{\frac{6\pi\beta}{\kappa(1+c)}} \exp\left[\frac{3i\gamma\omega^2}{2\kappa(1+c)(\Delta^2-\omega^2)}\right] \mathrm{H}(\Delta^2-\omega^2), \tag{3}$$

where ω is the deviation from ω_0, $c = \alpha\gamma/\beta\kappa$, $\Delta^2 = 3\sigma c/\alpha(2-c)$, and H is the Heaviside function. Other CDS parameters are $\psi = 3\gamma/\kappa(1+c)$, $T = 3\gamma/\kappa\Delta(1+c)$, and $P_0 = \beta\Delta^2/\gamma$. The solution (3) can be easily generalized for the more complicated nonlinear terms in the CGLE [4,10,13]. Another advantage of the solution (3) is that it allows developing an analytical perturbative theory of the solitonic sector of Eq. (1) in the spectral domain.

The equation for the Fourier image $f(\omega)$ of the small perturbation of $a(\omega)$ can be written as [2]

$$\begin{aligned} &[q-k(\omega)]f(\omega)+\frac{1}{\pi}\int_{-\infty}^{\infty}U(\omega-\omega')f(\omega')d\omega'+ \\ &+\frac{1}{2\pi}\int_{-\infty}^{\infty}V(\omega-\omega')f^*(\omega')d\omega' = S(\omega), \end{aligned} \tag{4}$$

where the linear wave wavenumber is $k(\omega) = \beta\omega^2 + i(\sigma + \alpha\omega^2)$, the kernels U and V are the Fourier images of $|a(t)|^2$ and $a(t)^2$, respectively ($a(t)$ is the time-dependent part of (2)). The source term $S(\omega)$ is the Fourier image of $\Gamma\{a(t)\}$ multiplied by i. Further, one may assume a phase matching between the soliton and its perturbation. This assumption allows U=V and Eq. (4) can be solved through the Neumann series

$$f_n(\omega) = \frac{S(\omega)}{q-k(\omega)} - \frac{3}{\pi[q-k(\omega)]}\int_{-\infty}^{\infty}U(\omega-\omega')f_{n-1}(\omega')d\omega', \tag{5}$$

where f_n is the n-th iteration and $f_0 = S(\omega)/[q-k(\omega)]$. The applications of the considered technique will be considered in the next section.

3. Perturbations of CDS

Let us consider some basic perturbation affecting the CDS. It will be seen, that the spectral perturbations can become irregular that induces a chaotic dynamics due to a feedback caused by the spectral dissipation.

3.1. Quintic nonlinearity

Let the quintic nonlinear term in Eq. (1) be $\Gamma\{A(z,t)\} = -\kappa\zeta P(z,t)^2 A(z,t)$. The tabletop spectrum of exact numerical solution is shown in Fig. 1 by solid black curve. The analytical approximations are shown by dashed and gray curves. They are very close to the numerical one, but have abrupt edges due to the Heaviside function in (3). One can conclude that the quintic nonlinear term can be treated as a perturbation in the case under consideration.

The successive terms of (5) are shown in Fig. 2. One can see that the tabletop spectrum becomes convex and some oscillating substructure develops. The last phenomenon explains the results of [5,6], where the cause of both regular and chaotic pulsations of CDS has been attributed to an excitation of the solitonic internal modes.

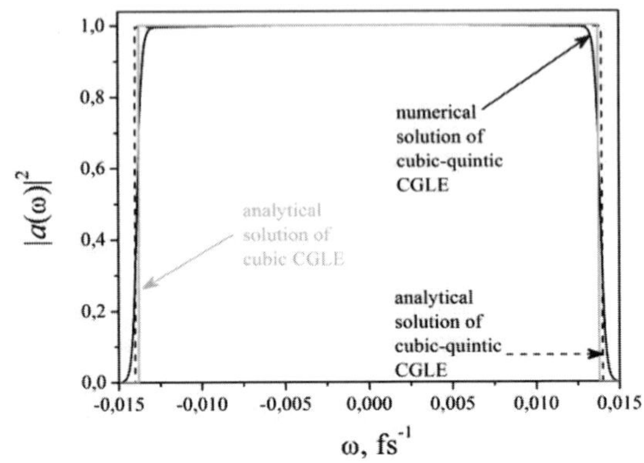

Figure 1. CDS spectra for i) numerical (solid black curve) and ii) approximate analytical solutions of the cubic-quintic CGLE (dashed black curve), as well as iii) the cubic CGLE (solid gray curve). Parameters correspond to a Cr:ZnSe mode-locked oscillator: α=16 fs^2, β=250 fs^2, κ=0.04γ; ζ=0.2γ (black curves) and 0 (gray curve).

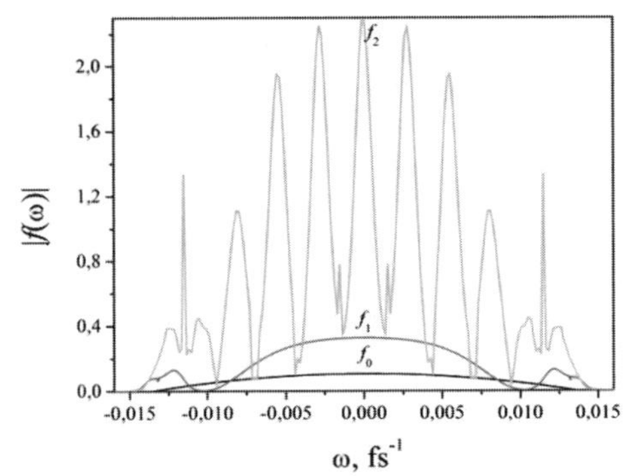

Figure 2. Absolute values of the successive terms in the Neumann series (5) for the quintic nonlinear perturbation.

3.2. Higher-order dispersions

Let us consider the third-order dispersion (TOD) term in Eq. (1), which can be expressed as $\Gamma\{A(z,t)\} = \theta\partial^3 A(z,t)/\partial t^3$ (θ is the TOD coefficient). The perturbed spectra (in zero- and first-orders of (5)) are presented in Fig. 3. An important feature of the perturbed CDS in presence of higher-order dispersion is an appearance of oscillating substructure growing to the spectrum edges. The phase changes dramatically in the vicinity of $\pm\Delta$ (Fig. 4) and this domain is very sensitive to perturbations due to their parametric amplification expressed by the term $1/[q-k(\omega)] \propto 1/(\Delta^2-\omega^2)$ in Eq. (5).

The induced substructure is fine and has a tendency to the growth of irregularities. On the other hand, the edges of spectrum are subjected to the maximum spectral loss described by the term $\alpha\partial^2/\partial t^2$ in Eq. (1). As a result of such loss, the spectral irregularities nearby the spectrum edges cause the irregular pulsations of soliton and, eventually, the chaotic mode-locking (Fig. 5), which has been reported in [6,9]. When the higher-order dispersions increase, the perturbations approach the spectrum center. Then the CDS becomes fragmented but remains localized (Fig. 5).

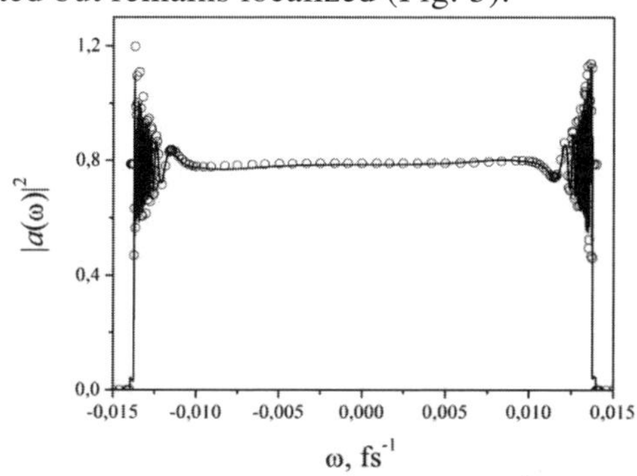

Figure 3. CDS spectra in presence of the TOD (θ=300 fs^3) in zero- (curve) and first- (circles) orders of (5).

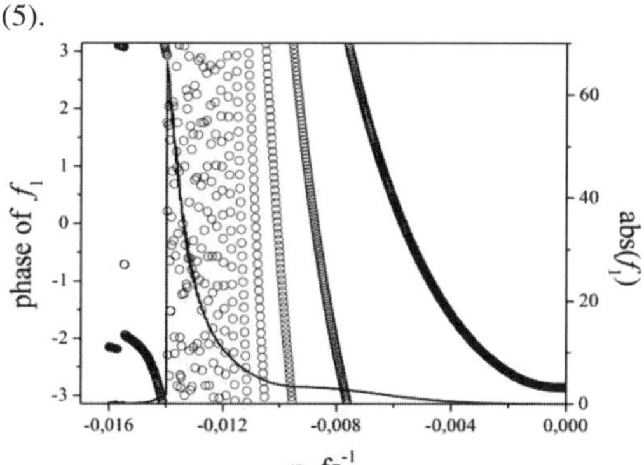

Figure 4. Absolute value (curve) and phase (circles) of f_1 corresponding to Fig. 3.

Another way to interpret the chaotic behavior in the presence of higher-order dispersions is to consider it as a result of the dispersive wave generation. Such a generation appears if the resonance condition is satisfied: $k(\omega)\equiv\beta\omega^2+\sum_{l=3}\theta_l\omega^l=q$. The resonant frequency ω_r providing such an condition is shifted inside the CDS spectrum (i.e. $|\omega_r|<\Delta$, Fig. 6) if the TOD parameter $\theta\neq0$. The source term $S(\omega)$ for such a perturbation is $\propto\theta\omega^3\,\mathrm{H}(\Delta^2-\omega^2)$ and increases towards the spectrum edges (Fig. 6) (unlike the case of the Schrödinger soliton [2]).

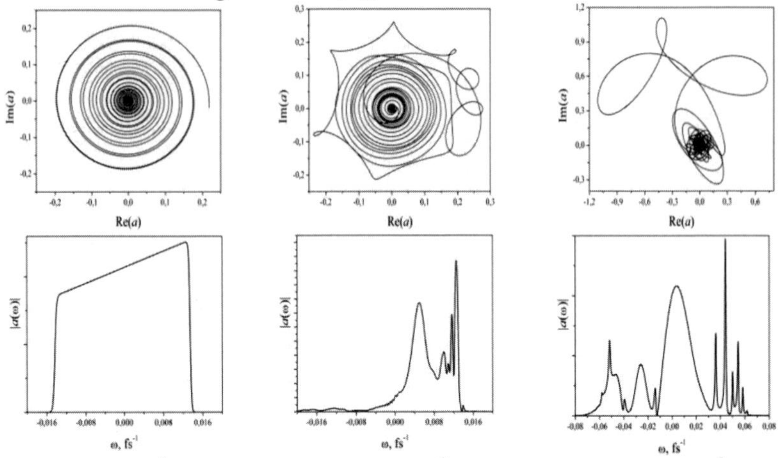

θ=1200 fs^3 θ=12000 fs^3 θ=15000 fs^3

Figure 5. Soliton chaotization with the TOD growth: phase portraits and spectra.

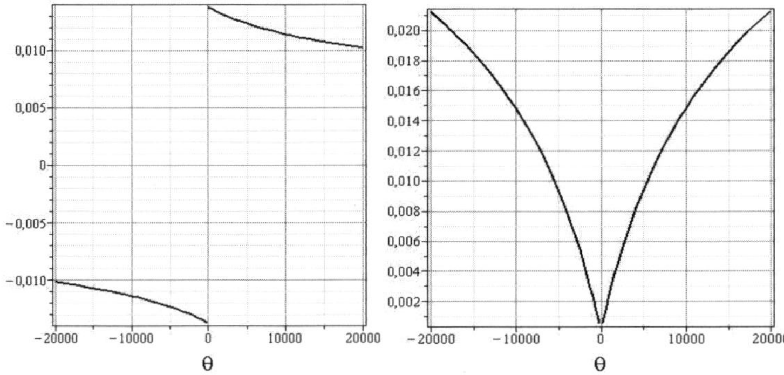

Figure 6. Left picture: resonant frequency ω_r [fs^{-1}] in dependence on θ [fs^3]; right picture: $\theta\omega_r^3$ in dependence on θ.

3.3. Narrowband losses: A gas filled oscillator or an impured fiber

Let us consider the case of solitonic propagation within a medium filled with impurities consisting of the l independent Lorentz lines centered at ω_l with the loss coefficients ε_l<0 and the linewidths Ω_l. Then

$$S(\omega) = ia(\omega)\sum_l \varepsilon_l \frac{1 - i(\omega - \omega_l)/\Omega_l}{1 + [(\omega - \omega_l)/\Omega_l]^2} \quad . \tag{6}$$

For a single line centered at $\omega_1 = \omega_0$, the central part of the CDS spectrum is shown in Fig. 7 for $a(\omega)$ corresponding to the approximate solutions of the cubic and cubic-quintic CGLE perturbed by zero and first terms of the Neumann series (5). One can see, that the perturbed spectral profile reproduces the real part of an

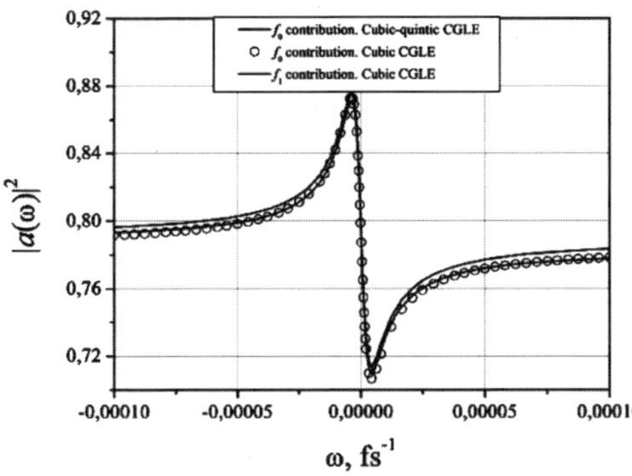

Figure 7. Central part of the CDS spectrum perturbed by a narrowband loss (Ω_1=4 GHz, ε_1=–0.005).

absorber permittivity like the case of the Schrödinger soliton [14]. However, in contrast to the perturbed Schrödinger soliton, the CDS perturbation spreads along the soliton spectral profile (Fig. 8; parameters of Fig. 7). That is the perturbation is not strongly bounded in the vicinity of ω_l and the higher-order terms of the Neumann series contribute. As a result, the CDS becomes perturbed also through the spectral loss growing towards the spectrum edges as it was in the case of perturbations induced by higher-order dispersions (see above). Thus, the spectrally localized perturbation distorts a whole CDS spectrum.

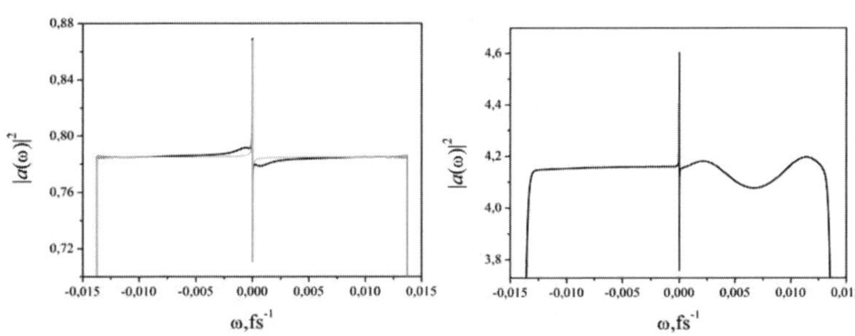

Figure 8. Left picture: the analytical CDS spectra (the cubic CGLE) perturbed by f_0 (gray) and f_1 (black). Right picture: the numerical spectrum (the cubic-quintic CGLE). $|a|^2$ is given in arbitrary units.

4. Conclusions

In this paper, the perturbative analysis of chirped dissipative solitons has been developed in the spectral domain. Three main types of perturbations have been considered: i) higher-order nonlinearity, ii) higher-order dispersions, and iii) narrowband losses. Unlike the perturbed Schrödinger soliton, the CDS has a nontrivial internal structure, which is strongly affected by perturbations. Even spectrally localized perturbations affect a whole CDS spectrum. Most sensitive parts of the CDS are its spectral edges, where the perturbations are resonantly enhanced and the irregularity develops. Since the spectral loss increases towards the spectrum edges, such a spectral irregularity becomes transformed into the irregular CDS dynamics.

Acknowledgements This work was supported by the Austrian Fonds zur Förderung der wissenschaftlichen Forschung, FWF project P20293.

References

[1] I.S.Aranson, L.Kramer. The world of the complex Ginzburg-Landau equation. *Rev. Mod. Phys.*, 74:99, 2002.

[2] N.N.Akhmediev, A.Ankiewicz. *Solitons: Nonlinear Pulses and Beams.* Chapman&Hall, 1997.

[3] F.X.Kärtner, Ed. *Few-cycle Laser Pulse Generation and its Applications.* Springer Verlag, Berlin, 2004.

[4] V.L.Kalashnikov. Chirped dissipative solitons of the complex cubic-quintic nonlinear Ginzburg-Landau equation. *Phys. Rev. E*, 80:046606, 2009.

[5] V. L. Kalashnikov, A. Chernykh. Spectral anomalies and stability of chirped-pulse oscillators. *Phys. Rev. A*, 75:033820, 2007.

[6] V.L.Kalashnikov. Chaotic mode-locking of chirped-pulse oscillators, in *Proc. 2nd Chaotic Modeling and Simulation International Conference*, 2009.

[7] V.L.Kalashnikov, E.Podivilov, A.Chernykh, S.Naumov, A.Fernandez, R.Graf, A.Apolonski. Approaching the microjoule frontier with femtosecond laser oscillators. Theory and comparison with experiment. *New Journal of Physics*, 7:217, 2005.

[8] S.Naumov, A.Fernandez, R.Graf, P.Dombi, F.Krausz, A.Apolonski. Approaching the microjoule frontier with femtosecond laser oscillators. *New Journal of Physics*, 7:216, 2005.

[9] V.L.Kalashnikov, A.Fernández, A.Apolonski. High-order dispersion in chirped-pulse oscillators. *Optics Express*, 16:4206, 2008.

[10] E.Podivilov, V.L.Kalashnikov. Heavily-chirped solitary pulses in the normal dispersion region: new solutions of the cubic-quintic complex Ginzburg-Landau equation. *JETP Letters*, 82:467 (2005).

[11] V.L.Kalashnikov, E.Podivilov, A.Chernykh, A.Apolonski. Chirped-pulse oscillators: theory and experiment. *Applied Physics B*, 83:503, 2006.

[12] V.L.Kalashnikov, A. Apolonski. Chirped-pulse oscillators: A unified standpoint. *Phys. Rev. A*, 79:043829, 2009.

[13] V.L.Kalashnikov. The unified theory of chirped-pulse oscillators, in *Proc. SPIE*, M. Bertolotti, Ed., 7354:73540T, 2009.

[14] V.L.Kalashnikov, E.Sorokin, J.Mandon, G.Guelachvili, N.Picque, I.T.Sorokina. Femtosecond lasers for intracavity molecular spectroscopy, in *Conference Abstract Vol. 3rd EPS-QEOD Europhoton Conf. on Solid-State,Fiber and Waveguided Light Sources*, 32G: TUoA.3, 2008.

Modified chaotic shift keying using indirect coupled chaotic synchronization for secure digital communication

Rupak Kharel[1], Krishna Busawon and Z. Ghassemlooy

University of Northumbria, Newcastle Upon Tyne,UK
Email: [1]rupak.kharel@unn.ac.uk

Abstract: In this paper, a modified chaotic shift keying method is proposed to transmit digital bits securely over a communication channel. The scheme is based upon encrypting the digital bits 0 and 1 into infinite levels by applying the keystream such that there is no recognisable pattern in the encoded transmitted signal. The encoded transmitting signal generated is shown to resist popular attack method therefore realizing a secure and trustworthy digital communication system.
Keywords: chaotic synchronization, chaotic shift keying, secure communication, return maps, encryption, chaotic simulation.

1. Introduction

For the last decade or so, the use of chaotic signals for transmitting messages securely over a channel have attracted a great deal of attention amongst researchers [1]. Chaotic signals have properties such as aperiodicity, sensitivity to initial conditions/parameter mismatches and have fundamentally broadband like nature. The main idea of using chaotic signal in secure communication is to bury the message spectrum/plaintext into the broad chaos spectrum; equivalent to burying inside the noise. At the receiver, chaotic synchronization is performed and the plain text message is revealed by performing some sort of inversion. Different methods are available in the literature which are Chaotic Masking, Chaotic Modulation, Chaotic Inclusion technique and Chaotic Shift Keying (CSK) [1]. Variations of these methods are also proposed but almost all of them fall on one or more of the above categories.

Chaotic masking is the first method proposed and is the simplest of all where the chaotic signal acts as a mask on which the message/plaintext is added to form the transmitted chaotic signal. In chaotic modulation, the message is used to modulate some parameter(s) of the chaotic system. In the so-called chaotic inclusion technique, the message is used to change the chaotic attractor directly at the phase space. The CSK method was proposed for transmitting binary signals and is fundamentally a special case of chaotic modulation. The transmitting bit either 0 or 1 determines the value of the parameter to be switched of the chaotic systems. At the receiver, chaotic synchronization is performed by setting parameter of the chaotic system equivalent to one of the bit value. By thresholding the synchronization error at the receiver, the transmitted binary value can easily be determined.

Chaos Theory: Modeling, Simulation and Applications
C. H. Skiadas, I. Dimotikalis and C. Skiadas (Eds)

Chaotic masking was shown to be insecure by various methods including Short's, return map analysis [1, 2]. The insecurity was shown for other methods, in particular, for CSK [2, 3], parametric modulation [2], and inclusion method [2]. Modification to these methods [4, 5] have also been proposed but to no avail [6, 7]. Other proposed methods based on the projective synchronization [8], phase synchronization [9], generalized synchronized were broken as well [10, 11]. Methods based on the time delay or the hyperchaos were also sought for increasing the security but they too were not entirely convincing (see eg. [12, 13]). Therefore, there is still a need to improve the existing methods if secure communication is to be realized using chaotic signals.
In this paper, we propose a methodology for improving the traditional CSK method by using the indirect coupling synchronization of the chaotic oscillators as developed in [14]. The main purpose of using indirect coupled synchronization is to generate synchronized keystream in the transmitter and receiver side. The binary signal to be transmitted will be modulated by an encrypted chaotic parameter using the keystream and encryption function into infinite levels between a given interval. This brings a considerable improvement on the traditional CSK method. In particular, we show that the proposed modified CSK method resists powerful decryption attacks such as return maps based attacks.
This paper is organized as follows. In Section 2, the modified CSK scheme for secure communication will be proposed. In Section 3, the implementation of the proposed technique using Lorenz and Chua's oscillator will be shown and simulation results to show successful message extraction will be presented. In Section 4, cryptanalysis of the proposed method using return map analysis will be discussed. Finally some concluding remarks will be made on Section 5.

2. Problem Statement

Consider a chaotic system with parameter β and output y described as follows:

$$\dot{x} = f(x, y, \beta)$$
$$y = h(x)$$

where $x \in R^n$ and $y \in R$. The function f and h are assumed to be smooth.
In CSK, the basic idea is to switch the parameter β of the chaotic system into two values depending on either 0 or 1 is to be transmitted, as follows:

$$\beta = \beta_m = \begin{cases} \beta_0 \text{ when } m = 0 \\ \beta_1 \text{ when } m = 1. \end{cases}$$

Upon receiving the modulated chaotic signal, the receiver is set with either of the parameters β_0 or β_1. Now, when the binary value m=0 is transmitted (β_0 used in receiver), synchronization is achieved; otherwise, synchronization is not possible, which constitute the basis for recovering the transmitted bits. Generally, in practice, some specific pattern associated with the transmitted chaotic signal can be observed when different binary values are sent. This is mainly due to some sort of switching involved in the process. Recovering the bits for intruders can simply amount to a classification problem, which can easily be done by methods such as return maps as explained in [2] or artificial neural network as in [3]. Thus, although CSK provides possibly the best way to transmit binary message using chaotic signals, it is not at all secure. Variations of CSK method have been reported such as in [15] but were not entirely convincing as shown in [16]. Therefore, in this paper, an effort is made to remove or blur any sort of pattern in the transmitted chaotic signal due to the switching of binary values.

3. Proposed Modified CSK Method

The proposed modified CSK method is illustrated in Figure 1.
We consider the following system to be used as transmitter

$$(\mathrm{T}):\begin{cases}\dot{x}=f(x,y_1,\beta_m)\\ y_1=g_1(x)\\ y_2=g_2(x)\end{cases}\quad \text{where } \beta_m=\begin{cases}\beta_0=\beta+e(0,k) \text{ if } m=0\\ \beta_1=\beta+e(\rho,k) \text{ if } m=1\end{cases}$$

and $x\in R^n$ and $y\in R$. The function f, g_1 and g_2 are assumed to be smooth.

Here β_m is the parameter to be modulated and $e(.)$ is an encryption algorithm, ρ is a constant scalar and k is the keystream for performing encryption. The function $e(.)$ is chosen such that its values (or image), $e(z,k)$, falls within the interval $[-h, h]$ where h is an encryption parameter. Therefore, β_m will always be in the interval $[-h+\beta, h+\beta]$. With the proper choice of h, it can be ensured that β_m always falls within the range such that (T) remains chaotic. y_1 is the transmitted signal for synchronization to the receiver described by:

$$(\mathrm{R}):\begin{cases}\dot{\hat{x}}=f(\hat{x},y_1,\hat{\beta}_m)\\ y_2=g_2(\hat{x}).\end{cases}\quad \text{where } \hat{\beta}_m=\beta_0=\beta+e(0,\hat{k}).$$

Here also we have $-h+\beta\leq\hat{\beta}_m\leq h+\beta$. If synchronization is achieved, it can be concluded that 0 is transmitted otherwise 1 is transmitted. The keystreams k and $\hat{k}$ are generated using indirect coupled synchronization and the systems for generating them are defined as

$$(\mathrm{A}):\begin{cases}\dot{u}=p(u,y_2)\\ k=q(u)\end{cases}\text{and } (\mathrm{B}):\begin{cases}\dot{\hat{u}}=p(\hat{u},\hat{y}_2)\\ \hat{k}=q(\hat{u}).\end{cases}$$

The systems defined by (A) and (B) are driven by y_2 and $\hat{y}_2$ respectively such that the keystreams are synchronized. It should be noted that y_2 and $\hat{y}_2$ are not always equal since the parameter β_m in the transmitter and the receiver are varying. In the transmitter, it is changing according to both binary values but in the receiver, it is changing only due to the binary value 0. But synchronization is still achievable for the keystream generating oscillator as shown later in the simulation results.

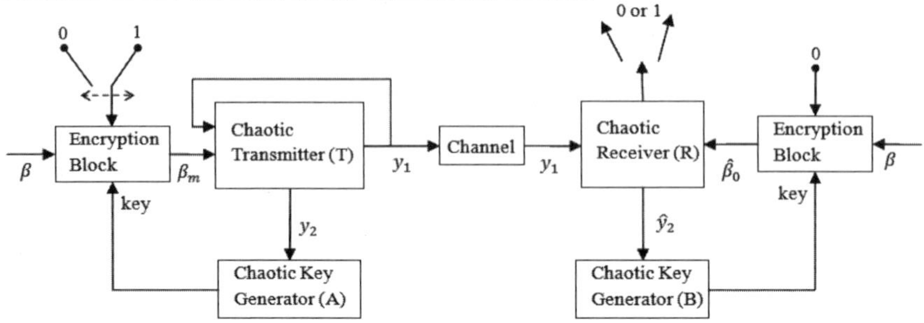

Figure 1. Modified CSK method.

4. Implementation using Lorenz and Chua's System

The proposed method is implemented using the Lorenz and Chua system. The Lorenz system acting as transmitter and receiver are given as

$$(T):\begin{cases}\dot{u}=-\sigma y_1+\sigma v\\ \dot{v}=-20y_1w+ry_1-v\\ \dot{w}=5y_1v-\beta_m w\\ y_1=u\\ y_2=v,\end{cases}\quad \text{and }(R):\begin{cases}\dot{\hat{u}}=-\sigma y_1+\sigma\hat{v}\\ \dot{\hat{v}}=-20\hat{u}\hat{w}+r\hat{u}-\hat{v}\\ \dot{\hat{w}}=5\hat{u}\hat{v}-\beta_0\hat{w}\\ y_2=\hat{v},\end{cases}$$

where parameter β_m in (T) modulates the binary signal $m(t)$ as mentioned previously. The encryption algorithm $e(z,k)$ is the n-shift cipher algorithm given as

$$e(z,k)=\underbrace{f_1(\ldots(f_1}_{n}(z,\underbrace{k),k),\ldots,k)}_{n}$$

and $f_1(z,k)$ given as

$$f_1(z,k)=\begin{cases}(z+k)+2h, & -2h\le(z+k)\le-h\\ (z+k), & h\le(z+k)\le h\\ (z+k)-2h, & 2h\le(z+k)\le 2h\end{cases}$$

where h is the encryption parameter. This means that that $-h+\beta\le\beta_m\le h+\beta$ and we have generated infinite levels of parameters between $-h+\beta$ to $h+\beta$ using the keystream k. The parameter h is chosen in such a way that β_m is always within the range where (T) is chaotic. In (R) the parameter β_0 is set as $\beta+e(0,k)$.

The keystream k and $\hat{k}$ in (T) and (R) are generated using the Chua's system defined as

$$(A):\begin{cases}\dot{p}=\alpha(q-p-\phi(y_2))\\ \dot{q}=y_2-q-s\\ s=-\delta q-\gamma s\\ k=d_0 p,\end{cases} \quad \text{and} \quad (B):\begin{cases}\dot{\hat{p}}=\alpha(\hat{q}-\hat{p}-\phi(\hat{y}_2))\\ \dot{\hat{q}}=\hat{y}_2-\hat{q}-\hat{s}\\ \dot{\hat{s}}=-\delta\hat{q}-\gamma\hat{s}\\ \hat{k}=d_0\hat{p}.\end{cases}$$

where d_0 is a scaling factor such that k and $\hat{k}$ lies within the interval [-*h*, *h*]. Note that here the signal y_2 and $\hat{y}_2$ is injected in the nonlinearity of the Chua's system. It means that the systems (A) and (B) are driven by y_2 and $\hat{y}_2$ respectively such that they synchronize with each other forming an indirect coupling.

The simulation of the proposed system is performed using Matlab/Simulink for the transmission of randomly generated digital bits. The different values used for systems (T), (R), (A) and (B) are taken as $\sigma = 16$, $r = 45.6$, $\beta = 4.2$, $\rho = 0.1$, $h = 0.2$, $\alpha = 10$, $\delta = 14.87$, $\gamma = 0$, $d_0 = 0.05$ and n for cipher shift algorithm is taken as 30.

Figure 2 shows the performance of the modified CSK method in successfully extracting the transmitted random bits. It can be seen that the random bits are successfully recovered at the receiver part using the proposed method exploiting the synchronization/desynchronization.

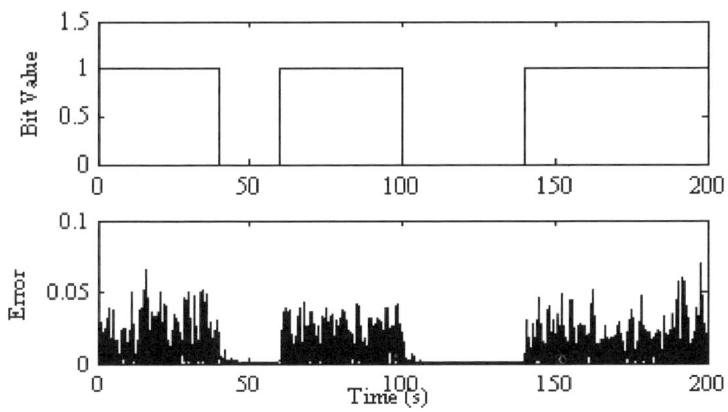

Figure 2. Transmission of random bits by CSK method using encryption and extraction.

5. Security Analysis

In this section security analysis of the proposed method using the most common decryption attack will be done. Not all of the attack methods are possible to be taken into consideration. However, Return Map (RM) being the common and the most powerful method for attacking CSK is used in this paper. RM sees the extraction of bits as a classification problem. In fact, extraction of either 0 or 1 in the transmitted signal is a classification problem, which is how digital equalization is also performed in digital communication. Other pattern classification tools can also be studied; however, if there is no pattern to classify for a method (RM in this case), it will generally mean that it will be valid for the remaining tools.

If x_i and X_i are the i-th minima and maxima for the transmitted signal y_1 respectively, then lets define $A_i = (X_i + x_i)/2$ and $B_i = (X_i + x_i)$. The plot of the B_i with respect to A_i is called RM of the signal y_1 [17].

The Figure 3 clearly illustrates the shortcoming of the classical CSK method where distinct two branches are seen when switched between only two parameters. Variation of CSK method was proposed as in [18] where authors have tried to confuse the intruders by including some switching parameters. However, there are branches in RM if the RM is zoomed in as shown in [19]. Methods like [4] has also been proposed but broken easily later [16]. As we shall show, the method proposed here, in this paper, can be employed to make the CSK method secure. The use of the keystream generated by indirect coupling synchronization at Tx and Rx will generate the parameter β_m of different values, in fact infinite possibilities within a range. The keystream generated is not part of the transmitted signal. Therefore, there is no way that the intruders will be able to generate it from only the knowledge of the transmitted carrier signal only. Without the knowledge of the keystream, it will be impossible for the intruders to find the change in the parameter $\hat{\beta}_m$ in the receiver to perform synchronization/desynchronization for extracting the bits.

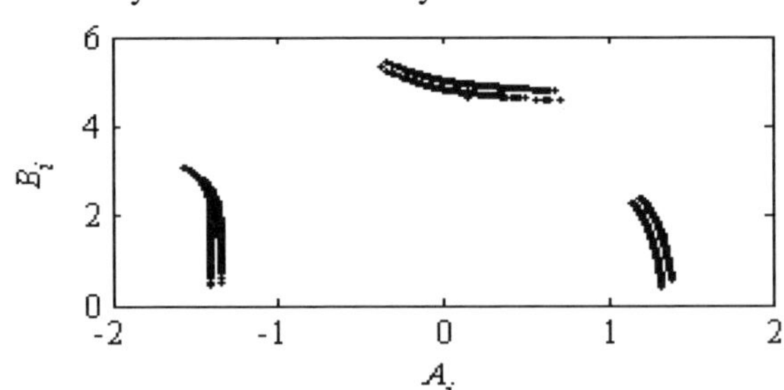

Figure 3. RM for traditional CSK scheme when β switched in two values.

Figure 4 shows the return map generated using the proposed method. It can clearly be seen that RM of the transmitted signal does not change according to the bit values being sent. This, therefore, will remove the possibility of seeing extraction of the bits as a classification problem.Consequently, by implementing the proposed methods a secure communication can be realized. For comparison, the RM of the transmitted signal is plotted when CSK is not implemented, i.e. when the transmitted bits do not modulate the parameter of the transmitter. It can be seen in Figure 5 that the RM is similar as in Figure 4 concluding that the proposed method does not necessarily change the return map of the transmitted carrier chaotic signal.

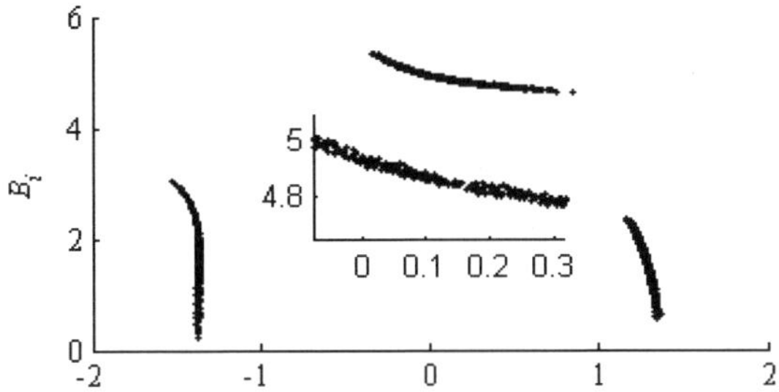

Figure 4. RM of the transmitted carrier signal using the proposed method.

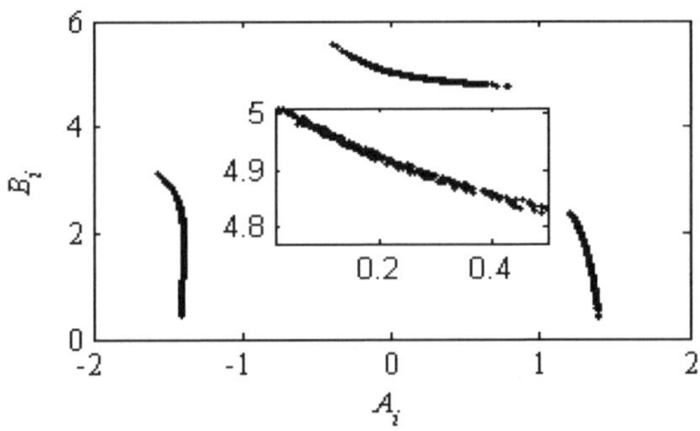

Figure 5: RM of the transmitted carrier signal when CSK is not implemented.

6. Conclusions

In this paper, the modified CSK method based on indirect coupled synchronization is proposed. The method is based on the use of an encryption algorithm whereby a chaotic keystreamb is generated using indirect coupled synchronization in the transmitter and receiver. The simulation showed that the method is able to extract the bit successfully at the receiver. Return map based attack is performed on the proposed method proving that the proposed method is secure. In fact, other pattern classification algorithm like ANN will also not be the suitable attack method because there is no pattern in the transmitted signal as depicted by the return map. The brute force attack will also not be an effective attack because of the large key space to choose from. Switch detection will not be an effective either because of the infinite levels generated by the encryption algorithm. Simulation confirms the effectiveness of the proposed method with regards to return map attacks.

References

[1] T. Yang, "A survey of chaotic secure communication systems," *International Journal of Computational Cognition,* vol. 2, pp. 81-130, 2004.

[2] T. Yang, L. B. Yang, and C. M. Yang, "Cryptanalyzing chaotic secure communication using return maps," *Physics Letters A,* vol. 245, pp. 495-510, 1998.
[3] T. Yang, L. B. Yang, and C. M. Yang, "Application of neural networks to unmasking chaotic secure communication," *Physica D,* vol. 124, pp. 248-257, 1998.
[4] S. Bu and B.-H. Wang, "Improving the security of chaotic encryption by using a simple modulating method," *Chaos Solitons & Fractals,* vol. 19, pp. 919-924, 2004.
[5] X. Wu, H. Hu, and B. Zhang, "Analyzing and improving a chaotic encryption method," *Chaos Solitons & Fractals,* vol. 22, pp. 367-373, 2004.
[6] S. Li, G. Alvarez, and G. Chen, "Breaking a choas-based secure scheme designed by an improved modulation method," *Chaos, Solitons and Fractals,* vol. 25, pp. 109-120, 2005.
[7] C. Y. Chee, D. Xu, and S. R. Bishop, "A zero-crossing approach to uncover the mask by chaotic encryption with periodic modulation," *Chaos Solitons & Fractals,* vol. 21, pp. 1129-1134, 2004.
[8] Z. Li and D. Xu, "A secure communication scheme using projective chaos synchronization," *Chaos, Solitons & Fractals,* vol. 22, pp. 477-481, 2004.
[9] J. Y. Chen, K. W. Wong, L. M. Cheng, and J. W. Shuai, "A secure communciation scheme based on the phase synchronization of chaotic systems," *Chaos,* vol. 13, pp. 508-514, 2003.
[10]G. Alvarez, S. Li, F. Montoya, M. Romera, and G. Pastor, "Breaking projective chaos synchronization secure communication using filtering and generalized synchronization," *Chaos Solitons & Fractals,* vol. 24, pp. 775-883, 2005.
[11]G. Alvarez, F. Montoya, G. Pastor, and M. Romera, "Breaking a secure communication scheme based on the phase synchronization of chaotic systems," *Chaos,* vol. 14, pp. 274-278, 2004.
[12]K. M. Short and A. T. Parker, "Unmasking a hyperchaotic communication scheme," *Physical Review E,* vol. 58, pp. 1159-1162, 1998.
[13]C. Zhou and C. H. Lai, "Extracting messages masked by chaotic signals of time-delay systems," *Physical Review E,* vol. 60, pp. 320-323, 1999.
[14]R. Kharel, K. Busawon, and Z. Ghassemlooy, "Indirect coupled oscillators for keystream generation in secure chaotic communication," in *Proceedings of the 48th IEEE conference on Decision and Control and 28th Chinese Control Conference 2009*, 2009.
[15]P. Palaniyandi and M. Lakshmanan, "Secure digital signal transmission by multistep parameter modulation and alternative driving of transmitter variables," *International Journal of Bifurcation and Chaos,* vol. 11, pp. 2031–2036, 2001.
[16]S. Li, G. Chen, and G. Alvarez, "Return-Map Cryptanalysis Revisited," *International Journal of Bifurcation and Chaos,* vol. 16, pp. 1557-1568, 2006.
[17]G. Perez and H. A. Cerdeira, "Extracting messages masked by chaos," *Phys. Rev. Lett.,* vol. 74, pp. 1970-1973,, 1995.
[18]D. Xu and C. Y. Chee, "Chaotic encryption with transient dynamics induced by pseudorandom switching keys," *International Journal of Bifurcation and Chaos,* vol. 14, pp. 3625–3631, 2004.
[19]D. Materassi and M. Basso, "Time Scaling of Chaotic Systems: Application to Secure Communications," *International Journal of Bifurcation and Chaos,* vol. 18, pp. 567-575, 2008.

Chaos problems in observer's mathematics

Boris Khots[1] and Dmitriy Khots[2]

[1]Compressor Controls Corp
4725 121st Street
Des Moines, Iowa, USA
Email: bkhots@cccglobal.com
[2]3710 S. 202nd Avenue
Omaha, Nebraska, USA
Email: dkhots@cox.net

Abstract. This work considers the solution of Cauchy problem (initial value problem) in a setting of arithmetic, algebra, and topology provided by Observer's Mathematics (see `www.mathrelativity.com`) and applies this solution to free wave equation, the linear (time-dependent) Schrodinger equation, the (time-dependent) Airy equation, the Korteweg-de Vries (KdV) equation, and quantum theory of two-slit interference. Certain results and communications pertaining to these problems are provided.
Keywords: Cauchy, Kowalevski, wave, Schrodinger, Airy, Korteweg-de Vries, two-slit interference, uncertainty, observer.

1 Introduction

In classical physics, it has been realized for centuries that the behavior of idealized vibrating media (such as waves on string, on a water surface, or in air), in the absence of friction or other dissipative forces, can be modeled by a number of partial differential equations, known collectively as dispersive equations. Model examples of such equations include the following:

- The free wave equation
$$u_{tt} - c^2 \Delta u = 0$$
where $u : R \times R^d \to R$ represents the amplitude $u(t,x)$ of a wave at a point in spacetime with d spatial dimensions, $\Delta = \sum_{j=1}^{d} \frac{\delta^2}{\delta x_j^2}$ is the spatial Laplacian on R^d, u_{tt} is short for $\frac{\delta^2 u}{\delta t^2}$, and $c > 0$ is a fixed constant.
- The linear Schrodinger equation
$$i\hbar u_t + \frac{\hbar^2}{2m} \Delta u = Vu$$

Chaos Theory: Modeling, Simulation and Applications
C. H. Skiadas, I. Dimotikalis and C. Skiadas (Eds)

where $u : R \times R^d \to R$ is the wave function of a quantum particle, $\hbar, m > 0$ are physical constants and $V : R^d \to R$ is a potential function, which we assume to depend only on the spatial variable x.

- The Airy equation

$$u_t + u_{xxx} = 0$$

where $u : R \times R \to R$ is a scalar function.

- The Korteweg-de Vries equation

$$u_t + u_{xxx} + 6uu_x = 0$$

which is a more refined version of the Airy equation in which the first nonlinear term is retained.

The theory of linear dispersive equations predicts that waves should spread out and disperse over time. However, it is a remarkable phenomenon, observed both in theory and practice, that once nonlinear effects are taken into account, solitary wave and soliton solutions can be created, which can be stable enough to persist indefinitely. In this paper we consider some properties of these equations from Observer's Mathematics point of view.

2 Cauchy-Kowalevski Theorem

The Cauchy-Kowalevski theorem is the main local existence and uniqueness theorem for analytic partial differential equations associated with Cauchy initial value problems. A special case was proved by Augustin Cauchy, [1], and the full result by Sophie Kowalevski, [6]. The first order Cauchy-Kowalevski theorem is about the existence of solutions to a system of m differential equations in n dimensions when the coefficients are analytic functions. The theorem and its proof are valid for analytic functions of either real or complex variables.

Let K denote either the fields of real or complex numbers and let $V = K^m$ and $W = K^n$. Let $A_1, \ldots, A_{n-1}$ be analytic functions defined on some neighborhood of $(0, 0)$ in $V \times W$ and taking values in the $m \times m$ matrices, and let b be an analytic function with values in V on the same neighborhood. Then there is a neighborhood of 0 in W on which the quasilinear Cauchy problem

$$\partial_{x_n} f = A_1(x, f)\partial_{x_1} f + \ldots + A_{n-1}(x, f)\partial_{x_{n-1}} f + b(x, f)$$

with initial condition $f(x) = 0$ on the hypersurface $x_n = 0$ has a unique analytic solution $f : V \to W$ near 0.

Lewy's example shows that the theorem is not valid for all smooth functions. The theorem can also be stated in abstract (real or complex) vector spaces. Let V and W be finite-dimensional real or complex vector spaces, with $n = dimW$. Let $A_1, \ldots, A_{n-1}$ be analytic functions with values in

$End(V)$ and b an analytic function with values in V, defined on some neighborhood of $(0,0)$ in $V \times W$. In this case, the same result holds.

The higher-order Cauchy-Kowalevski theorem can be stated as follows. If F and f_j are analytic functions near 0, then the non-linear Cauchy problem $\partial_t^k h = F(x,t,\partial_t^j,\partial_x^\alpha h)$, where $j < k$ and $|\alpha| + j \leq k$, with initial conditions $\partial_t^j h(x,0) = f_j(x)$, with $0 \leq j < k$, has a unique analytic solution near 0. This follows from the first order problem by considering the derivatives of h appearing on the right hand side as components of a vector-valued function.

3 Observer's Mathematics Arithmetic and Derivatives

Mathematics of Relativity - Observer's Mathematics was introduced by authors in [2], and then was published in [3],[4], and [5]. This work was created as an attempt to do away with the concept of infinity. The main concepts of Observer's Mathematics are as follows. Let W be the set of all real numbers, i.e. a set of all infinite decimal fractions. Let W_n be the set of all finite decimal fractions of length $2n$ (with n - positive integer). Visually W_n can be described as $W_n = \{\underbrace{\star\cdots\star}_{n}.\underbrace{\star\cdots\star}_{n}\}$. We consider a finite well-ordered (by n) system of observers, where each observer "lives" in W_n for some n, but "thinks" that he "lives" in W. We call this observer "naive" and denote by W_n-observer. We further assume that a W_n-observer, preserving his "naivety", sees that a W_k-observer actually sees W_k and deals with W_k, if $n > k$.

We now consider the arithmetic in W_n from the point of view of a W_m-observer with $m > n$.

For $c = c_0.c_1...c_n$, $d = d_0.d_1...d_n \in W_n$ addition and subtraction $(\pm_n)$ is defined as follows.

$$c \pm_n d = \begin{cases} c \pm d, \text{ if } c \pm d \in W_n \\ \text{not defined, if } c \pm d \notin W_n \end{cases}$$

(The $\pm$ sign is a standard arithmetic sign). We will write

$$((\dots(c_1 +_n c_2)\dots) +_n c_N) = \sum_{i=1}^{N}{}^{n} c_i$$

for $c_1,\dots,c_N$ iff the contents of any parenthesis are in W_n.

Next, multiplication is defined for $c = c_0.c_1...c_n$, $d = d_0.d_1...d_n \in W_n$.

$$c \times_n d = \sum_{k=0}^{n}{}^{n} \sum_{m=0}^{n-k}{}^{n} 0.\underbrace{0...0}_{k-1} c_k \cdot 0.\underbrace{0...0}_{m-1} d_m$$

where $c, d \geq 0$, $c_0 \cdot d_0 \in W_n$, $0.\underbrace{0...0}_{k-1} c_k \cdot 0.\underbrace{0...0}_{m-1} d_m$ is the standard product, and $k = m = 0$ means that $0.\underbrace{0...0}_{k-1} c_k = c_0$ and $0.\underbrace{0...0}_{m-1} d_m = d_0$. If either $c < 0$ or

$d < 0$, then we compute $|c| \times_n |d|$ and define $c \times_n d = \pm |c| \times_n |d|$, where the sign $\pm$ is defined as usual. Note, if the content of at least one parentheses (in previous formula) is not in W_n, then $c \times_n d$ is not defined.

The arithmetic that we introduced above coincides with the standard arithmetic until it meets the "allowed" boundaries of space. It turns out that the "boundary conditions" could cardinally affect the computational process in general.

From the point of view of W_n-observer (we will call such observers "naive", since they "think" that they "live" in W and deal with W) a real function y of a real variable x, $y = y(x)$, is called differentiable at $x = x_0$ if there is a derivative

$$y'(x_0) = \lim_{x \to x_0, x \neq x_0} \frac{y(x) - y(x_0)}{x - x_0}$$

What does the above statement mean from point of view of W_m-observer with $m > n$? It means that

$$|(y(x) -_n y(x_0)) -_n (y'(x_0) \times_n (x -_n x_0))| \leq 0.\underbrace{0\ldots01}_{n}$$

whenever

$$|y(x) -_n y(x_0)| = 0.\underbrace{0\ldots0y_l}_{l}\, y_{l+1}\ldots y_n$$

and

$$|(x -_n x_0)| = 0.\underbrace{0\ldots0x_k}_{k}\, x_{k+1}\ldots x_n$$

for $1 \leq k,\, l \leq n$, and x_k - non-zero digit. In [4], the following theorems have been proven:

Theorem 1. *From the point of view of a W_m-observer a derivative calculated by a W_n-observer ($m > n$) is not defined uniquely.*

Theorem 2. *From the point of view of a W_m-observer with $m > n$, $|y'(x_0)| \leq C_n^{l,k}$, where $C_n^{l,k} \in W_n$ is a constant defined only by n, l, k and not dependent on $y(x)$.*

Theorem 3. *From the point of view of a W_m-observer, when a W_n-observer (with $m > n \geq 3$) calculates the second derivative:*

$$y''(x_0) = \lim_{x_1 \to x_0, x_1 \neq x_0, x_2 \to x_0, x_2 \neq x_0, x_3 \to x_1, x_3 \neq x_1} \frac{\frac{y(x_3)-y(x_1)}{(x_3-x_1)} - \frac{y(x_2)-y(x_0)}{x_2-x_0}}{x_1 - x_0}$$

we get the following unequality:

$$(|x_2 -_n x_0| \times_n |x_3 -_n x_1|) \times_n |x_1 -_n x_0| \geq 0.\underbrace{0\ldots01}_{n}$$

provided that $y''(x_0) \neq 0$.

Analysis of concepts such as Free Wave equation, Schrodinger equation, two-slit interference, wave-particle duality for single photons, uncertainty principle, Airy and Korteweg-de Vries equations, and Schwarzian derivative shows that in Observers's Mathematics Cauchy-Kowalevski theorems become invalid. Instead, we have stochastic properties of partial (and ordinary) differential equations, both linear and non-linear.

4 Free Wave Equation

We consider the case when $d = 1$, i.e., $u : W_n \times W_n \to W_n$, from W_m-observer point of view, with $m > n$, where $W_n \times W_n$ means Cartesian product of W_n with itself. The free wave equation may be written as

$$u_{tt} -_n ((c \times_n c) \times_n u_{xx}) = 0$$

Then we have the following

Theorem 4. *Let*

$$c = c_0.c_1 \dots c_k c_{k+1} \dots c_n$$

and

$$u_{xx} = \pm u_0^{xx}.u_1^{xx} \dots u_l^{xx} u_{l+1}^{xx} \dots u_n^{xx}$$

with $2k < n$, $l < n$, $c_0 = c_1 = \dots = c_k = 0$, $c_{k+1} \neq 0$, $u_0^{xx} = u_1^{xx} = \dots = u_l^{xx} = 0$ *and* $u < k + l + 2$, *then* $u_{tt} = 0$.

Proof. We have $c \times_n c = d_0.d_1 \dots d_r d_{r+1} \dots d_n$, with $d_0 = d_1 = \dots = d_r = 0$, if $r = 2k$, $d_{r+1} \neq 0$. Thus, $(c \times_n c) \times_n u_{xx} = 0$ and $u_{tt} = 0$. QED.

Next, we have the following

Theorem 5. *If* $d_0 \geq \underbrace{9 \dots 9}_{p}$, *with* $0 < p \leq n$ *and* $u_0^{xx} \geq \underbrace{9 \dots 9}_{q}$, *with* $0 < q \leq n$ *and* $n < p + q$, *then there is no* u_{tt}, *such that* $u_{tt} = ((c \times_n c) \times_n u_{xx})$.

Proof. $(c \times_n c) \times_n |u_{xx}| > d_0 \times_n u_0^{xx}$, but $d_0 \times_n u_0^{xx}$ does not exist in W_n. QED.

5 Schrodinger Equation

Consider the following: $-(\hbar \times_n \hbar) \times_n \Psi_{xx} +_n ((2 \times_n m) \times_n V) \times_n \Psi = i((2 \times_n m) \times_n \hbar)\Psi_t$, where $\Psi = \Psi(x,t)$, $\hbar$ is the Planck's Constant, $\hbar = 1.054571628(53) \times 10^{-34}\ m^2 kg/s$. Then we have the following

Theorem 6. *Let* $36 < n < 68$, $m = m_0.m_1 \dots m_k m_{k+1} \dots m_n$, *with* $m \in W_n$, $m_0 = m_1 = \dots = m_k = 0$, $m_{k+1} \neq 0$, $k + 35 < n$, $V = 0$, *then* $\Psi_t = \Psi_t^0.\Psi_t^1 \dots \Psi_t^l \Psi_t^{l+1} \dots \Psi_t^n$ *and* $\Psi_t^0 = \dots \Psi_t^l = 0$, $\Psi_t^{l+1}, \dots, \Psi_t^n$ *are free and in* $\{0, 1, \dots, 9\}$, *where* $l = n - k - 36$, *i.e.,* Ψ_t *is a random variable, with*

$$\Psi_t \in \{(0.\overbrace{\underbrace{0 \dots 0}_{l} * \dots *}^{n})\},$$ *where* $* \in \{0, 1, \dots, 9\}$.

Proof. We have $-(\hbar \times_n \hbar) = 0$, $(2\times_n m)\times_n V = 0$, i.e., $i((2\times_n m)\times_n \hbar)\Psi_t = 0$ and $l = n - k - 36$. QED.

Corollary 1. *Let* $36 < n < 68$, $m = m_0.m_1 \ldots m_k m_{k+1} \ldots m_n$, *with* $m \in W_n$, $m_0 = m_1 = \ldots = m_k = 0$, $m_{k+1} \neq 0$. *Also, let* $V = v_0.v_1 \ldots v_s v_{s+1} \ldots v_n$, *with* $V \in W_n$, $v_0 = v_1 = \ldots = v_s = 0$, $v_{s+1} \neq 0$, *with* $\begin{cases} k+35 & < n \\ k+s+2 > n \end{cases}$, *then* $\Psi_t = \Psi_t^0.\Psi_t^1 \ldots \Psi_t^l \Psi_t^{l+1} \ldots \Psi_t^n$ *and* $\Psi_t^0 = \ldots \Psi_t^l = 0$, $\Psi_t^{l+1}, \ldots, \Psi_t^n$ *are free and in* $\{0, 1, \ldots, 9\}$, *where* $l = n - k - 36$, *i.e.,* Ψ_t *is a random variable, with*

$$\Psi_t \in \{(0.\overbrace{\underbrace{0\ldots0}_{l}*\ldots*}^{n})\}, \text{ where } * \in \{0, 1, \ldots, 9\}.$$

6 Two-Slit Interference

Quantum mechanics treats the motion of an electron, neutron or atom by writing down the Schrodinger equation:

$$-\frac{\hbar^2}{2m}\frac{\delta^2 \Psi}{\delta x^2} + V\Psi = i\hbar\frac{\delta \Psi}{\delta t}$$

where m is the particle mass and V is the external potential acting on the particle. As these particles pass through the two slits of any of the experiments they are moving freely; we, therefore, set $V = 0$ in the Schrodinger equation.

Now, consider the following:

$$-(\hbar \times_n \hbar) \times_n \Psi_{xx} +_n ((2 \times_n m) \times_n V) \times_n \Psi = i((2 \times_n m) \times_n \hbar)\Psi_t$$

where $\Psi = \Psi(x, t)$, $\hbar$ is the Planck's Constant, $\hbar = 1.054571628(53) \times 10^{-34}$ $m^2 kg/s$. Then we have the following

Theorem 7. *Let* $36 < n < 68$, $m = m_0.m_1 \ldots m_k m_{k+1} \ldots m_n$, *with* $m \in W_n$, $m_0 = m_1 = \ldots = m_k = 0$, $m_{k+1} \neq 0$, $k + 35 < n$, $V = 0$, *then* $\Psi_t = \Psi_t^0.\Psi_t^1 \ldots \Psi_t^l \Psi_t^{l+1} \ldots \Psi_t^n$ *and* $\Psi_t^0 = \ldots \Psi_t^l = 0$, $\Psi_t^{l+1}, \ldots, \Psi_t^n$ *are free and in* $\{0, 1, \ldots, 9\}$, *where* $l = n - k - 36$, *i.e.,* Ψ_t *is a random variable, with*

$$\Psi_t \in \{(0.\overbrace{\underbrace{0\ldots0}_{l}*\ldots*}^{n})\}, \text{ where } * \in \{0, 1, \ldots, 9\}.$$

Proof is given in [4]. The wave at the point of combination will be the sum of those from each slit. If Ψ_1 is the wave from slit 1 and Ψ_2 is the wave from slit 2, then $\Psi = \Psi_1 + \Psi_2$. The result gives the predicted interference pattern. Then by Theorem 1, we have

$$\Psi_{1t} = \Psi_{1t}^0.\Psi_{1t}^1 \ldots \Psi_{1t}^l \Psi_{1t}^{l+1} \ldots \Psi_{1t}^n$$

$$\Psi_{2t} = \Psi_{2t}^0.\Psi_{2t}^1 \ldots \Psi_{2t}^l \Psi_{2t}^{l+1} \ldots \Psi_{2t}^n$$

$$\Psi_{1t}^0 = \ldots = \Psi_{1t}^l = 0$$

Where $\Psi_{1t}^{l_1+1}, \ldots, \Psi_{1t}^n$ are free and in $\{0, 1, \ldots, 9\}$. and

$$\Psi_{2t}^0 = \ldots = \Psi_{2t}^l = 0$$

Where $\Psi_{2t}^{l_2+1}, \ldots, \Psi_{2t}^n$ are free and in $\{0, 1, \ldots, 9\}$ where $l = n - k - 36$.
Now we have the following

Theorem 8. *1. If $\Psi_{1t}^{l+1} + \Psi_{2t}^{l+1} > 9$, then $\Psi_1 + \Psi_2$ is not a wave.*
2. If $\Psi_{1t}^{l+1} + \Psi_{2t}^{l+1} < 9$, then $\Psi_1 + \Psi_2$ is a wave.
3. If $\Psi_{1t}^{l+1} + \Psi_{2t}^{l+1} = 9$, then $\Psi_1 + \Psi_2$ may or may not be a wave.

Proof. If $\Psi = \Psi_1 + \Psi_2$ is a wave, we have to have $\Psi = \Psi_0.\Psi_1 \ldots \Psi_l \Psi_{l+1} \ldots \Psi_n$ (which is necessary and sufficient) with $\Psi_0 = \Psi_1 = \ldots = \Psi_l = 0$. Thus, for the first statement in the theorem, we have $\Psi_l \neq 0$ and for the second statement, we must have $\Psi_l = 0$. For the third case both variants are possible. QED.

7 Airy and Korteweg-de Vries Equations

If $u : W_n \times W_n \to W_n$ then the Airy equation may be written as

$$u_t +_n u_{xxx} = 0$$

and Korteweg-de Vries equation may be written as

$$(u_t +_n u_{xxx}) +_n 6(u \times_n u_x) = 0$$

Then we have the following

Theorem 9. *Let*

$$u = u_0.u_1 \ldots u_k u_{k+1} \ldots u_n$$

and

$$u_x = u_0^x.u_1^x \ldots u_l^x u_{l+1}^x \ldots u_n^x$$

with $k < n$, $l \leq n$ and $u_0 = u_1 = \ldots = u_k = 0$ and $u_0^x = u_1^x = \ldots = u_l^x = 0$ and $k + l > n$, then Airy equation and Korteweg-de Vries equation have the solution.

Proof. In this case $u \times_n u_x = 0$ and $u_t +_n u_{xxx} = 0$. QED.

8 Acknowledgments

The authors thank Professor Christos H. Skiadas for his invitation to participate at the Chaotic Modeling and Simulation International Conference (CHAOS2010).

References

1.A.Cauchy. Comptes rend. *Reprinted in Oeuvres completes, 1 serie, Tome VII*, 15:1758, 1842.
2.B.Khots and D.Khots. Mathematics of relativity webbook, url: www.mathrelativity.com, 2004.
3.B.Khots and D.Khots. An introduction to mathematics of relativity. *Lecture Notes in Theoretical and Mathematical Physics*, 7:269306, 2006.
4.B.Khots and D.Khots. Quantum theory and observer's mathematics. *American Institute of Physics (AIP)*, 965:261264, 2007.
5.B.Khots and D.Khots. Physical aspects of observer's mathematics. *American Institute of Physics (AIP)*, 1101:311313, 2009.
6.S.Kowalevski. Zur theorie der partiellen differentialgleichung. Journal fr die reine und angewandte Mathematik, 80:132, 1875.

Freedom and necessity in computer aided composition: A thinking framework and its application

Johannes Kretz

Zentrum für innovative Musiktechnologie (ZiMT) der Universität für Musik und darstellende Kunst Wien, Vienna, Austria
Email: zimt@mdw.ac.at

Abstract: This paper presents some of the author's experiences with computer aided composition (CAC): the modeling of physical movements is used to obtain plausible musical gestures in interaction with constraint programming (rule based expert systems) in order to achieve precisely structured, consistent musical material with strong inner logic and syntax in pitch material. The "Constraints Engine" by Michael Laurson implemented in OpenMusic (IRCAM) or PWGL (Sibelius Academy) can be used to set up an interactive framework for composition, which offers a balance of freedom (allowing chance operations and arbitrary decisions of the composer) and necessity (through strict rules as well as through criteria for optimization). Computer Aided Composition is moving far beyond being "algorithmic" or "mechanical". This paper proposes an approach based on evolutionary epistemology (by the Austrian biologist and philosopher Rupert Riedl). The aim is a holistic synthesis of artistic freedom and coherent structures similar to the grown order of nature.
Keywords: Computer Aided Composition, CAC, physical modeling, music, composition, artificial intelligence, expert systems, evolutionary epistemology.

"Things derive their being and nature by mutual dependence and are nothing in themselves."
(Nagajuna, second century Buddhist philosopher)

"Nothing, in the development of this world, can be understood in isolated manner."
(Rupert Riedl, 1999)

1. Introduction

One of the particular challenges of composing is (and always was) finding the balance between simplicity and complexity, between freedom and necessity, and – which is even more difficult – the creation of a piece of art, that can be perceived at various levels of attention, for example in that way, that the most obvious level (the surface) can be understood and enjoyed easily without special pre-knowledge (even intuitively), but other, deeper, less obvious layers of the work provide more subtle contents satisfying the more sophisticated interests of the connoisseurs.

Chaos Theory: Modeling, Simulation and Applications
C. H. Skiadas, I. Dimotikalis and C. Skiadas (Eds)

Besides it turns out to be particularly difficult to formalize a sophisticated process of creation, where various structural layers of material from the surface to the inner structures have to interact permanently, and where various parameters (pitch, time) in various layers of complexity (melodies, chords, rhythmic cells, formal parts...) influence each other. In this paper we present a philosophical framework, allowing to represent musical thinking in an organic and meaningful way, giving the artist tools, where freedom, random choices and arbitrariness have its place equally to well structured organization of elements. Finally we will also give examples of the framework's practical application with the help of computers.

2. From Aristotle to evolutionary epistemology

2.1. *Humanities versus Natural Sciences*

In music, like in many other fields of human activity, the question of causality is essential. A music, which gives us the impression of absence of causes and effects is usually perceived as unsatisfying by listener. Therefore it might be worth looking in detail into a model of understanding causality in general.

Since the age of enlightenment we are confronted with two opposing thinking paradigms, fundamentally separating the scientific field of humanities from the natural sciences. The approach of humanities (fig.1) is mainly deductive, where truths are derived from abstract, fundamental certainties, which cannot be put into question. Based on these "truths" a hierarchical view on the world is unfolded by progressive specialization through deduction, providing a universe of nested classifications. On the other hand natural sciences (fig.2) build their view of the world empirically from the most elementary observations of physics, also establishing a hierarchical view of the world in nested classifications by progressive spezilalization, just starting from the other end. This leads – in the case of humanities – to a world, where everything has a final sense but lacks objectivity, since the fundamental "thruths" are given on the basis of religion, culture, ideology or coincidence, being not always the most objective guides through life. In the case of natural sciences on the other hand, we obtain a maximum degree of objectivity, but questions about the higher sense and goals can only be answered very vaguely.

Since the last decades of the 20th century it becomes more and more obvious that a more sophisticated multi-causal framework of thinking would be very useful, which possibly could help to overcome the splitting of our world into humanities and natural sciences.

Figure 1. Viewpoint of Humanities

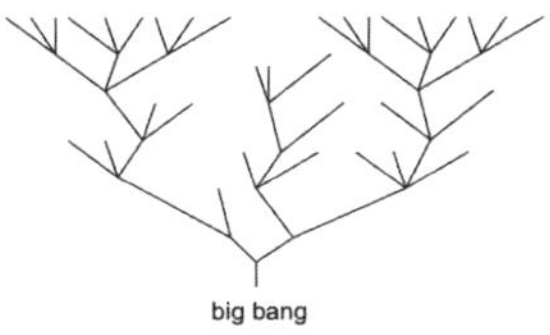

Figure 2. Viewpoint of Natural Sciences

2.2. *The four causes of Aristotle*

In this context the differentiated thinking model of the four causalities by Aristotle described around 350 B.C. in his book on Metaphysics [1] offers an approach, which we also find in the core of Evolutionary Epistemology [2] [3] [4] [5] [6] [7]. Aristotle mentions four causes, which interact in any development of complex entities:

- causa efficiens (energy)
- causa finalis (targets)
- causa materialis (bricks, elements)
- causa formalis (construction plan)

Causa efficiens represents effects of energy and matches mainly with the causality model of physics (fig 2.). *Causa finalis* on the other hand is the final aim or purpose of human activity and corresponds to the causality model of humanities (fig.1). Additionally there is another pair of antagonistic causes: *Causa materialis* describes the effects of the building material on the built objects, while *causa formalis* describes the effects of the construction plan on the built sub-entities.

This framework can be applied very well on composing music (see fig.3). Here *causa efficiens* is creative energy, the desire for experimentation, even the creative use of mistakes and errors (in the language of biology: mutations). On the other side *causa finalis* provides the aesthetic targets, the purpose of the music, (which can even be extra-musical like in dance music, opera, sacred music etc.). Nevertheless, composing would be extremely difficult, if it had to happen only between the causalities of the two causes, as would biological evolution be extremely slow in its success, if only mutation and selection would steer it. The chances are just to small that one can write a large work of music just by "try and error". Therefore both in evolution theory [2] and in music it is important to introduce a second pair of causes: The creation of materials, bricks, elements – in music: motives, patterns, chords, rhythmical cells etc. – makes the creative process more efficient. Of course it poses some constraints for the possible results, but nevertheless the increased efficiency and speed of creation outweighs the loss of possibilities by far, both, in biology and in composition. The hierarchical nested interaction of various layers of construction entities ranging from the level of materials (perspective of *causa materialis*) to the level of a construction plan (*causa formalis*) allows the creation of extended works with a reasonable amount of effort.

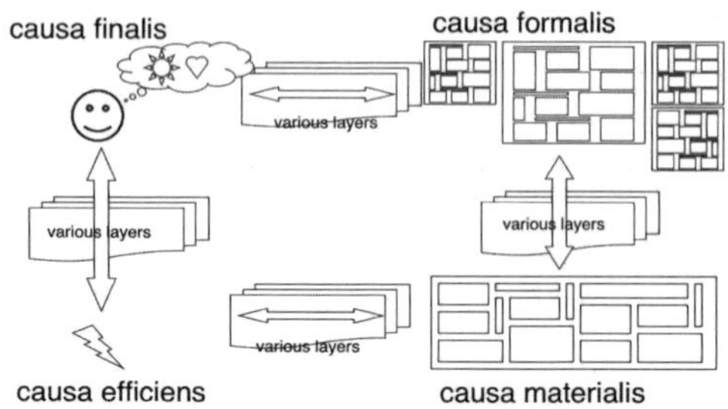

Figure 3. interaction of Aristotle's causes

In addition both, in traditional composition and in computer aided composition as well as in biological evolution [4], another aspect of causality has to be highlighted: If n elements (musical entities like notes, motives, chords etc.) are interacting, and each entity is interacting with each other, the number of possible interactions increases enormously for growing n. It is represented by the triangular number ("handshake problem"):

$$T_n = 1 + 2 + 3 + \ldots + (n-1) + n = \frac{n^2 - n}{2}$$

This means that interaction is getting too complex, when the number of entities is exceeding a certain number. (Compare fig.4).

A reasonable solution for this would be differentiating between internal and external interactions of entities forming groups of entities, interacting with other groups of entities, on various levels of hierarchy (see fig. 5 to 7).

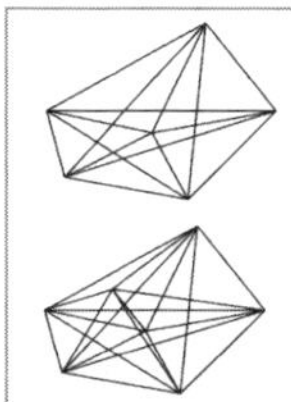

Figure 4. increasing complexity

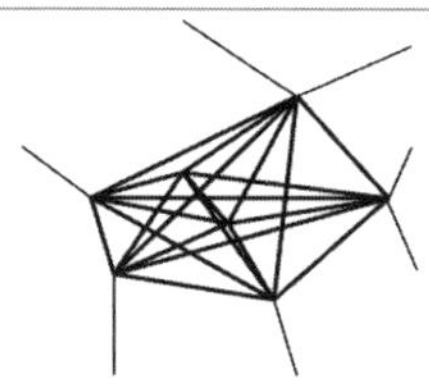

Figure 5. internal / external interaction

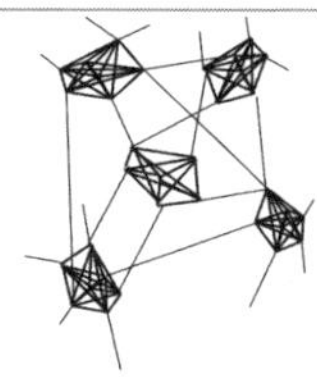

Figure 6. network of networks

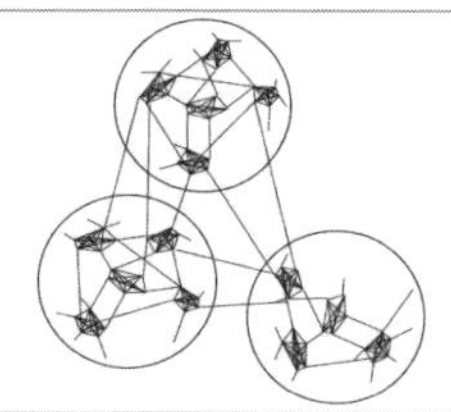

Figure 7. network of networks of networks

3. Musical application

Musical gestures representing physical movement have always played a certain role in music history. Also in contemporary computer aided composition the simulation of physical movements can serve as a main source of convincing plausible gestural data. Since we are used to "understand" this physical behavior of objects in the real world, the simulation of these movements with the means of music makes a lot of sense for our perception, too. These gestures can follow the known laws of physic (gravity, aerodynamics, etc. like in fig.8, where two bouncing rubber balls are simulated) or even add some "creativity" like in fig.9, where the two simulated rubber balls are "in love", attracting each other, so that their movements turn into a more interesting "dance", where each ball tries to meet the other, while their inertia mostly prevents an actual meeting.

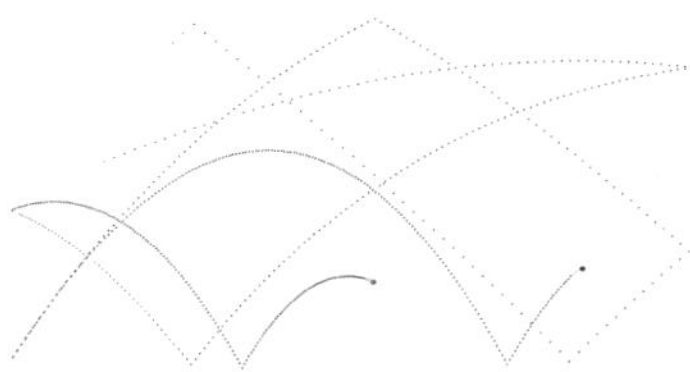

Figure 8. Physical modeling

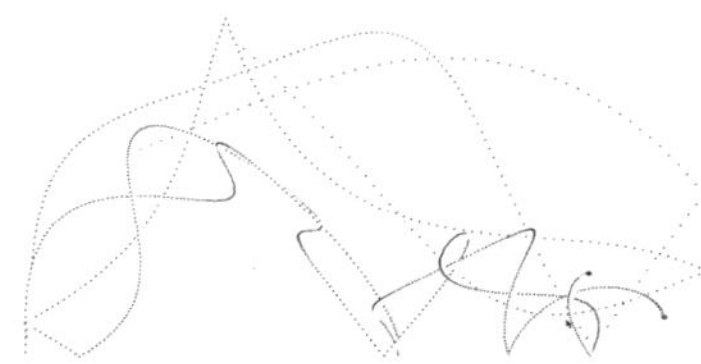

Figure 9. "Creative" modeling

When applying these gestures to musical material, it is evident that a direct translation of the position parameters into the domain of chromatic pitch would produce flat, rather unsatisfying musical results. While the curves can be useful to shape (or influence) the surface of music, other tools are needed to control the inner, structural side of music in interaction with these curves. (See fig. 10 and 11.)

One of the most promising methods for this is constraint programming. The user creates – in interaction with the computer – an expert system, where desired results are obtained by defining a search space (the basic material for possible solutions), a set of strict rules, allowing or forbidding certain characteristics of the solution, also (if needed) heuristic rules, which favor certain criteria of optimization. This approach originating from programming languages like PROLOG [8] was already implemented in IRCAM's Patchwork environment [9]. The same engine was then ported to OpenMusic [10] by Örjan Sandred, and is finally a core part of today's most advanced CAC environment PWGL [11] by Michael Laurson.

Note, that this approach of combining elements from a search space according to strict and heuristic rules corresponds very well with the interaction of *causa materialis* (the effects of the characteristics of the search space and the strict rules) and *causa formalis* (the steering towards certain desired global qualities with the help of heuristic rules).

The following examples are aimed to illustrate this approach. In the case of "second horizon" for piano and orchestra by Johannes Kretz the search space consisted of a set of allowed chords with all their inversions, octave permutations and transpositions. Additional constraints were applied: The range of the piano should not be exceeded. Two adjacent chords should have not more the three common notes, but at least one.

The shape of the rubber ball movement was implemented as heuristic rule, so that the engine always chose a solution as close as possible to the "ideal" shape with the help of a system of "bonus points". But the other (strict) rules had absolute priority. Fig.12 and 13 show possible results. It can be clearly seen, that the original shape is not represented perfectly, and that some of the inner – structural – necessities cause compromises of the shape. But exactly this compromise between outer and inner conditions of creation gives interesting, organic musical results. Compare [12] and [13] for details.

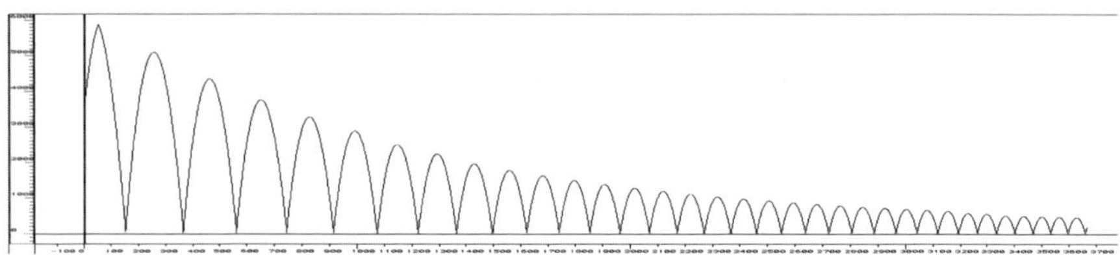

Figure 10. physical modeling of a bouncing ball

Figure 11. representation of fig. 8 in music notation

Figure 12. interaction of the shape of fig.8 with harmonic rules

Figure 13. Effect of additional rhythmical rules

Figures 14-20 show examples taken from the composition "full scope fancy" (2004) by Johannes Kretz. Here the search space was derived from a dodecaphonic row (fig. 14), which was used to generate a pool of chords (fig.15 and 16). The curves in fig. 17 and 18 should be approximated by the outer two voices of the final result, but since they were given only as heuristic rules, while the inner structure of the chords and melodies was defined by strict rules, we obtain both, an approximate physical behavior on the surface and a strict inner syntax in the details. (Compare fig.19 - 21).

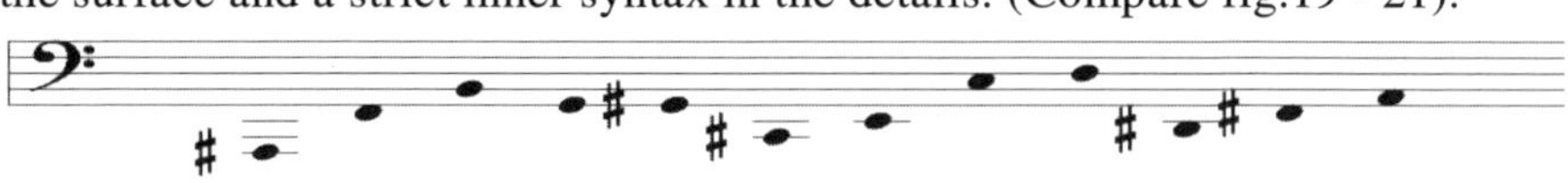

Figure 14: dodecaphonic row

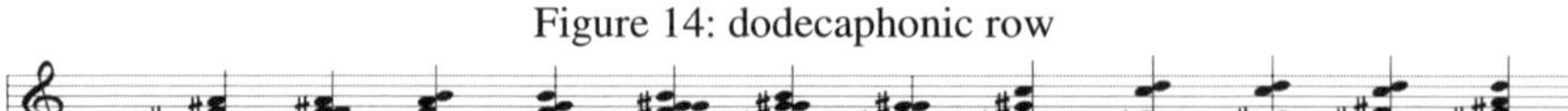

Figure 15. chord series derived from fig. 14.

Figure 16: extended chord series as search space

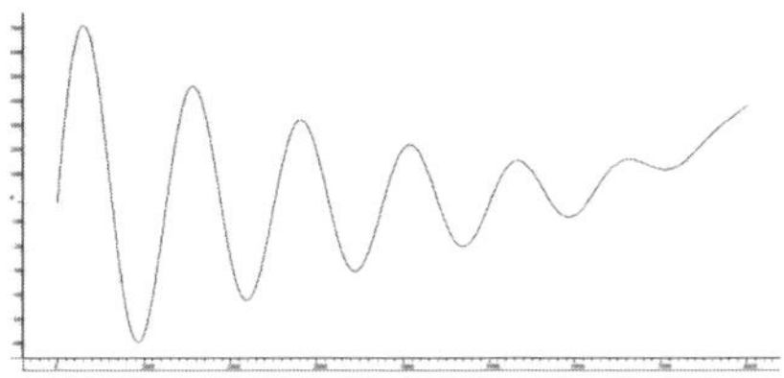

Figure 17. bass line: ballistic curve

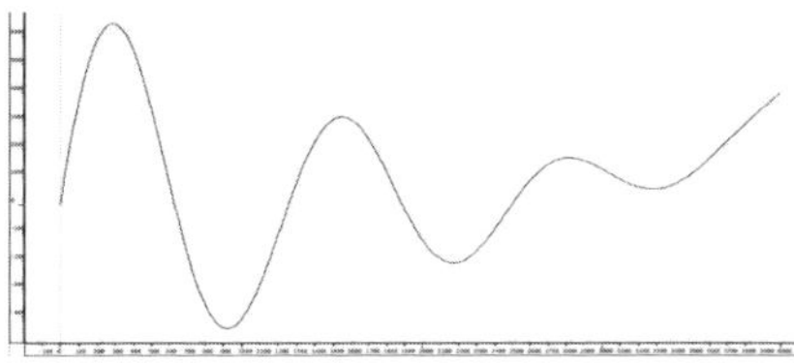

Figure 18: treble line: ballistic curve

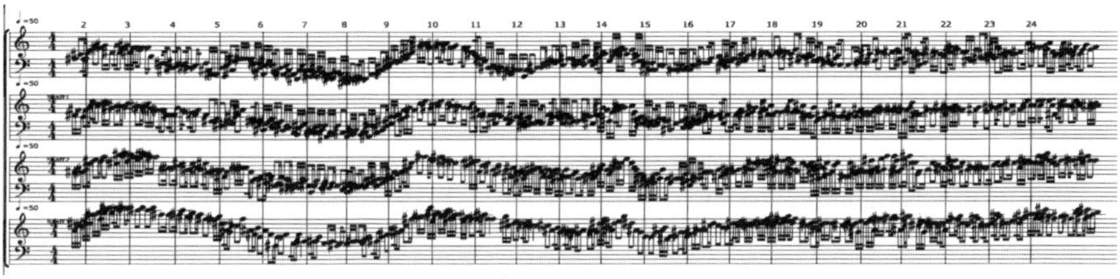

Figure 19. approximation through structured musical material

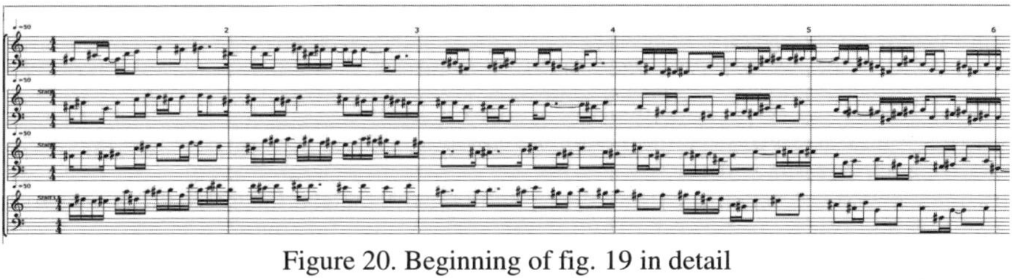

Figure 20. Beginning of fig. 19 in detail

Figure 21 again shows the difference between the result of physical modeling alone and the result of the interaction between physical modeling and syntactical strict rules.

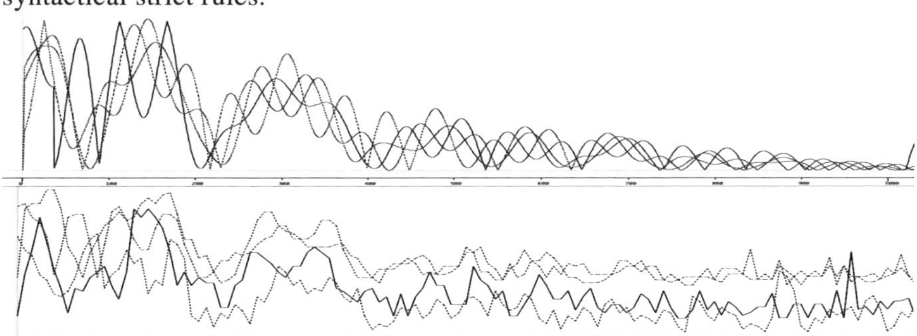

Figure 21. pure physical movement vs. structured approximation

4. Conclusions

The described method of computer aided composition differs significantly from more traditional approaches like algorithmic or stochastic composition. While a certain degree of randomness is allowed, when strict or heuristic rules don't suggest any difference between allowed decisions, the main focus here is the interaction of models in two levels: Firstly the interaction between strict rules and criteria of optimization has the potential to find solutions, that would be very hard to obtain by composing only with paper and pencil or other methods of computation. Secondly another iterative process – between the computer and the composer – controls the evolution of music: typically rules are added or modified after listening to the results of computation, until they match well with the aesthetics of the composer. In this way the results are approximating better and better to the desired results in an evolutionary process similar to the biological one.

References

[1] Aristotle: Methaphysics, book V 2, 350 B.C.
[2] Rupert Riedl: Evolution und Erkenntnis, Munich 1982.
[3] Rupert Riedl, (translated by R.P.S. Jefferies) „Order in Living Organisms: A Systems Analysis of Evolution", orginally: Die Ordnung des Lebendigen. Munich 1975.
[4] Rupert Riedl: Die Strategie der Genesis. Munich 1976.
[5] Rupert Riedl: Biology of Knowledge, the evolutionary basis of reason, Chichester, John Wiley & Sons 1984.
[6] Rupert Riedl: Zufall, Chaos, Sinn. Stuttgart 1999
[7] Rupert Riedl: Darwin, Zeus und Russels Huhn, Vienna 1994.
[8] Balaban, M. / Murray, N.V.: Machine Toungues X: Prolog. In: CMJ Vol.9 No. 3 Fall 1985, Cambridge 1985.
[9] Laurson, M. / Duthen, J.: Patchwork, a graphical language in Pre-Form. In: T.Wells and D. Butler (Ed.): „Proc. of the 1989 ICMC". San Francisco 1989: p.172-175
[10] Agon, C.: OpenMusic: Un langage visuel pour la composition musicale assistée par ordinateur, PhD Thesis, IRCAM -- Univ. Paris 6, 1998.
[11] Laurson, M. and Kuuskankare, M.: PWGL: A Novel Visual Language based on Common Lisp, CLOS and OpenGL. In Proc. of ICMC02, p.142-145, Gothenburg, Sweden 2002.
[12] Kretz, Johannes: Navigation of Structured Material in 'second horizon' for Piano and Orchestra (2002). Publ in: Agon C. et al.(eds.): The OM Composer's book 1. IRCAM, Paris 2006. p.107-127.
[13] Kretz, Johannes: Continuous Gestures of Structured Material, Experiences in Computer Aided Composition, publ. in: Baboni-Schinlingi, J. (ed.): „Prisma 01". Milano 2003. p.185-194.

A predator-prey model with the nonlinear self interaction coupling $x^k y$

I. Kusbeyzi[†*], O. O. Aybar[†*] and A. S. Hacinliyan[†‡*]

[†]Yeditepe University, Department of Information Systems and Technologies, Istanbul, Turkey
[*]Gebze Institute of Technology, Department of Mathematics, Kocaeli, Turkey
[‡]Yeditepe University, Department of Physics, Istanbul, Turkey
Emails: oaybar@yeditepe.edu.tr, ikusbeyzi@yeditepe.edu.tr, ahacinliyan@yeditepe.edu.tr

Abstract: A class of predator–prey models suggested by the continuous form of the following two dimensional map is studied

$$X_{n+1} = aX_n(1 - X_n - Y_n)$$
$$Y_{n+1} = aX_n Y_n$$

The continuous time form of this map generalizes the classical Lotka Volterra model by a quadratic self interaction term; an additional coupling of the form $x^k y$ in the prey equation is added.
Keywords: predator–prey models, bifurcation analysis, stability and normal form, time series analysis.

1. Generalization involving $x^k y$ Coupling

There are two motivations for generalizing Lotka Volterra (henceforth referred to as LV) type predator prey models. Firstly, there is a simple relation between quadratic and cubic self interactions [1, 2, 3, 4, 5]. If one lets $x= u^2$ and $y= v^2$, we get the interaction coupling uv^2 in the prey and u^2v in predator equation. It would therefore be of interest to study the simplest nontrivial generalization, namely couplings of the form xy^k in the prey equation [4, 5, 6, 7]. Secondly, Maxwell Bloch type models can be reduced to LV type models where both experimentally observed intermittency and limit cycle behavior not seen in the classical LV model appear in this generalization. The predator equation is that of the classical LV [7, 8, 9, 10]. For a given value of the number k, the prototype form of the model involving this generalization, after making the variable changes $x \to ax$ and $y \to ay$ is:

Chaos Theory: Modeling, Simulation and Applications
C. H. Skiadas, I. Dimotikalis and C. Skiadas (Eds)

$$\dot{x} = x(a-1) - x^2 - xy - a^{-k+1}x^k y$$
$$\dot{y} = -y + xy$$

The specialized choice of parameters is dictated by the requirement that the system be compatible with a resonant normal form for integer values of the folding parameter a.

The following observations can be made:
The general form of equilibrium points of the system given above are $(a-1,0)$ and $(1,\frac{a-2}{1+a^{-k+1}})$ for all k. In addition to these points the system acquires one more equilibrium point at the origin if $k \geq 1$. In this case, there are three saddle points whose linearized eigenvalues are $\{-1,a-1\},\{1-a,a-2\}$ and a third set of eigenvalues which start out as a saddle point for $a \leq 2$, degenerate into a sink and eventually into a stable focus for values of $a \approx 2.3$, the value is relatively insensitive to the value of k as long as $k > 2$ [2, 3, 10, 11, 12].

More specifically, if $\frac{-3a+a^2-a^k-2ak+a^2k}{2(a+a^k)}$ is defined as Δ then the eigenvalues of the k dependent equilibrium point are

$$\Delta \pm \frac{\sqrt{-4(-2+a)(a+a^k)^2+(a^k+a(3-2k)+a^2(-1+k))^2}}{2(a+a^k)}$$

By the change of variables $x = r\cos\mu$ and $y = r\sin\mu$ and averaging over a full period of μ, the following result which shows that there exists an infinite a family of periodic orbits near the parameter value $a = 2$ due to the crudity of the approximation independently of k [7, 8, 9, 13] .

$$\frac{dr^2}{dt} = (\frac{a}{2}-1)r^2$$

In order to understand this behavior, we pass to a rotating frame using $x = u(t)\cos\omega t + v(t)\sin\omega t$ and $y = -u(t)\sin\omega t + v(t)\cos\omega t$, where $u(t)$ and $v(t)$ are slowly varying functions. We can then average the system over a full period $0 \leq t \leq \frac{2\pi}{\omega}$. For $k = 2$ this yields the following system

$$\dot{u} = (\frac{a}{2} - 1)u - \omega v - \frac{u^2 v}{8a} - \frac{v^3}{8a}$$

$$\dot{v} = (\frac{a}{2} - 1)v + \omega u + \frac{uv^2}{8a} + \frac{u^3}{8a}$$

It is clear that for $a = 2$, the quadratic nonlinearity is the predominant one, the cubic terms drop oup if we form $\dot{u}u + \dot{v}v$. In particular, if $k = 2$, a family of periodic solutions are available, nonlinear terms persist. If $k = 3$ the same behavior occurs, but nonlinear terms vanish in the averaging stage. This explains both the behavior at $a = 2$ and the insensitivity of the results to the value of k [1,8,9, 10, 14].

The intermittency transition route has been observed in gain modulated interactively frequency doubled Nd:YAG lasers involve the second harmonics [14, 15, 16]. The Zeeman laser model with large cavity anisotropy also exhibits on-off intermittency [16]. The Maxwell - Bloch model that has been used in this work involve the fundamental mode. The $x^k y$ coupling modification was designed to overcome this limitation. Indeed intermittency has been observed.

When $a = 3$ and $k = 2$ there are three equilibrium points of the system in the linearized manner. Two of them, the origin and the point (2,0) are saddle points with the eigenvalues {2,-1} and {-2,1}. However the third equilibrium point (1,0.75) has the eigenvalues $\frac{1}{8}(-5 \mp i\sqrt{39})$ and gives rich dynamic behavior in its neighborhood. When k is chosen as bifurcation parameter around this point a subcritical Hopf point occurs at the point (1,0.2016) for $k \approx -0.25$ with eigenvalues $\{\mp i\}$. A family of limit cycles bifurcating from this Hopf point grows to achieve a Limit Point Cycle with double multiplier 1 indicating intermittency so the limit cycle manifold has a fold here.

2. Time Series Analysis

Time series derived from these models are examined for invariant parameters such as Lyapunov exponents, fractal dimension as a function of its parameters. The techniques used for this analysis is time series analysis [18, 19].

The generalized map for the LV system with quadratic self coupling in the prey equation is as follows:

$$X_{n+1} = aX_n - aX_n^2 - aX_nY_n - aX_n^kY_n$$
$$Y_{n+1} = bX_nY_n$$

For given values of k and *a=3* and *b=3.5*, the map diagrams are given as in Figure 1.

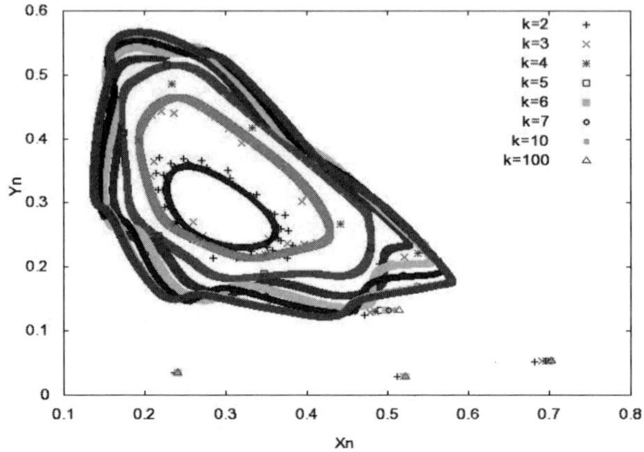

Figure 1. LV Maps

If a time series analysis approach is used, the results of the method of False Nearest Neighbors (FNN) indicate an embedding dimension of 3 irrespective of the value of *k*, as expected. This can be seen in Figure 2 [18, 19].

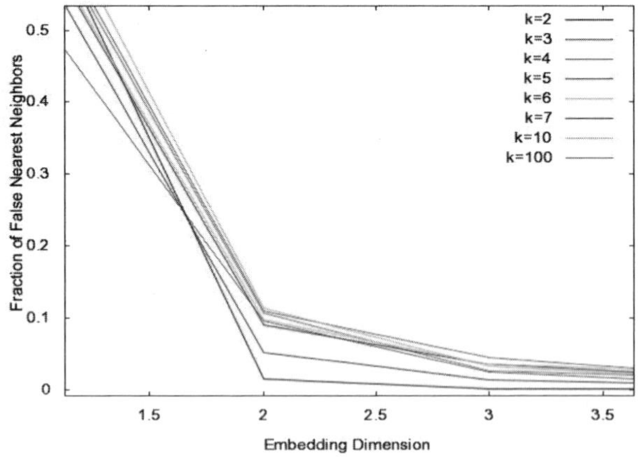

Figure 2. False Nearest Neighbors of LV Maps

On the other hand if the values of variables a and b are taken as 3 the map diagrams are given as in Figure 3.

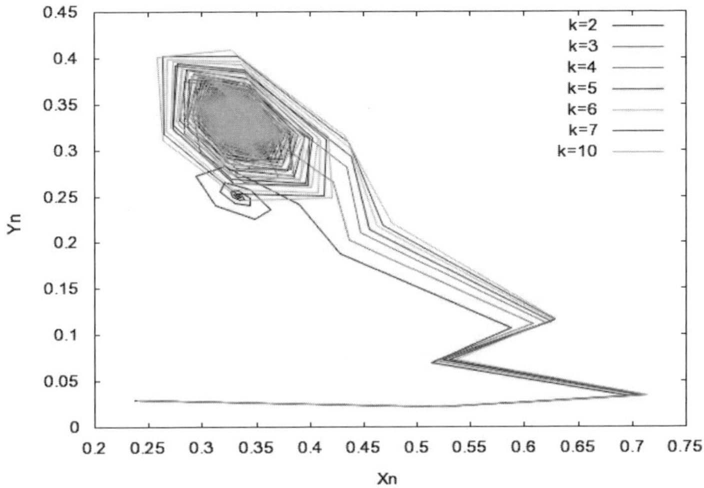

Figure 3. LV Maps

The mutual information curves are presented in Figure 4 below for this case. The maximal Lyapunov exponents are shown in Figure 5 below. The behavior near zero Lyapunov exponent has been enlarged in the inset [18, 19].

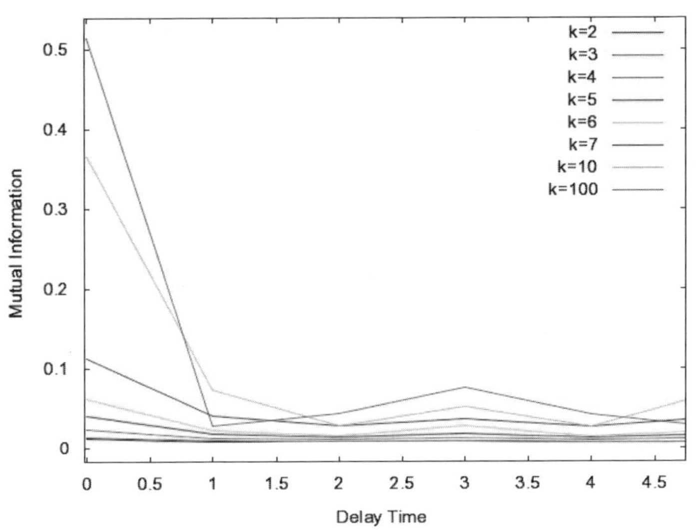

Figure 4. Mutual Information of LV Maps

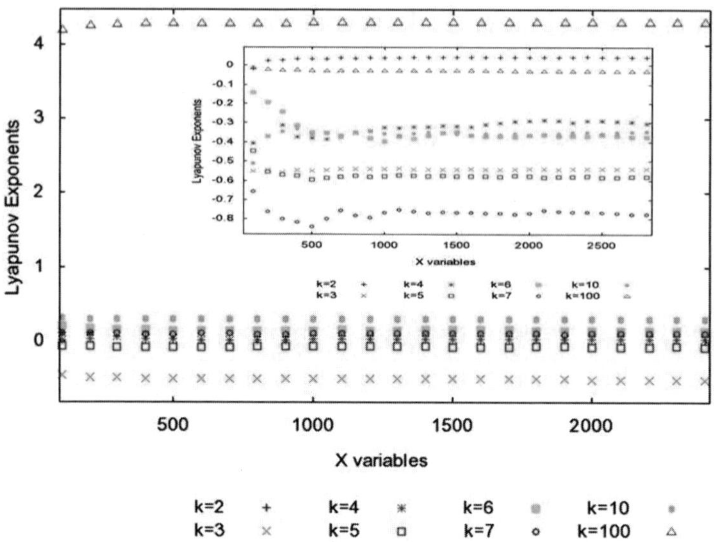

Figure 5 Maximal Lyapunov Exponents of LV Maps

3. Bifurcation Analysis

The corresponding LV differential equations for these maps are:

$$\dot{x} = (a-1)x - ax^2 - axy - ax^k y$$
$$\dot{y} = -y + bxy$$

In this generalization both parameters a and b can be used as bifurcation parameters. For starting values of $a=3$ and $b=3.5$ as analyzed in time series analysis there are three equilibrium points of the system which are (0,0), (0.666667,0) and (0.285714, 0.296719). The first two points are saddle points with the corresponding eigenvalues {2,-1} and {-2,1.33333} respectively, hence unstable, however later on we see that they are transcritical bifurcation points [1, 2, 3, 16, 17].

On the other hand the third equilibrium point is a stable one with the linearized eigenvalues $\{-0.555556 \pm i0.913354\}$. For bifurcation parameter a and $b=3.5$ there is a pitchfork bifurcation point for $a=1.4$ at (0.285714,0) indicated with one zero eigenvalue and a transcritical bifurcation point coinciding with the equilibrium point at the origin. Both the bifurcation diagram and the curve bifurcating from the pitchfork bifurcation point appear to be lines so that they form a shape of cross. The diagrams are given below in Figure 6 [2, 3, 15, 16].

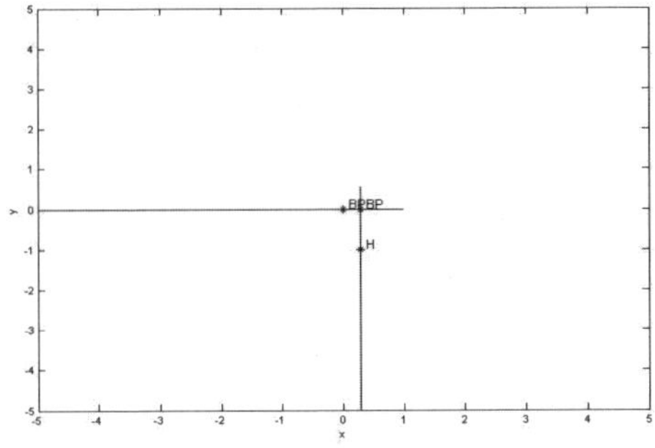

Figure 6. Bifurcation diagrams for the bifurcation parameter a

For the bifurcation parameter *b* the second equilibrium point is achieved to give a transcritical bifurcation point at *b=1.5*. The diagram is given below in Figure 7 [1, 2, 13, 14].

For starting values of *a=3* and *b=3* there are again three equilibrium points namely (0,0), (2/3,0), (1/3,1/4). The first two are again saddle points and the third a stable equilibrium with the eigenvalues $\{\frac{1}{8}(-5 \pm i\sqrt{39})\}$ [15, 16].

There exists two transcritical bifurcation points, one at (0.33333,0) for *a=1.5* although this is not an equilibrium anymore and one at (0,0) for *a=1*. No pitchfork bifurcation has been observed in this case.The bifurcation diagrams are given in Figures 6 and 7.

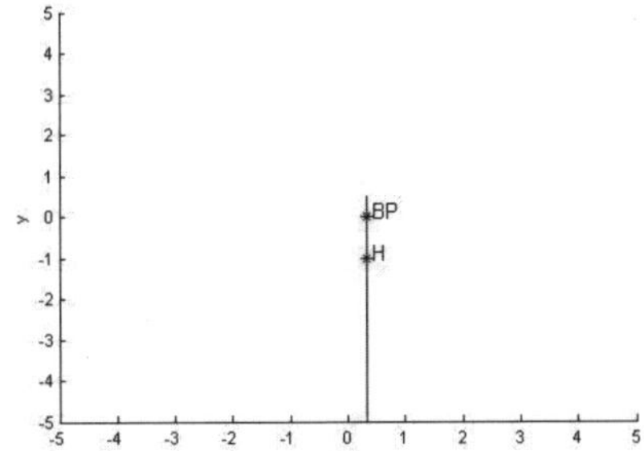

Figure 7. Bifurcation diagrams for the bifurcation parameter a

4. Conclusion

The Maxwell Bloch equations can be reduced to LV type equations, however they fail to deliver the whole picture (exponential depletion, no structurally stable oscilating regime, no limit cycle, no intermittency). Including the $x^k y + x^2$ terms adds both types of behavior to the laser equations. Results are insensitive to changes in *k* except for unusually large values of *k*, as long as *k*>2. The model is shown to have resonant normal forms corresponding to the added coupling terms for integer values of its parameter *a* [1, 2, 3, 14, 15, 18].

References

[1] A. S. Hacinliyan, O. O. Aybar, I. Kusbeyzi, I. Temizer and E. Akkaya. Maxwell–Bloch Eqautions as Predator-Prey System. *CHAOTIC SYSTEMS, Theory and Applications (CHAOS2009),* ISBN:978-981-4299-71-8. WorldScientific, 2010.

[2] I. Kusbeyzi, O. O. Aybar, A. S. Hacinliyan. Approximate solutions of Maxwell Bloch equations and possible Lotka Volterra type behavior. *Nonlinear Dynamics, Accepted for publication*, DOI: 10.1007/s11071-010-9695-5, 2010.

[3] I. Kusbeyzi and A. S. Hacınlıyan. Bifurcation scenarios of some modified predator-prey nonlinear systems. *J. Appl. Funct. Anal.* 4, 3, 519-527, 2009.

[4] A. J. Lotka. *Elements of Physical Biology*. Baltimore, Williams and Wilkins Company, 1925.

[5] Y. Nutku. Hamiltonian Structure Of The Lotka - Volterra Equations. *Physics Letters A*, I:145, 1990.

[6] D. Ghosh and A. R. Chowdhury. On the Bifurcation Pattern and Normal Form in a Modified Predator - Prey Nonlinear System. *Journal of Computational and Nonlinear Dynamics*, 2: 267-273, 2007.

[7] P. Yu and G. Chen. The simplest parametrized normal forms of Hopf and generalized Hopf bifurcations. *Nonlinear Dynamics*, 50:297-313, 2007.

[8] H. Zhu, S. A. Campbell and G. S. K. Wolkowicz. Bifurcation Analysis of A Predator-Prey System With Nonmonotonic Functional Response. *SIAM J.APPL. MATH.*, 2 63:636-682, 2002.

[9] H.W. Broer, V. Naudot, R. Roussarie and K. Saleh. Bifurcations of a Predator - Prey Model With Nonmonotonic Response Function. *R. Acad. Sci. Paris*, Ser. I 341:601-604, 2005.

[10] D. Xiao. Multiple Bifurcations in a Delayed Predator - Prey System with Nonmonotonic Functional Response. *Journal of Differential Equations*, 176:494-510, 2001.

[11] B. Hernndez-Bermejo and V. Fairn. Lotka-Volterra representation of general nonlinear systems. *Mathematical Biosciences*, 1 140:1-32, 1997.

[12] C. Christopher. Normalizable, Integrable and Linearizable Saddle Points in the Lotka-Volterra System. *Qualitative Theory Of Dynamical Systems*, 5:11-61, 2004.

[13] X. Huang, Y. Wang and A. Cheng. Limit cycles in a cubic predator-prey differential system. *J. Korean Math. Soc.,* 43:4:829-843, 2006.

[14] X. Huang, Y. Wang and L. Zhu. One and three limit cycles in a cubic predator-prey system. *Math. Meth. Appl. Sci.,* 30:501-511, 2007.

[15] E. C. Zeeman and M. L. Zeeman. An n-dimensional competitive Lotka-Volterra system is generically determined by the edges of its carrying simplex. *Nonlinearity*, 15:2019-2032, 2002.

[16] Gyu Ug Kim, Han Tae Choo, Dong Ik Kim, Young-Jai Park, Sung-Huan Gong, and Chil-Min Kim, Transition through ono® intermittency in Nd:YAG laser systems pumped by laser diodes. *JOSA B*, Vol. 20, Issue 2, pp. 302-306 doi:10.1364/JOSAB.20.000302, (2003).

[17] J. Redondo, Eugenio Roldn and G. J. de Valcrcel, On-off Intermittency in a Zeeman laser model. *Physics Letters A* Volume 210, Issues 4-5, Pages 301-306 doi:10.1016/0375-9601(95)00879-9, (1996).

[18] Atak K., O. O. Aybar, G. Sahin, A. Hacınlıyan and Y. Skarlatos. Chaoticity analysis of the current through pure, hydrogenated and hydrophobically modified PEG-Si thin films under varying relative humidity. *Central European Journal of Physics*. 7, 3, 568-574, 2009.

[19] Hacınlıyan A., Y. Skarlatos, G. Sahin, K. Atak and O. O. Aybar. Possible Stretched Exponential Parametrization for Humidity Absorption in Polymers. *The European Physical Journal E*. 28, 4, 369-376, 2009.

Evidence for deterministic chaos in aperiodic oscillations of acute lymphoblastic leukemia cells in long-term culture

George I. Lambrou[a], Aristotelis Chatziioannou[b], Spiros Vlahopoulos[a], Maria Moschovi[a] and George P. Chrousos[a]

[a]*1st Department of Paediatrics, University of Athens, 11527 Goudi, Athens, Greece*
[b]*Institute of Biological Research & Biotechnology, National Hellenic Research, Foundation, 48 Vassileos Constantinou Ave., Greece*
Emails: chrousog@mail.nih.gov, achatzi@eie.gr

Abstract: Biological systems are dynamic and possess properties that depend on two key elements: initial conditions and the response of the system over time. Conceptualizing this on tumor models will influence conclusions drawn with regard to disease initiation and progression. Alterations in initial conditions dynamically reshape the properties of proliferating tumor cells. The present work aims to test the hypothesis of *Wolfrom et al.*, that proliferation shows evidence for deterministic chaos in a manner such that subtle differences in the initial conditions give rise to non-linear response behavior of the system. Their hypothesis, tested on adherent Fao rat hepatoma cells, provides evidence that these cells manifest aperiodic oscillations in their proliferation rate. We have tested this hypothesis with some modifications to the proposed experimental setup. We have used the acute lymphoblastic leukemia cell line CCRF-CEM, as it provides an excellent substrate for modeling proliferation dynamics. Measurements were taken at time points varying from 24h to 48h, extending the assayed populations beyond that of previous published reports that dealt with the complex dynamic behavior of animal cell populations. We conducted flow cytometry studies to examine the apoptotic and necrotic rate of the system, as well as DNA content changes of the cells over time. The cells exhibited a proliferation rate of nonlinear nature, as this rate presented oscillatory behavior. The obtained data have been fit in known models of growth, such as logistic and Gompertzian growth.
Keywords: proliferation, deterministic chaos, aperiodic oscillations, non-linearity, CCRF-CEM.

1. Introduction

The scope of the present work is to test the hypothesis posed by *Wolfrom et al.* that proliferation shows evidence for deterministic chaos. Biological systems are dynamic systems. The knowledge on how to determine a present state from the previous ones is critical within many areas, or applications, varying from cancer to insect population control. However, it has been proven a very tedious work to discover laws underlying biological systems, since on one hand it is not easy to model such systems due to their complexity, and on the other hand, biological dynamical systems posses

Chaos Theory: Modeling, Simulation and Applications
C. H. Skiadas, I. Dimotikalis and C. Skiadas (Eds)
 (pp. 239 - 246)

significant adaptation capabilities. Their hypothesis was tested on adherent Fao rat hepatoma cells and it was found that these cells manifest aperiodic oscillations in their proliferation rate, giving evidence for deterministic chaos. We tested this hypothesis, adding specific modifications to the previously published experimental setup. We used the acute lymphoblastic leukaemia cell line CCRF-CEM since it provided an excellent substrate for modelling proliferation dynamics. Several studies have been occupied with the complex dynamic behaviour of animal populations [1-4]. However, very little is known about the dynamics of tumour cell proliferation [5] and even less is known about the state of proliferation dynamics during oncogenesis; that is until cells reach an adequate population to be diagnosed. The data that can be collected from tumours, regarding their dynamic nature, can only happen after a tumour has been diagnosed, which usually is too late for the patient, as all the progress-determining steps have taken place. Therefore, *in vitro* systems provide an excellent opportunity to study effects that are impossible to measure *in vivo*. Most importantly, they enable the study of long-term behaviour, which is required when it comes to reaching conclusions with regards to non-linearity and chaotic system behaviour. This, in particular, is impossible to happen, even with primary cultures of cells, since they are short-lived (15-20 days) when untransformed, and the only way is the use of established cell lines obtained from different organisms. For that reason we developed a modelling approach so as to simulate the *in vivo* conditions as best as possible. Cells were seeded at a low initial concentration of 20 cells/μl. Since they grew in suspension we assumed that they reached an even/equal distribution in the media solution. Measurements were taken at least at two-day intervals, thus obtaining more than 80 measurements in total, exceeding the *Wolfrom et al.* protocol, which took around 40 measurements. Also, taking a sample from a liquid culture has minimum effects on the total cell population, since it is not essential to trypsinize in order to take the sample; trypsinization, a requirement for obtaining samples from adherent cell cultures, stresses the cells and changes their proliferation dynamics. Cells were passaged at regular intervals. This practically removed the dead cells from the system and the remaining cells were allowed to grow again in a fresh medium. This allowed modelling of the growth of a tumour, such as leukaemia, in a space with finite capacity. Removal of cells modelled the circulation that removes dead cells from a particular position in the organism. Cells were grown for approximately 150 days (5 month period) while as previously reported, Fao cells were kept in culture for 200-240 days in total. The nature of proliferation dynamics may give insight into the way that cells not only proliferate but also differentiate.

2. Materials and Methods

2.1. *The CCRF-CEM cell line:*

The CCRF-CEM (T-ALL) cell line was used as the model, obtained from the European Collection of Cell Cultures (ECACC, United Kingdom). The CCRF-CEM cell line, a $CD4^+$ [6] and $CD34^+$ presenting cell line [7], was initially obtained from the peripheral blood of a 2 year old Caucasian female. She was diagnosed with lymphosarcoma which progressed to acute lymphoblastic leukaemia later on [8]. The child underwent irradiation therapy and chemotherapy prior to obtaining the cell line. Although remission was achieved at various stages, the disease progressed rapidly.[8] The cell line has been observed to undergo minor changes after long-term culture, except for the presence of dense granules in the nucleoli.[9] Finally, the CCRF-CEM cell line has been reported to manifest autocrine catalase activity, which participates to its mechanisms of growth and progression.[10]

2.2. *Cell culture conditions:* Cells were grown in RPMI-1640 medium, 10% FBS and 0.1× Streptomycin/Penicillin at 37 °C, 5% CO_2 and ~100% humidity. Cells were cultured in 75cm^2 in total medium volume of 25ml. Cells were seeded at an initial concentration of 20 cells/μl and ~200 cells/μl and were fed at regular intervals thereafter. Medium changes took place by centrifugation at 1000 rpm for 10min, the supernatant was discarded and the remaining cells were rediluted in 25ml media and were allowed to grow.

2.3. *Measurements, experimental setup and model*: The CCRF-CEM cells grow in suspension and therefore give an excellent model of avascular growth. In addition, the following assumptions have been made for its proliferation: a) extracellular signal transduction takes place autocrinaly, b) the cell distribution at seeding and thereafter is considered to be uniform and c) nutrient supply was considered to be stable since cells were fed at regular time intervals. All measurements have been performed in triplicates.

Wolfrom and collaborators (*2000*), had counted the cell population at the end of a time period varying from 5 to 7 days. At the end of this period, cells were trypsinized, measured and then seeded at an initial concentration of 10^5 cells per flask.

In the present study, before every measurement, flasks were gently shaken in order to assure that the sample taken consisted of a representative, equally distributed population size. For the growth dynamics study of the cell culture system, an experimental setup was developed, where cells were assayed at least every 48h and the media renewed every 3-5 days. For the measurements, 200μl from each flask was taken and measured with a NIHON KOHDEN CellTaq-α hematology analyzer.

In that way more than 80 measurements were obtained in a period of 150 days (5 months).

2.4. *Mathematical model and analysis*: We used a one-dimensional representation based on the assumption that the present state of our system is dependent upon the previous one. So, our system is better described by the logistic equation, as $f(x_{n+1}) = kx_n(1-x_n)$ (**1**) and with respect to time $\dot{x}_n = kx_n(1-x_n)$ (**2**) (the *logistic differential equation*). Both equations belong to the family of logistic equations of the form $f(x) = kx(1-x)$ (**3**), where k is the proliferation constant. For the analysis of the data collected we have utilized phase-space and return maps and used the geometrical representation proposed by *Wolfrom et al (2000)*. In addition, we have tested the dependence on initial conditions by using two different starting population sizes. For testing the chaotic behavior of the system, we have calculated the *Lyapunov* exponent and searched for strange attractors or sources. Many methods have been proposed for the calculation of *Lyapunov* exponents and it is considered to be a difficult task [11, 12]. In our case, given the function which we have based our model on, we used for the approximate estimation of *Lyapunov* parameters the following definition: Let f be a smooth map on $\Re$. The *Lyapunov* number $L(x1)$ for the orbit $\{x_1, x_2, \ldots, x_n\}$ is defined as $L(x_1) = \lim_{n \to \infty} \left(|f'(x_1)| \ldots |f'(x_n)| \right)^{\frac{1}{n}}$ (4) if the limit exists. In conjunction, the *Lyapunov* exponent $h(x_1)$ is defined as $h(x_1) = \lim_{n \to \infty} \frac{1}{n} \left(\ln|f'(x_1)| + \ldots + \ln|f'(x_n)| \right)$ (5).

3. Results

We have studied the proliferation dynamics of an acute lymphoblastic leukaemia cell line by developing a modelling approach, where cells from two different initial populations were allowed to grow with a periodic nutrient supply in order to sustain cell growth.

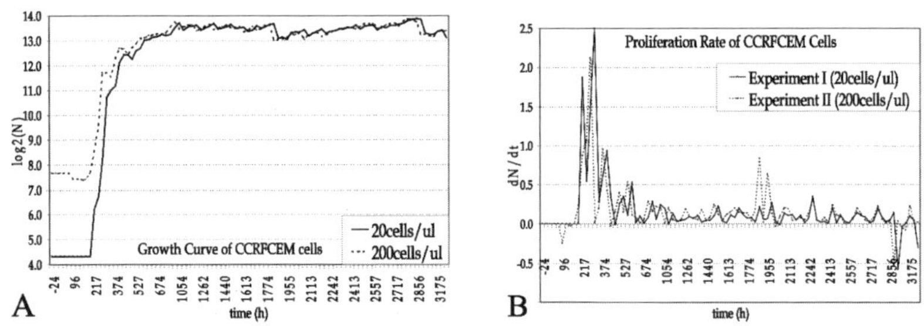

FIGURE 1. *Growth curve of the CCRF-CEM cells as a function of time shows the characteristic pattern of the iterated logistic equation (**A**). Proliferation rate of cells exhibits an almost unpredicted oscillatory behaviour (**B**). N is the actual measured cell population at time t. It is apparent that this oscillatory behavior has a declining tendency, something expected, since cells compete for space during their growth. Nutrients are considered to be abundant and equally distributed among the cells' environment.*

The time series produced from the experimental data showed a characteristic logistic pattern as described in (1), whereas proliferation rate with respect to time manifested an aperiodic oscillatory behavior (Fig. 1A and 1B). In fact, proliferation rate appears to manifest a saltatory pattern where cells after a period of "adjustment" to the environment start dividing rapidly. This was the first evidence that the dynamics of cell population can manifest complex behavior. Interestingly, when calculating the *Lyapunov exponent* of the two curves starting from different initial conditions, these gave different results (Fig. 2A). For the curve describing cell proliferation from 20cells/ul, the *Lyapunov exponent h* was calculated to be >0, whereas for the curve describing cells starting from 200cells/ul it was <0. Furthermore, for the orbit *Lyapunov exponent* was 0.13 for x_0=20 and -0.09 for x_0=200. The next criterion we investigated was whether the state-space manifested an asymptotically periodic pattern or not. Space-space representation has been proposed by Lorenz [13] and has been used by others in biological time-series [14]. State-space maps showed that irrespectively of the initial population (either 20 cells/ul or 200 cells/ul) both curves converged after a long period of cell culture (Fig. 2)

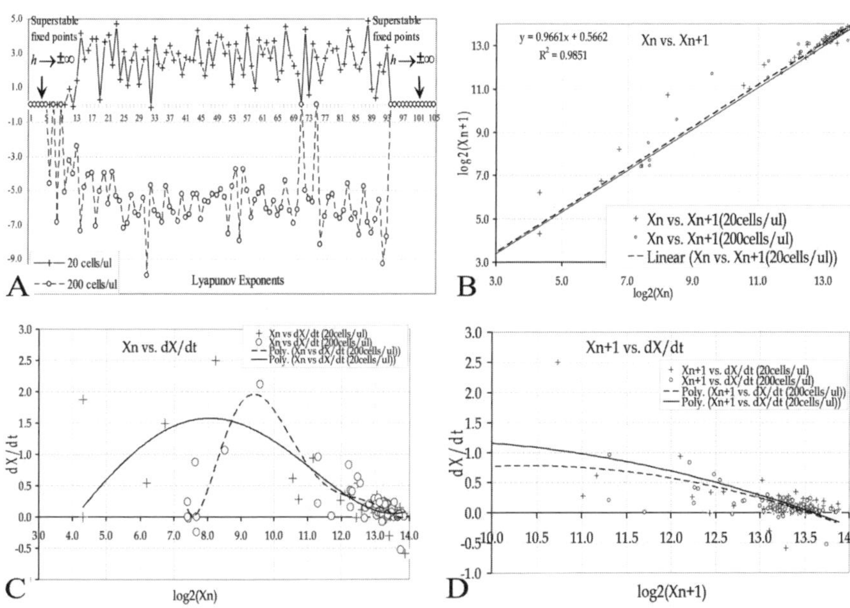

FIGURE 2. State-space of the growth curves of CCRF-CEM cells. *Cells were grown from two different initial populations: 20 cells/ul* (+) *and 200 cells/ul* (o). *Solid lines represent fitting curves for cells started from 200 cells/ul, while dashed lines represent fitting curves for cells started from 20 cells/ul.*

*When calculated for the two growth curves, Lyapunov exponents, gave opposite results (**A.**). Both orbits are asymptotically periodic. However, positive Lyapunov exponents were obtained only for x_0=20 cells/ul, hence it can be assumed that proliferation starting from 20cells/ul has chaotic orbits. Plotting the population x_n vs. x_{n+1} shows that both populations converge to a fixed plane (**B**) which, however, does not manifest asymptotical periodicity. Treating equation (2) as a two-dimensional function shows that the proliferation rate converges to a fixed plane as a function of x_n (**C**) and x_{n+1} (**D**).*

Probably our system reaches a steady-state in the long term (Fig. 2) or at least converges. Again, at first quite interesting was the fact that when we drew the return map of our initial data i.e. the population time-series, it appeared that it gave an almost perfect linear pattern. This was noteworthy because such a fact would mean that oscillatory behavior observed in the experimental measurements of the time-series was probably due to noise. To test this we performed a *Fast Fourier Transformation* (*FFT*) analysis (data not shown) which did not show the prevalence of noise. On the other hand, how does nonlinearity emerge? The logistic equation, on which we based our model, assumes that a population grows and then decays due to a limited amount of nutrients. In our model this was not the case since we kept a constant supply of nutrients in order to keep cells growing. This led us to the conclusion that the nonlinear factor was not the population *per se* but rather the proliferation constant k. In our initial assumptions we assumed that nutrients were abundant and constant, whereas the changing environmental factor was space for growth. Space becomes limited, as time progresses, activating inhibitory mechanisms for growth. Therefore, when drawing the populations at time t_n and t_{n+1} as a function of the proliferation rate (which is equal to $f'(x)$ for discrete time points) we observed that although the system converges towards its possible steady-state it manifests nonlinear behavior (Fig. 2B, 2C). Therefore, we analyzed the proliferation rates calculated for each discrete point in the same way. In other words, we used the return map of the proliferation rates, the first derivative of the proliferation curve at each discrete point. So, let $k_n=\{f'(x_0), f'(x_1),\ldots,f'(x_n)\}$ and $k_{n+1}=\{f'(x_1), f'(x_2),\ldots,f'(x_{n+1})\}$ where k is the proliferation rate from equations (1) and (2). K can also be written as $\frac{dX_n}{dt_n}$ and $\frac{dX_{n+1}}{dt_{n+1}}$ for t at n and $n+1$ respectively with $n \in \mathrm{N}$ as presented in Figure 3B. Interestingly, drawing k_n as a function of k_{n+1} gives us a curve with at least two fixed points for both initial cell populations indicating a period-3 orbit (Fig. 3A). This period-3 orbit implies chaos as it has been previously reported [15].

4. Conclusion

It has been reported previously that biological systems exhibit very complicated dynamics [3, 5]. The present work addressed the question as to whether chaotic dynamics could be detected in the proliferation of acute lymphoblastic leukaemia cells in long-term culture. This had been proposed previously for adherent cell lines [5]. To the best of our knowledge no reports have been published on the proliferation dynamics of suspension cell cultures and in general studies on proliferation dynamics *in vitro* are scarce. The question that arises at this point is whether such studies are meaningful, since *in vitro* systems represent *in vivo* systems only in part.

In vitro systems offer the capacity of performing long-term studies, and isolating the system under study to reduce noise, both of which are not possible in *in vivo* systems.

Cell cultures *in vitro* are considered to manifest a linear pattern of growth. Given the fact that the logistic equation takes into account limited nutrients and space, it predicts that a cell population would reach a steady-state within a certain time and eventually die out.

We have introduced a new constant to our model by making nutrient readily available but keeping space, in terms of total volume, constant. Under these conditions we draw the proliferation curve, which shows aperiodic oscillations, whereas the proliferation rate manifests these aperiodic oscillations most clearly. As it was shown in Fig. 2A, the return map of Fig.1A gave an almost perfectly linear function, where a "bell-shaped" geometry was expected.

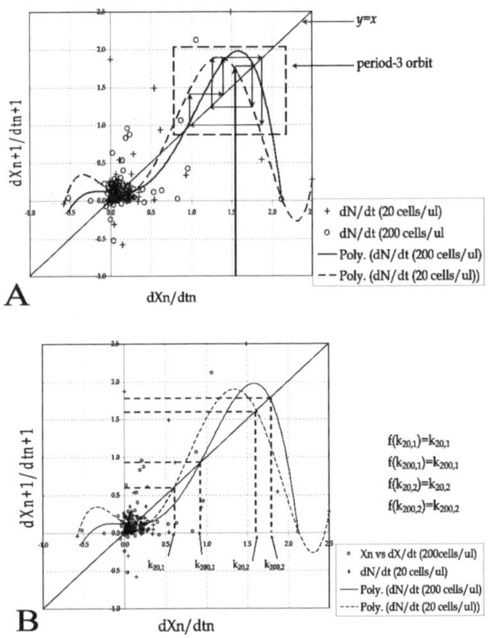

FIGURE 3. *Drawing the constant of proliferation* k_n $\left(\frac{dx_{n+1}}{dt_{n+1}}\right)$ *vs.* $k_{n+1}\left(\frac{dx_n}{dt_n}\right)$ *produces a bell-shaped curve (**B**) with at least two fixed points and period-3 orbit (**A**). The diagonal line is the y=x function, which high lights the fixed points. Solid curve line represents the polynomial fitting of cell population started from 200 cells/ul* (o). *Dashed curve line represents the polynomial fitting of the cell population started from 20 cells/ul* (+).

However, when we expanded our analysis to the return map of proliferation rate (Fig. 3B) we observed a nonlinear behaviour and we found geometrically that the fixed points on this curve manifest a period-3 orbit (Fig. 3A) which implies chaos dynamics. Our work shows evidence for deterministic chaos in the proliferative behaviour of leukaemia cells *in vitro*. Since this is a very complicated phenomenon it requires a lot more effort to understand the mechanisms underlying those dynamics. The implications from the understanding of these systems are tremendous. It will give us insight to the mechanisms of disease progression, such as in cancer, and enable building advanced models for the disease, which combine important features of both *in vitro* and *in vivo* systems. It is known that cancer starts and progresses slowly, at least before clinical presentation. Knowledge on the mechanisms of growth before the clinical symptoms become obvious may contribute to the early treatment of this and other diseases.

References

1. May, RM. Simple mathematical models with very complicated dynamics. *Nature* **261**(5560), p. 459-67(1976).
2. May, RM. Biological populations with nonoverlapping generations: stable points, stable cycles, and chaos. *Science* **186**(4164), p. 645-7(1974).
3. Mackey, MC and Glass, L. Oscillation and chaos in physiological control systems. *Science* **197**(4300), p. 287-9(1977).
4. Romond, PC, Rustici, M, Gonze, D, and Goldbeter, A. Alternating oscillations and chaos in a model of two coupled biochemical oscillators driving successive phases of the cell cycle. *Ann N Y Acad Sci* **879**, p. 180-93(1999).
5. Wolfrom, C, Chau, NP, Maigne, J, Lambert, JC, Ducot, B, Guerroui, S, and Deschatrette, J. Evidence for deterministic chaos in aperiodic oscillations of proliferative activity in long-term cultured Fao hepatoma cells. *J Cell Sci* **113 (Pt 6)**, p. 1069-74(2000).
6. Miranda, L, Wolf, J, Pichuantes, S, Duke, R, and Franzusoff, A. Isolation of the human PC6 gene encoding the putative host protease for HIV-1 gp160 processing in CD4+ T lymphocytes. *Proc Natl Acad Sci U S A* **93**(15), p. 7695-700(1996).
7. Naujokat, C, Sezer, O, Zinke, H, Leclere, A, Hauptmann, S, and Possinger, K. Proteasome inhibitors induced caspase-dependent apoptosis and accumulation of p21WAF1/Cip1 in human immature leukemic cells. *Eur J Haematol* **65**(4), p. 221-36(2000).
8. Foley, GE, Lazarus, H, Farber, S, Uzman, BG, Boone, BA, and McCarthy, RE. Continuous Culture of Human Lymphoblasts from Peripheral Blood of a Child with Acute Leukemia. *Cancer* **18**, p. 522-9(1965).
9. Uzman, BG, Foley, GE, Farber, S, and Lazarus, H. Morphologic variations in human leukemic lymphoblasts (CCRF-CEM cells) after long-term culture and exposure to chemotherapeutic agents. A study with the electron microscope. *Cancer* **19**(11), p. 1725-42(1966).
10. Sandstrom, PA and Buttke, TM. Autocrine production of extracellular catalase prevents apoptosis of the human CEM T-cell line in serum-free medium. *Proc Natl Acad Sci U S A* **90**(10), p. 4708-12(1993).
11. de Souza, SLT and Caldas, IL. Calculation of Lyapunov exponents in systems with impacts. *Chaos, Solitons & Fractals* **19**, p. 569-579(2004).
12. Mueller, PC. Calculation of Lyapunov Exponents for Dynamical Systems with Discontinuities. *Chaos, Solitons & Fractals* **5**(9), p. 1671-1681(1995).
13. Lorenz, N. Deterministic Nonperiodic Flow. *Journal of the Atmospheric Sciences* **20**(2), p. 130-157(1963).
14. Strizhak, P, Magura, IS, Yatsimirskii, K, and Masyuk, AI. Return Map Approach to Description of the Deterministic Chaos in Cytosolic Calcium Oscillations. *Journal of Biological Physics* **21**, p. 233-239(1995).
15. Li, TY and Yorke, JA. Period three implies chaos. *American Mathematical Monthly* **82**, p. 985-992(1975).

Scale invariance in chaotic time series: Classical and quantum examples

Emmanuel Landa[1], Irving O. Morales[1], Pavel Stránský[1,3], Rubén Fossion[1], Victor Velàzquez[2], J. C. López Vieyra[1] and Alejandro Frank[1]

[1]Institute of Nuclear Sciences, UNAM, México City, México
[2]Facultad de Ciencias, Universidad Nacional Autónoma de México, Mexico
[3]Institute of Particle and Nuclear Physics, Charles University in Prague, Czech Republic
Email: emmanuel.landa@nucleares.unam.mx

Abstract: Important aspects of chaotic behavior appear in systems of low dimension, as illustrated by the Map Module 1. It is indeed a remarkable fact that all systems tha make a transition from order to disorder display common properties, irrespective of their exacta functional form. We discuss evidence for $1/f$ power spectra in the chaotic time series associated in classical and quantum examples, the one-dimensional map module 1 and the spectrum of ^{48}Ca.

A Detrended Fluctuation Analysis (DFA) method is applied to investigate the scaling properties of the energy fluctuations in the spectrum of ^{48}Ca obtained with a large realistic shell model calculation (ANTOINE code) and with a random shell model (TBRE) calculation also in the time series obtained with the map mod 1. We compare the scale invariant properties of the ^{48}Ca nuclear spectrum sith similar analyses applied to the RMT ensambles GOE and GDE. A comparison with the corresponding power spectra is made in both cases. The possible consequences of the results are discussed.

Keywords: time series, fluctuation phenomena, $1/f$ noise, random processes and Brownian motion, nonlinear dynamic and chaos, quantum chaos, RMT.

1. Introduction

Classical chaos is understood as a non-linear phenomenon, which gives rise to an unpredictable time-evolution of the corresponding dynamical systems. In particular, it is characterized by an intrinsic instability in the orbits due to a high sensitivity to initial conditions. This is not to say that the physics describing the system is wrong, in fact, the physics predicts this chaotic motion, it is just extremely sensitive to the initial state of the system. In general, the dynamical instability of the orbits in a chaotic system is accompanied by the occurrence of strange attractors with a fractal structure in phase space e.g.in the Lorenz model.

The concept of *quantum chaos*, on the other hand, has no unique or precise definition as yet. Quantum chaos is often understood as the study of quantum manifestations of classical chaos. The sensitivity to the initial

Chaos Theory: Modeling, Simulation and Applications
C. H. Skiadas, I. Dimotikalis and C. Skiadas (Eds)

conditions is something which is not observed in quantum mechanics because it is a linear theory. However, both in classical and in quantum chaos "patterns of order in disorder" can be found: in classical chaos in terms of (strange) attractors in the dynamics, in quantum chaos universal laws have been proposed for the behaviour of fluctuations in excitation spectra. More in particular, it has been established that there is a relationship between energy-level fluctuation-properties of a quantum system and the dynamics of its classical analog. Quantum systems that have a classical analog that is chaotic, have spectral fluctuations that show a strong repulsion between energy levels and follow the predictions of Random-Matrix Theory, the Bohigas-Giannoni-Schmit-conjecture[1].

The notion of *scale invariance* appears in many different phenomena, in second order phase transitions it appears near the so called *critical points* where some physical quantities obey a power law behavior. In paricular, the correlation length, which is related the correlation function, near a critical point, it diverges and the system becomes scale invariant. Power law behavior has been observed in the study of chaotic time series,for example in the problem of a dripping faucet, in heartbeat and in many other phenomena. Recently, it was found that the power spectrum of the fluctuations of the eigenvalues of RMT ensembles and nuclear shell (TBRE) model calculations exhibit a power law behavior $1/f$ (with f being the frequency) whereas, for the case of integrable systems the corresponding power spectrum behaves as $1/f^2$ [2].

The aim of this contribution is to begin a study of the self-similar (or fractal) properties of the energy fluctuations in the spectrum of quantum chaotic systems and the fractal properties in classical chaotic time series that aim at a deeper understanding of the physical meaning of the exponent beta in power spectra that behave like the power law $1/f^\beta$. The contribution is organized as follows. Section 2 describes the scale invariance property with a elemental derivation. Seciton 3 is devoted to the discription of the method DFA in the study of discrete time series. Section 4 and Section 5 are devoted to the study of chaotic time series in classical (Map module 1) and quatum (Nuclear spectra of ^{48}Ca) examples respectively. Section 6 finally concludes.

2. Fractality and Scale Invariance

The outstanding feature of $1/f$ noise is that it is scale invariance, i.e. it looks the same for any choice of frequency or time unit, and for this reason it has been widely considered a prominent manifestation of the fractal character of many natural phenomena. The concept of a fractal is associated with geometrical objects satisfying two criteria: self similarity and fractional dimensionality. Self similarity means that an object is composed of sub-units and sub-sub-units on multiple levels that (statistically) resemble the structure of the whole object, but fractals are also presented in time: the fluctuations of a given quantity can appear the same when observed at different temporal resolutions. Thus, a fractal structure lacks any characteristic length scale. This fractal structure is seen, e.g., in the Lorenz attractor.

The *1/f* behavior of the power spectrum found in quantum fluctuations of the spectra of random hamiltonians[2] suggests the notion of scale invariance (the self-similarity is described by the power spectrum which measures the contribution of each frequency to the overall time series). In particular, the auto-correlation function of a *1/f* signal is scale independent, *i.e.*, the auto-correlation function is fractal. A demonstration in the continuum case is the following: suppose that the power spectrum[1] of a given time series has a *1/f* behavior, i.e.

$$S(f) = 1/f^{\beta}$$

Since the Fourier Transform of the power spectrum is identical to the autocorrelation function $C(\tau)$ (Wiener-Khinchin Theorem) and if we make a scale transformation in the time domain we have:

$$C(a\tau) = F^{-1}(\frac{1}{a}S(f/a)) = a^{\beta-1}F^{-1}(1/f^{\beta}) = a^{\beta-1}C(\tau).$$

Thus $C(a\tau)= C(\tau)$ with $\beta=1$. We have used the scaling property of Fourier Transforms, which is strictly valid only in the continuum case. For discrete time series, there are other tools for studying the scale invariant properties, including the DFA method [3]. The main result is then that a *1/f* behavior suggests scale invariance, this self-similar behavior is presented and the time correlations decay in a power law fhasion, systems with this behavior show a phase transition in the dynamic of the system. In the next section shows the correspondig discrete case study.

3. Detrended Fluctuation Analysis (DFA)

DFA is a method which allows the investigation of long range correlations and scaling properties in a random time series[3]. In the following we make a brief description of the DFA method. A time series $\delta(t)$ is self similar if the statistical properties of the full time series and the statistical properties of any rescaled subinterval of it, satisfy the scaling relation

$$\delta(t) \overset{PDF}{=} a^{\alpha}\delta(t/a),$$

where a is the scale factor in the *time* axis (a^{α} is the corresponding vertical scaling factor). The exponent alpha in the equation is defined as the self-similarity parameter. Let $\delta(i)$, $i=1 \ldots N$ be a time series, the DFA analysis of

---

[1] For a given time series δ_n, the power spectrum is defined as $S(f)=|F_f(\delta)|^2$, where $F_f(\delta)$ denotes the component of the discrete Fourier transform of δ, having frequency f.

$\delta(i)$ begins by defining an integrated time series

$$\psi(k) = \sum_{i=1}^{k} [\delta(i) - \langle \delta \rangle],$$

then the integrated time series is divided into boxes of equal length l, where a linear least-squares fit $\psi_l(k)$ (trend) is made. The difference (r.m.s.) between the integrated time series and the fit is measured by the *detrended fluctuation*

$$F(l) = \sqrt{1/N \sum_{k=1}^{N} [\psi(k) - \psi_l(k)]^2}.$$

This fluctuation can be calculated for all scale factors (or box sizes). In a *log-log* plot, a linear relationship between the fluctuation and the box size will indicate a scaling (power law) behavior. In this case the slope α_{DFA} in the *log[F(l)] vs log [l]* plot can be used to characterize the scaling properties of the original time series since $\alpha = \alpha_{DFA}$. As an example, if there is no correlation among the points in the original time series $\delta(i)$, i.e. the autocorrelation function $C(\tau) = 0$, for any time-lag $\tau \neq 0$, the time series behaves as white noise and the integrated time series corresponds to a random walk characterized by $\alpha_{DFA} = 0.5$[3]. Time series with short range (exponentially decaying) correlations $C(\tau) \sim e^{\tau/\tau 0}$, τ_0 being the characteristic scale, are also characterized by $\alpha_{DFA} \cong 0.5$ although some deviations from 0.5 may occur for small window sizes. Of special interest are the so called *persistent* (long time memory) time series for which the autocorrelation function has a power-like behavior. They are characterized by values $0.5<\alpha_{DFA}< 1$. The power spectrum of the corresponding time-series also displays a power-law behavior $S(f)\sim 1/f^{\beta}$ with $\beta=\alpha_{DFA}-1$, In particular, for time series with 1/f-noise (β=1) α_{DFA} =1[3].

4. Classical Case: Map Module 1

One of the characteristic properties of the strange attractors is a sensitive dependence on initial conditions. Discrete-time dynamical deterministic system, corresponding to one-dimensional maps of a given interval I,

$$f(x) = x + kx^2 (Mod1)$$

are perhaps the simplest models displaying a sensitive dependence on initial data. We study the dependence on initial conditions and related phenomena in the map module 1, repeated iteration of the above map equation yields a sequence that the value x_{n+1} at step $n+1$ depends explicitly only on the value x_n of the previous step n, it maps [0, 1] interval into itself as we see in the following discrete equation.

$$x_{n+1} = x_n + kx_n^2(Mod1) \quad x \in [0,1], \quad 0.9 < k \leq 1.10$$

k is called the control parameter, which plays a critical role to determine the dynamical behaviour of x_n. We study the dynamical behaviour of the Map Module 1 analyzing a global representation on the various regimes that are encountered as the control parameter k is varied. This can be done with the help of bifurcation diagrams and also with Lyapunov exponents, which are tools commonly used in nonlinear dynamic. Bifurcation diagrams display some characteristic properties of the asymptotic solution of a dynamical system as a function of a control parameter, allowing one to see at a glance where qualitative changes in the asymptotic solution occur. In our case we have a single dynamical variable, the bifurcation diagram is readily obtained by plotting a sample set of values of the sequence x_n as a function of the parameter k, similarly to the study has been done in the literature with the logistic map.

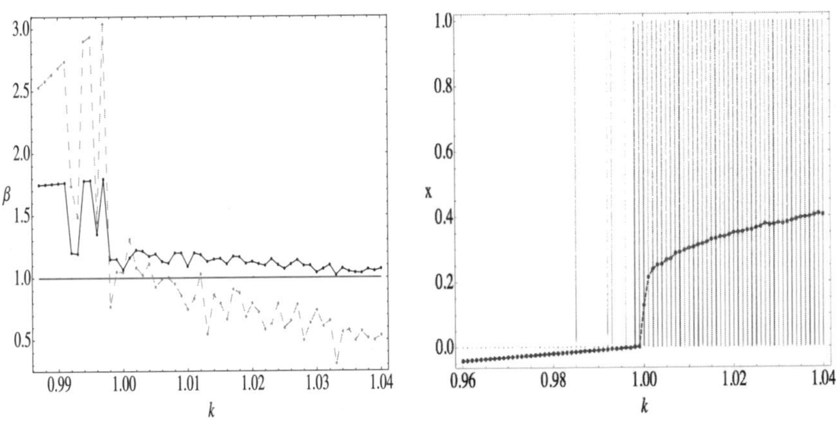

Figure 1. Fist panel of the figure on the vertical axis are the values of the slopes in the power spectrum ($S(f) \sim f^{\beta}$) blue dots, α_{DFA} values (red dots) and $\beta = \alpha_{DFA} - 1$ (green dots) and on the horizontal axis the value k that determines different time series. The second panel shows the Time series of the Map mod 1 plotted on the vertical axis as a function of the control parameter k, the time series are plotted vertically, the red points are the values of the Lyapunov exponents.

There are however, values of the control parameter in this range that lead to periodic behaviour of x_{n+1}. To exclude these periodic series we calculate for each value of k the value of the Lyapunov exponent λ_k, what follows, we will thus concentrate only on series x_{n+1} with $\lambda_k \geq 0$, a power spectrum study of the Map Module 1 has been carried out, we will apply the Detrended Fluctuation Analysis (DFA) technique on the *series of values* x_{n+1}. In literature, when *time series* are studied both in the classical as in the quantum world, what appears to be important are the *fluctuations* in the series and not the series itself. In the following, we want to study how to modify the behavior of the Map Module 1 as it approaches a critical value as reported in the literature [4], using k as a control parameter. To do so, the series of values x_n of the Map Module 1 must be "translated" into the fluctuations, of which the fluctuations can then be studied statistically. In our calculations; we used series of values x_{n+1} of dimension 3000. Once we have extracted the fluctuations we performed the calculations for the power spectrum and the DFA, we plot theses values in the first panel of the Fig. 1 showing blue values of the fit to the corresponding to the power spectrum, the slope values α_{DFA}(red dots) and the relation $\beta=\alpha_{DFA}-1$(green dots), where we see a correspondence of the beta exponent obtained with the fit in the power spectrum and the beta exponet found with the analysis of DFA when $k = 1$, also we can see in the second panel of the figure that the values of the Lyapunov exponents (red dots) see Fig. 2. Between these limiting values $-2<\beta<0$, β=-1 or $1/f$ noise, has been proposed as a particularly behaving system showing a fractal scale-invariant autocorrelation function. We must emphasize that the values in the power spectrum and DFA analysis after k=1 do not have good agreement being obtained with DFA analysis which agree with literature [4].

5. Quantum Case: Nuclear spectra

The fluctuations in a quantum spectrum are obtained by an unfolding procedure, *i.e.*, by substracting the gross features of the spectrum which can be modeled by a smooth function. In essence, this procedure consists in mapping the spectrum[2] E_i into a dimensionless spectrum ε_i, having a mean level density of 1:

$$E_i \rightarrow \varepsilon_i \equiv \bar{N}(E_i), \quad i = 1, \ldots, N.$$

where $N(E_i)$ is a smooth function fit of the staircase-like cumulative density function $N(E_i)$ (see e.g. [6]). In particular, the nearest neighbor spacing (NNS) is calculated as $s_i = \varepsilon_{i+1} - \varepsilon_{i+}$, $i=1,\ldots,N-1$ and <s>=1. The spectrum fluctuations can be defined by the quantity

$$\delta_n = \sum_{i=1}^{n} (s_i - \langle s \rangle) = [\varepsilon_{n+1} - \varepsilon_1] - n\langle s \rangle.$$

The stochastic discrete function δ_n measures the deviations of the distance between the first and the (n+1)-th unfolded states, with respect to the corresponding distance in a uniform (equally spaced) sequence having a unit level distance <s>=1. The sequence can be formally interpreted as a discrete

---

2 For the time being we assume a finite spectrum with N energy levels.

time series [2]. A more standard measure for the deviation from equal spacing is the Dyson-Metha *rigidity* function [5]. We have applied the DFA method and performed a spectral analysis to the energy fluctuations in the spectrum of ^{48}Ca. For comparison purposes we have also applied the analysis to the case of RMT ensembles GOE and GDE. In all cases the unfolding to the spectrum was done with a polynomial fit. After a careful analysis, a degree-7 polynomial fit was used in each case. However, the unfolding procedure is a delicate procedure when definining the energy fluctuations [6]. It can lead to wrong conclusions when not properly done.

^{48}Ca			
J^{π}	α	β	N
0^+	0.969	1.008	347
1^+	0.998	1.090	880
2^+	1.013	1.046	1390
3^+	1.020	1.183	1627
4^+	0.985	1.127	1755
5^+	0.916	1.198	1617
6^+	1.077	1.137	1426
7^+	1.095	1.180	1095
8^+	0.964	1.031	808

		α	β
^{48}Ca, $J^{\pi}=0^+$	(Shell Model)	0.969	1.008
	(TBRE)	1.003	0.987
GOE	$N=347$	0.951	0.998
	$N=1000$	0.942	1.069
GDE	$N=347$	1.338	1.604
	$N=1000$	1.398	1.786

Figure 2. Fist panel Table A, Self similarity parameter α obtained using a linear DFA method and the β exponent in the power spectrum of the energy fluctuations in the shell model calculations of ^{48}Ca with realistic interactions in different subspaces J^{π}. The dimension N of each subspace is also shown. Second panel Table B, comparison between shell model calculations with realistic interactions, random interactions (TBRE) and GOE and GDE calculations in the spectrum subspace $J^{\pi}=0^+$ of ^{48}Ca.

Large shell model calculations with realistic interactions (KB3) [7] were performed in the full *fp* shell for ^{48}Ca in the subspaces $J^{\pi}= 0^+,1^+,\ldots, 0^+$ by means of the ANTOINE code[7]. Within each subspace we calculated the energy fluctuations following the definition δ_n and applied a linear DFA analysis. The value of the self-similarity parameter α_{DFA} are found to be very close to 1 indicates an almost perfect non-trivial scale invariance, the largest deviations being 10% see Fig. 2 Table A.

In the present study we have also applied the DFA method to the energy fluctuations of the Two Body Random Ensemble (TBRE)[TBRE] shell model calculations for ^{48}Ca in the subspace $J^{\pi}=0^+$. For this calculations we have used 25 sets of energy levels. The self similarity parameter was calculated by an averaging procedure over the DFA results, and it was found to be α=1.01. This value is very similar to the value of the self similarity parameter obtained in the case of realistic case. For comparison purposes we applied the DFA method to the energy fluctuations in the case of GOE and GDE also a spectral analysis was made. In order to make a fair comparison we considered a GOE and GDE with the same dimension as the case of the subspace $J^{\pi}=0^{+}$, where both types of shell model calculations (with realistic and random interactions) were used. A set of 25 matrices in the ensemble was used. The results are reported in Table B in Figure 2.

6. Conclusions

In the clasical studies, Map Module 1, there are indications that a system carrying out a phase transition between an ordered and a disordered regime, the power spectrum of the fluctuations in the corresponding time series behaves like *1/f* near the critical point where chaos is setting in, $k=1$ and where the Lyapunov exponent value is set to near zero. We find in all cases that the DFA method is a more robust procedure than the direct calculation of spectral power when analyzing the data.
The DFA technique has been applied to the nuclear spectrum ^{48}Ca for different J^{π} states obtained with realistic and random Shell Model calculations. It was shown that the energy fluctuations in the spectrum of ^{48}Ca exhibit non-trivial scale invariance (fractal or self similar). This scaling behavior of the energy fluctuations was observed in random shell model (TBRE) calculations as well as in shell model calculations with realistic interactions.
Scale invariance was also observed in the case of GOE. Since scale invariance is observed in several classical chaotic phenomena, as well as in phase transitional critical points, these results suggest a possible underlying connection between classical and quantum chaos. This is an open question which we shall continue to investigate.

This work was supported by PAPIIT-UNAM (grant IN-113808), by DGAPA, UNAM, and by Conacyt, Mexico.

References

[1] Bohigas O., Giannoni M.-J., Schmit C., *Phys. Rev. Lett.* 52 (1984) 1.

[2] Relaño A., Gómez J.M.G., Molina R. A., Retamosa J. and Faleiro E., *Quantum Chaos and 1/f Noise*, *Phys. Rev. Lett.* 89, 244102 (2002)

[3] Peng, C.-K., Buldyrev, S.V., Havlin, S., Simons, M., Stanley, H. E. & Goldberger, A. L., *Phys. Rev.* E 49, 1685-1689 (1994).

[4] A. Ben-Mizrachi, I. Procaccia, N. Rosenberg, A. Schmidt, H. G. Schuster *Phys. Rev.* A 31, 1985.

[5] Mehta M.L., *Random Matrices*, 2nd ed., Academic Press, New York ñ London, 1991.

[6] E. Landa, I. Morales, C. hernández, V. Velàzquez, J.C López-Vieyra and A. Frank, *Rev. Mex. Fís.* S 54 (2008) 48.

[7] Kuo, T.T.S. and Brown G.E., *Nucl. Phys. A* 114, 241 (1968); Caurier E., computer code ANTOINE, CRN, Strasbourg (2000).

Acoustic emission within an atmospheric helium discharge jet

V. J. Law[1], C. E. Nwankire[2], D. P. Dowling[2] and S. Daniels[1]

[1]Dublin City University, National Center of Plasma Science and Technology, Collins Avenue, Glasnevin, Dublin 9, Dublin, Ireland
Email: vic.law@dcu.ie
[2]School of Electrical, Electronic and Mechanical Engineering, University College Dublin, Belfield, Dublin 4, Ireland

Abstract: This paper describes the thermal gas effluent and ion acoustic pressure wave interaction between the fundamental drive frequency and its harmonics within an atmospheric helium discharge. Deconvolution of the acoustic signal and the electrical signals reveal that the plasma jet undergoes a change in operational mode from chaotic (where the plasma is spatially and temporally inhomogeneous at the electrode surface) to stable (periodic in nature) when the plasma expands away from the electrodes and into the reactor cylinder. This effect is strongly influence by the helium flow and input power. In addition the generated acoustic signals is found to have a frequency response to that of a closed-end cylinder column which supports antinodes of n = 1 and 3. Decoding of the acoustic signal allows the helium thermal gas temperature to be obtained: T_{gas} ~ 290 K. The signal allows the axial gap distance between the jet nozzle and work surface to be estimated which has technology importance in terms of plasma metrology and in the basic understanding of atmospheric pressure plasma jet physics.
Keywords: corona discharge, phase-space diagrams, and acoustic emission.

1. Introduction

Atmospheric pressure plasmas are used in many engineering [1, 2], biological applications [3], and medicine [4]. In recent years multivariate analyses tools have been deployed to explore their plasma-surface interactions [5]. Acoustic metrology is one such promising plasma metrology tool that has been deployed on plasma welding processes [6] and reel-to-reel plasma process systems [7]. This paper is focused on the examination of a low temperature atmospheric pressure Corona discharge jet system called PlasmaStream™, which was developed by Dow Corning Plasma Solutions. Amongst the applications of this PlasmaStream™ source are: enhancing silicone adhesion to steel [8, 9], control of cell adhesion [10] antimicrobial applications [11] and the deposition of plasma polymerised coatings [12, 13, 14]. For this plasma device the initial gas breakdown produces a corona discharge which is spatially and temporally inhomogeneous, and then proceeds to stable and homogenous volume discharge at high flow rates and powers. Understanding this mode change in terms of electro-acoustic emission and gas flow dynamics shall help in the control of low temperature treatment of heat

Chaos Theory: Modeling, Simulation and Applications
C. H. Skiadas, I. Dimotikalis and C. Skiadas (Eds)

sensitive biological materials. The electro-acoustic being generated from the abundance of fast flow thermal molecules and rarer abundance of activated species (metastable and ions) that are rapidly quenched in the plume and at the surface. This paper reports on a systematic study of the non-invasive acoustic emission measurement of the PlasmaStream™ system as function of plasma nozzle to work-surface gap distance and helium gas flow. The analysis of which enables the physical geometry around the plasma-surface interaction to be understood in terms of gas flow dynamics, and surface response outcomes to discharge plume exposure. This paper is organised into the following sections. Section 2 provides the measurement methodology. Section 3 details the electrical measurements, and section 4 provides the acoustic measurements and deconvolution. Finally, section 5 forms the discussion of the data.

2. Measurement methodology

This work extends plasma acoustic metrology to an atmospheric helium discharge jet (see figure 1). The aim here is to correlate the plasma acoustic emission and electrical signals under varying spatial and temporal conditions by varying the applied power, gas flow rate and axial gap distance (d) between the jet nozzle and the work-surface.

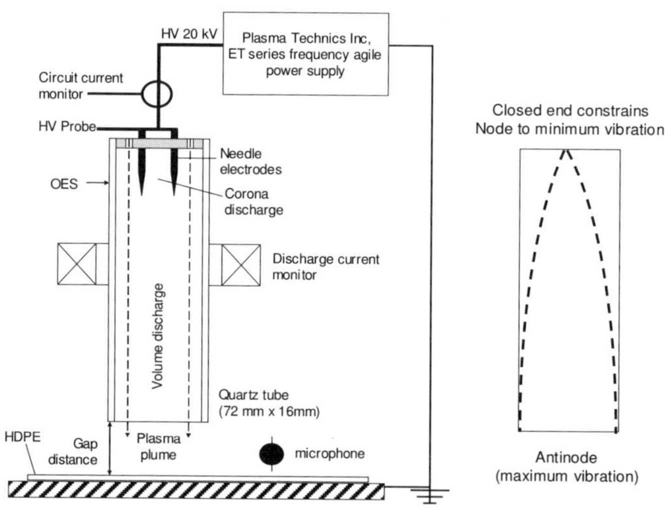

Figure 1: Schematic of Corona discharge atmospheric pressure jet, and model of closed cylinder column.

The approach employed here is to model the discharge plasma jet as an acoustic source that comprises a closed cylinder helium column of length (L = 72 mm + end correction), c_{sound} is the sound velocity of the gas and where the acoustic molecular / ion oscillation frequency (x_n) is modelled in equation 1. For this equation to hold true an end correction factor of 0.6 x the cylinder radius is used [15]. For this reactor the end correction is 0.6 x 16 mm = 4.8 mm.

$$x_n = \frac{c}{4L} \tag{1}$$

For a closed cylinder column, the aperture defines the maximum pressure vibration antinode, *n*, which is always an odd number. The initial value of c_{sound} is determined from equation 2.

$$c_{sound} = \sqrt{\frac{\gamma R T_{gas}}{M}} \tag{2}$$

In equation 2, γ is adiabatic constant for helium (1.6), *R* is the gas constant (8.123), T_{gas} is the helium gas temperature, and M is the molecular mass of helium (0.004 kg mol^{-1}).

For the Corona discharge cylinder jet geometries (see figure 1) reported here, equation 1 yields an acoustic frequencies of X_1 ~ 3.3 kHz, X_3 ~ 10 kHz and X_5 ~ 16.5 kHz when the helium column is not baffled. Here we note that when the jet nozzle is at an extended distance (~10 to 7 mm) from the work-surface the helium jet aperture (201 mm^2) is unrestricted. However as the jet nozzle is moved progressively closer to the work-surface the diminishing annular aperture ratio ((π x nozzle diameter) x d) between the jet nozzle and work-surface acts as a second aperture, resulting in an increased back-pressure in the processing zone. To investigate the effect of second aperture on the plasma acoustic emission, acoustic measurements of the plasma jet close to the equality of nozzle-surface aperture area / nozzle aperture area = 1. Here the values of d = 2.5 mm, 5 mm and 7.5 mm are used which correspond to aperture ratio values of 1.8, 1.25, and 0.62.

2.1. Electrical measurements

The current voltage phase-space diagram, for 1, 4 and 10 standard litre per minute (slm) of helium flow rates are shown in figure 2, where the current is plotted on the x-axis and the voltage on the y-axis. The discharge impedance is given along with the power; the data is plotted over 3 periods. For the 1 slm helium flow rate case (figure 2a), there is a chaotic, or strange, attractor, where one positive current spike at maximum negative voltage is generated along with a series of negative current spikes as the voltage becomes positive In the positive voltage region, the negative current spikes have a chaotic behaviour (that is for the three the periods the negative current spikes do not overlay each other). With increasing helium flow the phase space attractor become periodic, or cyclic, where a single negative current spike is produced and grows in magnitude and width. The discharge impedance and the drive frequency also falls. The associated digital image of each discharge is shown to the right of each phase space figure. These images illustrates that the discharge is constricted around the electrodes at 1 slm, and then progressively expands in the chamber as the helium flow is increased. This change in current waveform morphology is accompanied by an expansion of the discharge into the Quartz cylinder.

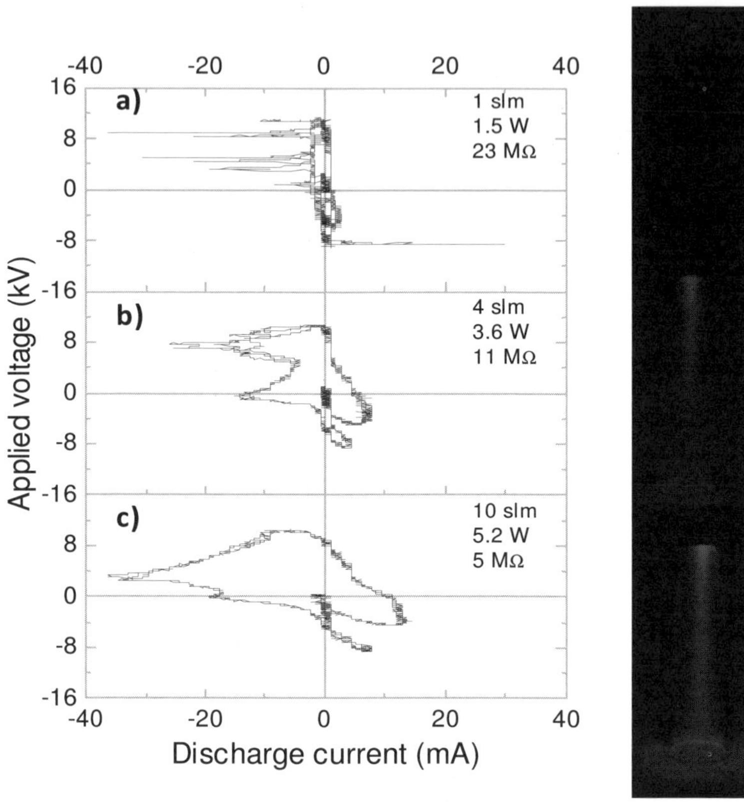

Figure 2: Current-voltage phase space diagram as a function of molecular helium flow rate for a fixed gap height of 2.5 mm. To the right of each phase-space portrait is the time average discharge image.

3. Acoustic measurements and deconvolution

An omni directional condenser microphone placed next to the work piece and within 20 mm of the plasma plume is used to capture the plasma sound pressure energy. The information is processed with the soundcard of a computer and analysed in the frequency-domain using LabVIEW 8.3 software [5 and 7]. For the experiments reported here the sound level of the discharge was measured to be between 86 to 88 dB(A). Figure 3 depicts the acoustic base-line of the system with no helium plasma and the power set to zero (black line) and with the plasma present (blue line) for a discharge power = 5.2 W and helium flow rate = 5 slm. In this figure the base-line acoustic power spectral density is confined to 0 to 5 kHz with discrete emission peaks at the drive frequency of the power supply 17.87 kHz and its harmonics. The helium plasma however exhibits the predicated acoustic emissions at n = 3.5 kHz and n = 9.5 kHz with -3db bandwidths of ~2 kHz. The fifth node may be present but is difficult to identify due to masking by the power supply frequency. Further features of note are: acoustic signal of the drive frequency is highly attenuated with harmonic number (-10dB per harmonic integer) and each consecutive bandwidth increases from 0.2 kHz at -10 dB at the fundamental to 1 kHz at the -10 dB level for the 4th harmonic.

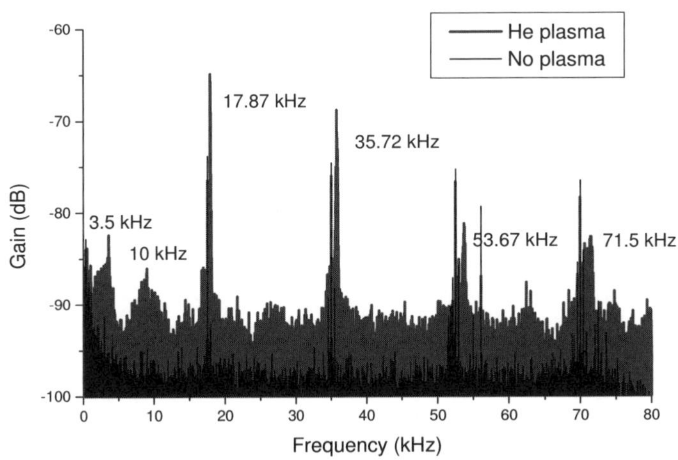

Figure 3: Base-line acoustic measurements identifying predicted helium acoustic nodes plus power supply drive frequency.

The following three figures (Figures 4, 5 and 6) survey the acoustic emission range from 0 to 70 kHz as a function of axial gap distance (2.5, 5 and 7.5 mm) for fixed helium flow rates of 2, 4, 6 and 8 slm. In this survey the discharge power is held constant at 5.2 W.

The analysis approach taken here is to deconvolve the complex acoustic emission into peaks which have specific origins. For this to be successful the plasma is required to have a degree of spatial and temporal stability, where the phase relation between signals of different origins is preserved. This intermodulation process is most likely to be present in the plasma volume expansion mode and to a lesser degree in the Corona discharge mode. This intermodulation between periodic time-vary signals is mathematically expressed in equation 3.

$$f_n \pm x \rightarrow f_{n-x}, f_{n+x} \quad (3)$$

Where f_n is the fundamental drive frequency or one of its harmonics, and X is the identified acoustic frequency due to molecule / ion oscillations.

The acoustic measurement (0 to 70 kHz) for d = 2.5 mm and an aperture ratio 0.62 is shown in Figure 4. The experimental measurement data reveals that the acoustic signals originating from the molecules / ion oscillations (3.5 and 10 kHz) increase in amplitude to +10 dB above the noise floor as the helium gas flow is increased from 2 to 8 slm. There is also a degree of intermodulation with the drive frequency and its harmonics. For example in the of the 8 slm case the flowing relationships are observed:

X_1 (3.5 kHz) mixes with f_1 (18 kHz) to produce sidebands at 15 kHz and 21 kHz

X_2 (9.5 kHz) mixes with f_1 (18 kHz) to produce sidebands at 8.5 kHz and 27.5 kHz

The result of both of these intermodulation processes generates the appearance of a double peak in the 8 to 10 kHz frequency region. Further to this, the process also occurs on each of the drive frequency harmonics, but this more difficult to observe due to the noise floor of the measurement.

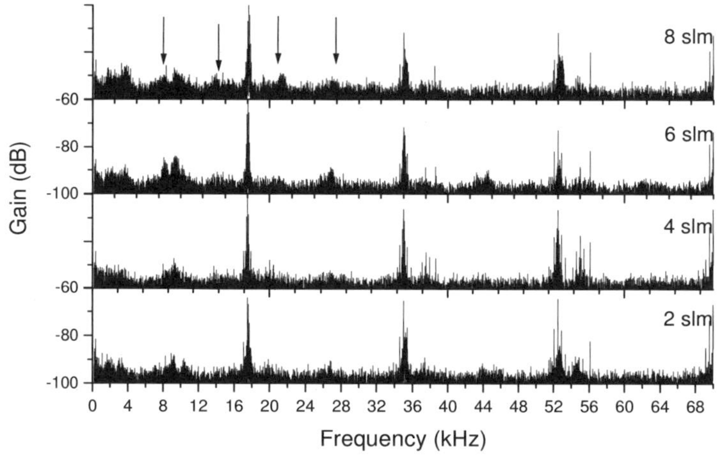

Figure 4: 2.5 mm gap distance at a discharge power of 5.2 W

Figure 5 shows the results for d = 5 mm and aperture ratio of 1.25. The first feature here to note is that drive frequency has been pulled to 17.6 kHz, indicting an increase in plasma capacitance to the work-surface [16]. Intermodulation process is also occurring. However the intermodulation of f_1 - $X_{n=3}$ product now overlays the original $X_{n=3}$.acoustic signal. This superposition generates strong sidebands around f_o and f_1 and f_2 at low helium flow rates.

Figure 6 provides further measurements for d = 7.5 mm, and aperture ratio of 1.8 at a discharge power of 5.2 W. Under these conditions the acoustic emission is found to be greater with respect to the small gap distance of 5 and 2.5 mm. The acoustic emissions are obtained some 15 dB above the noise floor at high helium flow rate (8 slm). Intermodulation between the molecular / ion acoustic signals and the drive and its harmonics are observed to be present at middle helium flow rates (6 and 4 slm). At low helium flow rates only X_1 interacts with the fundamental drive frequency.

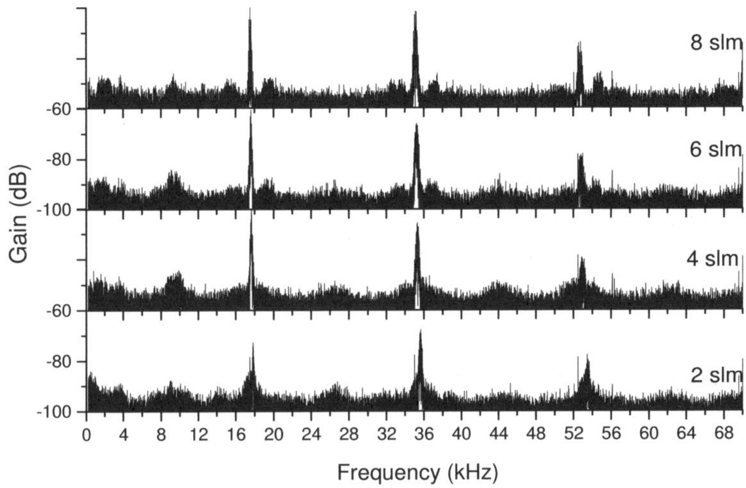

Figure 5: 5 mm gap distance at a discharge power of 5.2 W.

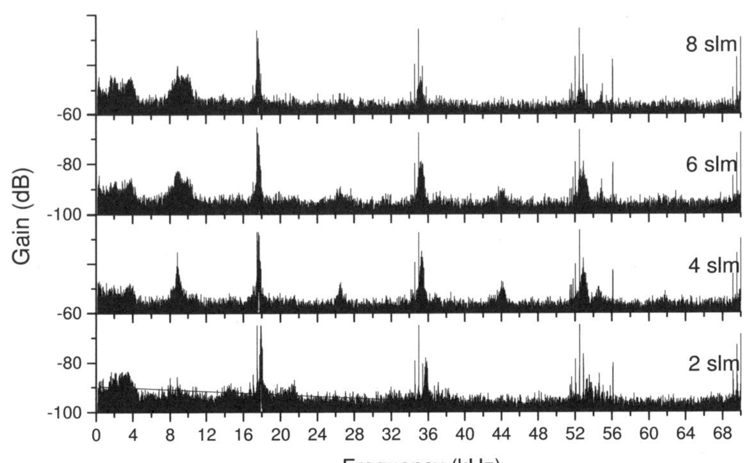

Figure 6: 7.5mm gap distance at 5.2 W discharge power.

4. Discussion

This work has examined the electro-acoustic emission from a low temperature discharge jet. In this configuration the jet exhibits a clarinet-like acoustic, or closed-end cylinder behaviour. The jet is found to have two operational modes: The electrical voltage-current phase-space diagrams map these two modes. The first mode is a Corona / Filamentary discharge [17] that has chaotic behaviour at low helium flow rates of 1-2 slm. Above this level of helium flow the voltage-current phase-space attractor becomes periodic (cyclic) with one major plasma ignition per cycle. Under these conditions, the discharge fills the cylinder with a visible plume issuing from the nozzle to the work-surface.

It is also found that there is a correlation of these modes with the electro-acoustic emission. When the discharge is in the Corona / Filamentary mode [17], the acoustic fundamental is present and the acoustic 3rd harmonic (overtone) is weakly supported. When the discharge fills the cylinder generating only one current event per cycle both the acoustic fundamental and it 3rd harmonic (overtone) under go an intermodulation with the power supply drive frequency.

The strongest acoustic measurement of the sound velocity is found at $X_{n=3}$ frequency (~10 kHz) at high flow rates when the plasma jet is at extended distance from the work-surface with an aperture ratio of 1.8. Under these conditions, the influence of the work-surface can be considered to be minimal.

As the axial gap distance is decreased to d = 5 mm and aperture ratio of 1.25, the first acoustic node ($X_{n=1}$ = 3.5 kHz) signal increase in strength and interacts with the drive frequency. Under these conditions, the second aperture can be considered to be influencing the gas flow properties within the plume-surface interaction zone.

Finally as the axial gap distance is reduced further (d = 2.5 mm) where the aperture ratio is 0.62, the $X_{n=3}$ frequency signal increase with helium gas flow.

Based on the electro-acoustic measurements made, it can be concluded that the reactor geometries provide a unique electro-acoustic emission signature. Within the acoustic signature an intermodulation process is identified with the power source drive frequency and its harmonics. A simple acoustic model (equations: 1 and 2) based on chamber geometry is used to describe the thermal molecular acoustic emission. The model predicts a series of odd number acoustic nodes (n = 1, 3, 5 ...). The value of n = 1 and 3 are confirmed by experimental measurements. The thermal gas temperature of the discharge effluent is found by using equation (2). Consider first the acoustic emission model (equation 2) for this reactor where $L = 0.0768$ m (72 mm + end correction of 4.8 mm) and a resonant fundamental frequency value of 3.2 kHz. From these values a gas velocity of 983 m s^{-1} is obtained which equates to a gas temperature of ~290 K (~17^0C). This value agrees well with the thermal gas temperature (18°C) reported in reference [14] using an alcohol thermometer. As the even 2nd harmonic (overtone) is not present (as predicted by the closed-end cylinder model (equation 1) the 3rd harmonic (overtone) may be used and does produce a similar gas velocity and helium gas temperature.

For the intermodulation process, electromagnetic interaction between activated low temperature (400 ±50 K) nitrogen ions and radicals [17] and drive frequency could be a possible mechanism. This relationship is mathematically in equation (3) as a RF mixing mechanism. Undoubtedly the intermodulation model must be more complex as the two electromagnetic sources have different origins: the electrodes for the drive frequency signal and the nozzle aperture for the acoustic. The coherence of these two signals will therefore be function of axial distance and position of the microphone. Nevertheless the electro-acoustic emission measurement proves a means of process monitoring by linking the acoustic signature to gas temperature and gap distance and mode of operation.

The above experimental observations lead to a number of non-invasive acoustic plasma metrology innovations to reference [7] and patent [18] in atmospheric pressure plasma processing.

5. Acknowledgements

This work is partly supported by Enterprise Ireland grant CDT/7/IT/304 and the Science foundation Ireland under grant 08/SRC/1141.

6. References

[1] K. Baba, T. Okada, T. Kaneko, R. Hatakeyama, H. Yoshiki. Investigation of gas–liquid interface in atmospheric-pressure micro plasma with solution. Thin Solid Films **515** 4308, 2007.

[2] B. Twomey, D. P. Dowling, G. Byrne, L. O'Neill and L O'Hare. Properties of siloxane coatings deposited in a reel-to-reel atmospheric pressure plasma system. Plasma Process and Polymers. **4,** 450-454, 2007.

[3] G. Fridman, A. D. Brooks, M. Balasubramanian, A. Fridman, A. Gutsol, V. N. Vasilets, H. Ayan, G. Friedman. Comparison of direct and indirect effects of non-thermal atmospheric-pressure plasma on bacteria. Plasma Process. Polym. **4**, 370, 2007.

[4] M. G. Kong, G. Kroesen, G. Morfill, T. Nosenko, T. Shimizu, J. van Dijk and J L Zimmermann. Plasma medicine: an introductory review. New Journal of Physics **11,** 11501 2009.

[5] V. J. Law, N. O'Connor, B. Twomey, D. P. Dowling and S. Daniels. Visualization of atmospheric pressure plasma electrical parameters. Topics of Chaotic Systems: Selected Papers of Chaos 2008 International Conference. 204-213, Ed. C H Skiadas, I Dimotikalis & C Skiadas. (World Scientific Publishing 2009). ISBN 978-981-4271-33-2

[6] Z. Szymanski, J. Hoffman and J. Kurzyna. Plasma plume oscillations during welding of thin metal sheets with a Co_2 laser. J Phys. D: Appl, Phys. 34 189-199, 2001.

[7] J. Tynan, V. J. Law, P Ward, A. M. Hynes J Cullen, G. Byrne, D. P. Dowling and S. Daniels. Comparison of pilot and industrial scale atmospheric pressure glow discharge systems including a novel electro-acoustic technique for process monitoring, PSST, **19**, 015015, 2010.

[8] C. E. Nwankire, M. Ardhaoui, and D. P. Dowling. The effect of plasma-polymerised silicon hydride-rich polyhydrogenmethylsiloxane on the adhesion of silicone elastomers Polym. Int. **58** 996-1001, 2009.

[9] L. O'Neill, N, Shephard, S. R. Leadley, and L. A. O'Hare. Atmospheric pressure plasma polymerised primer to promote adhesion of silicones. J. Adhesion **84** 562-77, 2008.

[10] M. Ardhaoui, M. Nassiri, C. Brugha, A K Keenan, M. Al-Rubeai, and D. P. Dowling. Evaluation of cell behaviour on atmospheric plasma deposited siloxane and fluorosiloxane coatings. Submitted to J. Adhesion Science and Technology. 2009.

[11] L. A. O'Hare, L. O'Neill, and A. J. Goodwin. Anti-microbial coatings by agent entrapment in coatings deposited via atmospheric pressure plasma liquid deposition J. Surf. Interface Anal. **38**, 2006.

[12] L. O'Neill, and C. O'Sullivan. Polymeric coatings deposited from an aerosol-assisted non-thermal plasma jet. Chemical Vapor Deposition. **15** 21-6, 2009.

[13] C. E. Nwankire and D. P. Dowling. Influence of nm thick atmospheric plasma pressure deposited coatings on silicon elastomer to stainless steel. Submitted to J. Adhesion Sci, Technol, 2009.

[14] P. A. F. Herbert, L. O'Neill and J Jarosynska-Wolinska. Soft plasma polymerization of gas state precursors from an atmospheric pressure corona plasma discharge. Chem, Mater, **21**, 4401-4407, 2009.

[15] H. Levine and J. Schwinger. On the radiation of sound from an unflanged circular pipe. Physical Review. **73**(4) 383-406, 1948.

[16] V. J. Law. Process induced oscillator frequency pulling and phase noise within plasma systems. Vacuum **82**(6) 630-638, 2008.

[17] C. E. Nwankire, V. J. Law, A Nindrayog, B. Twomey, K. Niemi, W. G. Graham, D. P. Dowling. Electrical, Thermal and optical diagnostics of the plasmastream atmospheric plasma jet system. Accepted for Plasma Chemistry, Polymer Processing, May 2010.

[18] Patent: GB 2459858 A (published: 11 November 2009), and Patent: WO 209/135919 A1 (published: 12 November 2009).

Experimental investigation of the enhancement factor for irregular undirected and directed microwave graphs

Michał Ławniczak, Szymon Bauch, Oleh Hul and Leszek Sirko

Institute of Physics, Polish Academy of Sciences, Al. Lotników 32/46, 02-668 Warszawa, Poland; Email: sirko@ifpan.edu.pl

Abstract: Quantum graphs are widely used to investigate properties of quantum chaos. Experimentally, quantum graphs are simulated by microwave graphs (networks) consisting of joints, microwave cables and other microwave components such as attenuators and circulators. This is possible due to an equivalency of the one-dimensional Schrödinger equation describing a quantum system and the telegraph equation describing an ideal microwave network. We present the results of the experimental study of the enhancement factor $W_{S,\beta}$ for irregular fully connected hexagon undirected and directed microwave graphs in the presence of absorption. We measured the two-port scattering matrix $\hat{S}$ for the undirected microwave graphs which statistical properties of eigenfrequencies, typical for the systems with time reversal symmetry (TRS), can be described by Gaussian Orthogonal Ensemble (GOE) in Random Matrix Theory (RMT), and for the directed graphs possessing the properties of the systems with broken time reversal symmetry, which can be described by Gaussian Unitary Ensemble (GUE) in RMT. The measurements were performed as a function of absorption, which was varied by microwave attenuators.

Keywords: quantum chaos, microwave networks, enhancement factor, scattering matrix, time reversal symmetry, broken time reversal symmetry.

1. Introduction

Quantum graphs consisting of one-dimensional wires were introduced by Pauling, more than seventy years ago [1]. His idea was used later to describe organic molecules [2] by free electron models. Quantum graphs can be considered as idealizations of physical networks in the limit where the widths of the wires are much smaller than their lengths, i.e. assuming that propagating waves remain in a single transversal mode. There are many systems modeled by quantum graphs - mesoscopic systems [3, 4], quantum wires [5, 6], excitation of fractons in fractal structures [7, 8], electromagnetic optical waveguides [9, 10], and microwave networks [11, 12].

Quantum graphs are perfect systems for studying quantum chaos [11–22] and localization phenomena [23–25]. One may study ideal graphs, without absorption as well as more realistic systems, open graphs with leads [14, 15].

In this paper we present results of experiments in which quantum graphs were simulated by microwave networks. For small absorption this is

Chaos Theory: Modeling, Simulation and Applications
C. H. Skiadas, I. Dimotikalis and C. Skiadas (Eds)

possible due the equivalence of the Schrödinger equation describing quantum systems and the telegraph equation describing microwave circuits [11]:

$$\frac{\partial^2}{\partial x^2}\Psi_{ij}(x)+k^2\Psi_{ij}(x)=0 \ , \tag{1a}$$

$$\frac{\partial^2}{\partial x^2}U_{ij}(x)+\frac{\omega^2\varepsilon}{c^2}U_{ij}(x)=0 \ . \tag{1b}$$

Assuming that $\hbar = 2m = 1$ there exist the following equivalencies: $\Psi_{ij}(x) \Leftrightarrow U_{ij}(x)$ and $k^2 \Leftrightarrow \frac{\omega^2\varepsilon}{c^2}$. In the case of bigger absorption microwave attenuators, which are mainly responsible for the losses in the system, can be described by the means of the optical potentials [12].

2. Experiment

A microwave network is a structure of N vertices connected by bonds e.g., coaxial cables. A coaxial cable consists of an inner conductor of radius r_1 surrounded by a concentric conductor of inner radius r_2. The space between the inner and the outer conductors is filled with a homogeneous material having a dielectric constant ε. For a frequency ν below the onset of the next TE_{11} mode only the fundamental TEM mode can propagate inside a coaxial cable. In our case the cut-off frequency of the TE_{11} mode is $\nu_c = \frac{c}{\pi(r_1+r_2)\sqrt{\varepsilon}} \approx 32.9$ GHz [26], where r_1 = 0.05 cm is the inner wire radius of the coaxial cable (SMA-RG402), while r_2 = 0.15 cm is the inner radius of the surrounding conductor, and ε = 2.08 is the Teflon dielectric constant [27, 28].

Primary, relatively weak, absorption of networks is due to losses in SMA cables. However, to change the absorption in the wider range, in the fully controllable way, we used microwave attenuators. The absorption strength is given by $\gamma = \frac{2\pi\Gamma}{\Delta}$, where Γ is the mean resonance spectrum width and Δ is the mean level spacing.

Microwave networks may simulate undirected quantum graphs with preserved time reversal symmetry (TRS) and the directed ones with broken TRS. The system without TRS is obtained by introducing to the network microwave circulators. A microwave circulator is a non-reciprocal three port passive device. A wave which enters port 1 of a circulator exits port 2, a wave into port 2 exits port 3, and finally a wave into port 3 exits port 1. In the experiment Anritsu PE8403 microwave circulators with the operating frequency range 7 – 14 GHz were used

We have measured the two-port scattering matrix

$$\hat{S} = \begin{bmatrix} S^{aa} & S^{ab} \\ S^{ba} & S^{bb} \end{bmatrix} \tag{2}$$

The elements S^{aa}, S^{bb} are connected with the signals reflected from the ports *a* and *b*, respectively, while the elements S^{ab}, S^{ba} are connected with the transmitted signals.

Figure 1 shows the experimental setup for measuring the two-port scattering matrix $\hat{S}$. We used Agilent E8364B microwave vector network analyzer to measure the scattering matrix $\hat{S}$ of fully connected hexagon microwave networks in the frequency window: 7–14 GHz. The networks were connected to the vector network analyzer through the leads - HP 85133-616 and HP 85133-617 flexible microwave cables - connected to the 6-joint vertices. The 5-joints were the other four vertices of the networks. The scattering matrices were measured for the networks with moderate absorption (without attenuators), and with strong absorption. In the latter case 7 or all fifteen bonds of the networks contained microwave attenuators. For measuring the two-port scattering matrix $\hat{S}$ of directed microwave networks simulating quantum graphs with broken TRS we used four Anritsu PE8403 microwave circulators.

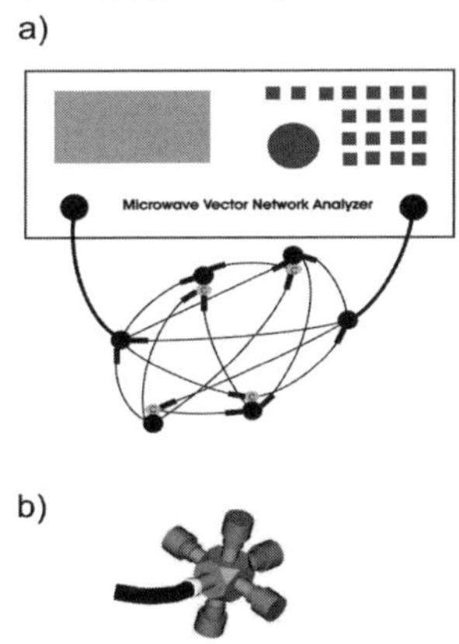

Figure 1. (a) The scheme of the experimental setup for measuring the two-port scattering matrix $\hat{S}$. Each bond of the network presented in this figure contains a microwave attenuator. To construct the directed networks, simulating systems with broken time reversal symmetry, we used four microwave circulators. (b) The scheme of the setup for measuring the radiation scattering matrix s^{kk} (see the formula (5)). In this case 50 Ω loads were connected to the six-joint vertex instead of the microwave network.

3. The enhancement factor

Knowing the scattering matrix $\hat{S}$ one can determine the elastic enhancement factor $W_{S,\beta}$ defined by the following relation [29, 30]

$$W_{S,\beta} = \frac{\sqrt{\operatorname{var}(S^{aa})\operatorname{var}(S^{bb})}}{\operatorname{var}(S^{ab})}, \tag{3}$$

where $\operatorname{var}(S^{ab}) = \langle |S^{ab}|^2 \rangle - |\langle S^{ab} \rangle|^2$ denotes the variance of the scattering matrix element S^{ab}, and β is a symmetry index, equal 1 for the quantum graphs with preserved TRS – Gaussian orthogonal ensemble (GOE) and 2 for the those with broken TRS. – Gaussian unitary ensemble (GUE) in random matrix theory (RMT).

It is worthy to notice that for $\gamma >> 1$ $W_{S,\beta}$ should not depend on the direct processes present in the system and equals $2/\beta$ [29-31]. Calculations performed within RMT indicate that for the other values of the parameter γ the enhancement factor $W_{S,\beta}$ might depend both on the coupling to the system and the parameter γ itself [31].

The variances of the fluctuations of multi-port impedance parameters, which are necessary to estimate the enhancement factor $W_{S,\beta=1}$ were also studied within electromagnetic scattering theory in [32].

The experimental values of the parameter $\gamma = \frac{\left(\gamma^{(a)} + \gamma^{(b)}\right)}{2}$ were estimated for each realization of a network by adjusting the theoretical mean reflection coefficients

$$\langle R \rangle_{th}^{(k)} = \int_0^1 dR P(R), \tag{4}$$

to the experimental ones $\langle R \rangle^{(k)} = \left\langle s^{kk} s^{kk^*} \right\rangle$ obtained after eliminating the direct processes [12]. Here the index k = a, b denotes the port a and b. In the impedance approach [33, 34] the scattering matrix s^{kk} of a network for the perfect coupling case (no direct processes present) can be extracted from the formula

$$s^{kk} = \frac{\left(1 - z^{kk}\right)}{\left(1 + z^{kk}\right)}, \tag{5}$$

where the normalized impedance z^{kk} of a chaotic microwave network is given by

$$z^{kk} = \frac{\mathrm{Re}\, Z^{kk} + i\left(\mathrm{Im}\, Z^{kk} - \mathrm{Im}\, Z_r^{kk}\right)}{\mathrm{Re}\, Z_r^{kk}}. \tag{6}$$

In the formula (6) $Z^{kk} = \frac{Z_0\left(1 + S^{kk}\right)}{\left(1 - S^{kk}\right)}$ and $Z_r^{kk} = \frac{Z_0\left(1 + S_r^{kk}\right)}{\left(1 - S_r^{kk}\right)}$ are the network and the radiation impedances expressed, respectively, by the network S^{kk} and the radiation S_r^{kk} scattering matrices. Z_0 is the characteristic impedance of the transmission lines feeding the 6-joint vertices. The radiation scattering matrix S_r^{kk} is the scattering matrix measured at the input of the coupling structure for the same coupling geometry, but with the vertices (walls) of the system removed to infinity. The scheme of the setup used to measure the radiation scattering matrix S_r^{kk} of the 6-joint vertex is shown in Figure 1(b). The five 50 loads are connected to the microwave joint to simulate the vertices removed to infinity.

For the systems with preserved TRS ($\beta = 1$) the explicit analytic expression for the distribution *P(R)* of the reflection coefficient *R* is given in

[35]. The explicit form of the distribution *P(R)* for systems with broken TRS ($\beta = 2$) is presented in [36, 37].

In figure 2 we show the enhancement factor $W_{S,\beta}$ of the microwave networks simulating quantum graphs with preserved and broken TRS, in the function of the absorption strength γ.

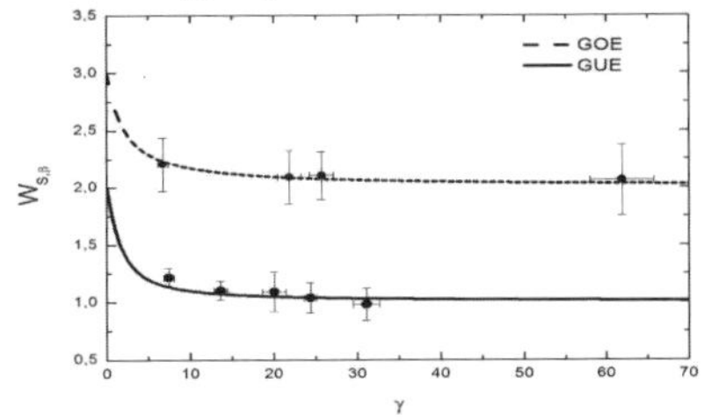

Figure 2. The enhancement factor $W_{S,\beta}$ for the undirected microwave networks simulating the graphs with TRS (dashed line) and for the directed networks which simulate the graphs with broken TRS (solid line). The measurements were performed in the frequency range: 7-14 GHz.

The measurements of $W_{S,\beta=1}$ were done for the undirected networks containing no attenuators (the smallest value of the parameter $\gamma = 6.5$), for the ones containing fifteen 1 dB attenuators ($\gamma = 22$), nine 1 dB and six 2 dB attenuators ($\gamma = 25.7$) and finally fifteen 2 dB attenuators ($\gamma = 61.9$). In all of the cases our results were averaged over 60 microwave network configurations. The total "optical" length of the networks including joints and attenuators was varied for different network configurations from 538 cm to 681cm. The experimental results for the undirected networks simulating graphs with preserved TRS are in general good agreement with the theoretical ones predicted by [29, 30]. Even for moderate absorption $\gamma = 6.5$ the experimental result $W_{S,\beta=1} = 2.2 \pm 0.24$ is very close to the theoretical one. In order to show the spread of the results obtained for different graph configurations the assigned error was calculated using the definition of the sample standard deviation. Because of absorption of microwave cables (network bonds) we could not test experimentally predicted by the theory [29,30,38] increase of the enhancement factor $W_{S,\beta=1} \rightarrow 3$ for small values of the parameter γ. The measurements of the enhancement factor for this range of the parameter have been recently reported in [39]. In the experiment a flat microwave cavity with ohmic losses was used. However, due to relatively large spread of the experimental results and their big uncertainties the limit $\lim_{\gamma\rightarrow 0} W_{S,\beta} = 3$ still remains to be tested.

The measurements for the directed networks simulating graphs with broken TRS were performed in the range of the parameter $7 < \gamma < 32$. The five experimental points were obtained for the networks containing four microwave circulators and different number of microwave attenuators. Beginning from the lowest absorption, the measurements were done for the microwave networks containing no attenuators ($\gamma = 7.4$), seven 1 dB attenuators ($\gamma = 13.6$), fifteen 1 dB attenuators ($\gamma = 20.1$), nine 1 dB and six 2 dB attenuators ($\gamma = 24.4$) and finally for the ones containing fifteen 2 dB attenuators ($\gamma = 31.1$). In all of the cases our results were averaged over 80 microwave network configurations. The total "optical" length of the networks including joints, circulators and attenuators, was varied for different network configurations from 528 cm to 699 cm.

Also in this case the experimental results are in good agreement with the theoretical values of $W_{S,\beta=2}$ for moderate and strong absorption predicted by [29–31]. Only for the smallest value of the parameter $\gamma = 7.4$ the experimental value of $W_{S,\beta=2} = 1.21 \pm 0.08$ lies slightly above the theoretical one while for the largest value $\gamma = 31.1$ the experimental value of $W_{S,\beta=2}$ is below the theoretical one.

4. The properties of the directed microwave networks

In order to characterize the directed and undirected networks the cross-correlation function $c_{12}(\nu)$ and the level spacing distribution *P(s)* were used. The cross-correlation function $c_{12}(\nu)$ defined by the following relation [11, 39]

$$c_{12}(\nu) = \frac{\left\langle S_{12}(\nu+\Delta\nu) S_{21}^{*}(\nu+\Delta\nu) \right\rangle_{\Delta\nu}}{\sqrt{\left\langle \left|S_{12}(\nu+\Delta\nu)\right|^2 \right\rangle_{\Delta\nu} \left\langle \left|S_{21}(\nu+\Delta\nu)\right|^2 \right\rangle_{\Delta\nu}}} \tag{7}$$

determined from our measurements of the scattering matrix is shown in figure 3. The averaging $\langle \ldots \rangle_{\Delta\nu}$ was performed over the frequency window $\Delta\nu = [-500{,}500]$ MHz.

The cross-correlation function $c_{12}(\nu)$ is equal 1 for quantum systems with TRS and should be visibly smaller than 1 for the quantum systems even with partially broken TRS.

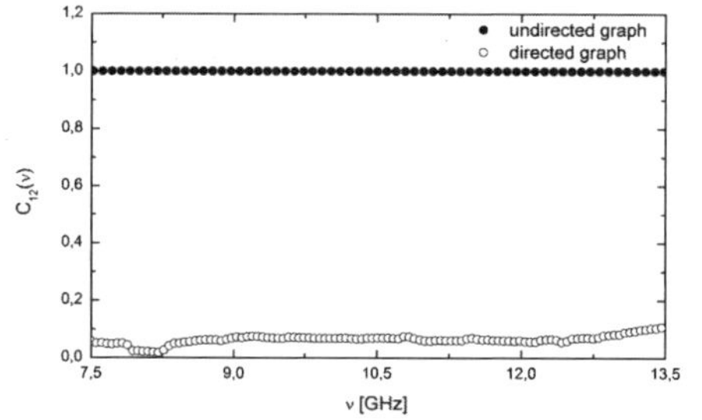

Figure 3. The cross-correlation function $c_{12}(\nu)$ for the directed microwave network without direct process (open circles) with absorption strength $\gamma = 5.63$ in the function of microwave frequency. For the comparison the results for the undirected network with absorption strength $\gamma = 4.31$ (full circles) are also shown.

The results shown in Figure 3 confirm that the directed microwave networks consisting microwave circulators can be used to simulate quantum graphs with broken TRS. It is important to pointing out that the microwave circulators introduce only weak absorption to the microwave networks in contrast to the Faraday isolators, which are often used for the same purpose [11].

Properties of the directed networks with microwave circulators - but without microwave attenuators, were investigated using the level spacing distribution *P(s)* [22]. The distribution *P(s)* is shown in Figure 4.

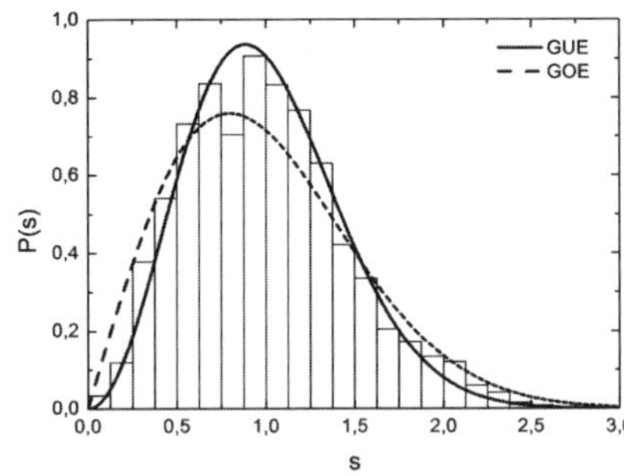

Figure 4. The distribution *P(R)* averaged for 20 realizations of the directed microwave networks

In order to minimize absorption of the networks bonds (microwave cables) the distribution *P(s)* was obtained only for the lower frequency range 7–9 GHz. It should be stressed out that the level spacing distribution can be determined only in the case of weak absorption, when the resonance spectrum is well resolved.

The distribution *P(s)* was averaged over 20 microwave network configurations. In this way 1378 eigenfrequencies of the networks were used in the calculation of the distribution *P(s)*. Figure 4 shows that the experimental distribution *P(s)* is relatively close to the theoretical prediction for GUE in RMT (full line), characteristic for the systems with broken TRS.

5. Summary

We studied experimentally the elastic enhancement factor $W_{S,\beta}$ for the undirected and directed microwave irregular networks simulating, respectively, quantum graphs with preserved (symmetry index $\beta = 1$) and with broken time reversal symmetry ($\beta = 2$) in the presence of moderate and strong absorption. We show that the experimental results for the undirected and directed networks in the presence of strong absorption are, respectively, in good agreement with the theoretical ones $W_{S,\beta=1} = 2$ and $W_{S,\beta=2} = 1$ predicted by [29–31] within the framework of random matrix theory (RMT). Some deviations of the experimental results from the theoretical ones visible for moderate and strong absorption might suggest that the measurements of the enhancement factor could be influenced by some nonuniversal processes, such as coupling of the measuring system to the microwave graphs.

We also presented the experimentally obtained cross-correlation function $c_{12}(\nu)$ and the level spacing distribution *P(s)* for the directed microwave networks. These results together with the result for the enhancement factor $W_{S,\beta}$ show that the microwave circulators can be successfully used to construct the directed microwave networks simulating systems with broken TRS. Moreover, our results clearly show that all these quantities can be successfully used as tools to investigate properties of chaotic quantum systems with broken TRS.

Acknowledgments. This work was partially supported by the Polish Ministry of Science and Higher Education grant No. N N202 338838.

References

[1] L. Pauling, J. Chem. Phys. 4, 673 (1936).
[2] H. Kuhn, Helv. Chim. Acta, 31, 1441 (1948).
[3] Y. Imry, Introduction to Mesoscopic Systems (Oxford, New York, 1996).
[4] D. Kowal, U. Sivan, O. Entin-Wohlman, Y. Imry, Phys. Rev. B 42, 9009 (1990).
[5] E. L. Ivchenko, A. A. Kiselev, JETP Lett. 67, 43 (1998).
[6] J.A. Sanchez-Gil, V. Freilikher, I. Yurkevich, and A. A. Maradudin, Phys. Rev. Lett. 80, 948 (1998).
[7] Y. Avishai and J.M. Luck, Phys. Rev. B 45, 1074 (1992).
[8] T. Nakayama, K. Yakubo, and R. L. Orbach, Rev. Mod. Phys. 66, 381 (1994).
[9] C. Flesia, R. Johnston, and H. Kunz, Europhys. Lett. 3 , 497 (1987).
[10] R. Mitra and S. W. Lee, Analytical techniques in the Theory of Guided Waves (Macmillan, New York, 1971).
[11] O. Hul, S. Bauch, P. Pakonski, N. Savytskyy, K. Życzkowski, and L. Sirko, Phys. Rev. E 69, 056205 (2004).
[12] M. Lawniczak, O. Hul, S. Bauch, P. Seba, and L. Sirko, Phys. Rev. E 77, 056210 (2008).
[13] T. Kottos and U. Smilansky, Phys. Rev. Lett. 79, 4794 (1997).
[14] T. Kottos and H. Schanz, Physica E 9, 523 (2003).
[15] T. Kottos and U. Smilansky, J. Phys. A 36, 3501 (2003).
[16] F. Barra and P. Gaspard, Journal of Statistical Physics 101, 283 (2000).
[17] G. Tanner, J. Phys. A 33, 3567 (2000).
[18] P. Pakonski, K. Życzkowski and M. Kus, J. Phys. A 34, 9303 (2001).
[19] P. Pakonski, G. Tanner and K. Życzkowski, J. Stat. Phys. 111, 1331 (2003).
[20] R. Blumel, Yu Dabaghian, and R.V. Jensen, Phys. Rev. Lett. 88, 044101 (2002).
[21] O. Hul, P. Seba, and L. Sirko, Physica Scripta T135 014048 (2009).
[22] M. Lawniczak, S. Bauch, O. Hul, and L. Sirko, Physica Scripta T135 014050 (2009).
[23] H. Schanz and T. Kottos, Phys. Rev. Lett. 90, 234101 (2003).
[24] L. Kaplan, Phys. Rev. E 64, 036225 (2001).
[25] O. Hul, P. Seba, and L. Sirko, Phys. Rev. E 79, 066204 (2009).
[26] D. S. Jones, Theory of Electromagnetism (Pergamon Press, Oxford, 1964), p. 254.
[27] K. H. Breeden and A. P. Sheppard, Microwave J. 10, 59 (1967); Radio Sci. 3, 205 (1968).
[28] N. Savytskyy, A. Kohler, S. Bauch, R. Blumel, and L. Sirko, Phys. Rev. E 64, 036211 (2001).
[29] Y.V. Fyodorov, D.V. Savin, and H.-J. Sommers, J. Phys. A 38, 10731 (2005).
[30] D.V. Savin, Y.V. Fyodorov, and H.-J. Sommers, Acta Physica Polonica A 109, 53 (2005).
[31] X. Zheng, S. Hemmady, T. M. Antonsen, Jr., S. M. Anlage, and E. Ott, Phys. Rev. E 73, 046208 (2006).
[32] B. Michielsen, F. Isaac, I. Junqua, and C. Fiachetti, arxiv:math-ph/0702041v1 13 Feb. 2007.
[33] S. Hemmady, X. Zheng, T.M. Antonsen Jr., E. Ott, and S.M. Anlage, Acta Physica Polonica A 109, 65 (2006).
[34] S. Hemmady, X.Zheng, E. Ott, T.M. Antonsen, and S.M. Anlage, Phys. Rev. Lett. 94, 014102 (2005).
[35] D.V. Savin, H.-J. Sommers, and Y.V. Fyodorov, JETP Letters 82, 544 (2005).
[36] C. J. W. Beenakker and P. W. Brouwer, Physica E 9, 463 (2001).
[37] Y.V. Fyodorov, JETP Letters 78, 250 (2003).
[38] M. Lawniczak, S. Bauch, O. Hul, and L. Sirko, Phys. Rev. E **81**, 046204 (2010).
[39] B. Dietz, T. Friedrich, H. L. Harney, M. Miski-Oglu, A. Richter, F. Schafer,
[40] J. Verbaarschot,and H. A. Weidenm¨uller, Phys. Rev. Lett. 103, 064101 2009).

Algorithmic sound composition using coupled cellular automata

Jaime Serquera and Eduardo R. Miranda

ICCMR, Interdisciplinary Centre for Computer Music Research
University of Plymouth, UK
Emails: jaime.serquera@plymouth.ac.uk, eduardo.miranda@plymouth.ac.uk

Abstract: In this paper we introduce a new approach to algorithmic sound composition using a bespoke technique combining coupled Cellular Automata (CA) and Histogram Mapping Synthesis. Two CA are used: a hodge podge machine and a growth model. The latter serves as control of the former. The hodge podge machine can exhibit different kinds of behaviour depending on the values of a set of rule parameters. Our method explores the fact that different simultaneous behaviours can be evolved within the same automaton if we bring into play different sets of parameter values. However, we restrict the number of parameter sets to two. Therefore, the CA growth model will have only two states and will delimit two dynamic zones in the hodge podge machine, each of which governed by a different set of parameter values. The predictable evolution of the two zones will produce a controlled dynamic sound spectrum. Among all the possibilities that this process affords for the composition of a variety of sounds algorithmically, we highlight its application to the attack portion of a sound, making it dynamically more complex than the rest of the sound.

Keywords: sound synthesis, cellular automata, histogram mapping synthesis, additive synthesis.

1. Introduction

Computers have been used as a tool for composition since the dawn of Computer Science in the 1950's. The *Illiac Suite for String Quartet*, composed in the USA in late 1950's by Lejaren Hiller (composer) and Leonard Isaacson (mathematician), is often cited as the first piece of music involving materials generated by a computer; e.g., the fourth movement was generated using a Markov chain [5]. Whilst the computer can afford composers the design of very complex abstract structures by using a plethora of modelling methods (including chaotic systems), most of composers have been focusing on using these systems to process information at the so called "note level". In a nutshell, the computer is used here to process musical notes represented symbolically. In tandem to these developments, computing technology also fostered the development of digital signal processing tools, which is increasingly giving composers access to very fine control of sounds. In this case, composers are no longer limited to the capabilities of acoustic musical instruments to play their music. That is, now composers have the possibility to compose "their instruments". Nowadays, the practice of

Chaos Theory: Modeling, Simulation and Applications
C. H. Skiadas, I. Dimotikalis and C. Skiadas (Eds)

musical composition may also include the synthesis of sounds at the micro-level; that is at the level of their spectral components and split second variations. In this paper we focus on using cellular automata to compose sounds at the crossroads of the art of algorithmic composition and the art of sound synthesis.

Cellular automata are mathematical/computational models normally implemented as a regular grid of cells in one or more dimensions. Each cell may assume any state from a finite set of *V* values. CA evolve in successive generations at every time unit. For each generation, the values of all cells change simultaneously according to a set of transition rules that takes into account the states of the neighbouring cells. The states of the cells may represent different colours and therefore, the functioning of a two-dimensional cellular automaton may be displayed on the computer screen as a sequence of images, like an animated film.

Cellular automata have been of interest to computer musicians because of their emergent structures –patterns not created by a single rule but through the interaction of multiple units with relatively simple rules. This dynamic process leading to some order allows the musician to explore new forms of organization. The fundamental motivations for sound synthesis research are on the one hand the imitation of sounds produced by acoustic instruments and on the other hand the search for new sounds. In the latter respect, even though novelty is a goal, it is a common practice to model and simulate certain properties of acoustic instrument sounds. In sound synthesis CA are normally used for controlling over time the parameters of a synthesis instrument. Many of the synthesis techniques demand enormous amounts of control data for obtaining interesting results, making it difficult to be controlled manually. CA represent a solution to this problem because with few parameter specifications it is obtained massive amounts of structured data. The goal is to transfer the structured evolution of CA onto the sound synthesis domain. This is always done through a mapping, a set of correspondences between different domains.

There have been different mapping attempts [2] ranging from direct assignments of CA values, like in Lasy [3], to higher-level approaches intending to map the overall CA behaviour, like in Chaosynth [6]. We are interested in the second type of approach. Our research strategy is based on the analysis of CA evolutions by means of digital signal processing techniques in order to discover structural information of their organization. Then we proceed with the mapping of the analysis results onto appropriate synthesis parameters. Histogram Mapping Synthesis (HMS) is a fruit of this approach.

The control is the main problem of CA. An automaton is defined as a device operating under its own hidden power. The term is derived from the Greek "automatos" which means acting of one's own will or independly, self-acting, self-moved. But all this does not mean that they are uncontrollable. In this paper we provide a mechanism for the control over time of a hodge podge machine by another cellular automaton, a growth model.

2. Histogram Mapping Synthesis (HMS)

Histogram Mapping Synthesis is a recently new sound synthesis technique. It was originally conceived working with CA but it is not restricted to them. Concisely speaking, the method involves a mapping from histograms onto sound spectrograms.

The functioning of a two-dimensional automaton is considered as a sequence of digital images and it is analysed by histogram measurements of every CA image. Such a CA analysis gives a histogram sequence.

The histogram of a grey level digital image is a graphical representation of the number of occurrences of each grey level[1] in the image. By dividing the number of occurrences by the total number of pixels of the image, the histogram is expressed in probabilistic terms giving an estimate of the probability of occurrence of each grey level in the image.

In general terms our mapping method works as follows: the bins of the histogram sequence are considered to be bins of a spectrogram. With an appropriate automaton, in the histogram sequence it is possible to find structural elements resembling spectral components of a sound. For example, from a histogram analysis of the hodge podge machine we discovered structural elements similar to sinusoidal components, others similar to noise components and others similar to transients [7]. This makes such a mapping process distinctive; in most other cases there is not an intuitive correspondence between the components of the automaton and the components of a sound.

With these structural elements we can design the time varying frequency content of a sound; we can build a spectrogram. This spectrogram can be rendered into sound using different synthesis techniques –the structural elements of the histogram sequences become control data for the synthesis program.

3. Two Cellular Automata

In this research two CA are coupled, a hodge podge machine and a CA growth model.

3.1 Hodge Podge Machine

The hodge podge machine is a mathematical model of the oxidation of carbon monoxide. During the numerical investigation of such a cellular automaton, it turned out that this automaton not only describes the typical behaviour of the CO oxidation, but that it leads to a self-sustained organization of fascinating spatial patterns, such as circular and spiral waves. These are very similar to those observed in excitable media, e.g. the Belousov-Zhabotinsky reaction [4].

---

[1] Apart from this definition, in this paper we refer to colours instead of grey levels because we usually display the CA in the computer screen using a palette of different colours.

In the hodge podge machine, the states of a cell can be interpreted metaphorically as follows: the state characterized by a minimum value 0 is called "healthy". The state given by a maximum value *V*-1 is called "ill". All other states in between are called "infected". The original hodge podge rules can be found in [4]. Our investigations include the experimentation with modifications in the rules in order to explore new musical possibilities [7]. For this research we have implemented the following transition rules:

$$m_{x,y}[t+1] = \begin{cases} round(A/r_1) + round(B/r_2) & for\ m_{x,y}[t] = 0 \\ \min\{round(S/A) + K,\ \ V-1\} & for\ 0 < m_{x,y}[t] < V-1 \\ 0 & for\ m_{x,y}[t] = V-1 \end{cases}$$

where the state of a cell at a time step t is denoted by $m_{x,y}$[t]; x and y are the horizontal and vertical coordinates of the location of the cell in the automaton; A and B represent respectively the number of "infected" and "ill" cells in the neighbourhood; S stands for the sum of the states of all cells in the neighbourhood; K, r_1 and r_2 are constants $\in \mathbb{N}$; *round* stands for the nearest integer; and V is the number of possible states that a cell can adopt. For this research we have considered the Moore neighbourhood and periodic boundary conditions, i.e., a torus.

The hodge podge machine can exhibit different kinds of behaviour depending on the values of its rule parameters K, r_1, r_2 and V. For sound synthesis purposes we identified three suitable behaviours [8], two of which are considered in this research: the quasi-synchronic and the spiral waves. Each of these behaviours can be obtained from a variety of different combinations of parameter values.

The quasi-synchronic behaviour (in which all the cells reach the maximum state almost simultaneously) generates a type of histogram sequences from which we can obtain structures similar to sinusoidal and noise components. After the maximum state is reached, patterns of distorted circumferences emerge. These shapes will create narrow bands or peaks in the histogram which are similar to sinusoidal components. From here, the cell values grow towards the maximum state and the boundaries of the distorted circumferences become less defined, creating wide bands in the histogram which are similar to noise bands. At each cycle of the automaton, this process is repeated but with slightly different distorted circumferences shapes, creating structures in the histogram sequences with time varying amplitudes. In the histogram sequences we differentiate a number of bin-zones according the structures that we can find in them. In this research we focus on the structures similar to sinusoidal components (or partials) that are located in a zone close to the Kth bin.

The spiral wave behaviour creates a type of histogram sequence that is of great interest for sound synthesis. When the spiral waves start to develop, they create structures in the histograms that resemble sound partials with increasingly high amplitudes. Once the spiral waves have completely evolved, they often expand themselves covering the whole CA grid. This creates sequences of images which are cyclic and stable. At this point, the amplitudes of the partials often stop increasing and settle to relatively stable values.

Broadly speaking, the most relevant fact derived from the histogram analysis of our hodge podge adaptation exhibiting these two behaviours is that starting from a uniform random distribution of colours the automaton self-organizes through a very specific set of predominant colours.

3.2 CA Growth Model

The growth model used in this research is the one we developed in [9] as an extended version of the multitype voter model.

We implemented the multitype voter model with this transition rule: a number between 0 and 1 is chosen as to be the update probability for all cells. Then, for each cell in the grid, a random number between 0 and 1 is generated at every time step. If the random number generated for the given cell is higher than the update probability, then the state of the cell changes to that of one of its neighbours selected uniformly at random. Neighbour is defined as the four orthogonally adjacent cells: north, east, south and west. The boundary conditions are periodic.

The CA growth model springs out of the following modifications to the previous rule: we define only two cell states, empty and occupied. The functioning is the same (the same rule), but in order to ensure a growth we impose that occupied cells can not become empty. With this, the occupied cells will grow covering the whole automaton.

4. Simultaneous Behaviours

We conducted an experiment delimiting two halves in a hodge podge machine each of which governed by a different set of parameter values in order to evolve two simultaneous behaviours. We only established different values for r_1, r_2 and K; V was the same on both halves.

From that experiment we observed that since the two behaviours are not isolated they are not independent, and it is visually noticeable that they influence each other. On many occasions it seems that one of the two behaviours has a stronger influence over the other. For example, if we run a quasi-synchronic behaviour in one half and a spiral wave behaviour in the other, it usually happens that the spiral waves emerging on one side of the automaton propagate to the other side but with a noticeable visual change so that we can differentiate two halves in the automaton (Figure 1).

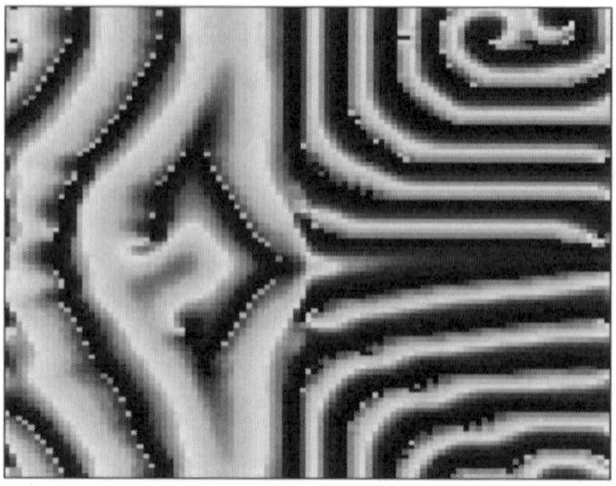

Figure 1. Simultaneous behaviours in a hodge podge machine. The left half has the rule parameters adjusted to produce a quasi-synchronic behaviour whereas the right half has the rule parameters adjusted to produce a spiral wave behaviour.

However, independently of the visual effect, what is interesting for us from the fact that each half is governed by a different set of parameter values is that each half will evolve through a different set of predominant colours. And that fact is well reflected in the histograms. For instance, if we evolve the same type of behaviour in both halves, for example the quasi-synchronic, but with different sets of parameter values, it can happen that at the beginning we can differentiate visually the two halves but after a while, the influence of one side over the other makes that the automaton has only one visual appearance as a whole; we can not distinguish two halves. However, from a histogram analysis we can always identify two superposed histogram sequences.

5. Adding Dynamism

The next step to further explore the possibility of producing different simultaneous histogram sequences is to add dynamism to the process.

The CA growth model (GM) will control the hodge podge machine (HPM) by determining the parameter values that govern each of its cells. We use the GM to create and evolve different regions of parameter values for the HPM. We bring into play only two different sets of parameter values, one associated to empty cells of the GM (set of parameter values E) and the other associated to occupied cells (set of parameter values O). Both two CA have the same size and therefore if a GM cell is empty that means that the cell located in the same place in the HPM will be governed by the set of parameter values E. On the contrary, if one cell of the GM is occupied, the corresponding cell in the HPM will be governed by the set of parameter values O. Both two sets of parameter values will be adjusted to produce the same type of behaviour in order to work with the same type of histograms.

The GM will evolve in the form of predictable transitions. Such an automaton with two states allows two types of transitions. The most immediate one is to go from the set of parameter values E to O (Figure 2).

That will produce a smooth transition between two histogram sequences and will lead to the production of a cross-fade sound effect.

Figure 2. A CA growth model evolution starting with all the cells empty. At some point it is introduced an occupied cell in the middle of the automaton which will grow covering the whole automaton.

The second type of transition is more interesting for sound synthesis. It consists in going from two simultaneous sets of parameter values, O & E, to O. An example of such a transition is illustrated in Figure 3.

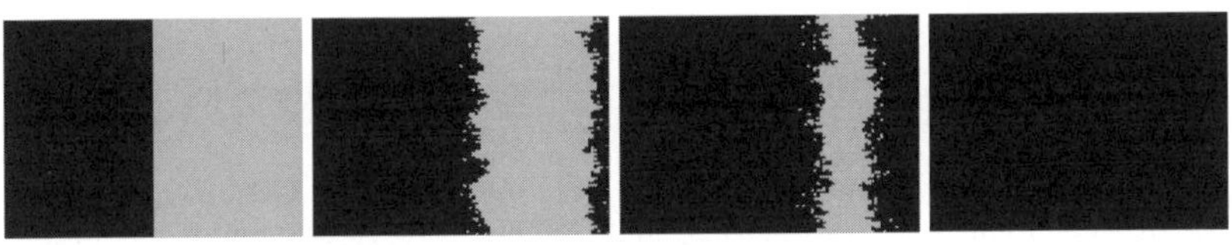

Figure 3. A CA growth model evolution starting with occupied cells in the left half and empty cells in the right half.

That means starting with two superposed histogram sequences and ending up with just one. With this, working with the quasi-synchronic behaviour we can simulate an important characteristic of acoustic instruments: they usually produce more partials in the attack portion than in the rest of the sound. In addition, our system models another process observed in acoustic instruments in which energy passes between the various modes of vibration, some increasing in amplitude, some decreasing [1] (Figure 4).

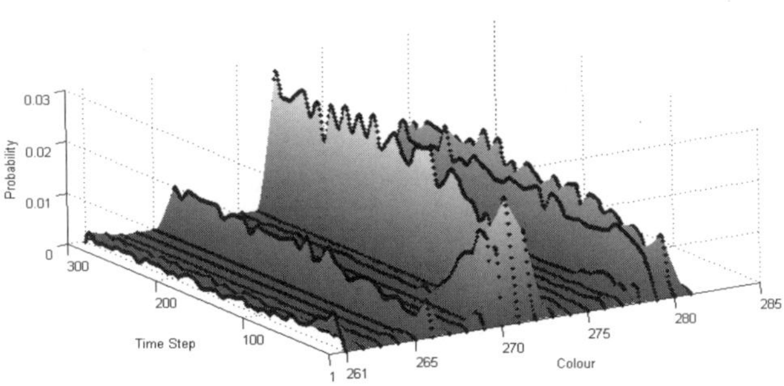

Figure 4. Zoom of a histogram sequence of a hodge podge machine exhibiting two simultaneous quasi-synchronic behaviours and controlled over time by a CA growth model. The bins of the histogram sequence can represent the time varying amplitudes of sound partials, some of which are short lived in the attack portion. We can also observe amplitude compensations between the disappearing "partials" and the permanent ones.

6. Conclusions

In this paper we have provided a coupling method for controlling over time an adaptation of the hodge podge machine by another automaton, a CA growth model.

The choice of the latter automaton has been determined by the musical applications that it can afford with the HMS technique. We have achieved the synthesis of sounds with more complex spectra in the attack portion than in the rest of the sound, an important feature found in sounds produced by acoustic instruments.

Other CA can be suitable for controlling the hodge podge machine with our method. For instance the multitype voter model can produce dynamic areas evolving with more unpredictability than the growth model.

References

[1] Aldoshina, I. et al.: The analysis of peculiarities of Russian bells acoustic parameters. 114th AES Convention. Amsterdam (2003)

[2] Burraston, D. and Edmonds, E.: Cellular Automata in Generative Electronic Music and Sonic Art: A Historical and Technical Review. Digital Creativity 16, 165–185 (2005)

[3] Chareyron, J.: Digital Synthesis of Self-Modifying Waveforms by Means of Linear Automata. Computer Music Journal 14, 25–41 (1990)

[4] Gerhardt, M. and Schuster, H.: A cellular automaton describing the formation of spatially ordered structures in chemical systems. Physica D 36, 209–221 (1989)

[5] Hiller, L. A. and Isaacson, L. M.: Experimental Music: Composition with an electronic computer. McGraw Hill. New York (1959).

[6] Miranda, E. R.: At the Crossroads of Evolutionary Computation and Music: Self-Programming Synthesizers, Swarm Orchestras and the Origins of Melody. Evolutionary Computation Journal 12, 137–158 (2004)

[7] Serquera, J. and Miranda, E. R.: Spectral Synthesis and Control with Cellular Automata. Proceedings of the International Computer Music Conference. Belfast (2008)

[8] Serquera, J. and Miranda, E. R.: Cellular Automata Sound Synthesis: From Histograms to Spectrograms. Automata 2008: Theory and Applications of Cellular Automata. Luniver Press, Frome (2008)

[9] Serquera, J. and Miranda, E. R.: Evolutionary Sound Synthesis: Rendering Spectrograms from Cellular Automata Histograms. Proceedings of EvoApplications / EvoMUSART, Lecture Notes in Computer Science. Springer-Verlag, Berlin (2010)

Efficient large-scale forcing in finite-difference simulations of steady isotropic turbulence

Ryo Onishi, Yuya Baba and Keiko Takahashi

Earth Simulator Center / Japan Agency for Marine-Earth Science and Technology, Yokohama, Japan
Email: onishi.ryo@jamstec.go.jp

Abstract: This study proposes a new forcing scheme suitable for massively-parallel finite-difference simulations of steady isotropic turbulence. The proposed forcing scheme, named reduced-communication forcing (RCF), is based on the idea of the conventional large-scale forcing, but requires much less data communication, leading to a high parallel efficiency. It has been confirmed that the RCF works intrinsically in the same manner as the conventional large-scale forcing. Comparisons have revealed that the fourth-order finite-difference model run in combination with the RCF (FDM-RCF) is as good as the spectral model, while requiring less computational costs.
Keywords: steady isotropic turbulence, finite-difference simulation, high-performance computing, intermittency.

1. Introduction

Rapid developments in computational facilities enable us to investigate, for example, the intermittent nature of turbulence across a wide range of Reynolds numbers. However, there is still a need for numerical data at even higher Reynolds numbers. For example, there is currently a debate concerning the power-laws that govern the Reynolds dependence of the flatness factor of the longitudinal velocity gradient $F=\langle(\partial u/\partial x)^4\rangle/\langle(\partial u/\partial x)^2\rangle^2$ (< > represents the mean value over the domain.), which is a measure of intermittency. Tabeling and Willaime (2002) [1] suggest there may be a transition in the power-law at around Re_λ~700, where Re_λ is the Taylor microscale-based Reynolds number defined as $Re_\lambda=u'l_\lambda/\nu$, where u' is the RMS of the velocity, l_λ is the Taylor microscale and ν is the kinematic viscosity. Specifically, they suggest that the flatness factor increases up to Re_λ=700, but then decreases before eventually increasing again. However, the wind-tunnel experiments by Gylfason et al (2004) [2] show no such transition. Direct numerical simulation (DNS) of turbulent flows has the potential to settle this argument because it is free from experimental ambiguities such as the effects of using Taylor's hypothesis, one dimensional

Chaos Theory: Modeling, Simulation and Applications
C. H. Skiadas, I. Dimotikalis and C. Skiadas (Eds)
 (pp. 281 - 288)

surrogates and so on. Using a cubic grid with 4096 points along each dimension, Ishihara et al. (2007) [3] achieved Re_λ=675 with $k_{max}l_\eta$~2, where k_{max} is a maximum effective wavenumber and l_η is the Kolmogorov length. Unfortunately, the Reynolds number for this large-size computation was still not sufficient to settle the argument. Ideally, we would like to increase Re_λ by a factor 2, which demands approximately $(2^{3/2})^4=2^6$ times. (The power of 4 comes from the sum of three spatial dimensions and one time dimension.) Massive parallel computing is a must for such a large computation.

Numerical simulations of isotropic turbulence are most often formulated in spectral space since the periodic boundary conditions are then easily implemented. The spectral models are, however, facing a technical problem in massively-parallel computing. Although parallel vrsions of the 3D FFT algorithm are available in several scientific libraries, heavy data communication makes it difficult for spectral models to achieve a high parallel efficiency in highly parallel distributed-memory calculations. (Generally, these use the Message Passing Interface (MPI) library.) For this reason, finite difference models (FDMs) have become an attractive alternative to spectral models.

Because of their inferior accuracy, FDMs have generally not been less favored than spectral models. For example, Herring et al. (1974) [4] reported that FDMs require twice the resolution of spectral models to achieve the same accuracy. This is partly because many FDM simulations use upstream schemes, which contain numerical diffusion. For example, the FDM of Rai and Moin (1991) [5] uses the fifth-order upstream scheme. Past FDMs also suffered from inappropriate fomulations of the finite-difference scheme, leading to non-conservation of mass, momentum and kinetic energy. Morinishi et al. (1998) [6] corrected this deficiency and formulated fully conservative high-order accurate finite-difference schemes. Kajishima et al. (1998) [7] then showed their FDM with fourth-order central difference FDM was as good as a spectral model in simulations of decaying turbulence. In the first part of this paper we will re-examine the reliability of FDMs in comparison with spectral models.

One particular computational problem arises when the finite-difference approach is applied to steady isotropic turbulence (which is preferable to decaying turbulence in terms of statical analysis). The most common method for forcing the flow is to apply a forcing to the Navier-Stokes equations for all modes in the wavenumber sphere $|\boldsymbol{k}|<k_f$, often refered to as large-scale forcing. Michioka and Komori (2004) [8] employed this approach in their 512^3 gridpoint FDM. In order to appy the forcing, they performed Fourier transformations on the complete velocity field in order to obtain the low-wavenumber components. however, this direct application of the large-scale forcing technique reduces the parallel efficiency in larger-size computations since it requres communication of the whole velocity field.

This study proposes a new forcing scheme suitable for massively-parallel FDMs. The proposed scheme is based on the idea of the conventional large-scale forcing scheme, but requires much less data communication. We call the scheme *reduced-communication forcing* (RCF). The RCF scheme performs volume-averaging on the velocity fields before Fourier transforming to extract large-scale motions. Small-scale information is removed by the volume-averaging, but this is not an issue when forcing large-scales. The size of the volume-averaged data is, of course, smaller than that of the full data, and this reduces the data communication required and leads to a high parallel efficiency. The main aims of this study are to compare the properties of the RCF scheme with those of the conventional large-scale forcing and linear forcing schemes ([9],[10]), and to achieve high-performance computations using a fourth-order FDM with the RCF.

2. Finite-Difference Model for Incompressible Navier-Stokes Equation

2.1. *Governing equations*

We solve the three-dimensional continuity and Navier-Stokes equations for incompressible flows. In our finite difference model (FDM), spatial derivatives are calculated using fourth-order central differences. We employed the conservative scheme by [6] for the advection term, and the second-order Runge-Kutta scheme for time integration. The HSMAC scheme was used to solve the Poisson equation arising from the continuity equation. Velocities and pressure were modified interatively until the the resudiuals for the continuity equation became smaller than a threshold. The governing equations were discretized on a cubic domain of length $2\pi L_0$, and periodic boundary conditions were applied in all three directions. The flow cube was discretized uniformly into N^3 grid points.

2.2. *Conventional large-scale forcing*

The most common way to maintain a statistically steady state is to force the low wavenumber components. All Fourier modes with wavenumber components $|\boldsymbol{k}|$ between 0.5 and 1.5 (e.g., [11]) or 0.5 and 2.5 (e.g., [12]) are usually forced. This restriction of the forcing to large scales allows the development of a natural inertial range at smaller scales; a white-noise random forcing may influence the intrinsic chaotic behavior of the turbulent flows.

At each timestep, the equations are first integrated without forcing, with the resulting Fourier modes denoted by $u_i^*(\mathbf{k},t+\Delta t)$. Those Fourier modes with wavenumber $|\boldsymbol{k}|$ between n-0.5 and n+0.5 are then multiplied by a constant:

$$u_i(\mathbf{k},t+\Delta t) = u_i^*(\mathbf{k},t+\Delta t)\sqrt{E_n(0)/E_n(t+\Delta t)}\,, \quad (1)$$

where $E_n(t+\Delta t)$ is the energy of the modes with wavenumbers between n-0.5 and n+0.5, calculated from $u_i^*(\mathbf{k},t+\Delta t)$. $u_i(\mathbf{k},t+\Delta t)$ is the updated Fourier amplitude. When forcing the Fourier modes with $|\boldsymbol{k}| < 2.5$, the modes n=1 and 2 are treated similarly.

The conventional large-scale forcing scheme is formulated in Fourier space and is therefore easily implemented in spectral models. For FDMs however, which are formulated in gridpoint space, large-scale forcing requires transformations to and from Fourier space. If full Fourier transformations are used, covering the full spectrum, this requires data-gathering from the whole computational domain. This is not an issue for single-processor calculations [8], but may lead to inefficiencies when a large number of processors is used. The efficiency of the scheme can be improved substantially, however, by restricting the transformation to those Fourier modes being forced. In this case, it is no longer necessary to gather the full velocity field from each process; all that is required is the projection of that field onto the modes being forced.

For example, if (20,20,20) processes are used for a $2{,}000^3$ grid, and the forcing is for $0.5<|\boldsymbol{k}|<2.5$, the number of values to be communicated for each velocity component (u, v or w) is $20^3 \times 19 \times 2$, where the factor of 2 comes from phase calculation and that of 19 comes from the number of modes; i.e., (1,0,0), (1,1,0), (1,1,1), (2,0,0), (2,1,0), (2,1,1) and their permutations. This data size is much less than the $2{,}000^3$ values required for communicating the complete set of data. Note, however, that for this case the scheme to be introduced in section 2.3 requires the communication of only 20^3 values.

2.3. *Reduced-communication forcing (RCF)*

We propose a new forcing scheme suitable for massively-parallel FDMs. The proposed scheme is based on the idea of the conventional large-scale forcing scheme, but requires much less data communication. We call the scheme *reduced-communication forcing* (RCF). The basic idea is to volume-average the velocity fields onto a coarse grid before Fourier transforming to extract the large-scale motions. This reduces the data size to be communicated to N_c^3, where N_c^3 is the number of gridpoints on the coarse grid.

For the forcing of modes $0.5<|\boldsymbol{k}|<1.5$, the calculation procedure for the RCF scheme is as follows

(i) Each process volume-averages the time-integrated local velocities obtained without forcing $u_i{*}(\mathbf{k},t+\Delta t)^{\mathrm{local}}$, giving $\overline{u_i^*}(\mathbf{x}^c,t+\Delta t)^{\mathrm{local}}$, where $\mathbf{x}^c$ represents the filtered coarse-grid system with size N_c^3.

(ii) An arbitrary process collects $\overline{u_i^*}(\mathbf{x}^c,t+\Delta t)^{\mathrm{local}}$ and constructs the volume-averaged velocities for the whole domain $\overline{u_i^*}(\mathbf{x}^c,t+\Delta t)$.

(iii) The process distributes $\overline{u_i^*}(\mathbf{x}^c,t+\Delta t)$ to all the processes.

(iv) Each process calculates the Fourier components $\overline{u_i^*}(\mathbf{k}^1,t+\Delta t)^{\mathrm{local}}$, where $0.5<|\mathbf{k}^1|<1.5$, using forward Fourier transformations, and then it calculates the energy for the modes $\mathbf{k}^1$, $\overline{E_1}(t+\Delta t)$.

(v) Velocities for the modes $\mathbf{k}^1$ are transformed to gridpoint space using inverse Fourier transformations, giving $\overline{u_i^*}(\mathbf{x}^1,t+\Delta t)$.

(vi) Each process updates the velocities in order to maintain the energy in the mode.

$$u_i(\mathbf{x},t+\Delta t) = u_i(\mathbf{x},t+\Delta t) + \overline{u_i^*}(\mathbf{x}^1,t+\Delta t)\left(\sqrt{\overline{E_1}(0)/\overline{E_1}(t+\Delta t)}-1\right), \qquad (2)$$

When forcing $0.5<|\mathbf{k}|<2.5$, the modes for which $1.5<|\mathbf{k}^2|<2.5$ are treated in a similar way in steps (iv), (v) and (vi). $\overline{E_1}$ is smaller than E_1 because the volume-average filtering is not an exact spectral filtering. Assuming $E(k)=Ak^{-3/5}$, $\overline{E_1}=0.88E_1$ if $N_c=5$. However, note that this reduction applies to both numerator and denominator in the square-rooted term on the RHS of equation (2), so the effect cancels.

When forcing $0.5<|\mathbf{k}|<1.5$, the dimension N_c^3 of $\mathbf{x}^c$ is given by N_c=MAX(n_x, n_y, n_z, 3), where n_x, n_y and n_z are the numbers of processes in x-, y- and z-directions. When forcing $0.5<|\mathbf{k}|<2.5$, N_c^3 is given by (n_x, n_y, n_z, 5).

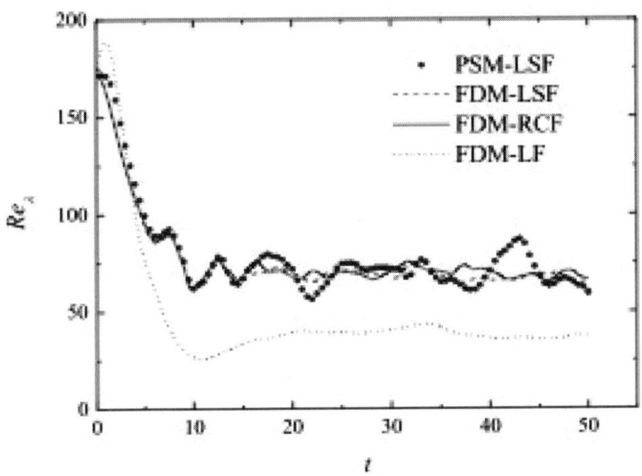

Figure 1. Temporal evolutions of Reynolds number Re_λ for the case N=128

3. Results and Discussion

3.1. *Steady isotropic turbulence*

Steady isotropic turbulence was attained four combinations of forcing scheme and flow calculation model: (I) conventional large-scale forcing in the pseudo-spectral model (PSM) with two-thirds dealiasing [13], hereafter referred to as PSM-LSF, (II) conventional large-scale forcing in the FDM; FDM-LSF, (III) reduced-communication forcing in the FDM; FDM-RCF, and (IV) linear forcing in the FDM; FDM-LF. The first three combinations are based on the same idea of forcing only large-scale motions. In this study, large-scale forcing inject energy in the large-scale motions with wavenumber $|\mathbf{k}|$ between 0.5 and 1.5. In FDM-LF, the forcing coefficient was set so as the mean energy dissipation rate $<\varepsilon>$ to converge towards 0.2, which was about the values from the other three combinations.

Figure 1 shows the time evolutions of Taylor-microscale based Reynolds number Re_λ for the case N=128. The horizontal axes show the nondimensional time. It shows that all four simulations achieve statistically steady state after around t=15, which is about three times the eddy turnover time $3T_e$. The FDM-RCF, PSM-LSF and FDM-LSF simulations in which only the large-scale motions are forced show similar trends, while FDM-LF produces a somewhat different behaviour. One should note that for a given resolution FDM-LF reaches significantly smaller values of Re_λ than the other simulations. This is a deficiency of the linear forcing scheme, consistent with the findings of [10]. The results have confirmed that the using the RCF scheme in conjuction with the FDM works intrinsically in the same manner as the conventional large-scale forcing scheme, succeeding in producing typical steady isotropic turbulence. This is achieved at a smaller computational cost (35% less floating-operations) than the conventional combination of LSF with the pectral model.

3.2. *Parallel computing for higher Reynolds numbers*

The flatness factor of the longitudinal velocity gradient F, which is a measure of intermittency, defined as

$$F = \frac{\left\langle (\partial u / \partial x)^4 \right\rangle}{\left\langle (\partial u / \partial x)^2 \right\rangle^2}. \tag{3}$$

Figure 2 shows the flatness factor F obtained from FDM-RCF using N=128, 256, 512 and 1,000 at high resolution; $k_{N/2} l_\eta \sim 2$, where $k_{N/2}(=N/2)$ is the effective wavenumber in FDM and l_η is the Kolmogorov scale. Results from the earlier studies are also drawn. The power-law increase of our flatness factors is $F\sim Re_\lambda^{0.34}$, which are consistent with the past studies in the range $80< Re_\lambda<340$: For example, $F\sim Re_\lambda^{0.31}$ (atmospheric observations by [14]), $F\sim Re_\lambda^{0.32}$ (empirical fomula of [15]), $F\sim Re_\lambda^{0.39}$ (wind-tunnel experiments by [2]) and $F\sim Re_\lambda^{0.34\pm0.03}$ (DNS of [3]). Unfortunately, the calculated range of Re_λ is not large enough to settle the debate on the power-law transition of the flatness factor at around Re_λ=700. Larger size computations are currently under going.

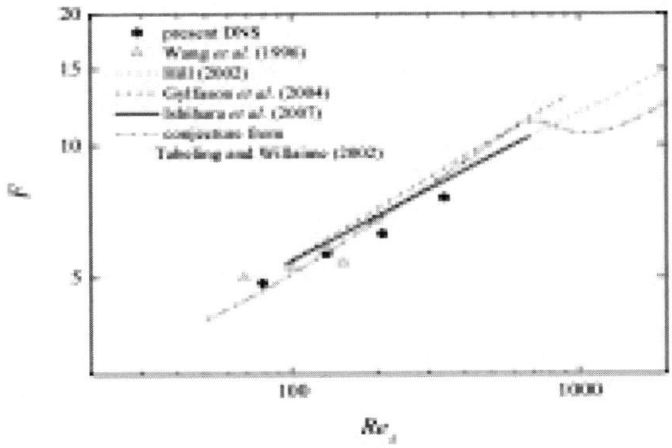

Figure 2. Flatness factor F of the longitudinal velocity gradient in different Reynolds numbers Re_λ

4. Conclusions

This study proposes a new simple forcing scheme suitable for massively-parallel finite-difference simulations of steady isotropic turbulence. The proposed forcing scheme, named *reduced-communication forcing* (RCF), is based on the idea of the conventional large-scale forcing scheme, but requires much less data communication. The RCF performs volume-averaging on the velocity fields before applying Fourier transforms to extract the large-scale motions. Small-scale information is removed by volume-averaging the data, but this is not an issue when forcing the large scales. The size of the volume-averaged data is of course smaller than that of the full set of data values, which reduces the data communications required for parallel computing and leads to a high parallel efficiency. It is confirmed that a finite-difference model (FDM) adopting a conservative fourth-order scheme is as good as a conventional spectral model, and that the RCF scheme works intrinsically in the same manner as the conventional large-scale forcing scheme. We therefore conclude that the combination of this FDM with the RCF scheme (FDM-RCF) is a promising tool for massively-parallel simulations of steady isotropic turbulence, which will contribute to a better understanding of small-scale structures such as intermittency.

Acknowledgments

This research was supported by a Grant-in-Aid from the Ministry of Education, Culture, Sports, Science and Technology (MEXT) of Japan for Young Scientists (B) 21760143, and was partially supported by Core Research for Evolutional Science and Technology (CREST) Program "Advanced Model Development and Simulations for Disaster Countermeasures" of Japan Science and Technology Agency (JST). Large-size numerical simulations presented were carried out on the Earth Simulator 2 in Japan Agency for Marine-Earth Science and Technology.

References

[1] P. Tabeling and H. Willaime, Transition at dissipative scales in large Reynolds number turbulence, *Phys. Rev. E*, 65:066301, 2002.

[2] A. Gylfason, S. Ayyalasomayajula and Z. Warhaft, Intermittency, pressure and acceleration statistics from hot-wire measurements in wind-tunnel turbulence, *J. Fluid Mech.*, 501:213-29, 2004.

[3] T. Ishihara, Y. Kaneda, M. Yokokawa, K. Itakura and A. Uno, Small-scale statistics in high-resolution direct numerical simulation of turbulence: Reynolds number dependence of one-point velocity gradient statistics, *J. Fluid Mech.*, 592:335-366, 2007.

[4] R. Herring, S. A. Orszag, R. H. Kraichnan and D. G. Fox, Decay of two-dimensional homogeneous turbulence, *J. Fluid Mech.*, 66:417-444, 1974.

[5] M. M. Rai and P. Moin, Direct simulation of turbulent flow using finite-difference schemes, *J. Comput. Phys.*, 96:15-53, 1991.

[6] Y. Morinishi, T. S. Lund, O. V. Vasilyev and P. Moin, Fully Conservative Higher Order Finite Difference Schemes for Incompressible Flow, *J. Comput. Phys.*, 143:90-124, 1998.

[7] T. Kajishima, T. Ohta, K. Okazaki and Y. Miyake, High-order Finite-difference method for incompressible flows using collocated grid system, *JSME Intl. J.*, 41B:830-839, 1998.

[8] T. Michioka and S. Komori, Large-Eddy Simulation of a Turbulent Reacting Liquid Flow, *AIChE J.*, 50:2705-2720, 2004

[9] T. S. Lundgren, Linearly forced isotropic turbulence, *CTR Annu. Res. Briefs*, 461-473, 2003

[10] C. Rosales and C. Meneveau, Linear forcing in numerical simulations of isotropic turbulence: Physical space implementations and convergence properties, *Phys. Fluids*, 17:095106, 2005.

[11] R. M. Kerr, High-order derivative correlations and the alignment of small-scale structures in isotropic numerical turbulence, *J. Fluid Mech.*, 153:31-58, 1985.

[12] Y. Kaneda, T. Ishihara, M. Yokokawa, K. Itakura and A. Uno. Energy dissipation rate and energy spectrum in high resolution direct numerical simulations of turbulence in a periodic box, *Phys. Fluids*, 15:L21-24, 2003.

[13] R. Onishi, K. Takahashi and S. Komori, Influence of gravity on collisions of monodispersed droplets in homogeneous isotropic turbulence, *Phys. Fluids*, 21:125108, 2009.

[14] R. A. Antonia, A. J. Chambers and B. R. Satyaprakash, Reynolds number dependence of high-order moments of the streamwise turbulent velocity derivative. *Boundary-layer Met.*, 21:159-171, 1981.

[15] R. J. Hill, Scaling of acceleration in locally isotropic turbulence, *J. Fluid Mech.*, 452:361-370, 2002.

Rendering statistical significance of information flow measures

Angeliki Papana and Dimitris Kugiumtzis

Aristotle University of Thessaloniki
Department of Mathematical, Physical and Computational Sciences
Faculty of Engineering
54124 Thessaloniki, Greece
Emails: agpapana@gen.auth.gr, dkugiu@gen.auth.gr

Abstract: Information causality measures, i.e. transfer entropy and symbolic transfer entropy, are modified using the concept of surrogate data in order to identify correctly the presence and direction of causal effects. The measures are evaluated on multiple bivariate time series of known coupled systems of varying complexity and on a range of embedding dimensions. The proposed modifications of the causality measures are found to reduce the bias in the estimation of the measures and preserve the zero level in the absence of coupling.
Keywords: information flow, causality, transfer entropy.

1 Introduction

Given a set of time series observations, it is essential to assess whether they originate from coupled or uncoupled systems, detect the hidden causal dependencies between them and understand which system is the driver. For these purposes, many model-free measures have been developed based on phase synchronization [8], geometry in state space [1,7] and information theory [9,11,10].

There have been recently comparative studies on the different causality measures [2,5]. In a recent evaluation of different causality measures [6], we observed in many cases significant measure values in the absence of coupling that can be misinterpreted as weak coupling. Here, we investigate this on information measures. Their estimation involves the estimation of entropies that can have bias depending on the time series length, state space reconstruction and system complexity.

We attempt to establish the zero level of information measures when the systems are not coupled and for this we use surrogate data. The suggested surrogates are extracted by randomly shuffling the reconstructed points of the driving time series, so that the dynamical properties of each system are

Chaos Theory: Modeling, Simulation and Applications
C. H. Skiadas, I. Dimotikalis and C. Skiadas (Eds)

preserved in the point representation and only the coupling, if present, is destroyed. We believe that shuffling randomly the reconstructed points is more appropriate than shuffling the samples of the driving time series, as proposed in the so-called effective transfer entropy [4]. The original, effective, and our corrected information measures are tested on simulated systems of varying complexity. We also study their performance for a range of embedding dimensions (different for the driving and the driven system) and in the presence of noise.

The structure of the paper is as follows. In Section 2, we present the information causality measures and the proposed corrections using surrogate data. In Section 3, we report the results of our simulation study, and in Section 4, we discuss the results and draw conclusions.

2 Information causality measures and proposed corrections

Let $\{x_t\}$ and $\{y_t\}$, $t = 1, \ldots, n$, denote two simultaneously observed univariate time series derived from the dynamical systems X and Y, respectively. The reconstructed vectors of the two state spaces are defined as $\mathbf{x}_t = (x_t, x_{t-\tau_x}, \ldots, x_{t-(m_x-1)\tau_x})'$ and $\mathbf{y}_t = (y_t, y_{t-\tau_y}, \ldots, y_{t-(m_y-1)\tau_y})'$, where $t = 1, \ldots, n'$, $n' = n - \max\{(m_x-1)\tau_x, (m_y-1)\tau_y\}$, m_x and m_y are the embedding dimensions, and τ_x and τ_y the delays for X and Y, respectively.

We consider measures of the information flow from X to Y, denoted $X \to Y$. Transfer entropy (TE) quantifies the amount of information explained in Y at one step ahead from the state of X, accounting for the concurrent state of Y [9]. In terms of entropy, TE is defined as

$$\mathrm{TE}_{X \to Y} = H(\mathbf{x}_t, \mathbf{y}_t) - H(y_{t+1}, \mathbf{x}_t, \mathbf{y}_t) + H(y_{t+1}, \mathbf{y}_t) - H(\mathbf{y}_t), \tag{1}$$

where $H(\mathbf{x})$ is the entropy of $\mathbf{x}$. We consider the differential entropy and estimate it from the correlation sum $C(\mathbf{x})$ as $H(X) \simeq \ln C(\mathbf{x}) + m \ln r$, where $C(\mathbf{x})$ is the estimated cumulative density of inter-point distances at embedding dimension m and a given distance r [3]. To account for the different dimensions in each entropy term in (1), we use the standardized Euclidean norm in the computation of $C(\mathbf{x})$. Thus TE is estimated by the correlation sums as

$$\mathrm{TE}_{X \to Y} = \log \frac{C(y_{t+1}, \mathbf{x}_t, \mathbf{y}_t) C(\mathbf{y}_t)}{C(\mathbf{x}_t, \mathbf{y}_t) C(y_{t+1}, \mathbf{y}_t)}. \tag{2}$$

The so-called effective transfer entropy (ETE), is defined as the difference of TE computed on the original bivariate time series from the TE computed on a surrogate bivariate time series, where the driving time series X is randomly shuffled [4]. In order to gain better stability of the measure, we consider instead of a single surrogate an ensemble M surrogates and ETE is then defined as

$$\mathrm{ETE}_{X\to Y} = \mathrm{TE}_{X\to Y} - \frac{1}{M}\sum_{l=1}^{M}\mathrm{TE}_{X_l\to Y}, \tag{3}$$

where $l = 1,\dots,M$ and $\mathrm{TE}_{X_l\to Y}$ is the TE for the l-th surrogate.

The random shuffling of the driving time series aims at destroying any possible coupling structure between systems X and Y. Thus the mean of the TE on the surrogate bivariate time series estimates the significance level of TE and thus positive values of $\mathrm{ETE}_{X\to Y}$ indicate true coupling.

There is a serious shortcoming in ETE. The random permutation of the samples in $\{x_t\}$ destroys not only the time correspondence between $\{x_t\}$ and $\{y_t\}$, but also the dynamics in X, represented in the reconstructed points $\{\mathbf{x}_t\}$. The dynamics of the individual systems may fool the coupling measures and give false positive values in the absence of coupling. This form of bias in TE cannot be rendered by ETE and the significance level (for zero coupling) is not attained by the TE on these surrogate data. The remaining bias can be both positive and negative depending on, among others, the state space reconstruction. Therefore we suggest to shuffle the reconstructed points $\{\mathbf{x}_t\}$ instead. Moreover, to gain stability in the estimation of the significance level of TE, we propose to subtract the entropy on these surrogate data at each of the two terms in (1), where $\mathbf{x}_t$ occurs, i.e. $H(\mathbf{x}_t,\mathbf{y}_t)$ and $H(y_{t+h},\mathbf{x}_t,\mathbf{y}_t)$. Thus considering the estimation of TE using correlation sums in (2), the terms $C(y_{t+1},\mathbf{x}_t,\mathbf{y}_t)$ and $C(\mathbf{x}_t,\mathbf{y}_t)$ are replaced by the average on M replications of the random shufflings of the reconstructed vectors of X

$$C_s(y_{t+1},\mathbf{x}_t,\mathbf{y}_t) = \frac{1}{M}\sum_{l=1}^{M} C(y_{t+1},\mathbf{x}_{t_l},\mathbf{y}_t),\quad C_s(\mathbf{x}_t,\mathbf{y}_t) = \frac{1}{M}\sum_{l=1}^{M} C(\mathbf{x}_{t_l},\mathbf{y}_t)$$

where $t_l = 1,\dots,n'$ are randomly chosen for each $t = 1,\dots,n'$ and this is repeated M times. Then, the TE in the absence of coupling between X and Y is

$$\mathrm{TEs}_{X\to Y} = \log\frac{C_s(y_{t+1},\mathbf{x}_t,\mathbf{y}_t)C(\mathbf{y}_t)}{C_s(\mathbf{x}_t,\mathbf{y}_t)C(y_{t+1},\mathbf{y}_t)}.$$

The suggested corrected transfer entropy (CTE) is thus derived as

$$\mathrm{CTE}_{X\to Y} = \mathrm{TE}_{X\to Y} - \mathrm{TEs}_{X\to Y} = \log\frac{C(y_{t+1},\mathbf{x}_t,\mathbf{y}_t)C_s(\mathbf{x}_t,\mathbf{y}_t)}{C_s(y_{t+1},\mathbf{x}_t,\mathbf{y}_t)C(\mathbf{x}_t,\mathbf{y}_t)}. \tag{4}$$

The symbolic transfer entropy (STE) is a recently suggested version of TE using rank points instead of reconstructed points [10]. For each vector, say $\mathbf{y}_t$, the ranks of its components assign a rank-point $\hat{\mathbf{y}}_t = [r_1, r_2, \dots, r_{m_y}]$, where $r_j \in \{1, 2, \dots, m_y\}$ for $j = 1,\dots,m_y$. Following this sample-point to rank-point conversion, the sample y_{t+1} in Eq.(1) is taken as a rank point at time $t+1$, $\hat{\mathbf{y}}_{t+1}$, and STE is defined in terms of entropy as

$$\mathrm{STE}_{X\to Y} = H(\hat{\mathbf{x}}_t,\hat{\mathbf{y}}_t) - H(\hat{\mathbf{y}}_{t+h},\hat{\mathbf{x}}_t,\hat{\mathbf{y}}_t) + H(\hat{\mathbf{y}}_{t+h},\hat{\mathbf{y}}_t) - H(\hat{\mathbf{y}}_t). \tag{5}$$

The estimation of the Shannon entropy terms is straightforward by estimating the probabilities from the relative frequencies of the observed ranks. The effective symbolic transfer entropy (ESTE) and corrected symbolic transfer entropy (CSTE) are defined analogously to ETE and CTE, respectively.

3 Results

In all simulations, the time series were normalized to have mean zero and standard deviation one and the distance r in the calculation of correlation sums was set to 0.2. The first simulation system is comprised of two Henon maps coupled in one direction

$$\begin{aligned} x_{t+1} &= 1.4 - x_t^2 + 0.3x_{t-1} \\ y_{t+1} &= 1.4 - cx_t y_t + (1-c)y_t^2 + 0.3y_{t-1} \end{aligned} \tag{6}$$

with coupling strengths $c = 0$, 0.15, 0.1, 0.2, 0.3, 0.4 and 0.5. The second system is comprised by two Mackey-Glass systems with unidirectional coupling

$$\begin{aligned} \frac{\mathrm{d}x}{\mathrm{d}t} &= \frac{0.2x_{t-\Delta_1}}{1+x_{t-\Delta_1}^{10}} - 0.1x_t \\ \frac{\mathrm{d}y}{\mathrm{d}t} &= \frac{0.2y_{t-\Delta_2}}{1+y_{t-\Delta_2}^{10}} + c\frac{0.2x_{t-\Delta_1}}{1+x_{t-\Delta_1}^{10}} - 0.1y_t. \end{aligned} \tag{7}$$

The two systems can have different complexity determined by Δ_1 and Δ_2 taking all combinations of the values 17, 30, 100. The coupling strength is $c = 0$, 0.05, 0.1, 0.15, 0.2, 0.3, 0.4 and 0.5.

The time series lengths for the coupled Henon maps are $n = 512, 1024, 2048$ and for the Mackey-Glass systems $n = 2048$. Gaussian white noise is added on the time series with mean zero and standard deviation (SD) 0.05 and 0.2. We also investigate the influence of varying embedding dimensions m_x and m_y on the performance of the measures. For each system, coupling strength and condition (n, noise SD, m_x and m_y), all measures presented in Section 2 are computed on 100 realizations for both directions, $X \to Y$ (the correct dependence direction) and $Y \to X$.

For the coupled Henon maps, all TE measures (including ETE and CTE) correctly detect the direction of the information flow for all combinations of m_x and m_y, even for small time series lengths (see Fig.1a and b). Only for $m_x = 1$ and $m_y > 1$, the three measures spuriously increase with the coupling strength in the direction $Y \to X$. TE is positively biased and ETE is negatively biased for $c = 0$ for a number of combinations of the embedding dimensions, whereas CTE attains always the zero level. ETE is affected by the selection of the embedding dimensions more than TE and CTE and has also larger variance. With the addition of noise, the direction of the information flow is still correctly detected (see Fig.1c for noise SD 0.2),

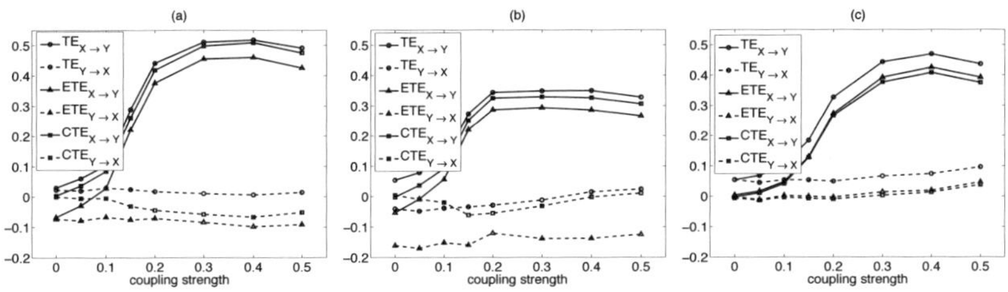

Fig. 1. (a) Mean estimated values of TE, ETE and CTE for both directions from 100 realizations of the noise-free unidirectionally coupled Henon maps, with $n = 512$ and $m_x = m_y = 2$. (b) As in (a) but for $m_x = 2, m_y = 4$. (c) As in (a) but for noise SD 0.2.

with ETE being more affected than the other measures especially for large embedding dimensions. The variance of all information measures gets also much larger. Also for the noisy data, CTE improves the performance of TE giving values at the zero level for $c = 0$ and for varying embedding dimensions and time series lengths. The best results for all measures are obtained when the embedding dimensions are equal.

STE, ESTE and CSTE correctly detect the direction of information flow for $m_x \geq m_y > 2$, for all time series lengths (see Fig.2). STE tends to increase

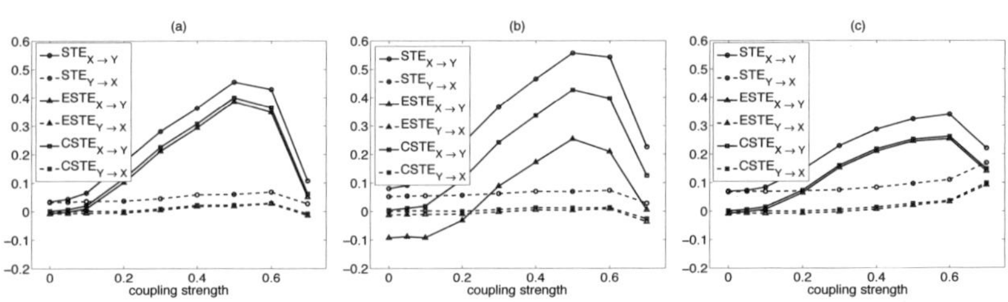

Fig. 2. (a) Mean estimated values of STE, ESTE and CSTE for both directions from 100 realizations of the noise-free unidirectionally coupled Henon maps, with $n = 512$ and $m_x = m_y = 3$. (b) As in (a) but for $m_x = 4, m_y = 3$. (c) As in (a) but for noise SD 0.2.

in both directions with the embedding dimensions m_x and m_y, while ESTE and CSTE tend to decrease. Thus, for embedding dimensions larger than 3, STE is positively biased, ESTE is negatively biased, whereas CSTE preserves again the zero level for $c = 0$. As for the TE measures, STE measures are also stable to noise (see Fig.2c).

The embedding dimension affects more the information causality measures on the unidirectionally coupled Mackey-Glass systems. TE and CTE detect correctly the direction of information flow only for $m_x \geq m_y$, while ETE varies a lot with m_x and m_y and takes negative values for small c (see Fig.4a and b). For example, in Fig.4a where $m_x = 2, m_y = 5$ all measures

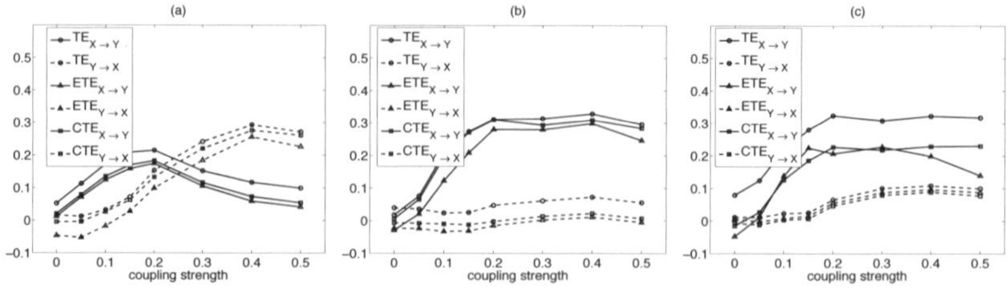

Fig. 3. (a) Mean estimated values of TE, ETE and CTE for both directions from 100 realizations of the noise-free unidirectionally coupled Mackey-Glass systems with $\Delta_1 = 30, \Delta_2 = 100$, with $n = 2048$ and $m_x = 2, m_y = 5$. (b) As in (a) but for $m_x = 5, m_y = 2$. (c) As in (b) but for 20% noise level.

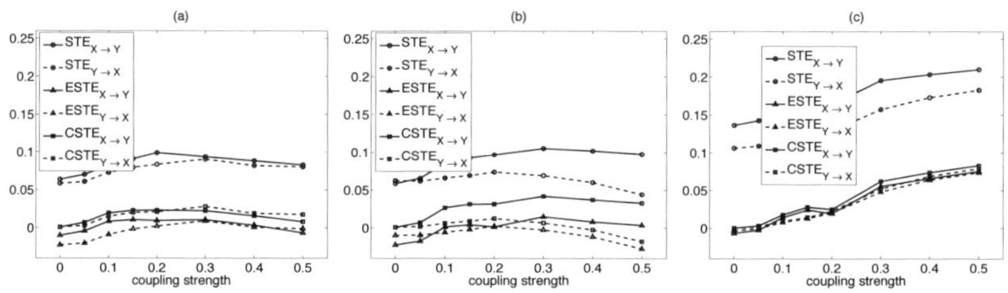

Fig. 4. (a) Mean estimated values of STE, ESTE and CSTE for both directions from 100 realizations of the noise-free unidirectionally coupled Mackey-Glass systems with $\Delta_1 = 30, \Delta_2 = 100$, with $n = 2048$ and $m_x = 3, m_y = 4$. (b) As in (a) but for $m_x = 4, m_y = 3$. (c) As in (a) but for 20% noise level.

vary with c, but in the same way, giving larger values for the correct direction when c is small and the opposite for larger c. This contradicting behavior was found for all combinations of Δ_1 and Δ_2. However, in all cases CTE maintains the zero level in both directions for $c = 0$, indicating that there are other sources of bias for $c > 0$, probably due to the mixed dynamics that cannot be accounted for by the random shuffling of the reconstructed points. When noise is added to the time series, TE is more sensitive to the selection of the embedding dimensions and gives positive values for $c = 0$, whereas CTE stays close to zero for all but very large m_x and m_y (see Fig.4c).

The problem of inconclusive estimation of coupling direction for $m_x < m_y$ holds also for STE, as shown in Fig.4a and b. For all scenarios of complexity of the coupled Mackey-Glass systems, STE obtains significant positive values also for $c = 0$ when $m_x < m_y$, while ESTE obtains often negative values. CSTE follows the same dependence on c as STE but displaced so that CSTE falls to zero for $c = 0$. It is interesting that in the presence of noise, STE gets even larger value for $c = 0$ and increases faster for larger c, and CSTE has the same course with c as STE but starts at the zero level for $c = 0$ (see Fig.4c). We illustrate this nice property of CTE and CSTE for $c = 0$ in Fig.5 as a function of $m_x = m_y$. First, we note that STE measures have much less

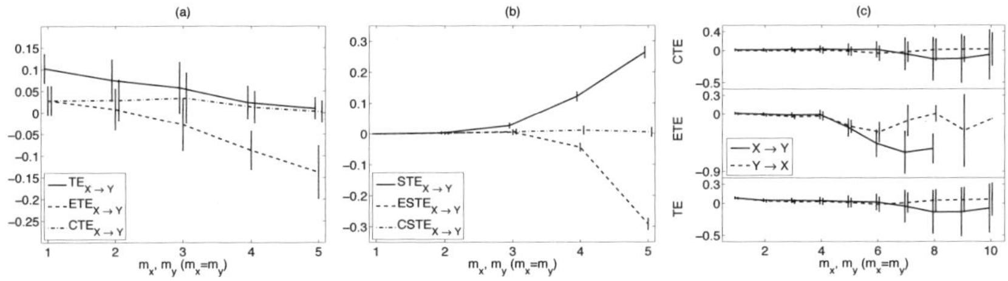

Fig. 5. (a) Mean TE, ETE and CTE (error bars denote the standard deviation) from 100 realizations of the noise-free uncoupled Mackey-Glass systems with $\Delta_1 = \Delta_2 = 17$, $n = 2048$, and for varying $m_x = m_y$. Only the one direction is shown as the two systems are the same. (b) As in (a) but for STE, ESTE and CTE. (c) As in (a) but for $\Delta_1 = 30$, $\Delta_2 = 100$ and the two directions are shown, as given in the legend, with a measure at each panel.

variance than the respective TE measures and attain the zero level for small $m_x = m_y$ (in Fig.5a only the distribution of CTE contains zero for varying $m_x = m_y$). In Fig.5c, where the systems have different complexity, ETE gets more affected by the individual system complexity as the embedding dimension increases, TE and CTE do not differ much in the two directions, but only CTE is at the zero level.

4 Discussion

We propose to correct the entropy terms in the definition of transfer entropy (TE) and symbolic transfer entropy (STE) by subtracting the average entropy computed on an ensemble of the trajectories of the two systems, where for the driving system the points of the reconstructed trajectory are randomly shuffled. We have found that this correction gives more stable estimates than when taking the overall average over all entropy terms. This can be explained by the variation in the estimation of each entropy term on the surrogate data. The corrected TE and STE, termed CTE and CSTE, respectively, attain the correct statistical significance, i.e. their values are within the zero level when the two systems are uncoupled. This has been illustrated with simulations on the unidirectionally coupled Henon map and Mackey-Glass system of varying complexity for the driving and the driven system. In the same simulations, we compared the so-called effective TE (ETE), and introduced also effective STE (ESTE), which use random shuffling of the samples of the time series instead of the reconstructed points. Both ETE and ESTE are found to be rather unstable and very sensitive to variations of the embedding dimensions. It is interesting that the correct significance of CTE and CSTE holds for a range of varying embedding dimensions and in the presence of noise, as well as for different data sizes.

In all works on coupling measures, the embedding dimension of the driving system m_x and driven system m_y are always equal and often the two systems X and Y are of the same complexity (though not identical in parameters). One would expect that different complexity of X and Y would suggest different m_x and m_y. Our simulation results confirm that the popular choice $m_x = m_y$ gives indeed the best results and identifies best the correct direction of information flow even when the systems differ in complexity.

Acknowledgments This research project is implemented within the framework of the "Reinforcement Programme of Human Research Manpower" (PENED) and is co-financed at 90% jointly by E.U.-European Social Fund (75%) and the Greek Ministry of Development-GSRT (25%) and at 10% by "The Epilepsy Centre - SSE", Oslo University Hospital, Oslo, Norway.

References

1. J. Arnhold, P. Grassberger, K. Lehnertz, and C.E. Elger. A robust method for detecting interdependences: Application to intracranially recorded EEG. *Physica D*, 134(4):419–430, 1999.
2. T. Kreuz, F. Mormann, R.G. Andrzejak, A. Kraskov, K. Lehnertz, and P. Grassberger. Measuring synchronization in coupled model systems: A comparison of different approaches. *Physica D*, 225(1):29–42, 2007.
3. S. Manzan and C. Diks. Tests for serial independence and linearity based on correlation integrals. *Studies in Nonlinear Dynamics and Econometrics*, 6(2):1005, 2002.
4. R. Marschinski and H. Kantz. Analysing the information flow between financial time series. *European Physical Journal B*, 30:275–281, 2002.
5. M. Palus and M. Vejmelka. Directionality of coupling from bivariate time series: How to avoid false causalities and missed connections. *Physical Review E*, 75:056211, 2007.
6. A. Papana and D. Kugiumtzis. Detection of directionality of information transfer in nonlinear dynamical systems. In *Topics on Chaotic Systems, Selected Papers from CHAOS 2008 International Conference, 3-6 June, Chania, Crete*, 2008.
7. M.C. Romano, M. Thiel, J. Kurths, and C. Grebogi. Estimation of the direction of the coupling by conditional probabilities of recurrence. *Physical Review E*, 76:036211, 2007.
8. M.G. Rosenblum and A.S. Pikovsky. Detecting direction of coupling in interacting oscillators. *Physical Review E*, 64(4):045202, 2001.
9. T. Schreiber. Measuring information transfer. *Physical Review Letters*, 85(2):461–464, 2000.
10. M. Staniek and K. Lehnertz. Symbolic transfer entropy. *Physical Review Letters*, 100(15):158101, 2008.
11. M. Vejmelka and M. Palus. Inferring the directionality of coupling with conditional mutual information. *Physical Review E*, 77(2):026214, 2008.

Complexity theory and physical unification: From microscopic to macroscopic level

G. P. Pavlos[1], A. C. Iliopoulos[1], L. P. Karakatsanis[1], V. G. Tsoutsouras[1] and E. G. Pavlos[2]

[1]Department of Electrical and Computer Engineering, Democritus University of Thrace, 67100 Xanthi, Greece
[2]Department of Physics, Aristotle University of Thessaloniki, 541 24 Thessaloniki, Greece
Email: gpavlos@ee.duth.gr

Abstract: During the last two decades, low dimensional chaotic or self-organized criticality (SOC) processes have been observed by our group in many different physical systems such as space plasmas, the solar or the magnetospheric dynamics, the atmosphere, earthquakes, the brain activity as well as in informational systems. All these systems are complex systems living far from equilibrium with strong self-organization and phase transition character. The theoretical interpretation of these natural phenomena needs a deeper insight into the fundamentals of complexity theory. In this study, we try to give a synoptic description of complexity theory both at the microscopic and at the macroscopic level of the physical reality. Also, we propose that the self-organization observed macroscopically is a phenomenon that reveals the strong unifying character of the complex dynamics which includes thermodynamical and dynamical characteristics in all levels of the physical reality. From this point of view, macroscopical deterministic and stochastic processes are closely related to the microscopical chaos and self-organization. In this study the scientific work of scientists such as Wilson, Nicolis, Prigogine, Hooft, Nottale, El Naschie, Castro, Tsallis, Chang and others is used for the development of a unified physical comprehension of complex dynamics from the microscopic to the macroscopic level.
Keywords:. chaoticity, quanticity, far from equilibrium phase transition, scale invariance, self-organization, fractal spacetime, path integrals amplitude.

1. Introduction

The great challenge of complexity theory emerges from old and essential problems such as: the time arrow, the existence or not of a simple and fundamental physical level for a unified description of macroscopic and microscopic levels, the relation between the observer and the examined object, etc. In general, as far as the complexity theory and every new level of the physical reality are concerned, new concepts and new classifications are required. Initially, the first principles of the physical theory modeled the entire cosmos, using a reductionist point of view, as a deterministic, integrable, conservative, mechanistic and objectifying whole. However, Boltzmann's probabilistic interpretation of entropy as well as the non-integrability of the large Poincare systems rose for the first time serious doubts concerning the universality and plausibility of the primitive classical physical theory. Moreover, two great revolutions reconfigured the absolute objectifying character of the classical physical theory. The first one, concerning the macrocosm, was the Relativity Theory of Einstein and the second one, concerning the microcosm, was the Quantum Theory

Chaos Theory: Modeling, Simulation and Applications
C. H. Skiadas, I. Dimotikalis and C. Skiadas (Eds)

of Heisenberg, Schrodinger and Dirac. Finally, the reductionist character of the physical theory has been disputed after the scientific job of Feigenbaum, Poincare, Lorenz, Prigogine, Nicolis, Ruelle, Takens and other scientists who founded the chaos and complexity theory.

Recently, Ord, Nottale, El Naschie, Tsallis and other scientists introduced new theoretical and fertile concepts for the complex nature of the physical reality and the unification of the physical theory at the microscopic and macroscopic levels. In particular, complexity theory includes: chaotic dynamics in finite or infinite dimensional state space, far from equilibrium phase transitions, long range correlations and pattern formation, self-organization and multi-scale cooperation from the microscopic to the macroscopic level, fractal processes in space and time and other significant phenomena.

In this study, we provide a comprehensive description of the novel concepts included in the complexity theory from microscopic to macroscopic level. Some of the modern concepts that can be used for a unified description of complex systems and for the understanding of modern complexity theory, as it is manifested at the macroscopic and the microscopic level, are the fractal geometry and fractal space-time, scale invariance and scale relativity, phase transition and self-organization, path integral amplitudes, renormalization group theory, stochastic and chaotic quantization and E-infinite theory, etc.

2. Complexity as a new theory

Complexity is considered as a new and independent physical theory which was developed after the Relativity Theory and Quantum mechanics. It is related to far from equilibrium dynamics and concerns the creation and destruction of spatiotemporal patterns, forms and structures. According to Prigogine, Nicolis [1-3] and others, complexity theory corresponds to the flow and development of space-time correlations instead of the fundamental local interactions. Moreover, according to Sornette [4], systems with a large number of mutually interacting parts and open to their environment can self-organize their internal structure and their dynamics with novel and sometimes surprising macroscopic emergent properties. These general characteristics make the complexity theory, a fundamentally probabilistic theory of the non-equilibrium dynamics.

Until now complexity concerns mainly macroscopic physical systems. The traditional point of view is to explain macroscopic complex dynamics by the basic microscopical physical theory which is a fundamental Quantum Field Theory (QFT) with all its modern manifestations like supersymmetry and gauge symmetries or superstring theory (ST) including the M-theory (superstring) or D-branes theories [5]. Also, such a type of fundamental field theory belongs to the general class of time reversible and deterministic physical theory. However, the quantum probabilism is not included inherently in the fundamental theory but as an external interpretation of its mathematical formulation. It concerns the observer and the action measurement (observation), but not the dynamics of the objective physical system which corresponds to an objective and deterministic process in a general state space. At the macroscopic level, we are obliged as observers or as knowing subjects to use statistical or probabilistic methods. From this point of view, chaos with its famous characteristic of sensitivity to initial conditions (chaotic instability) or the older Heisenberg's uncertainty principle do not cause any trouble to the theoretical or dogmatic determinism rooted to the Cartesian metaphysic realism.

After all it is significant to search for a unifying route through complexity of the physical theory at the microscopical and macroscopical or cosmological level even of it is too early for such a dream. In the following, we present significant evidence for such a route.

3. Complexity as a kind of macroscopic quanticity

The central point of complexity theory is the possibility for a physical system, which includes a great number of parts or elements, to develop internal long-range correlations leading to macroscopic ordering and coherent patterns. These long-range correlations can also appear at the quantum level. In particular, according to the general entanglement character of the quantum theory, the quantum mechanical states of a system with two or more parts cannot be expressed as the conjunction of quantum states of the separate parts. This situation generally reflects the existence of non-local interactions and quantum correlations while the measurements bearing on either part correspond to random variables which are not independent and can be correlated independently of the spatial distance of the parts [6]. This means that the quantum density operator cannot be factored while the quantum state corresponds to the global and undivided system. The macroscopic manifestation of the quantum possibility for the development of long range correlations is the spontaneous appearance of ordered behavior in a macroscopic system examples of which are phenomena like superfluidity and superconductivity or lasers [7].

These quantum phenomena display coherent behavior involving the collective cooperation of a huge number of particles or simple elements and a vast number of degrees of freedom. They correspond also to equilibrium phase transition processes which constitute the meeting point of quantum theory and complexity. Here the development of quantum long range correlations leads to a macroscopic phase transition process and macroscopic ordering. It is not out of logic or physical reality to extend the unifying possibility of quantum process for development of long-range correlations or the quantum entanglement character to a macroscopic self-organizing factor causing also the far-from equilibrium symmetry breaking and macroscopic pattern formation. From this point of view we can characterize complexity as a kind of a macroscopic quanticity [8-10].

4. Quantum theory as a kind of microscopic chaoticity or complexity

Bohm [11] imagined that the quantum theory must be the manifestation of subquantum complex dynamics. During the last years, we observe the productive onset or the impetus invasion of chaos and complexity from macroscopic to the microscopic quantum level [12]. In the following we present some novel concepts in this direction:

- Analytical continuation and fractal space-time can convert an ordinary diffusion equation into a Schrödinger equation and a telegraph equation into a Dirac's equation. From this point of view analytical continuation is a short of cut quantization.
- Positive Lyapunov exponents of non-abelian gauge fields reveal the significance of chaos for the quantum field theory [13].
- Coupled map lattices with spatiotemporal chaotic profile can be used to simulate quantum field theories in an appropriate scaling limit.
- Kaneko coupled map lattices including chaotic strings provide the background for the Parisi-Wu stochastic quantization of ordinary string and quantum field theories [14, 15]. Chaotic strings can be used also to provide a theoretical argument why certain standard model parameters are realized in nature reproducing numerical values of the electroweak and strong coupling constants masses of the known quarks and leptons neutrino, W boson and Higgs mass.
- Renormalization group (RG) flows on the superstring world sheet becomes chaotic and leads to non-Markovian Fokker-Planck equation with solutions describing the transition from order to chaos and revealing the Feigenbaum universal constant [16]. The appearance of this constant reveals the scaling of space-time curvatures at the fixed points of the RG flow which becomes chaotic near singularities where the curvature is very large [17, 18].
- The Parisi-Wu [19] stochastic quantization theory relates the quantum field theory in D-dimensions to a classical Langevin equation in $D+1$ -dimensions where the Parisi-Wu fictitious time plays the role of an extra dimension. In this picture there exists a short of classical stochasticity and quantum theory duality

[20]. The stochastic quantization can be transform to chaotic quantization, similar to chaotic deterministic dynamical systems which can generate Langevin dynamics in an appropriate scaling limit. In this approach, quantum field theories can be simulated by chaotic dynamics.

- Non-extensive statistics [21], fractal string and branes, fractal statistics, fractons and anyons particles as well as chaotic M(atrix) theory indicate the establishment of chaos and complexity at the microscopic and the quantum level [5]. In this direction, Gerardus 't Hooft raised the conjecture that quantum theory can be derived as the low-energy limit of a microscopically deterministic but dissipative theory [22]. According to this concept classical Perron-Frobenius operators or deterministic automata can produce quantum states in Hilbert spaces as well as the Schrödinger equation [23-26].

5. The road of complexity for the physical theory unification

In this way, the Feynman rules and diagrams become common tool from the estimation of probabilistic processes at the microscopic quantum level or the macroscopic level of continuous media as they are being described by the Ginzburg-Landau model [28, 29]. In this direction, we could imagine Feynman rules and renormalization group theory as the universal characteristics of probabilistic processes at the microscopic and the macroscopic level. The renormalization group equations have many common features with non-linear dynamical systems, so that besides the existence of isolated fixed points, the coupling in a renormalizable field theory may flow also towards more general even fractal attractors. This could lead to Big Mess scenarios in application to multiphase systems, from spin-glasses and neural networks to fundamental string theory [17]. In this direction Cristopher Hill [30] introduced the fractal theory space where the key idea is that the Feynman path integral is invariant under a sequence of renormalization group transformations that map the kth lattice into the k-1 lattice. In the continuum limit these models produced quantum field theories in fractal dimensions $D=4+\varepsilon$. These theories are connected to the scaling behavior of fractal strings (branes), while the couplings oscillate on a limit cycle. Moreover the concept of fractal space-time can be used for the foundation of an extended Einstein's Relativity Principle unifying the micro and macro levels [9, 31].

In this direction, Ord [32] showed that fractal trajectories in space with Hansdorff dimension $D=2$ exhibit both an uncertainty principle and a De Broglie wave - particle duality relations. Furthermore, L. Nottale introduced the principle of Scaled Relativity according to which the laws of physics are scale invariant. This theory is related also to the concept of fractal space-time. According to Nottale, the consequence of scale invariance principle and space-time fractality opens the door for a grand unification of cosmos, from the microscopic quantum level to the macroscopic and cosmological level. The starting point of the theory is the refusing of the unjustified assumption of the differentiability of the space-time continuum.

The non-differentiable space-time continuum is necessarily fractal. The development of the theory starts by making the various physical quantities explicitly dependent on the space-time scale while the fundamental laws become also scale dependent. In this frame of theory the non-differentiability of space-time implies the breaking of time reversibility, and the global unification of microscopical and macroscopical laws.

6. Mathematical traces of complexity theory

Quantum theory or Relativity theory apart from their conceptual basis they correspond to unambiguous or an all-embracing mathematical form. Also these theories include as a limit the classical mechanistic or field theory. On the contrary, for complexity theory there is does not exist such a complete mathematical formulation. However, in the following we are going to present some of the mathematical expressions of the complexity theory. We believe that for a fully developed theory of complexity there must be also a fast developed consistory of mathematical expressions related logically to the existing physical theory.

6.1 *The thermodynamics point of view*

Complexity theory becomes effective when dynamical system exists far from equilibrium, where the entropy follows the equation

$$S = S_0 + \delta S + \delta^2 S \tag{1}$$

Far from equilibrium, according to Tsallis theory, the Boltzmann-Gibbs entropy:

$$S_{BG} = -\kappa \cdot \sum P_i \cdot \ln P_i = -\kappa \cdot \int P(x) \cdot \ln P(x) dx \tag{2}$$

is the limit of a more general and non-extensive entropy. The non-extensive Tsallis entropy S_q, given by the equation:

$$S_q = \frac{1 - \sum P_i^q}{q-1} = \frac{1 - \int [P(x)]^q \, dx}{q-1} \tag{3}$$

For $q = 1$ the new and non-extensive statistical mechanics is transformed to the known Boltzmann-Gibbs statistical mechanics [21].

Every thermodynamical system is also a dynamical system with internal dynamical states and this point of view exists as the propabilistic character of the internal dynamics. Thus, the entropy corresponds to the information that can be obtained by knowing the internal microscopic state of the system.

Near equilibrium $\delta S = 0$ and $\delta^2 S$ is a Lyapunov function of positive time derivative, $\frac{\partial}{\partial t} \delta^2 S > 0$. The time derivative corresponds to entropy productions. However far from equilibrium the time derivative of $\delta^2 S$ may be negative corresponding to decreasing of entropy and development of order [1]. Significant concepts of Thermodynamics developed for the near equilibrium phenomena can be used for the description of far from equilibrium phenomena. Such concepts are: thermodynamical potentials, phase-transition, order parameters, nucleation phenomena. Also the Ginzburg-Landau theory can be extended for the description of the far from equilibrium phase transition processes and pattern formation in many different dynamical systems [33] (Haken, 1983).

6.2 *The dynamical point of view*

Complex systems can be at the same time thermodynamical and dynamical systems. The far from equilibrium dynamics can be described by the generalized free energy H function which depends generally upon the macroscopic field $\varphi(\vec{x}, t)$ (order parameter field). The generalized free energy function can be estimated by using generalized Hamiltonians and Lagrangian, while the dynamics is given by a Langevin type equation:

$$\frac{\partial \varphi(\vec{x}, t)}{\partial t} = -\frac{\delta H(\varphi)}{\delta \varphi(\vec{x}, t)} + Df \tag{4}$$

where f is a stochastic component.

The Langevin equation (t) is equivalent to a functional Fokker-Planck equation

$$\frac{\partial P}{\partial t} = \frac{\delta}{\delta \phi} \cdot \left(\frac{\delta H}{\delta \phi} \cdot P + \frac{\delta}{\delta \phi} \cdot (DP) \right) \tag{5}$$

where $P[\varphi(\vec{x}, t), t]$ is the probability density functional [34].

The probability density functional can be calculated by the method of math integral formalism. According to Chang [35], for nonlinear stochastic systems near criticality the correlations among the fluctuations of the random dynamical fields are extremely long-ranged and there exist many correlations states. The far from equilibrium generalization of the renormalization group transformation includes the existence of fixed points (singular point) for the Lagrangian scaling flow on the parameter stochastic Lagrangian affine space. As the stochastic nonlinear-complex system is perturbed from a particular fixed point (meta-equilibrium state) it may be attached to another critical state revealing symmetry breaking and phase transition processes.

Anomalous diffusion and non-Gaussian distributions, low dimensional chaos, self organized critically, spatiotemporal chaos, spatiotemporal patterns, directed percolation processes, scaling and other macroscopic complex phenomena can be described by the application of the generalized renormalization group theory. In the above description the Langevin equations can be of more general type corresponding to a general type of master equations and discrete spatio-temporal stochastic or deterministic chaotic processes. According to the relation:

$$\partial_t \left| P_t \right\rangle = L \left| P_t \right\rangle \tag{6}$$

Where $\left| P_t \right\rangle$ is the probability distribution vector and L the so-called Liouville operator which generates the temporal evolution [36]. It is a truth that from the microscopic to the macroscopic level, the modern physical theory includes non-linear dynamical processes in finite-dimensional or infinite-dimensional state spaces. The dynamical processes can be of different and independent form at every level of dynamical description of nature. Superstrings, elementary particles or other physical forms and patterns observed at the macroscopic or cosmological level can be produced as solutions of non-linear mathematical equations including Hamiltonians, Lagrangians, and other mathematical forms corresponding to continuing or discrete fields, quantum or classical.

The great dream of theoretical scientists is to develop some kind of organizing physical principles valid at all levels. The strong simultaneous presence of chaos and fractal self similarity at the macroscopic and at the quantum level indicates the possibility for the existence of such global organizing principles. Some significant concepts in this direction are described in the following.

6.3 *Fractal theory space and renormalization group theory*

According to Nottale [30] theory space can be endowed with certain geometrical symmetries based essentially on the renormalization group (RG) transformations. Fractal lattices, known already in condensed matter physics, are used to construct classes of new theory spaces. Theory space lattices are defined recursively by replacing each site with a new simplex of S sites, while this process is iterated k times. In the limit $K \to \infty$ this lattice describes a continuum theory whose properties are determined be certain scaling laws of critical systems. At finite critical temperature the Feynman path integral is invariant under the *RG* transformations. Wilson's approach is based on an effective action

$$F_A(t/\varphi) \sim \log Z_A(t/\varphi) \tag{7}$$

$$Z_A(t/\phi) = \int_A D\phi \cdot \exp\left(\sum_i t^i O_i(\phi) \right) \tag{8}$$

where Z_A is the partition function and the integration extends over a functional space of fields. According to Morozov and Niemi [17], a *RG* flow could indeed tend towards a nontrivial attractor with even a fractal structure corresponding to a chaotic flow in the space of couplings. This could lead to a Big Mess scenario in applications to multiphase systems, from spin-glasses and neural networks to fundamentals string theories. The concept of the fractal theory space can be extended in a general unifying scheme to include Langevin processes or directed percolation processes corresponding to macroscopic far from equilibrium processes [28, 29, 37].

6.4 ***Scale Relativity Theory***

According to Nottale [30] the Scale Relativity theory extends the Einstein Relativity theory to scale transformations of resolution [9, 31]. In this theory the axiom of space-time differentiability is given up and the space-time corresponds to a fractal set. This is in accordance with Feynman's path integral approach to quantum mechanics, which at small length and time scales reveals continuous and non-differentiable paths with fractal dimension $D=2$. The fractal character of space-time indicates that the space-time variables can be described as depending also on resolutions $\Phi=\phi(x,\varepsilon)$. Nottale extended the principle of scale Relativity states that the laws of nature should apply to all reference systems whatever their state if scale. This is a scale covariance principle in the spirit of renormalization group theory corresponding to the typical form:

$$\frac{\partial\Phi(x,\varepsilon)}{\partial\ln\varepsilon}=\beta(\varphi) \tag{9}$$

In this theory the probabilistic character of quantum theory is derived naturally by the fractal nature of space-time while the microscopic quantum laws and the macroscopic stochastic processes are the different manifestation of the same physical law under the scale transformation in the fractal space-time. This led us to consider the macroscopical chaos and complexity in natural systems as the manifestation of the large scale fractal nature of space-time. From this point of view, the Feynman path integral formulation of physical theory in a fractal space time is the unifying tool of the microscopic and macroscopic complexity. In this way, the path probability density, the correlation function and the generating functional are the common tools in a unified point of view of microscopic and macroscopic stochastic processes according to the relation

$$G^n(x_1,x_2,\dots,x_n;J)=\frac{\delta}{\delta J(x_1)}\cdots\frac{\delta}{\delta J(x_n)}\log Z[J] \tag{10}$$

where the $Z_G=\int D\varphi\cdot e^{-H(J)}$ is the partition function of the system and G^n are the correlation functions. Classical macroscopic distributed far from equilibrium systems and microscopic quantum field systems are connected through stochastic and chaotic quantization theory. According to the Parisi-Wu [19] approach the classical field Φ described by an action $S[\Phi]$ can be quantized by means of a stochastic Langevin equation:

$$\frac{\partial\Phi(x,t)}{\partial t}=-\frac{\delta S(\phi)}{\delta\phi}\cdot(x,t)+f(x,t) \tag{11}$$

where f denotes spatio-temporal Gaussian white noise.

The classical stochastic process corresponds to a non-equilibrium $(D+1)$-dimensional space-time system while its corresponding Fokker-Planck distributed equation at the equilibrium state is identified with the theory of the quantum field Φ in the D-dimensional space-time. The classical stochastic dynamics in the $(D+1)$-dimensional space-time is described by the correlators $\langle\Phi(x_1,t_1)\dots\Phi(x_n,t_n)\rangle$ where $\Phi(x,t)$ are the solutions of equation (10). The classical stochastic correlators in the limit $t\to\infty$ reproduce the Feynman graphs of the corresponding D-dimensional quantum theory.

On the other hand according to Beck [14, 15] there are deterministic chaotic dynamical systems that can generate the Langevin dynamics of equation (10). This approach corresponds to the chaotic quantization theory as it simulates quantum fields by chaotic dynamics. Moreover Beck introduced a class of coupled map lattices, by using coupled third – order Tschebyscheff polynomials, which can simulate quantum field theories. The stochastic and chaotic quantization theories reveal the mathematical similarity of the classical and quantum dynamics but more than this it indicates the deep unity of physical theory at the microscopic and the macroscopic level. This point of view can be supported by modern theoretical constructions which can produce the quantum theory as the manifestation of a subquantum chaotic dynamics.

G.'t Hooft [12, 22] conjectured that quantum mechanics can logically arise as the low energy limit of a microscopically deterministic but dissipative theory. Biro, Mueller and Matinyan [13] showed that quantum field theory can emerge in the infrared limit of a high-dimensional quantum field theory as Yang- Mills fields can quantize themselves.

7. Becoming before Being

Quantum theory and Relativity theory (special and general) introduced the notion of Becoming in the heart of Being showing that Being is the production of Becoming. Complexity theory is a theory of Being as Becoming in its essence. From this point of view we believe that complexity theory, as the essential theory of becoming and pattern formation, shows the route for a global unification of physical theory. Particularly, the quantum states include becoming and potentiality in an objective way. Because of this the physical magnitudes are related probabilistically with the quantum states. Entropy and complexity are inherent in the quantum states. Similarly in relativity material particles are condensed energy and space and time are continuously. In this point of view the great Hellenic philosophers Heraclitus, Plato and Aristotle meet one another. For Heraclitus all the cosmos is one. For Aristotle matter is the potentiality of the being and for Plato the essence of things is the form.

According to the modern theoretical concepts, the essence of the physical theory is the mathematical structure-equations with solutions corresponding to the observed phenomena. All this state of knowledge can be included in the general form of Feynman path integral formulation of the probability amplitudes of dynamics:

$$\langle \Phi_2, t_2 || \Phi_1, t_1 \rangle \approx N \int_{\Phi_1}^{\Phi_2} D\Phi \, e^{i \cdot S / \hbar} \qquad (12)$$

where Φ is the dynamical state and S the action functional of the dynamics. The unification of micro and macro dynamics passes through the fractal theory, and the scale invariance of Feynman amplitudes. The Feynman principle must be related somehow to the general probabilistic Liouville dynamical theory. As we show in the following the probabilistic character of modern theory reveals the fundamental character of becoming or process in the physical theory. According to the radical conjecture of Prigogine becoming is prior to being in a fundamental way including even the physical laws. According to Kovacs [38] many of the pattern formation processes observed in systems of quite different genesis exhibit astonishing analogies independently of the composition of the system and of the interactions between individual components. As the formation of structures is a geometrical event, the observed qualitative analogies indicate common topological origin. From this point of view, physics is geometry and topology.

In this direction bifurcation theory and symmetry breaking processes are the common base of the far from equilibrium dynamics in micro and macro cosmos. In the bifurcation theory, bifurcation phenomena include topological equivalence while the bifurcations are largely determined by geometrical constraints as the dimension and shape or inherent symmetries. The structural stability of physical patterns at every level of physical reality is inherent in the fundamental law (laws) of physics derived by geometrical invariance principles such as space-time symmetries or gauge symmetries. The link between geometry and physics is due to topological invariance principles which cause inherent stability properties. Structural stability proves includes the interaction and correlation physical phenomenology of physical systems. On the other hand, self-organization is a far from equilibrium thermodynamical process revealing entropy decrease and order development according to the equation

$$dS = d_e S + d_i S \tag{13}$$

where $d_e S$ is the reversible entropy transfer and $d_i S$ is the irreversible produced entropy within the system. Thermodynamics is coupled to dynamic through the general Liouville equation

$$\partial_t f(t) = Lf(t) \tag{14}$$

where $f(t)$ is the infinite dimensional distribution vector $f(t)=\{f_0, f_1(x_1), f_2(x_1,x_2)\ldots\}$ including the Bogolinbov, Born, Green, Kirkwood, Yvon (BBGKY) [39] correlations hierarchy. The dynamics of correlations corresponds to a functional Hilbert space H^{∞}. In order the dynamical theory to be in agreement with the thermodynamical theory, the Liouville equation (eq.14) must be solved in an extended functional space called rigged Hilbert space or Gelfand space according to Petrosky and Prigogine [40, 41]. For the integrable classical or quantum dynamical systems the dynamical theory does not lead to thermodynamics. Chaotic dynamical systems are non-integrable Poincare systems. For chaotic or non-integrable dynamical systems, the Liouville equation can be solved in an extended functional space which includes the thermodynamical states and the irreversible Markov stochastic processes underlying to the far from equilibrium thermodynamics. The extended physical space in a Rigged Hilbert space or Gelfald space, in which the infinite dimensional correlation hierarchy and the corresponding extended Liouville equation must include as general solutions the observed physical forms or states in the microscopic or the macroscopic level. Also, the physical states in the Rigged Hilbert space correspond to possibilities or potentialities. In this way, the extended in the Rigged Hilbert space Liouville equation can be transformed to a Fokker-Planck equation including dynamics and thermodynamics simultaneously with breaking of time reversal symmetry and entropy production. According to Prigogine the flow of correlations in the extended Riggen Hilbert space creates the observable physical states and the observed dynamics. In this way, the process as becoming is before being and correlations as potentialities are the fundamental physical reality.

8. Chaos, fractals and unification of dynamics to thermodynamics

In this chapter, we conjecture that the above mathematical construction must be related somehow to the infinite dimensional fractal space included in the E-infinite theory of El Naschie [42]. Von Neumann understood that the quantum dynamics includes non-commutative geometry of the quantum space which has no points at all. This means that the physical reality is not created by geometrical points or by simple material points without internal structure. At every level physical reality is complex including forms, structures and patterns. Non-commutative geometry corresponds also to fuzzy spaces. The E-infinite theory of El Naschie introduces randomness and scale invariance to the very concept of space-time geometry. Space-time is an infinite-dimensional, hierarchical and random geometrical manifold with infinite numbers of equivalent paths. In this space, any so-called point will always reveal a structure on a close examination. Cantorian space-time E^{∞} of the El Naschie theory is a form of non-commutative geometry and it can be constructed using an infinite number hierarchy of Cantor fractal sets mixed together in all possible forms of union and intersections.

Let us remember that in the Euclidean theory of space, the curved surface is included as a subset of Euclidean points. In the Riemannian theory the Euclidean space is a local characteristic of the general Riemannian manifold. For Einstein, the space-time Riemannian manifold is the primitive or fundamental physical reality, which includes every other physical form as a geometrical characteristic. Of course the dynamical cosmology and the expanded space included in the equations of general relativity of Einstein was for him an undesirable trouble. For modern complexity theory the Mandelbrot fractal theory, the Euclidean or Riemannian manifolds are the smoothed out character of a primitive fractal manifold with scale dependent dimensions and other topological invariant properties. For El Naschie and other modern scientists every physical form in the universe is a scale dependent geometric characteristic of the infinite dimensional fractal Cantor's manifold. In El Naschie's theory, space-time is an infinite dimensional fractal that happens to have $D=4$ as the expectation value for the topological dimension. The topological dimension 3+1 means that in our low energy resolution the world appears to us as it were four dimensional. According to Iovane, Ahmed, and Agop [43-45] the difference between micro and macro physics depends only on the resolution in which the observer looks at the world. Nature shows us structures with scaling rules where hierarchy clustering properties are revealed from cosmological to microscopical objects. The universe is self-similar from the quantum to the cosmological level according to the scale invariance law:

$$R(N)=\frac{h}{Mc}N^{\alpha} \tag{15}$$

where R is the radius of the astrophysical structures, M is the total Mass of the self-gravitating system, h is the Planck constant, N is the number of nucleons, c the speed of light and $a=3/2$. For $N=1$, R is equal to the Compton wavelength. There is no breaking point between microscopic and large scale universe. This is in agreement with El Naschie's E-infinite theory of cantorian space-time, the Golden Mean and the Fibonacci numbers, as well as to the general theory of the stochastic self-similar processes in a fractal space. Also it appears that the universe has a memory of its quantum origin as was suggested by Penrose. Consequently the universe with its structures as all scales from the quantum level to the organic cell and human to super clusters or galaxy and the cosmological level is a self-similar complex system and includes self-similar processes at every level. In this theory the fractal space-time includes in a fundamental way the probabilistic character of the physical dynamics as well as the symmetry braking of time reversal.

9. Epilogue: Complexity theory and unification of cosmos

We live in an infinite dimensional and continuously created cosmos. Anaximander thought that every finite physical form is born in the infinite. Also, Heraclitus thought that every physical form is protogenic (primitive) motion ordered by the cosmic (common) reason. Cantor understood every finite set as subset or finite projection of the transfinite Cantor set hierarchy. El Naschie, Nottale and others, thought the E-infinite Cantor's fractal space-time theory as the unifying theory of micro and macro physical forms in a continuation of Einstein's geometrization of physics or as well as a continuation of the quantum or Popper's and Prigogine's ontological probabilism. Fractal geometry and fractal space-time couples, long range correlations, symmetry breaking and scaling invariance with the quantum uncertainty principle and non-commutativity, all together in an ontological and dynamical whole can be described through fuzzy logic and Godel's theory. All these new and old concepts reveal that complexity can be the strong framework for the physical cosmos unification. According to the general scale relativity, based on the fractal space-time concept, the physical laws at every physical level must be self-similarly related. In this direction, we introduce the following conjecture. Non-linearity at every level creates chaoticity, fractality, bifurcation and symmetry breaking and finally pattern formation. Physical forms correspond to topological invariance and structural stability at every level of physical reality. The physical laws are also created as physical forms included, after a general symmetry breaking process, in the new concept of the dynamical physical reality. Quantum states are also complex patterns and are created by a microscopic self-organization process similar to macroscopic self-organization which creates the macroscopic physical states. From this point of view, determinism and stochasticity are two faces of the same nonlinear and complex unique dynamics. The observed self-organization process is self-similar and self-consistent at every level including even the level of conscience and mind. According to Prigogine and Nicolis the flow of the dynamics of correlations is the deeper truth of complexity. The flow of correlations seems to be the most primitive «non-material» dynamical and infinite dimensional «matter» of the physical theory producing nonlinearity, determinism, probabilism and stochasticity at every level of the physical reality. After all, we move at the limits of scientific thought which may be not the last word of the truth.

References

[1] I. Prigogine, Time, structure and fluctuations, Nobel Lecture, 1977.
[2] G. Nicolis and I. Prigogine, Exploring Complexity: An Introduction, San Francisco: W.H. Freeman, 1989.
[3] G. Nicolis, Irreversible thermodynamics, Rep. Prog. Phys., 42, 225, 1979.
G. Nicolis, Physics of Far-From-Equilibrium Systems and Self-organisation, in Davies, Cambridge University Press, 1989.
[4] D. Sornette, Critical Phenomena in Natural Science, Springer, 2006.
[5] C. Castro, On non-extensive statistics, chaos and fractal strings, Physica A 347, 184, 2005.
[6] A. Shimony, Conceptual foundations of quantum mechanics, in Davies, Cambridge University Press, 1989.
[7] A. Leggett, Low temperature physics, superconductivity and superfluidity, in Davies, Cambridge University Press, 1989.
[8] K. Wilson, The renormalization group and critical phenomena, Rev. Mod. Phys., **55**, 583, 1983.
[9] L. Nottale, Scale-relativity and quantization of the universe, Astron. Astrophys. 327, 867, 1997.
[10] P. Davies, The Cosmic Blueprint, William Heinemann Ltd., Great Britain, 1987.
[11] D. Bohm and B. Hiley, The undivided universe, Routledge, London, 1993.
[12] G. 't Hoof, How Does God Play Dice? (Pre-)Determinism at the Planck Scale, Quantum [Un]speakables: From Bell to Quantum Information, 307, 2001.
[13] T. Biro, B. Muller, and S. Matinyan, Chaotic Quantization: Maybe the Lord plays dice, after all?, Lecture Notes in Physics, 633,164, 2004.
[14] C. Beck, Higher correlation functions of chaotic dynamical systems-a graph theoretical approach, Nonlinearity 4, 1131, 1991.
[15] C. Beck, Chaotic quantization of field theories, Nonlinearity 8, 423, 1995.
[16] I. Kogan and D. Polyakov, Chost-Matter Mixing and Feigenbaum Universality in String Theory, Physics of Atomic Nuclei, 66, 2062, 2003.
[17] A. Morozov and J. Niemi, Can Renormalization Group Flow End in a Big Mess?, Nuclear Physics B, 666, 311, 2006.
[18] D. Polyakov, Ads / CFT correspondence, critical strings and stochastic quantization, Class. Quantum Grav. 18, 1979, 2001.
[19] G. Parisi and Y. Wu, Perturbation theory without gauge fixing, Scientia Sinica, Vol. 25, No. 4, 484, 1981.
[20] P. Damgaard and H. Huffel, Stochastic Quantization, Phys. Rev. Let. 152, 227, 1987.
[21] C. Tsallis, Introduction to Nonextensive Statistical Mechanics, Springer, 2009.
[22] G. 't Hooft, Determinism and Dissipation in Quantum Gravity, Basics and highlights in fundamental physics: proceedings of the International School of Subnuclear Physics, 37, 397, 2001.
[23] B. De Wit, On the quantum mechanics of supermembranes, Nuclear Physics B 305, 545, 1988.
[24] T. Banks, W. Fischler, S.H. Shenker and L. Susskind, M theory as a matrix model: A conjecture, Phys. Rev. D, 55, 5112, 1997.
[25] J. Frohlich and J. Hoppe, On Zero-Mass Ground States in Super-Membrane Matrix Models, Commun.Math.Phys., 191, 613, 1998.
[26] I.Ya. Aref'eva, P.B. Medvedev, O.A. Rytchkov and I.V. Volovich, Chaos in M(atrix) Theory, Chaos, Solitons & Fractals, 10, 213, 1999.

[27] J. Binney, N.J. Dowrick, A.J. Fisher and M.E.J. Newman, The Theory of Critical Phenomena: An introduction to the renormalization group, Oxford University Press, 1992.
[28] R. Bausch, H.K. Jansshen and H. Wagner, Renormalized Field Theory of Critical Dynamics, Z.Physik B24, 113, 1976.
[29] P. Hohenberg and B. Halperin, Theory of dynamic critical phenomena, Rev. of Mod. Phys. 49, 435, 1977.
[30] C. Hill, Fractal theory space: Spacetime of noninteger dimensionality, Phys. Rev. D 67, 085004, 2003.
[31] L. Nottale, Fractal Space-Time and Micro-physics, Towards a Theory of Scale Relativity, Edit. World Scientific, 1993.
L. Nottale, Scale Relativity: First Steps toward a Field Theory, invited conference in "Relativity in General", Eds. J. Diaz Alonso and M. Lorente Paramo, 121, 1993.
L. Nottale, Emergence of Structures from chaos, Colloquium on Cellular Automata, 268, 1993.
L. Nottale, Scale Relativity, Fractal Space-time and Quantum Mechanics, Chaos, Solitons & Fractals, 4, 361, 1994.
L. Nottale, The Scale-Relativity Program, Chaos, Solitons & Fractals, 10,459, 1999.
[32] G.N. Ord, Fractal space-time: a geometric analogue of relativistic quantum mechanics, J. Phys. A: Math. Gen. 16, 1869, 1983.
[33] H. Haken, Synergetics: Introduction and Advanced Topics, Springer, 1983.
[34] T. Chang, D.D. Vvedensky and J.F. Nicoli, Differential renormalization-group generators for static and dynamic critical phenomena, Phys. Rep. 217, 279, 1992.
[35] T. Chang, Low-Dimensional Behavior and Symmetry Breaking of Stochastic Systems near Criticality-Can These Effects be Observed in Space and in the Laboratory?, IEEE Trans. On Plasma Science, 20, 691, 1992.
[36] H. Hinrichsen, Non-equilibrium phase transitions, Physica A 369, 1, 2006.
[37] M. Henkel, H. Hinrichsen and S. Lubeck, Non-Equilibrium Phase Transitions, Volume I: Absorbing Phase Transitions, Springer, 2008.
[38] A. Kovacs, Nonlinear Dynamics of Spatio-Temporal Processes in Complex Systems, Mathl. Comput. Modelling, 19,47, 1994.
[39] R. Balescu, Equilibrium and nonequilibrium statistical mechanics, A Wiley-Interscience publication, United States of America, 1975.
[40] T. Petrosky and I. Prigogine, Poincare Resonances and the Extension of Classical Dynamics, Chaos, Solitons & Fractals, 7, 441, 1996.
[41] T. Petrosky and I. Prigogine, The Liouville space extension of quantum mechanics, Advances in Chemical Physics, Vol. XCIX, Ed. I. Prigogine and S. Rice, 1997.
[42] M.S. El Naschie, Superstrings, Knots, and Noncommutative Geometry in $E^{(\infty)}$ Space, Int. Journal of Theoretical Physics, 37, 2935, 1998.
M.S. El Naschie, Complex vacuum fluctuation as a chaotic "limit" set of any Kleinian group transformation and the mass spectrum of high energy particle physics via spontaneous self-organization, Chaos, Solitons & Fractals, 17, 631, 2003.
M.S. El Naschie, A review of E-infinity theory and the mass spectrum of high energy particle physics, Chaos, Solitons & Fractals, 19, 209, 2004.
M.S. El Naschie, The Concepts of E Infinity: An elementary introduction to the Cantorian-fractal theory of quantum physics, Chaos, Solitons & Fractals, 22, 495, 2004.
M.S. El Naschie, Einstein's dream and fractal geometry, Chaos, Solitons & Fractals, 24, 1, 2005.
M.S. El Naschie, Elementary prerequisites for E-infinity (Recommended background readings in nonlinear dynamics, geometry and topology), Chaos, Solitons & Fractals, 30, 579, 2006.
[43] G. Iovane, E. Laserra and F.S. Tortoriello, Stochastic Self-Similar and Fractal Universe, Chaos, Solitons & Fractals, 20, 415, 2003.
G. Iovane, Varying G, accelerating Universe, and other relevant consequences of a stochastic self-similar and fractal Universe, Chaos, Solitons & Fractals, 20, 657, 2004.
G. Iovane, Mohamed El Naschie's $\varepsilon^{(\infty)}$ Cantorian space-time and its consequences in cosmology, Chaos, Solitons & Fractals, 25, 775, 2005.
[44] N. Ahmed, Cantorian small world, Mach's principle and the universal mass network, Chaos, Solitons & Fractals, 21, 773, 2004.
[45] M. Agop, El Naschie's $\varepsilon^{(\infty)}$ space-time, hydrodynamic model of scale relativity theory and some applications, Chaos, Solitons & Fractals, 34, 1704, 2007.

Regular variation, Paretian distributions, and the interplay of light and heavy tails in the fractality of asymptotic models

Dinis D. Pestana[1], Sandra M. Aleixo[2] and J. Leonel Rocha[3]

[1] FCUL, Universidade de Lisboa, and CEAUL, Lisboa, Portugal
Email: dinis.pestana@fc.ul.pt
[2] Mathematics Unit, DEC, Instituto Superior de Engenharia de Lisboa, and CEAUL, Lisboa, Portugal; Email: sandra.aleixo@dec.isel.ipl.pt
[3] Mathematics Unit, DEQ, Instituto Superior de Engenharia de Lisboa, and CEAUL, Lisboa, Portugal; Email: jrocha@deq.isel.ipl.pt

Abstract: Classical central limit theorems, culminating in the theory of infinite divisibility, accurately describe the behaviour of stochastic phenomena with asymptotically negligible components. The classical theory fails when a single component may assume an extreme protagonism. The early developments of the speculation theory didn't incorporate the pioneer work of Pareto on heavy tailed models, and the proper setup to conciliate regularity and abrupt changes, in a wide range of natural phenomena, is Karamata's concept of regular variation and the role it plays in the theory of domains of attraction, [8], and Resnick's tail equivalence leading to the importance of generalized Pareto distribution is the scope of extreme value theory, [13].
Waliszewski and Konarski discussed the applicability of the Gompertz curve and its fractal behaviour for instance in modeling healthy and neoplasic cells tissue growth, [15]. Gompertz function is the Gumbel extreme value model, whose broad domain of attraction contains intermediate tail weight laws with a wide range of behaviour.
Aleixo et al. investigated fractality associated with *Beta* (p,q) models, [1], [2], [10] and [11]. In this work, we introduce a new family of probability density functions tied to the classical beta family, the $Beta^*(p,q)$ models, some of which are generalized Pareto, that span the possible regular variation of tails. We extend the investigation to other extreme stable models, namely Fréchet's and Weibull's types in the General Extreme Value (GEV) model.

Keywords: power laws, extension of the logistic parabola, Gompertz growth model and Gumbel law, extreme value laws, growth models.

1. Introduction

Power laws have a natural connection with fractals and chaotic dynamics, as explained in detail in [14]. But on the other hand, power laws have an important smoothing effect, since power laws like behaviour is at the root of Karamata's regular variation, [8], a theory that became an important tool in many branches of Mathematics, [5]. Since Feller, [6], observed that Doeblin's and Gnedenko theory of the domains of attraction of the additive stable laws could be simply rephrased in Karamata's framework, slow variation and

Chaos Theory: Modeling, Simulation and Applications
C. H. Skiadas, I. Dimotikalis and C. Skiadas (Eds)

regular variation became the election tool in characterizing domains of attraction, letting De Haan, [7], achieve a complete characterization of domains of attraction for stable extreme value laws, Bingham solve a more general problem of domains of attraction of generalized convolution algebras, [4], and Kozubowski and Rachev characterize geometric-stable domains of attraction, [9]. No one, as far as we are aware, pointed out that weak convergence of types problems in Probability theory are connected to the renormalization group theory in Mathematical Physics, and that the appropriateness of the regular variation theory to deal with domains of attraction is a side-effect of the self-similarity resulting from power laws and scaling.

In here we shall try to bring together similar ideas that, having been developed in diverse fields, seem so far to be strangers to each other. Namely, we shall use an extension of the beta function and of beta densities to develop an extension of Verhults growth model leading to the Gompertz function — which, in fact, plays an important role in extreme value theory, where it is known as Gumbel distribution function.

The Gumbel extreme stable law has in its domain of attraction laws both with infinite right endpoint and with finite right endpoint. In fact, laws in the Gumbel model domain of attraction aren't heavy tailed, and this can be a serious drawback for the Gompertz growth model proposed by Waliszewski and Konarski as a cancer growth model, [15]. A simple extension of the differential equation whose solution is the Gompertz function leads to the Fréchet extreme value model, which for some values of the malthusian parameter can be very heavy-tailed.

Observe that the introduction of a generalized Pareto densities family, first described by Pickands, [12], is at the core of a presentation of extreme value theory in a POT (Peaks over Thresholds) setting; the paretian laws are the stable elements in this approach, and once again quasi-power functions are substantive in characterizing domains of attraction, and Resnick's tail equivalence is the key concept tying the two theories, [13].

In section 2, we introduce a family of probability density functions tied to the classical beta family. In fact, a $Beta(p,2)$ probability density function can be viewed as the truncation of the MacLaurin series expansion for the models $Beta^*(p,2)$ we investigate. We focus our interest in the $Beta^*(2,2)$ model, constructing an extension of the Verhulst model, that show that stable laws for maxima provide a natural framework to study growth when some external behaviour is expected as in the case of cancer growth.

In section 3, we present a bare sketch of the dynamical behaviour of the models based on $Beta^*(p,q)$ laws, focusing once again on the $p=q=2$ case, due to space restrictions. Finaly, in section 4, we point out how these models are connected to generalized mixtures of power functions, a fact that seems to be at the core of the fractal and ultimately chaotic behaviour of the iterates of the difference growth model for non-overlaping generations.

2. An extension of the Beta densities, and Gompertz growth model

The Verhulst growth model has a prestigious history in population dynamics, and the developments of fractal geometry in the XXth Century enlightened some curious features, such has chaotic behaviour followed by extinction, when the malthusian reproduction rate r is "large" in the logistic parabola

$$f(x)=r\,x(1-x)\ \mathrm{I}_{(0,1)}(x).$$

Observing that $f(x)=r\,x(1-x)\mathrm{I}_{(0,1)}(x)=f_{r,2,2}(x)$ is proportional to the *Beta(2,2)* density, the authors investigated, [1], [2], [3], [10], and [11], the dynamical behaviour of functions

$$f_{r,p,q}(x)=r\,x^{p-1}(1-x)^{q-1}\ \mathrm{I}_{(0,1)}(x)$$

proportional to $Beta(p,q)$ densities, discovering namely that interesting chaotic patterns of behaviour were closely related to the generalized skewness and kurtosis functions — i.e., to the standardized third and fourth central moments of the model, [3].

In here we use a non-trival extension of Euler's Beta function

$$\mathrm{B}(p,q)=\int_0^1 x^{p-1}(1-x)^{q-1}\mathrm{d}x,\quad p,\,q>0,$$

namely

$$\mathrm{B}^*(p,q)=\int_0^1 x^{p-1}(-\ln x)^{q-1}\mathrm{d}x,\quad p,q>0,$$

to extend the Verhults growth model.

Observe that $-\ln x=\sum_{k=1}^{\infty}\frac{(1-x)^k}{k}$ when $x\approx 0$, and hence $\mathrm{B}(p,q)$ may be viewed has a first order approximation of $\mathrm{B}^*(p,q)$.

Note that, while $\mathrm{B}(p,q)=\frac{\Gamma(p)\Gamma(q)}{\Gamma(p+q)}$, where $\Gamma(\alpha)=\int_0^{\infty} x^{\alpha-1}\mathrm{e}^{-x}\mathrm{d}x,\ \alpha>0$, is the Euler's gamma function, $\mathrm{B}^*(p,q)=\frac{\Gamma(q)}{p^q}$. In particular,

$$\mathrm{B}^*(p,2)=\sum_{k=1}^{\infty}\frac{\mathrm{B}(p,k+1)}{k}=\frac{1}{p^2}.$$

Hence

$$h_{p,q}(x)=\frac{1}{B^*(p,q)}x^{p-1}(-\ln x)^{q-1}\ \mathrm{I}_{(0,1)}(x)=\frac{p^q}{\Gamma(q)}x^{p-1}(-\ln x)^{q-1}\ \mathrm{I}_{(0,1)}(x)$$

is a new probability density function, which we designated by $Beta^*(p,q)$ density, and in particular, the $Beta^*(2,2)$ density

$$h_{2,2}(x)=4\,x\,(\text{-}\ln x)\,\mathrm{I}_{(0,1)}(x)$$

is a mixture of the $Beta(2,k+1)$ densities, with weights

$$w_k = \frac{4}{k\,(k+1)(k+2)}\,.$$

Now, we consider a natural extention of the logistic parabola growth model inspired in this new $h_{2,2}$ probabilistic model, namely

$$G'(x)=r\,G(x)(\text{-}\ln G(x)),\;\; r>0,\;\; G(x)\in(0,1)$$

whose solution $G(x)=\mathrm{e}^{\mathrm{e}^{-rx+C}}$, C constante, is the Gompertz growth model. A similar observation is made in [15]. Observe, also, that $\Lambda(x)=\mathrm{e}^{\mathrm{e}^{-rx}}\mathrm{I}_{\mathrm{IR}}(x)$ is the widely used extreme value Gumbel model.
Further observe that the solution of

$$G'_\nu(x)=r\,x^\nu\,G_\nu(x)(\text{-}\ln G_\nu(x)),\, r>0,\;\; G_\nu(x)\in(0,1)$$

is $G_\nu(x)=\mathrm{e}^{\mathrm{e}^{-\frac{rx^{\nu+1}}{\nu+1}+C}}$, C constante.

Hence, when $\nu\to-1$, with $C=\dfrac{r}{\nu+1}$, using L'Hôpital rule we get

$$\lim_{\nu\to-1}\frac{-r\left(x^{\nu+1}-1\right)}{\nu+1}=-r\ln x$$

and we obtain the solution $G_{-1}(x)=\mathrm{e}^{-x^{r}}$ which is a Fréchet-r extreme value distribution. Laws in the extreme value Fréchet domain of attraction for maxima must have infinte right endpoint, and can be severely heavy-tailed.

3. Dynamical study of the models proportional to $Beta^*(p,q)$ densities — a brief sketch

Consider the family of unimodal maps with a parameter $r\in[m,M]$, $m\in\mathrm{IR}_0^+$, where $g_{r,p,q}:\,]0,1]\to[0,1]$, such that

$$g_{r,p,q}(x)=r\,x^{p-1}(-\ln x)^{q-1}\;\text{ with }\; p,q>1\;\text{ and }\; r>0$$

satisfies the following conditions:

- $g_{r,p,q}\in C^3(\,]0,1[\,)$;
- $g'_{r,p,q}(x)\neq 0, \forall\, x\neq c$ ($c=\mathrm{e}^{(1-q)/(p-1)}$ is the critical point of $g_{r,p,q}$);
- $g'_{r,p,q}(c)=0$ and $g''_{r,p,q}(c)<0$;

- $g_{r,p,q}(0^+) = g_{r,p,q}(1) = 0$; $g_{m,p,q}(c) = 0$ and $g_{M,p,q}(c) = 1$;
- The Schwarz derivative of $g_{r,p,q}(x)$ is given by

$$S(g_{r,p,q}(x)) = \frac{g'''_{r,p,q}(x)}{g'_{r,p,q}(x)} - \frac{3}{2}\left(\frac{g''_{r,p,q}(x)}{g'_{r,p,q}(x)}\right)^2$$

$$= -\frac{q(q-1)^2(q-2) + 4(p-1)q(q^2-3q+2)\ln x + (6p^2-12p+5)(q-1)^2\ln^2 x}{2x^2\ln^2 x(q-1+(p-1)\ln x)^2}$$

$$-\frac{2(2p^3-6p^2+5p-1)(q-1)\ln^3 x + p(p-2)(p-1)^2\ln^4 x}{2x^2\ln^2 x(q-1+(p-1)\ln x)^2}.$$

It is verified that $S(g_{r,p,q}(x)) < 0$, for $x \in]0,1] \setminus \{c\}$, except for some values at the beginning or at the end of the interval $]0,1]$ where $S(g_{r,p,q}(x)) \geq 0$, but it does not disturb the dynamical behaviour of $g_{r,p,q}$, see Figs. 1 and 2, and [2], [3]. Note that, $S(g_{r,p,q}(c)) = -\infty$.

The $g_{r,p,q}(x)$ maps are proportional to the probability density function $h_{p,q}(x)$.

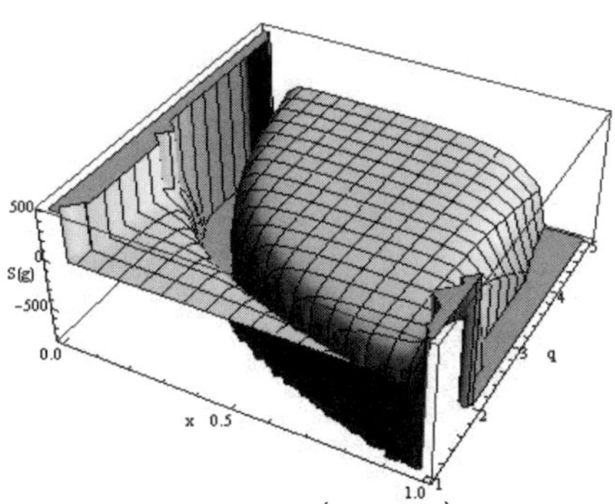

Graphic of $S(g_{r,2,q}(x))$

Figure 1

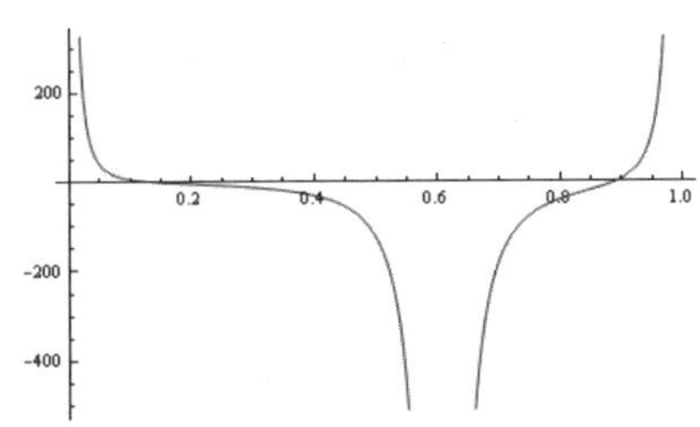

Graphic of $S(g_{r,2,1.5}(x))$

Figure 2

In Fig. 3 below, we exemplify the bifurcation diagrams corresponding to the unimodal family $g_{r,p,q}$, using the case $p = q = 2$. For different values of $p, q > 1$, diagrams are in essence similar, with some non-utterly dramatic changes in shape and scale. In Fig. 4 we can see a 4-periodic orbit for the map proportional to the $Beta^*(2,2)$ model.

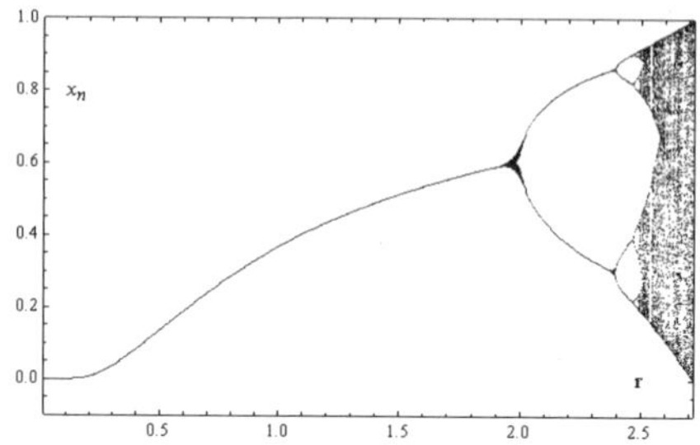

Bifurcation Diagram for $g_{r,2,2}$

Figure 3

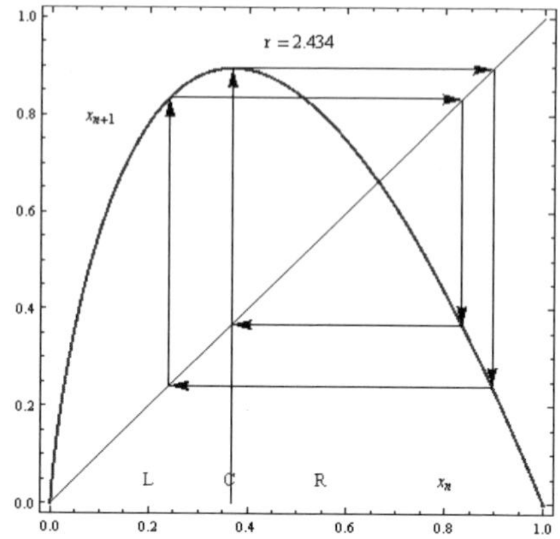

4-period orbit for $g_{r,2,2}$

Figure 4

Period	Sequence	$Beta^*(2,3)$	$Beta^*(2,2)$	$Beta^*(3,2)$	h_{top}
2	$(CR)^\infty$	1.230	2.218	4.412	0
4	$(CRLR)^\infty$	1.481	2.434	4.678	0
8	$(CRLR^3LR)^\infty$	1.535	2.478	4.736	0
6	$(CRLR^3)^\infty$	1.607	2.535	4.812	0.241
8	$(CRLR^5)^\infty$	1.643	2.564	4.847	0.304
7	$(CRLR^4)^\infty$	1.689	2.599	4.886	0.382
5	$(CRLR^2)^\infty$	1.722	2.624	4.927	0.414
7	$(CRLR^2LR)^\infty$	1.749	2.645	4.967	0.442
8	$(CRLR^2LR^2)^\infty$	1.769	2.660	4.998	0.468
3	$(CRL)^\infty$	1.786	2.673	5.038	0.481
	CRL^∞	1.847	2.718	5.437	ln 2

Table 1: Values of the parameter r of some symbolic sequences and topologycal entropy.

In Table 1, we can observe the values of the parameter r for which are obtained the several periodic orbits correspondent to the respective kneading symbolic sequences. This values were obtained from the construction of symbolic lists for increasing values of the parameter *r*, using suitable computacional programs implemented in *Mathematica* 7.0. As a consequence of all this maps $g_{r,p,q}$ are unimodal, the admissible sequences are exactly the same, as well as the respective values of topological entropy, cf. [1] and [2].

4. $Beta^*(p,q)$ models as signed mixtures of power functions

In section 2, we have seen that the probability density function $h_{2,2}$ is a convex mixture of the $Beta(2,k+1)$ densities. On the other hand, using the binomial expansion of the factor $(1-x)^k$, we observe that each $Beta(2,k+1)$ probability density function is a signed mixture — in the sense that the sum of the weights is 1,

$$\sum_{j=0}^{k} w_j^* = \frac{1}{\mathrm{B}(2,k+1)} \sum_{j=0}^{k} (-1)^j \frac{k!}{(j+2)\, j!\,(k-j)!} = 1,$$

although odd indexed weights are negative and even indexed weights are positive — of $Beta(j+2,1)$ densities:

$$\frac{x\,(1-x)^k}{\mathrm{B}(2,k+1)} = \sum_{j=0}^{k} \frac{(-1)^j}{j+2} \binom{k}{j} \frac{1}{\mathrm{B}(2,k+1)} (j+2) x^{j+1}.$$

Therefore, the $Beta^*(2,2)$ density function $h_{2,2}(x) = 4\,x\,(-\ln x)\,\mathrm{I}_{(0,1)}(x)$ is a signed mixture

$$4\,x\,(-\ln x)\,\mathrm{I}_{(0,1)}(x) = \sum_{k=1}^{\infty} \sum_{j=0}^{k} (-1)^j \frac{4\,k!}{k\,(j+2)\, j!\,(k-j)!} (j+2)\, x^{j+1}\, \mathrm{I}_{(0,1)}(x)$$

of $Beta(j+2,1)$ densities, in other words a signed mixture of power laws with natural exponents, with $\sum_{k=1}^{\infty} \sum_{j=0}^{k} (-1)^j \binom{k}{j} \frac{4}{k\,(j+2)} = 1$.

This is at the core of the duality observed in the growth of populations — for instance populations of neoplasic cells in tumours — as modeled by regularly varying tailed distributions and the correponding asymptotic extreme value laws: each positive even component influencing the exponential growth rate is moderated by the retroaction of the next negative odd term. The relevance of power laws in fractality and chaos is well-known [14], and the connection with regular variation is evident. Our more versatile family of model inherits and amplifies the interplay between malthusian growth term and retroactive term of the logistic parabola, which results in an interesting heavy and light tail equilibrium result.

Acknowledgements

This research has been partially supported by FCT/OE, CEAUL and ISEL.

References

[1] S. M. Aleixo, J. L. Rocha and D. D. Pestana. Populational growth models proportional to beta densities with Allee effect. *Am. Inst. of Phys.,* 1124: 3-12, 2008.

[2] S. M. Aleixo, J. L. Rocha and D. D. Pestana. Dynamical behaviour in the parameter space: new populational growth models proportional to beta densities. In Luzar-Stiffler, V., Jarec, I. and Bekic, Z. (eds.), *Proc. of the ITI 2009, 31th Int. Conf. on Information Technology Interfaces*, 213-218, 2009.

[3] S. M. Aleixo, J. L. Rocha and D. D. Pestana. Dynamical analysisin probabilistic models: random Beta(p,2)-Cantor sets, *Notas e Comunicações do CEAUL*, nº6/2010.

[4] N. H. Bingham. Factorisation theory and domains of attraction for generalised convolution algebras, *Proc. London Mathem. Soc.*, (3) 23: 16-30, 1971.

[5] N. H. Bingham, C. M. Goldie and J. L.Teugels. *Regular Variation*, Cambridge University Press, Cambridge, 1987.

[6] W. Feller. On regular variation and local limit theorems, in J. Neyman (ed.) *Proc. Fifth Berkeley Symp. Math. Stat. Prob.*, vol II, Part I: 373-388, 1967.

[7] L. De Haan. *On Regular Variation and its Applications to the Weak Convergence os Sample Extremes*, Math. Centre Tract 32, Amsterdam, 1970.

[8] J. Karamata. Sur un mode de croissance régulière des fonctions, *Mathematica* (Cluj), 4: 38-53, 1930.

[9] T. J. Kozubowski and S. T.Rachev. Univariate geometric stable distributions, *J. Computational Analysis Appl.*, 1: 177-217, 1999.

[10] D. D. Pestana, S. M. Aleixo and J. L. Rocha. Hausdorff dimension of the random middle third Cantor set. In Luzar-Stiffler, V., Jarec, I. and Bekic, Z. (eds.), *Proc. of the ITI 2009, 31th Int. Conf. on Information Technology Interfaces*, 279-284, 2009.

[11] D. D. Pestana, S. M. Aleixo and J. L. Rocha. The Beta(p,1) extensions of the random (uniform) Cantor sets. *Disc. Math. Prob. Stat.,* 29: 199-221, 2009.

[12] J. Pickands. Statistical inference using extreme order statistics, *Ann. Satatist.*, 3: 119-131, 1975.

[13] S. I. Resnick. Tail equivalence and its applications. *J. Appl. Prob.*, 8: 136-156, 1971.

[14] M. Schroeder. *Fractals, Chaos, Power Laws: Minutes from an Infinite Paradise*, Dover, New York, 2009.

[15] J. Waliszewski and J. Konarski. A Mystery of the Gompertz Function, in G.A. Losa, D. Merlini, T. F .Nonnenmacher and E.R. Weibel (eds.), *Fractals in Biology and Medicine*, Birkhäuser, Basel, 277-286, 2005.

Tools for investigation of dynamics of DC-DC converters within Matlab/Simulink

Dmitry Pikulin

Riga Technical University, Riga, Latvia
Email: pikulin03@inbox.lv

Abstract: In this paper the study of complex phenomenon in buck converter under voltage mode control, operating in discontinious current mode, within Matlab/Simulink simulation environments is provided. To perform simulations different types of models are used: based on discrete-time maps, differential equations and real elements (including different nonidealities). The main goal of this paper is to detect the ability of various Matlab/Simulink models to identify and to explore different types of complex behaviour, such as chaos and bifurcataions, in switch-mode DC-DC converters, as system parameters are changed, as well as to estalish the possibilities of each model in this kind of investigation. Simulations are carried out by means of Matlab/Simulink simulation environment that provides wide range of blocks and elements for complete investigation procedure, including the implementation of all types of mentioned models and appropriate result data postprocessing and visualization. The verification of accuracy of developed models is based on the detection of Feigenbaum numbers. All models with definite level of precision are able to reveal that under certain circuit parameters period doubling route to chaos is observed.
Keywords: bifurcation diagram, buck converter, chaos, simulation tools, subharmonics.

1. Introduction

Modern electronic devices are equipped with switch mode DC-DC converters, which have several advantages over linear ones: smaller weight, size and much higher efficiency. It is well known, that these converters have always been designed to operate in the specific type of periodic operation, known as period-1 mode, in which all waveforms repeat at the same rate as driving clock. However, under certain conditions, the circuit may operate in the period-n regime (in which the periods of all waveforms are exactly n times larger then that of driving clock), or in random-like fashion - *chaotic regime* (in which all waveforms become aperiodic). As some circuit parameters are slightly varied, the operation of converter can toggle between different operating regimes, sometimes in abrupt manner. Such a phenomenon, where one mode fails to operate and other picks up, is termed as *bifurcation*. The existence of this type of nonlinear phenomenon shows

Chaos Theory: Modeling, Simulation and Applications
C. H. Skiadas, I. Dimotikalis and C. Skiadas (Eds)

that even if the converter is appropriately designed to work in period-1 regime, it could fail to operate as expected, if some parameters are varied, causing it to assume another, undesirable mode of operation.

During the last three decades a significant amount of work in analysis of different chaotic effects and control of nonlinear phenomena (such as chaos and bifurcations) in switch mode DC–DC converters has been reported by power electronics specialists in course of studying and designing power electronic circuits [1,2,3].

These researchers established that the DC–DC converters under feedback control may exhibit various types of nonlinear phenomena, such as period doubling, Hopf bifurcation, quasi-periodic and chaotic operation. The already mentioned findings highlight that the investigation of these phenomena is important and vital, if reliable power supplies are to be designed.

Subharmonic and chaotic operation modes could be avoided and the reliable DC-DC converters could be designed only if the specialists have some background knowledge in nonlinear dynamics. Unfortunately the linear methods used by power electronics specialists alone have limited possibilities in analyzing nonlinear phenomena in switch mode DC-DC converters. The modern simulation software, providing designers and researchers with powerful tools for analysis of power electronic circuits, could help to overcome the mentioned drawback.

In this paper the simulation aspects of power electronic converters are addressed and different simulation approaches are evaluated and compared. The paper is organized as follows. The second section of the paper provides the information about the system under test – the voltage-mode controlled DC–DC buck converter operating in discontinuous inductor current mode (DCM). This section also presents the brief description of the Matlab/Simulink models used to investigate the period doubling route to chaos. In Section 3 the analysis of simulation results is provided. Finally concluding comments are presented in the Section 4.

2. Models of Voltage Mode Controlled Buck Converter

2.1. *Basic operation principles of buck converter*

The circuit diagram and all parameters of the voltage-mode controlled DC-DC buck converter under test are shown in the Figure 1. It is assumed that the converter operates in the DCM, so that the inductor current falls to zero for the part of switching period.

The feedback loop keeps track of the output voltage variation and adjusts accordingly the value of the duty cycle. In this control scheme the difference between output voltage and reference voltage is processed by the compensation network, which generates a control signal v_{con}:

$$v_{con}(t) = g(V_{ref} - v_c), \tag{1}$$

where g(.) – the function determined by the compensation network and including also the parameter k – small signal feedback gain.

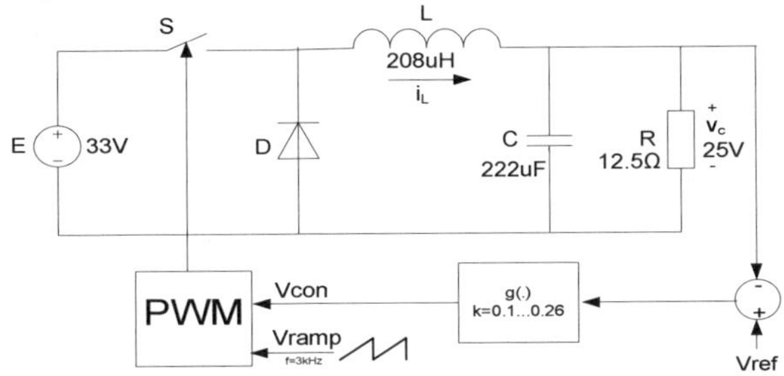

Figure 1. Voltage mode controlled buck converter

The control signal (1) efficiently determines the value of duty cycle in order to minimize the output voltage ripples and achieve fast dynamic response. The obtained signal v_{con} is compared with periodic ramp signal, to generate pulse-width modulated signal that drives the switch S.

2.2. *Switched state-space (SSS) model*

The first simulation model under test is the switched state-space model and it is based on the precise system descriptive differential equations. Depending on the state of the switch S and the value of inductor current, there are three possible topologies that are described by the following state-space equations:

$$\begin{cases} \dfrac{dv_C}{dt} = \dfrac{1}{C}\left(i_L - \dfrac{v_C}{R}\right) \\ \dfrac{di_L}{dt} = \dfrac{1}{L}(E - v_C) \end{cases} \quad \text{Switch is ON} \qquad (2)$$

$$\begin{cases} \dfrac{dv_C}{dt} = \dfrac{1}{C}\left(i_L - \dfrac{v_C}{R}\right) \\ \dfrac{di_L}{dt} = \dfrac{1}{L}(-v_C) \end{cases} \quad \text{Switch is OFF and } i_L>0; \qquad (3)$$

$$\begin{cases} \dfrac{dv_C}{dt} = \dfrac{1}{C}\left(- \dfrac{v_C}{R}\right) \\ \dfrac{di_L}{dt} = 0 \end{cases} \quad \text{Switch is OFF and } i_L=0. \qquad (4)$$

It could be seen, that the DC-DC converter is the example of hybrid system, which includes the discrete event (DE) and the continuous event (CE) systems. The DE system processes events that receive and emit control signals, but the CE system evolves continuously in time, in agreement with some physical laws and based on signals received from DE system.

To simulate the dynamics of the converter, the system is separated into two subsystems: the continuous time system implements the equations (2), (3), (4) and the discrete event system provides switching between different topologies.

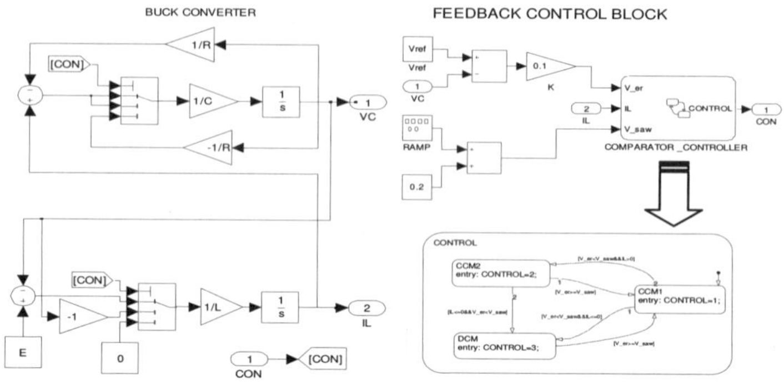

Figure 2. Simulink block diagram of the SSS model and the control block

The voltage feedback circuit block includes the sawtooth generator, reference voltage source and the logic block which produces the output of the control block. As buck converter is operating in DCM and there are three possible topologies, the signal provided by control block is in the range [1,2,3]. The most suitable way for the implementation of the logic block is the use of Simulink Stateflow chart, representing the DE system, which makes the transition from one state to another prescribed state, provided that the condition defining this change is true. The control signal is than fed to the MULTIPORT SWITCH block that operates as multiplexer and accordingly to the control signal, passes to the output the appropriate input signal, thus defining the topology of the converter. Figure 2 shows the switched state-space model of the buck converter and the appropriate control block.

2.3. *Power System Blockset (PSB) model*

The next model under test is the DCM buck converter implemented by means of Power System Blockset which allows creating the simulation model

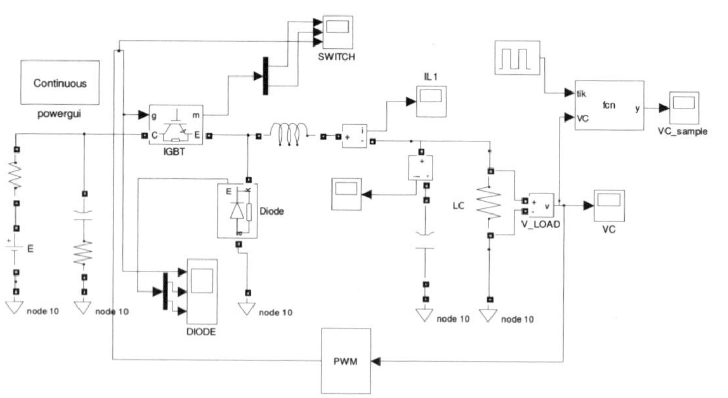

Figure 3. PSB schematic diagram of the voltage – mode controlled buck converter

of DC-DC converter, based on the precise circuit, composed from the real element models. As it is possible to integrate the PSB with standard Simulink blocks, there is the possibility to implement complex control circuitry.

Figure 3 shows that in the PSB model the physical structure of the circuit is retained in the contrast to the block diagram of the SSS model. It should be noted that the design based on PSB would save significant amount of time in model building as it doesn't require any additional mathematical or logical preprocessing. On the other hand it works a number of times slower, than the SSS model, as it takes a lot of time for Simulink to obtain the difference equations, based on the circuit diagram created.

2.4. *The model based on iterative mappings*

Although switch mode DC-DC converters are modeled using switched state-space models, discrete-time models are also found to be useful in analytical investigation of nonlinear phenomena [4].

The most widely used discrete-time model for DC-DC converters is called *iterative map (F)* that provides the relationship that links the state variable samples:

$$x_{n+1} = F(x_n)$$

The iterative maps are used as analytical tools for locating bifurcation points and constructing bifurcation diagrams. C.K.Tse [5] has derived one – dimensional map in closed form for the capacitor voltage of the buck converter, operating in DCM under voltage – mode control:

$$v_{C,n+1} = \frac{\alpha v^2{}_{C,n} + \beta (H(D - k(v_C - V_{ref})))^2}{v_{C,n}} \times \frac{E(E - v_{C,n})}{v_{C,n}} \qquad (5)$$

where $D = \sqrt{\dfrac{(1-\alpha)V_C^2}{\beta E(E-V_C)}}$; $\alpha = 1 - \dfrac{T}{RC} + \dfrac{T^2}{2C^2R^2}$; $\beta = \dfrac{T^2}{2LC}$; $H(x) = \begin{cases} 0, \_if\_x<0 \\ 1, \_if\_x>1 \\ x, otherwise. \end{cases}$.

D - steady-state duty cycle; H(.) – accounts to limit the range of duty cycle between 0 and 1;V_{ref} – reference voltage; k – small signal gain.

The iterative map shown allows evaluating the dynamics of buck converter, when one or more system parameters are changed. The drawback of this mapping is that parasitics of the inductor, diode, switch and capacitor are not taken into account.

In order to facilitate the investigation of buck converter operating in DCM by means of iterative maps, the Matlab Graphical User Interface (GUI) was designed. The main window of this GUI (see. Figure 4) is divided into several logical regions. At the beginning the user has to select the type of the converter (buck or boost) and define the input data. After the confirmation of parameters, the user is able to generate iteration diagram, bifurcation diagram or the graph of Lapunov exponents in the selected range of parameter k values.

The bifurcation diagram, based on the iterative mapping, was generated by means of this GUI.

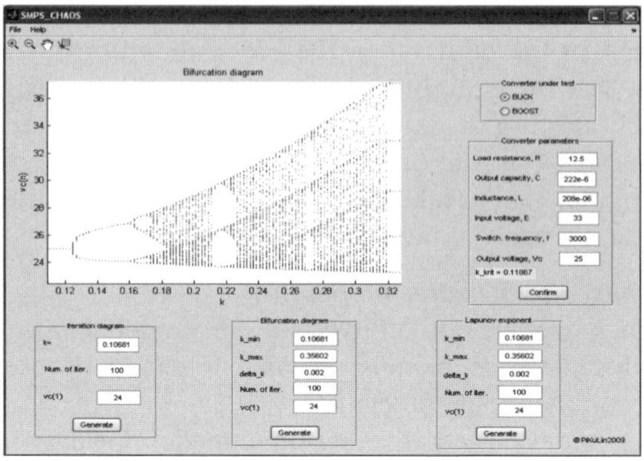

Figure 4. The designed GUI

3. Simulation Results

One of the most convenient ways to study the dynamics of the system is to obtain a summary chart of the different types of its behavior when some parameters are varied. The simplest case of bifurcation diagram, when there is the variation of only one parameter – small signal feedback gain k, is observed in this investigation. Thus the bifurcation diagram consists of an x-y plot, where the samples of output voltage are plotted against the chosen bifurcation parameter k. This bifurcation diagram clearly shows the behavioral change of the converter within the parameter range of interest.

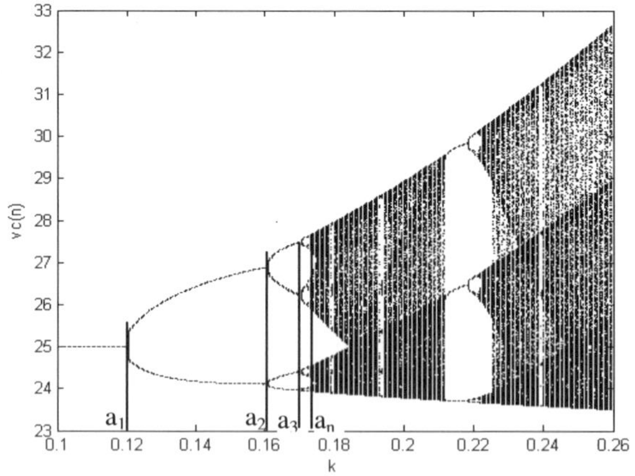

Figure 5. Bifurcation diagram obtained from SSS model

The general appearance of the bifurcations diagrams generated by all models under test are very similar, so the Figure 5 shows an example diagram obtained only from iterative map.

The obtained bifurcation diagrams show that, as the value of k is varied, the behavior of the system changes from stable period-1 to chaotic mode of operation. More precisely, the period doubling is the characteristic route for the voltage mode controlled buck converter to follow it changes from periodic motion to complex aperiodic motion.

It is well known, that Feigenbaum revealed, that there is universal solution common to all systems undergoing period doubling and found the coefficient of geometric convergence that allows linking the parameter values of system, at which the period doubling occurs. This coefficient has now been named "Feigenbaum constant" $\delta = 4.66920161\ldots$.

The above mentioned universal property gives us a possibility to evaluate simulation results, calculating the value of Feigenbaum number for each model and comparing it to Feigenbaum constant. The model, in which the difference between the calculated Feigenbaum number and the Feigenbaum constant is the smallest, is the most precise.

The comparison of different models, taking into account the accuracy of calculated Feigenbaum numbers and time required for the simulation, as well as some advices for most suitable application areas for each model are summarized in the Table 1.

Table 1.The comparison of different simulation models

Model	Simulation speed	Difference from Feigenbaum constant δ, %	The most suitable application
SSS model	1	-0.05%	Theory/ Experiments
PSB model	3	-2.09%	Experiments
Iterative map	2	-4.82%	Theory

4. Conclusions

This investigation includes the evaluation of different simulation approaches in order to create the models of switch-mode DC-DC converters. The suitability of these models for simulation of complex behavior of buck converter was considered.

As it has been mentioned in Section 3, all models allowed distinguishing the variety of complex behavior exhibited by buck converter, as the small signal feedback gain was varied. Simulation results were compared with each other, obtaining the bifurcation diagrams, calculating the Feigenbaum number for each model and evaluating simulation times.

The results summarized in the Table 1 demonstrate the good agreement of Feigenbaum number, obtained from the bifurcation diagram of SSS model with the Feigenbaum constant, so it could be concluded, that this model provides the most accurate description of nonlinear phenomena in buck converter under study. The drawback of this model is that the inclusion of non – ideal element parameters would lead to noticeable complexity of the system. As this model is based on the precise differential equations, it could be used in order to verify some analytical calculations.

The fact, that the validity of the iterative mapping relies on the level of approximation, and on the number of elements of Taylor series used to obtain the simplified solutions of switch state – space equations, explains the big difference between δ and Feigenbaum number obtained from this model. Despite the mentioned drawback, the designed GUI provides practical tool for investigation of complex behavior of buck converter, operating in DCM.

The Feigenbaum number obtained from the PSB model differs from δ, as this model is based on equations generated automatically by Simulink environment, and also includes some additional parasitics.

In order to obtain clear bifurcation diagrams, depicting the real behavior of the system, sufficiently large number of data points should be generated and the initial transients after the change of bifurcation parameter should be discarded. These preconditions determine the simulation speed and the resolution of the obtained bifurcation diagram. As PSB model takes into account parasitics of real elements, transient times are really noticeable and this model provides the lowest simulation speed and resolution of bifurcation diagram. On the other hand, the fact that this model is capable of taking into account the parasitics of real elements makes it possible to use this model as the evaluation tool for experimental results.

References

[1] Tse C.K. Complex Behaviour of Switching Power Converters. CRC Press, 2004.-262p.

[2] Banerjee S., Verghese G. Nonlinear Phenomena in Power Electronics: Attractors, Bifurcations, Chaos, and Nonlinear Control. New York: IEEE Press, 2001.-441p.

[3] Hasler M. Electrical circuits with chaotic behavior // Proceedings of IEEE, vol. 75, no. 8, August 1987, pp. 1009–1021.

[4] Hammil D.C., Deane J.H.B., Jefferies D.J. Modeling of chaotic dc/dc converters by iterative nonlinear mappings // IEEE Transactions on Circuits and Systems Part I. - Vol.35, no.8,1992, pp. 25-36.

[5] Tse C.K. Chaos from a buck regulator operating in discontinuous conduction mode // International Journal of Circuit Theory and Applications, vol. 22, 1994, pp. 263–278.

Chaos as compositional order

Eleri Angharad Pound

Interdisciplinary Centre for Scientific Research in Music
University of Leeds, UK
Email: eleri.pound@gmail.com

Abstract: Composition is a combination of determined combinations of notes, durations and timbres usually decided upon in advance by a composer who plans carefully the sounds she desires. There is also always an element of chance present in acoustic music due to the 'human' element of the performance in that the performers will add their own interpretation of the dynamics and errors in terms of precise durations and pitches. Some composers have exploited this chance element more than others, allowing more space within the composition for the performers to make choices during the course of the piece. Composers such as Cage and Bussotti offer varying degrees of freedom within pieces resulting in unpredictability of the resulting sound of the composition. Other composers attempt to control as far as possible every parameter of the music as seen in serialist composers such as Webern and Boulez.

This paper is delivered from the point of view of a composer who is intrigued by the relationship between the notation and the resultant sound, specifically, in terms of the relationship between the written elements determined by the composer and the unpredictability that arises due to those elements which cannot or are deliberately not written. These elements are then left to the interpretation and/or choice of the performer during the performance resulting in a composition which differs sonically from performance to performance. Chaos offers this combination of determination and the appearance of disorder: a clear structure within which are a number of elaborate chaotic-appearing options. The paper will focus on a composition-in-progress for voices which will offer the performers some choices based on the idea of sensitivity on initial conditions. Each singer will be provided with a set of headphones through which they will be fed a choice of pitches, the choices made for the first few pitches will determine the choices provided to the singer later on in the composition.

The paper will also outline the concepts behind the piece and how the use of chaos will provide compositional parameters with the overall development of the piece being determined by choices made by the performers. The paper will also briefly explain the unusual tuning systems set in place for the piece which is a system proposed by Dr Rob Sturman based on non-linear dynamics.

Chaos Theory: Modeling, Simulation and Applications
C. H. Skiadas, I. Dimotikalis and C. Skiadas (Eds)

1. The composer-performer balance

Composition is a combination of determined combinations of notes, durations and timbres usually decided upon in advance by a composer who plans carefully the sounds she desires. Often there will be much pre-compositional planning, contributing to the final selection of notes and the structure of the piece. There is, however, also an element of unpredictability always present in acoustic music. This is due to the 'human' element of the performance in that the performers will 'interpret' the score presented to them. In addition to this errors occur in terms of precise durations and pitches throughout a performance.

This element of imprecision has been demonstrated or exploited by some composers in various ways. Conlon Nancarrow is a composer who wrote mainly for player piano due to the complex rhythmic patterns and ratios central to his music, therefore no human element is introduced into his pieces during performance as there is no performer as such. However, Nancarrow recognised the importance of this human 'error' in rhythm as something that made music sound more 'natural' rather than a purely mechanical sound. After analysing the exact rhythms that were more likely to be played, he then built those imprecisions into his player piano scores. This means that Nancarrow's music on player piano sounds closer to a human performer than other mechanical instruments. As Kyle Gann explains in his book *The Music of Conlon Nancarrow*:

> *...Nancarrow has from the very beginning used the player piano to recreate rhythmic liberties taken in performance that no notation could convey.* (Gann 1996. p.9)

Other composers have explored the chance element of music; building chance into the compositions themselves. An obvious example to take here is John Cage, who used chance in various ways throughout his life. One way was to create a piece using chance operations, famously in this case the use of the I Ching to determine elements of the piece. This removes control of the music from the composer, however the resulting score is actually extremely deterministic, and to a performer appears much the same as any other piece of music. An example of this is *Music of Changes* (1951), composed using the I Ching, the hexagrams being used to determine the parameters of each musical event (Pritchet 1996, p.78). This method in some ways correlates more closely with the ideas of the integral serialists who set out to determine each parameter within music in advance, the method of pitch generation and aesthetic behind the two composition techniques are entirely different, however the results can be much the same. Another way in which Cage explored chance in music, was to allow more space within the composition for the performers to make choices during the course of the piece. This means that John Cage again loses control over the final sonic result of a performance; however he will have placed certain restrictions in place to assist the performers in the choices that they make. Cage's piece for 20 harps *Postcards from Heaven* presents each harpist with three ragas and associated talas, a set structure and timing guideline, along with guidelines and ideas of how to use the material given. The performers are given information and the space within which to interpret them. Cage explains in the performance notes in the score:

> *Improvisation may be "melodic" and/or "percussive." "Melodic" means proceeding stepwise, leaping only in the opposite direction, following a leap by a step or steps in the opposite direction, continually establishing, that is, the character of the raga.* (Cage 1982)

As a composer I am interested in the relationship between the determined elements of a composition and those which are less predictable and also how to control the freedoms given to a performer while retaining ownership of each piece. I am intrigued by the relationship between notation and the sonic result, and have become more interested in exploring other means of notating music other than traditional notation to achieve the desired results. I have written pieces with more freedom recently, particularly in terms of rhythm and tempo. The first piece of mine to explore the use of indeterminacy is a piece called *Pryderi* for Bass clarinet, fretless bass guitar and drum kit. More than in previous pieces, *Pryderi* sets out to allow more freedom in the performance of the music, using unspecified rhythmic values throughout and allowing for improvisation in the bass guitar part. The piece was written to explore how these elements of improvisation and freedom of time could be combined with a strict numerical process which I commonly used in other compositions. The intention was to create a composition which included this element of freedom in performance, but also to determine enough parameters that the piece would be recognisable in each performance as *Pryderi*. The important issue to me is that when I allow for these indeterminacies, that I create a piece with freedoms while still retaining ownership and control of the sonic result.

I am interested in using the idea of chaos theory, in the form of fractals or sensitivity to initial conditions as a basis for composition that incorporates chance. Chaos theory looks at dynamical systems that seem initially to be random and chaotic, but are however completely deterministic if one knows the exact initial conditions. Chaos often describes beautiful images and can be seen in the form of weather systems, coast lines, branches on trees and so on, there have been various discussions of music being fractal in nature (see for example Hsu & Hsu 'Self-similarity of the "1/f noise" called music'). In Lochhead's article 'Hearing Chaos' she describes the Lorenz Attractor; 'it represents the unpredictable behaviour of a system, but at the same time shows an implicit order' (Lochhead 2001. p. 220). These concepts can be used to create new ways of exploring elements of choice while retaining 'control' as a composer due to the essentially deterministic nature of the mathematical concept.

2. Arnold tongues and tuning systems

I am in the process of composing a piece for voices which uses various concepts derived from chaos theory. The piece will use a new tuning system developed in collaboration with Dr Rob Sturman from the School of Mathematics, University of Leeds. This tuning system is based on the use of a circle map:

$$\theta \rightarrow \theta + \Omega - \frac{k}{2\pi} Sin(2\pi\theta)$$

A perfect circle (where $k = 0$) can be used to represent equal temperament; it is divided into twelve equal parts (for the tuning commonly used in western music) each representing a semitone. To move from one note to the next a rotation takes place within the circle by the same angle each time. For twelve-tone equal temperament, this angle is 1/12 of the circle, meaning after twelve rotations the circle is complete, reaching the starting point once more.

Using values of $k \neq 0$ normally won't produce a perfect circle, however for certain irrational values of k, the circle map equation will still produce a perfect circle, a circle made up of unequal parts. Translating this into a tuning system provides us with a scale containing a limited number of pitches and containing intervals of varying sizes within the octave.

The first time this tuning system was used was during my PhD, when I explored the use of a 21-division tuning to create a microtonal triple harp. The tuning system provided a new set of intervals to work with, the majority not found in conventional tuning. One of the pieces written for the tuning and instrument combination is *Pluck 8* for solo retuned triple harp, a piece which again allows for freedom in terms of rhythm, leaving the performer to decide on the tempo, and individual note spacing through the piece. In terms of pitch, the piece explores each of the possible intervals between adjacent strings, growing from the smallest interval to the largest, which is in this case from 32.7 cents to 99.59 cents.

3. Music for voices – Circles 7

Following this project with the harp, I took some four division versions of the tuning system to use on string instruments, by retuning each individual string on the instruments. The piece I am currently working on, which will be title *Circles 7* is a composition for voices which will use multiple versions of the 4-division tuning as follows:

4-div-A

Cents	Pitch in Hertz
0.0000	440.0000
318.8484	528.9790
588.7185	618.2123
871.2414	727.7980
1200.0000	880.0000

4-divB

Cents	Pitch in Hertz
0.0000	440.0000
325.788881	531.103950
565.600232	610.011758
840.006516	714.784799
1200	880

4-div C

Cents	Pitch in Hertz
0.0000	440.0000
266.478807	513.217084
469.034957	576.917885
761.435204	683.069730
1200.0000	880.0000

4-div-D

Cents	Pitch in Hertz
0.0000	440.0000
360.58227	541.885767
521.434672	594.646555
759.505627	682.308827
1200.0000	880.0000

4-div-A

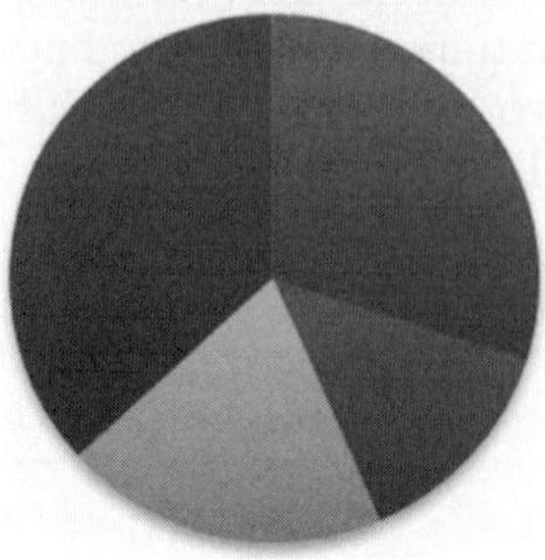

4-div-B

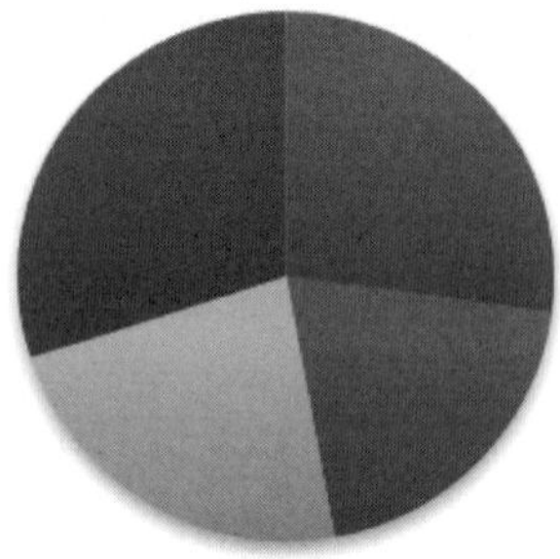

4-div-C

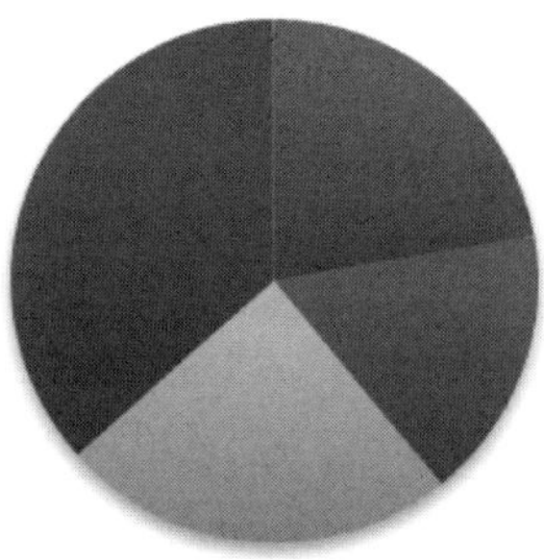

4-div-D

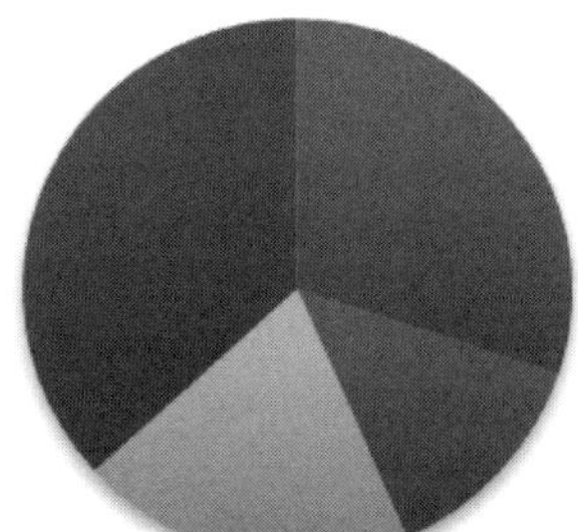

The initial concept of the piece was to begin from the idea of sensitivity of initial conditions. A group of five singers would start from similar points but then follow different paths depending on the choices made by the performers. Each singer would be provided with a choice of pitches generated by an electronics programme Max/MSP, the pitch chosen by the singer would then be used by the programme to generate a new set of pitches.

The piece will use the original 21-division tuning system as the starting point, offering to the singers a choice of four pitches which are close together in pitch as follows:

797.23 Hz, 812.67 Hz, 828.17 Hz and 844.20 Hz

These pitches are spaced 33.21 cents, 32.7 cents and 33.19 cents apart respectively; that is, approximately a sixth of a tone each time. Each singer will make a choice between these pitches which they will then sing, potentially creating a very close harmony between the voices. The singers will then be asked to sing another pitch of their choice. This interval will be analysed to find the nearest interval in one of the tuning systems listed above. From this point on the Max/MSP patch will generate a choice of four pitches, for each singer using the tuning system closest to the initial interval sung. Each time the singer makes a choice, the sung pitch will be used to generate a new version of the tuning system using the pitch as a randomly chosen degree of the scale. This means that there may be a large number of different pitches used during the course of a performance; however there will be a limited number of intervals between those pitches. The choices each singer at the very beginning of the piece will determine the development of their own line through the composition, which will (potentially) be very different to the other singers in the performance. Each singer has a semi-determinate path to follow with limited intervals, however the combination of notes between singers and so the harmony created is indeterminate with a larger number of options possible depending on the exact pitches used at the beginning by the singers.

4. Conclusion

This composition is still a work in progress: the conceptualisation of the piece has now been more or less completed and so the next step is to work on the realisation and programming of the electronics part. The concept of the piece has developed through many stages before arriving at the point where the tuning system encompasses the entire piece while still not offering the listener an opportunity to hear the harmony within an individual tuning system, but rather a harmony between the systems. The way that the choices affect the development of the piece has become similar to Markov Chains where 'consecutive transitions link up so that the destination for one transition serves as the source for the next transition' (Ames: 1989, p.175). After the first two choices, which will determine the tuning system to be used initially for each singer, every subsequent choice made by the singer affects only the next set of pitches offered each time.

This method of pre-composition will ensure that the piece retains a unifying identity, allowing the composer to keep an element of control over the ‘character’ of the sonic result. I will not be able to predict the exact harmonies or melodic line; however, I have set in place limits for the singers to work within and so the control is not completely relinquished – the balance is still a creative tension between the composer and the performer.

References

1. Ames, Charles. (1989). 'The Markov Process as a Compositional Model: A survey and Tutorial' *Leonardo* 22:2, 175-187.

2. Boulez, Pierre. (1975) *Conversations with Celestien Deliege.* London: Eulenberg Books.

3. Cage, John. (1982) *Postcard from Heaven* London: Edition Peters.

4. Fauvel, John, Raymond Flood, and Robin Wilson, eds. (2003). *Music and Mathematics: from Pythagoras to fractals.* Oxford: Oxford University Press.

5. Gann, Kyle. (1995). *The Music of Conlon Nancarrow.* Cambrige: Cambridge University Press.

6. Lochead, Judy (2001). 'Hearing Chaos' *American Music* 19 (2), 210-246.

7. Pritchett, James. 1996. *The Music of John Cage* Cambridge: Cambridge University Press

8. Sturman, Rob. (2005). Tuning: A dynamical Systems approach. Poster presented at the Royal Institution of Great Britain Conference, 18th March 2005.

Beta(*p*,*q*)-Cantor sets: Determinism and randomness

J. Leonel Rocha[1], Sandra M. Aleixo[2] and Dinis D. Pestana[3]

[1]Mathematics Unit, DEQ, Instituto Superior de Engenharia de Lisboa, and CEAUL, Lisboa, Portugal
Email: jrocha@deq.isel.ipl.pt

[2]Mathematics Unit, DEC, Instituto Superior de Engenharia de Lisboa, and CEAUL, Lisboa, Portugal
Email: sandra.aleixo@dec.isel.ipl.pt

[3]FCUL, Universidade de Lisboa, and CEAUL, Lisboa, Portugal
Email: dinis.pestana@fc.ul.pt

Abstract: Usually randomness appears as a sophisticated extension of deterministic models, that are then presented as expectation of some class of random models (this approach is exceedingly well managed in the classical Barucha-Reid's treatise on random functions and stochastic processes). The works [1], [2], [3] and [5] summarize previous studies by the authors, using stochastic definitions of extensions of Cantor's fractal to put forward appropriate deterministic models, that in a precise sense are the expectation of a structured class of models, and investigated bifurcations, Allee' effect, and the Hausdorff dimension. $Beta(p,q)$ models, with either $p=1$ or $q=1$, or the classical Verhulsts model $(p=q=2)$, proportionate interesting computable models for which computations both of Hausdorff dimension and probabilities can be explicitly evaluated, either analytically or using the Monte Carlo method.
The present extension, axed on arbitrary symbolic dynamical systems, further develops new fundamental classes of geometric constructions, and exploits the interplay of determinism and randomness on the richness of the limit fractal set, in a recursive construction. This sheds new light on the concept of Hausdorff dimensionality. We show that the dependence of the random order statistics is at the core of the apparent anomaly of consistently smaller Hausdorff dimensions of the random sets, when compared with the corresponding "expected" deterministic counterparts. We also recover Falconner's, Pesin's and Weiss' (among others) ideas on recursive geometric constructions as a straightforward approach to important issues in fractality and chaos.
Keywords: $Beta(p,q)$ densities, random and deterministic $Beta(p,q)$-Cantor sets, Hausdorff dimension.

Chaos Theory: Modeling, Simulation and Applications
C. H. Skiadas, I. Dimotikalis and C. Skiadas (Eds)

1. Introduction

Our research has been aimed at defining and characterizing a new structure of random Cantor sets when the middle sets removed at each step have a $Beta(p,q)$ law, with $q=1$, [5], and with $q=2$, [3]. In this work, we generalize these ideas to any other values of the parameter $q>0$. In section 2, we formally define random and deterministic $Beta(p,q)$-Cantor sets. In section 3, we developed methods to determine the Hausdorff dimension of these new sets, and in last section we present some numerical results which show that the Hausdorff dimension of a certain random $Beta(p,q)$-Cantor set is smaller than the Hausdorff dimension of the correspondent deterministic $Beta(p,q)$-Cantor set, as a consequence of Jensen's inequality.

2. Random and deterministic *Beta*(*p*,*q*)-Cantor sets

Consider that we remove the middle subinterval $S_1^{(0)}=\left]X_{1:2}^{(0,1)},X_{2:2}^{(0,1)}\right[$, from the interval $[0,1]$, where $X_{1:2}^{(0,1)}$ and $X_{2:2}^{(0,1)}$ are the minimum and maximum, respectively, of two independent observations of a random variable $X_1^{(0)}\cap Beta(p,q)$. After that, in each step n, we remove the middle subinterval $S_k^{(n\text{-}1)}$ in each one of the $k=1,2,\ldots,2^{n-1}$ intervals remaining from the n-1 previous step. This procedure corresponds to a new geometric construction of random type using the distribution beta with shape parameters p and q, with adequated support in each interval remaining. In this work, as a similar way to the one used to define the random middle third Cantor set, random $Beta(p,1)$-Cantor sets, [5], and random $Beta(p,2)$-Cantor sets, [3], we present the general definition of the random $Beta(p,q)$-Cantor sets.

Definition 1: Let $X_1^{(0)}$ be a $Beta(p,q)$ random variable defined in the interval $[0,1]$, $X_{1:2}^{(0,1)}$ and $X_{2:2}^{(0,1)}$ be the minimum and the maximum, respectively, of a random sample of dimension two from $X_1^{(0)}\cap Beta(p,q)$. Let,

$F_0=[0,1]=J_1^{(0)}$;

$F_1=F_0\setminus\left]X_{1:2}^{(0,1)},X_{2:2}^{(0,1)}\right[=\left[0,X_{1:2}^{(0,1)}\right]\cup\left[X_{2:2}^{(0,1)},1\right]=J_1^{(1)}\cup J_2^{(1)}$;

$F_{n-1}=\bigcup_{k=1}^{2^{n-1}}J_k^{(n-1)}$ and $F_n=\bigcup_{k=1}^{2^{n}}J_k^{(n)}$, where for each $k=1,2,\ldots,2^{n-1}$,

$J_{2k-1}^{(n)}\cup J_{2k}^{(n)}=J_k^{(n\text{-}1)}\setminus\left]X_{1:2}^{(n-1,k)},X_{2:2}^{(n-1,k)}\right[$

with $X_{1:2}^{(n-1,k)}$ and $X_{2:2}^{(n-1,k)}$ the minimum and the maximum of a random sample of dimension two of $X_k^{(n-1)} \cap Beta\left(p, q, J_k^{(n-1)}\right)$, respectively. The random fractal connected to a general random variable *X*, i.e., the random $Beta(p,q)$-Cantor set, is

$$F_{p,q} = \bigcap_{n=0}^{\infty} F_n .$$

In a correspondent deterministic approach, we consider the intersection of the sets obtained starting from the interval $[0, 1]$, and removing iterativelly the middle expected subinterval $\left]\boldsymbol{E}\left[X_{1:2}^{(n\text{-}1,k)}\right], \boldsymbol{E}\left[X_{2:2}^{(n\text{-}1,k)}\right]\right[$.

Therefore, formally, we can defined the deterministic $Beta(p,q)$-Cantor set as follows:

Definition 2: Let $X_1^{(0)}$ be a $Beta(p,q)$ random variable defined in the interval $[0, 1]$, i.e., $X_1^{(0)} \cap Beta(p,q)$, where $\boldsymbol{E}\left[X_{1:2}^{(0,1)}\right]$ and $\boldsymbol{E}\left[X_{2:2}^{(0,1)}\right]$ are the expected values of the minimum and the maximum of a random sample of dimension two of $X_1^{(0)}$, respectively. Let,

$H_0 = [0, 1] = \boldsymbol{J}_1^{(0)}$;

$H_1 = H_0 \setminus \left]\boldsymbol{E}\left[X_{1:2}^{(0,1)}\right], \boldsymbol{E}\left[X_{2:2}^{(0,1)}\right]\right[= \left[0, \boldsymbol{E}\left[X_{1:2}^{(0,1)}\right]\right] \cup \left[\boldsymbol{E}\left[X_{2:2}^{(0,1)}\right], 1\right] = \boldsymbol{J}_1^{(1)} \cup \boldsymbol{J}_2^{(1)}$;

$H_{n-1} = \bigcup_{k=1}^{2^{n-1}} \boldsymbol{J}_k^{(n\text{-}1)}$ and $H_n = \bigcup_{k=1}^{2^n} \boldsymbol{J}_k^{(n)}$, where for each $k = 1, 2, \ldots, 2^{n-1}$,

$\boldsymbol{J}_{2k-1}^{(n)} \cup \boldsymbol{J}_{2k}^{(n)} = \boldsymbol{J}_k^{(n-1)} \setminus \left]\boldsymbol{E}\left[X_{1:2}^{(n-1,k)}\right], \boldsymbol{E}\left[X_{2:2}^{(n-1,k)}\right]\right[$

with $\boldsymbol{E}\left[X_{1:2}^{(n-1,k)}\right]$ and $\boldsymbol{E}\left[X_{2:2}^{(n-1,k)}\right]$ the expected values of the minimum and the maximum of a random sample of dimension two of $X_k^{(n-1)} \cap Beta\left(p, q, \boldsymbol{J}_k^{(n-1)}\right)$, respectively. The "mean" fractal or the deterministic $Beta(p,q)$-Cantor set, is

$$C_{p,q} = \bigcap_{n=0}^{\infty} H_n .$$

Note that, the deterministic $Beta(p,q)$-Cantor sets $C_{p,q}$ satisfy the open set condition, [4].

Remark 3: Considering $X_k^{(n-1)} \cap Beta\left(p, q, \boldsymbol{J}_k^{(n-1)}\right)$, its distribution function is given by:

$$F_{X_k^{(n-1)}}(x)=\int_{a_k^{(n-1)}}^{x}\frac{t^{p-1}(1-t)^{q-1}}{B\left(p,q,\boldsymbol{J}_k^{(n-1)}\right)}\,\mathrm{d}t\ \mathrm{I}_{\boldsymbol{J}_k^{(n-1)}}(x)+\mathrm{I}_{\left(b_k^{(n-1)},+\infty\right)}(x)$$

with $\boldsymbol{J}_k^{(n-1)}=\left[a_k^{(n-1)},b_k^{(n-1)}\right]$. The mean values of the minimum and maximum of a two dimensional random sample, can be calculated as follows:

$$\boldsymbol{E}\left[X_{1:2}^{(n-1,k)}\right]=\int_{a_k^{(n-1)}}^{b_k^{(n-1)}}(1-F_X(x))^2\,\mathrm{d}x=\int_{a_k^{(n-1)}}^{b_k^{(n-1)}}\left(1-\int_{a_k^{(n-1)}}^{x}\frac{t^{p-1}(1-t)^{q-1}}{B\left(p,q,\boldsymbol{J}_k^{(n-1)}\right)}\,\mathrm{d}t\right)^2\mathrm{d}x \quad (1)$$

and

$$\boldsymbol{E}\left[X_{2:2}^{(n-1,k)}\right]=\int_{a_k^{(n-1)}}^{b_k^{(n-1)}}\left(1-(F_X(x))^2\right)\mathrm{d}x=\int_{a_k^{(n-1)}}^{b_k^{(n-1)}}\left(1-\left(\int_{a_k^{(n-1)}}^{x}\frac{t^{p-1}(1-t)^{q-1}}{B\left(p,q,\boldsymbol{J}_k^{(n-1)}\right)}\,\mathrm{d}t\right)^2\right)\mathrm{d}x. \quad (2)$$

3. The Hausdorff dimension of the random and deterministic *Beta*(*p*,*q*)-Cantor sets

The definition of random fractals presented in previous section, preserves one of the main features of fractality, namely self-similarity, but more sophisticated, in essence that the self-similarity of deterministic fractals:

- In the random $Beta(p,q)$-Cantor set $F_{p,q}$, we have $[0,1]=F_0\supset F_1\supset\cdots\supset F_n\supset\cdots$, a decreasing sequence of closed intervals, where F_n is the union of 2^n closed and pairwise disjoint intervals $J_k^{(n)}$.
- Each interval $J_k^{(n-1)}$ of F_{n-1} contains two intervals of F_n, since the middle interval from the three intervals with (dependent) random lengths in which F_{n-1} is divided, is always eliminated in the following step n. These intervals are denoted by $J_{2k-1}^{(n)}$ and $J_{2k}^{(n)}$, respectively. The left endpoint of $J_{2k-1}^{(n)}$ coincides with the left endpoint of $J_k^{(n-1)}$ and the right endpoint of $J_{2k}^{(n)}$ coincides with the right endpoint of $J_k^{(n-1)}$.
- The lengths of the intervals $J_{2k-1}^{(n)}$ and $J_{2k}^{(n)}$, denoted by $\tilde{J}_{2k-1}^{(n)}$ and $\tilde{J}_{2k}^{(n)}$, respectively, are random variables, and we inforce random self-similarity

requiring the ratios $C_{2k-1}^{(n)} = \dfrac{\tilde{J}_{2k-1}^{(n)}}{\tilde{J}_k^{(n-1)}}$ to have the same probability distribution throughout, for any of the steps n-1 and n, and for any interval $J_k^{(n-1)}$ of F_{n-1}, and also that the ratios $C_{2k}^{(n)} = \dfrac{\tilde{J}_{2k}^{(n)}}{\tilde{J}_k^{(n-1)}}$ have the same probability distribution, for any of the steps n-1 and n, and for any interval $J_k^{(n-1)}$ of F_{n-1}. Note that, the ratios $C_{2k-1}^{(n)}$ and $C_{2k}^{(n)}$ do not necessarily have the same probability distribution and they are not independent.

As we assume that, for all steps n, with $n = 1, 2, \ldots$, all the ratios $C_{2k-1}^{(n)}$ have the same probability distribution, we can use in particular the ratio:

$$C_1 = C_1^{(1)} = \frac{\tilde{J}_1^{(1)}}{\tilde{J}_1^{(0)}} = \frac{\tilde{J}_1^{(1)}}{1} = \tilde{J}_1^{(1)} = X_{1:2}^{(0,1)} - 0 = X_{1:2}^{(0,1)} \quad (3)$$

and similarly, as we assume that in each step the ratios $C_{2k}^{(n)}$ do have the same probability distribution, the ratio is:

$$C_2 = C_2^{(1)} = \frac{\tilde{J}_2^{(1)}}{\tilde{J}_1^{(0)}} = \frac{\tilde{J}_2^{(1)}}{1} = \tilde{J}_2^{(1)} = 1 - X_{2:2}^{(0,1)}. \quad (4)$$

In view of the geometric construction of the above random sets, and of Falconer's Theorem 15.1, [4], we compute the Hausdorff dimension of the random $Beta(p,q)$-Cantor set using the equation $\boldsymbol{E}\left[C_1^s + C_2^s\right] = 1$.

For $X_1^{(0)} \cap Beta(p,q)$, the probability density functions of the minimum and of the maximum of a two dimensional random sample are given by:

$$f_{X_{1:2}^{(0,1)}}(x) = 2\left(1 - F_{X_1^{(0)}}(x)\right) f_{X_1^{(0)}}(x) \text{ and } f_{X_{2:2}^{(0,1)}}(x) = 2\, F_{X_1^{(0)}}(x)\, f_{X_1^{(0)}}(x),$$

where $F_{X_1^{(0)}}(x)$ and $f_{X_1^{(0)}}(x)$ are:

$$F_{X_1^{(0)}}(x) = \int_0^x \frac{1}{B(p,q)} t^{p-1}(1-t)^{q-1} \mathrm{d}t \; \mathrm{I}_{(0,1)}(x) + \mathrm{I}_{(1,+\infty)}(x)$$

and

$$f_{X_1^{(0)}}(x) = \frac{1}{B(p,q)} x^{p-1}(1-x)^{q-1} \mathrm{I}_{(0,1)}(x).$$

The expected values of the transformations of the random variables C_1^s and C_2^s are given by:

$$\boldsymbol{E}\left[\left(X_{1:2}^{(0,1)}\right)^s\right] = \int_0^1 x^s f_{X_{1:2}^{(0,1)}}(x)\,\mathrm{d}x \quad \text{and} \quad \boldsymbol{E}\left[\left(1 - X_{2:2}^{(0,1)}\right)^s\right] = \int_0^1 x^s f_{X_{2:2}^{(0,1)}}(1-x)\,\mathrm{d}x.$$

Now, we can state the following result:

Theorem 4: With probability 1, the random $Beta(p,q)$-Cantor set $F_{p,q}$ has Hausdorff dimension $Dim_H F_{p,q}$ equal to *s*, where *s* is the solution of the equation

$$\boldsymbol{E}\left[\left(X_{1:2}^{(0,1)}\right)^{s}\right] + \boldsymbol{E}\left[\left(1 - X_{2:2}^{(0,1)}\right)^{s}\right] = 1.$$

On the other hand, to determine the Hausdorff dimension of the deterministic $Beta(p,q)$-Cantor set, we use an analogous version of the Theorem 9.3 of Falconer, [4]. Nevertheless, it is necessary to determine the similarity ratios used in each step of the deterministic $Beta(p,q)$-Cantor set construction. Considering that the randomness of the $Beta(p,q)$-Cantor sets comes from the minimum and the maximum of a random variable with beta distribution, we have only two similarities.

Having in mind that, in each step n the ratios $C_{2k-1}^{(n)}$ have the same probability distribution, and the ratios $C_{2k}^{(n)}$ also have the same probability distribution, as explained before, and using (3) and (4), then:

$$\boldsymbol{E}[C_1] = \boldsymbol{E}\left[X_{1:2}^{(0,1)}\right] \text{ and } \boldsymbol{E}[C_2] = 1 - \boldsymbol{E}\left[X_{2:2}^{(0,1)}\right].$$

Therefore, the similarity ratios are $c_1 = \boldsymbol{E}\left[X_{1:2}^{(0,1)}\right]$ and $c_2 = 1 - \boldsymbol{E}\left[X_{2:2}^{(0,1)}\right]$ to the left and right intervals, respectively. The expressions of $\boldsymbol{E}\left[X_{1:2}^{(0,1)}\right]$ and $\boldsymbol{E}\left[X_{2:2}^{(0,1)}\right]$ were calculated from Eqs. (1) and (2). So, we can state the following result:

Theorem 5: Let S_i denote the similarities defined on IR, with ratios c_i, with $i = 1,2$. If $C_{p,q}$ is the invariant set satisfying $C_{p,q} = \bigcup_{i=1}^{2} S_i\left(C_{p,q}\right)$ then $Dim_H C_{p,q} = s$ where *s* is the solution of the equation

$$\left(\boldsymbol{E}\left[X_{1:2}^{(0,1)}\right]\right)^{s} + \left(1 - \boldsymbol{E}\left[X_{2:2}^{(0,1)}\right]\right)^{s} = 1.$$

4. Numerical simulations

At first sight the intuitive (but misguided) idea is that the random fractal $F_{p,q}$ should have a bigger Hausdorff dimension than the correspondent "mean fractal" $C_{p,q}$; In fact, the distribution of the retained spacings is skewed, and thus, the probability of retaining larger portions than in the deterministic fractal is higher than 0.5. But as a consequence, the removed portion in later steps is big, and the overall effect is that the random fractal is less dense, as explained below.

To gain a deeper insight, we are going to evaluate the probability that the sum of the lengths of the intervals removed until the step n in the construction of the random fractal, which we shall denoted by $S_{2,R}^{(n)}$ in what follows, is greater than the sum of the lengths of the intervals removed until the step n in the construction of the correspondent "mean fractal", denoted by $S_{2,D}^{(n)}$ in what follows. This evaluation cannot be done analytically, but the evaluation is readily performed using Monte Carlo methods.
To make the Monte Carlo simulation for determining these probabilities and the correspondent 95% confidence intervals, we used in each case 5000 runs.
On the other hand, in order to compute $S_{2,D}^{(n)}$ of the "mean fractal" $C_{p,q}$, observe that in the first step we obtain $\boldsymbol{J}_1^{(1)} \cup \boldsymbol{J}_2^{(1)} = [0, a] \cup [b, 1]$, where $a = a_1^{(0)} = \boldsymbol{E}\left[X_{1:2}^{(0,1)}\right]$ and $b = b_1^{(0)} = \boldsymbol{E}\left[X_{2:2}^{(0,1)}\right]$. A straightforward extension is stated in the theorem that follows:

Theorem 6: The length of the sum of the intervals removed in the construction of a "mean fractal" $C_{p,q}$, until the step n, is given by

$$S_{2,D}^{(n)} = 1 - \left(a + (1 - b)\right)^{n}$$

with $n = 1, 2, \ldots$, where $a = \boldsymbol{E}\left[X_{1:2}^{(0,1)}\right]$ and $b = \boldsymbol{E}\left[X_{2:2}^{(0,1)}\right]$.

The proof of this theorem can be seen in [2].

In Table 1, we compute the probability that the accumulated length of the random middle sets removed in the recursive construction of the random $Beta(p,q)$-Cantor set $F_{p,q}$ be greater than the accumulated length of removed subintervals in the construction of the corresponding deterministic $Beta(p,q)$-Cantor set $C_{p,q}$, for some values of p and q. While in the first steps this probability is less than 0.5, in the next steps, for small values of p, the odds are in favour that the length of the removed random set exceeds the length of what has been removed in its deterministic counterpart. In fact, at each step of the recursive construction of the random fractal and of its deterministic correspondence, this pattern will apply: the probability that the accumulated length of the removed intervals in the random case exceeds the accumulated length of the removed intervals in the corresponding deterministic fractal increases steadily.
The dependence structure of order statistics, skewness and the consequent unequal mean and median contribute to this surprising reversal, and this deeper analysis of the situation shows that we should indeed expect that the random fractal be less dense in $[0, 1]$. Thus, smaller Hausdorff dimension is indeed a coherent result. This is a consequence of Jensen's inequality. Although the Hausdorff dimensions of both the random and the

corresponding deterministic $Beta(p,q)$-Cantor sets increase with *p*, we have $Dim_H F_{p,q} < Dim_H C_{p,q}$.

Table 1: Estimated probability of $\boldsymbol{P}\left[S_{2,R}^{(n)} > S_{2,D}^{(n)}\right]$ and respective confidence interval

	***Beta*(1,1)**		***Beta*(2,1)**		***Beta*(3,1)**	
Step	**Est.Prob**	**95% Conf. Int.**	**Est.Prob**	**95% Conf. Int.**	**Est.Prob**	**95% Conf. Int.**
1	0.4570	(0.4432;0.4708)	0.4252	(0.4115;0.4389)	0.4222	(0.4085;0.4358)
2	0.5148	(0.5010;0.5287)	0.4834	(0.4696;0.4973)	0.4890	(0.4751;0.5028)
3	0.5516	(0.5378;0.5654)	0.5176	(0.5038;0.5315)	0.5182	(0.5043;0.5320)
4	0.5744	(0.5607;0.5881)	0.5354	(0.5216;0.5492)	0.5246	(0.5107;0.5384)
5	0.5776	(0.5639;0.5913)	0.5430	(0.5292;0.5492)	0.5402	(0.5263;0.5540)
6	0.5936	(0.5800;0.6072)	0.5462	(0.5324;0.5600)	0.5476	(0.5338;0.5613)
7	0.6002	(0.5866;0.6138)	0.5512	(0.5374;0.5649)	0.5524	(0.5386;0.5661)
8	0.6048	(0.5913;0.6184)	0.5496	(0.5358;0.5633)	0.5556	(0.5418;0.5693)
9	0.6084	(0.5949;0.6219)	0.5528	(0.5390;0.5665)	0.5616	(0.5478;0.5753)
10	0.6098	(0.5963;0.6233)	0.5500	(0.5362;0.5637)	0.5630	(0.5492;0.5767)
	***Beta*(1,2)**		***Beta*(2,2)**		***Beta*(3,2)**	
Step	**Est.Prob**	**95% Conf. Int.**	**Est.Prob**	**95% Conf. Int.**	**Est.Prob**	**95% Conf. Int.**
1	0.4398	(0.4260;0.4536)	0.4300	(0.4163;0.4437)	0.4230	(0.4093;0.4367)
2	0.5012	(0.4873;0.5151)	0.4898	(0.4759;0.5037)	0.4836	(0.4697;0.4975)
3	0.5252	(0.5114;0.5390)	0.5128	(0.4989;0.5277)	0.5034	(0.4895;0.5173)
4	0.5374	(0.5236;0.5512)	0.5222	(0.5084;0.5360)	0.5190	(0.5052;0.5328)
5	0.5454	(0.5316;0.5592)	0.5232	(0.5094;0.5370)	0.5282	0.5144;0.5420)
6	0.5574	(0.5436;0.5712)	0.5266	(0.5128;0.5404)	0.5346	(0.5207;0.5484)
7	0.5638	(0.5501;0.5775)	0.5254	(0.5116;0.5392)	0.5388	(0.5250;0.5526)
8	0.5636	(0.5499;0.5773)	0.5266	(0.5128;0.5404)	0.5470	(0.5332;0.5608)
9	0.5666	(0.5529;0.5803)	0.5258	(0.5120;0.5396)	0.5492	(0.5354;0.5630)
10	0.5684	(0.5547;0.5821)	0.5274	(0.5136;0.5412)	0.5502	(0.5364;0.5640)

Acknowledgements

This research has been partially supported by FCT/OE, CEAUL and ISEL.

References

[1] S. M. Aleixo, J. L. Rocha and D. D. Pestana. Populational growth models proportional to beta densities with Allee effect. *Am. Inst. of Phys.,* 1124: 3-12, 2008.

[2] S. M. Aleixo, J. L. Rocha and D. D. Pestana. Dynamical behaviour in the parameter space: new populational growth models proportional to beta densities. In Luzar-Stiffler, V., Jarec, I. and Bekic, Z. (eds.), *Proc. of the ITI 2009, 31th Int. Conf. on Information Technology Interfaces*, 213-218, 2009.

[3] S. M. Aleixo, J. L. Rocha and D. D. Pestana. Dynamical analysisin probabilistic models: random Beta(p,2)-Cantor sets, *Notas e Comunicações do CEAUL*, nº6/2010.

[4] K. Falconer. *Fractal Geometry, Mathematical Foundations and Applications*, John Wiley & Sons, New York, 1990.

[5] D. D. Pestana, S. M. Aleixo and J. L. Rocha. The Beta(p,1) extensions of the random (uniform) Cantor sets. *Disc. Math. Prob. Stat.,* 29: 119-221, 2009.

Predicting chaos with second method of Lyapunov

Vladimir B. Ryabov

Future University Hakodate, 116-2 Kamedanakano-cho, Hakodate, Hokkaido, Japan 041-8655
Email: riabov@fun.ac.jp

Abstract: We overview several analytic methods of predicting the emergence of chaotic motion in nonlinear oscillatory systems. A special attention is given to the second method of Lyapunov, a technique that has been widely used in the analysis of stability of motion in the theory of dynamical systems but received little attention in the context of chaotic systems analysis. We show that the method allows formulating a necessary condition for the appearance of chaos in nonlinear systems. In other terms, it provides an analytic estimate of an area in the space of control parameters where the largest Lyapunov exponent is strictly negative. A complementary area thus comprises the values of controls, where the exponent can take positive values, and hence the motion can become chaotic. Contrary to other commonly used methods based on perturbation analysis, such as e.g., Melnikov criterion, harmonic balance, or averaging, our approach demonstrates superior performance at large values of the parameters of dissipation and nonlinearity. Several classical examples including mathematical pendulum, Duffing oscillator, and a system of two coupled oscillators, are analyzed in detail demonstrating advantages of the proposed method compared to other existing techniques.
Keywords: direct method of Lyapunov, Lyapunov exponents, Melnikov method, harmonic balance, averaging, saddle-node bifurcation, local expansion rates, pendulum, Duffing.

1. Introduction

One of the classical problems in the analysis of nonlinear systems is predicting the properties of motion from the functional form of the governing differential equations. In particular, the arising of various instabilities leading to bifurcations of regular motions and, finally to the appearance of strange attractors in the phase space, have been at the focus of active research in the last decades [1,2]. In general terms, a typical problem posing of this type looks like the following one.

Given a dynamical system described by a set of differential equations

$$\frac{d\mathbf{x}}{dt} = \mathbf{F}(\mathbf{x}, t, \mathbf{c}), \qquad \mathbf{x} \in \Re^n; \mathbf{c} \in \Re^m \tag{1}$$

where $\mathbf{c}$ is vector of m control parameters, specify the areas in the $\mathbf{c}$-space where the behavior is periodic, quasiperiodic, or chaotic. The borders of such areas in a two-dimensional cross-section of the $\mathbf{c}$-space present bifurcation lines, and the cross-sections themselves constitute state diagrams. Despite

Chaos Theory: Modeling, Simulation and Applications
C. H. Skiadas, I. Dimotikalis and C. Skiadas (Eds)

considerable efforts aimed at designing analytic techniques for predicting different bifurcations and conditions for the transition to chaos in dynamical systems, the number of tools available for this kind of analysis remains limited, and each method deals with a very restricted class of differential equations.

In comparatively simple cases like, e.g, saddle-node bifurcation in periodically forced systems or Andronov-Hopf bifurcation in autonomous oscillators, the bifurcation conditions can be found by a two-step procedure, when an approximation to a solution (typically, periodic or quasiperiodic) is found first, followed by the stability analysis. The bifurcation lines in this approach are the borders of stability areas found at the final step of analysis.

Several approaches have been proposed for finding the conditions for the transition to chaos. The methods are known by the names of their inventors, the most famous ones being those of Chirikov, Melnikov, and Shilnikov. It should be noted, however, that all of these methods can be called "indirect" ones, since they do not find the conditions of chaotization of motion, but analyze the phenomena accompanying chaos. In particular, in the case of Hamiltonian systems treated by Chirikov method, the analysis predicts the phenomenon of overlap of resonances, Melnikov method finds the position of homoclinic bifurcation (intersection of stable and unstable manifolds of a saddle state), whereas Shilnikov method looks for special trajectories, saddle-focus loops, prerequisite of a chaotic motion.

On the other hand, it appears natural to use the very definition of chaotic motion, i.e. exponential instability of chaotic trajectories with respect to variations in initial conditions, for predicting its appearance. The sensitivity to initial conditions is described by the spectrum of Lyapunov exponents λ_i that characterize average growth rates of perturbations in orthogonal directions along a trajectory. It has been recently demonstrated [3] that a necessary condition for chaos can be efficiently formulated in terms of Temporally Local Lyapunov Exponents (sometimes also called Instantaneous Lyapunov Exponents, ILE), μ_i, that represent the local time-dependent expansion rates at a given point on the trajectory. The standard Lyapunov exponents λ_i are simply long time averages of corresponding ILE. Requirement for all ILE to be strictly negative is equivalent to imposing a condition of asymptotic stability on the fiducial trajectory. In this paper, we show that analysis in terms of ILE is equivalent to that of direct (second) method of Lyapunov. Since chaos usually appears in dynamical systems as a result of a loss of stability by some periodic orbits, its prediction requires using a variant of the Lyapunov method developed for non-autonomous systems [4]. The conclusion on the instability of motion that can lead to chaos can be derived from very similar arguments to those formulated in Chetaev theorem [4, 5].

2. Lyapunov exponents and direct (second) method

Lyapunov exponents of an arbitrary system of type (1), are usually calculated by the analysis of solutions in linearized system

$$\frac{d\mathbf{y}}{dt} = \hat{\mathbf{J}}\left(\mathbf{x}^*(t)\right)\mathbf{y}\,, \tag{2}$$

where $\hat{\mathbf{J}}\left(\mathbf{x}^*(t)\right) \equiv \partial\mathbf{F}\left(\mathbf{x}^*(t)\right)/\partial\mathbf{x}$ is n×n time dependent Jacobian matrix, $\mathbf{y}$ is an n-vector of perturbations around the trajectory $\mathbf{x}^*(t)$. The standard algorithm of calculating the spectrum of Lyapunov exponents [6] consists in solving the equations (2) simultaneously with (1) for a set of mutually orthonormal vectors $\{\mathbf{y}_k\}$, (k = 1, 2, ..., n) and estimating expansion rates (ILE)

$$\mu_k(t) = \frac{d\left[\ln \rho_k(t)\right]}{dt} \tag{3}$$

for the lengths $\rho_k = \left\|\mathbf{y}_k\right\|$ of the vectors $\{\mathbf{y}_k\}$. Then, estimates of Lyapunov exponents can then be derived as time averages

$$\lambda_k = \lim_{T\to\infty} \frac{1}{T}\int_0^T \mu_k(t)dt \tag{4}$$

Stability of the solution $\mathbf{x}^*(t)$ is thus defined by the time-dependent function $\mu_1(t)$, i.e. it is asymptotically stable if $\mu_1(t) < 0$ for all times, and hence the largest Lyapunov exponent is negative. As it is discussed in [3, 4], the conditions of stability can be often derived by analysing the properties of Eqs.(1-2), without finding their exact solutions. In particular, it had been demonstrated that the condition for stability in a wide class of nonlinear oscillatory systems is defined by the amplitude of motion only, i.e. the motion becomes unstable when the trajectory crosses certain threshold. The estimate of the critical amplitude can be obtained by some other technique and formulate the stability (instability) criteria in terms of control parameters.

On the other hand, the stability of motion can also be analysed by the direct method of Lyapunov [7], which establishes the stability conditions from the analysis of so-called Lyapunov function, $V(\mathbf{y})$. In order to be stable, a solution $\mathbf{x}^*(t)$ should evolve in a way that ensures that the function $V(\mathbf{y})$ always keeps different sign compared to its time derivative, $\dot{V}(\mathbf{y})$. Selecting, for example, the Lyapunov function to be definite positive, one can predict the stability of solutions in the areas of the phase space and the space of controls $\mathbf{c}$ where the condition $\dot{V}(\mathbf{y}) < 0$ is satisfied.

We are now in a position of demonstrating equivalence of the stability analysis in terms of ILE and direct method of Lyapunov. Indeed, selecting the Lyapunov function in the form that guarantees its positive definiteness

$$V(\mathbf{y}) = \sum_{i=1}^{n} y_i^2\,, \tag{5}$$

one can see that it coincides with the squared norm $\rho_1 = \left\|\mathbf{y}_1\right\|$ of an arbitrary vector from the set $\{\mathbf{y}_k\}$. Then considering the ratio

$$\frac{\dot{V}}{V} = \frac{\frac{d}{dt}\left(\rho_1^2\right)}{\rho_1^2} = 2\frac{\dot{\rho}_1}{\rho_1} \equiv 2\mu_1 \tag{6}$$

we conclude that the demand for the opposite signs of the functions $V(\mathbf{y})$ and $\dot{V}(\mathbf{y})$ is equivalent to the requirement of $\mu_1(t) < 0$. Therefore, the problem of stability is reduced to the analysis of sign of the function $\dot{V}(\mathbf{y})$, and the problem consists in finding the areas in the space of controls where the condition $\dot{V}(\mathbf{y}) < 0$ is satisfied in certain range of variation for the variable $\mathbf{x}$: $x_{\min} < \mathbf{x}^*(t) < x_{\max}$.

The interval of values $[x_{\min}; x_{\max}]$ that locates the solution $\mathbf{x}^*(t)$ can be estimated by other methods, such as e.g., averaging, harmonic balance, or multiple time scale analysis [8]. As it is shown in [3], the demand of $\dot{V}(\mathbf{y}) < 0$ imposed by Lyapunov method leads to the necessity of introducing a linear coordinate transform that converts the function (5) to that of Lyapunov-type and hence makes feasible the stability analysis. In the next chapter, we demonstrate the efficiency of the method with several examples of classical nonlinear oscillators.

3. Examples of nonlinear oscillators

3.1. *Single-well cubic potential*

Differential equation describing the time evolution of Duffing oscillator with nonlinearity of hardening type ($\varepsilon > 0$) reads as

$$\frac{d^2x}{dt^2} + \delta\frac{dx}{dt} + \omega_0^2 x + \varepsilon x^3 = e(t) \tag{7}$$

It is known that this system has chaotic solutions, even in the simplest case of one-frequency excitation $e(t) = F\cos(\omega t)$, at certain combination of the parameters F and ω [9]. For hardening type of nonlinearity, no analytic methods exist for predicting the appearance of chaotic motion. To apply Lyapunov stability analysis, one needs to consider a notation $(x; \dot{x}) \to (x_1; x_2)$ and a linear coordinate transform [3]

$$z_1 = \frac{\delta}{\delta + 2\omega_0}x_1; \quad z_2 = x_1 + \frac{\delta}{2}x_2 \tag{8}$$

that leads to the following expression for the ratio $\dot{V}/V$

$$\frac{\dot{V}}{V} = -\delta + \left(\delta - \frac{6\beta x^2}{\delta + 2\omega_0}\right)\sin\Phi \tag{9}$$

and allows estimating maximal stable amplitude as

$$x_{\max} = \sqrt{\frac{\delta(\delta + 2\omega_0)}{3\varepsilon}} \tag{10}$$

If the value of x exceeds the threshold given by Eq.(10) the motion of the system may (although not necessarily) become unstable. In order to estimate the amplitude of motion we use the simplest approach by calculating the response function of linear oscillator, i.e. Eq. (7) at $\varepsilon = 0$

$$x_{\max} = F \Big/ \sqrt{\left(\omega_0^2 - \omega^2\right)^2 + \delta^2 \omega^2} \tag{11}$$

Combining Eqs. (10-11), we obtain the threshold line in the parameter plane $F - \omega$. In Fig. 1 we plot the result, together with saddle-node bifurcation lines obtained with averaging method [8]. As one can see, the combination of the proposed criterion with linear response theory works well, accurately predicting the onset of saddle-node bifurcation, i.e. the lowest order instability with respect to the parameter *F*. Other bifurcations, including period-doubling cascades and chaos, occur at much higher values of *F*.

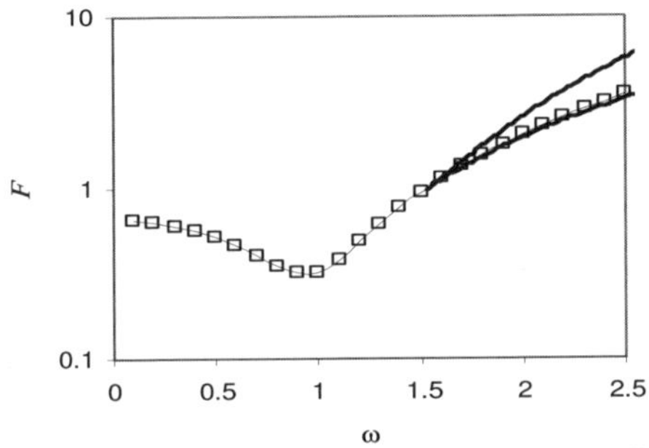

Fig.1. Lyapunov stability of single–well Duffing-type oscillator with hardening nonlinearity (squares); saddle-node bifurcation (heavy line) in the vicinity of principal resonance $\omega \approx \omega_0$.
Parameters: $\delta = 0.5; \omega_0 = 1; \varepsilon = 1$.

3.2. *Mathematical pendulum*

Equation describing the dynamics of a simple pendulum can be written as

$$\frac{d^2x}{dt^2} + \delta \frac{dx}{dt} + \sin(x) = f(t) \tag{12}$$

Performing the analysis similar to the case of Duffing oscillator, we come to the two cases that should be considered, depending on the value of dissipation parameter δ

1. If $\delta < 1$, then the maximal stable displacement is defined by

$$|x|_{\max} = \cos^{-1}\left[(1-\delta)^2\right] \tag{13a}$$

2. If $\delta > 1$, then the size of stability area does not depend on dissipation and is limited by the value of

$$|x|_{\max} = \frac{\pi}{2} \tag{13b}$$

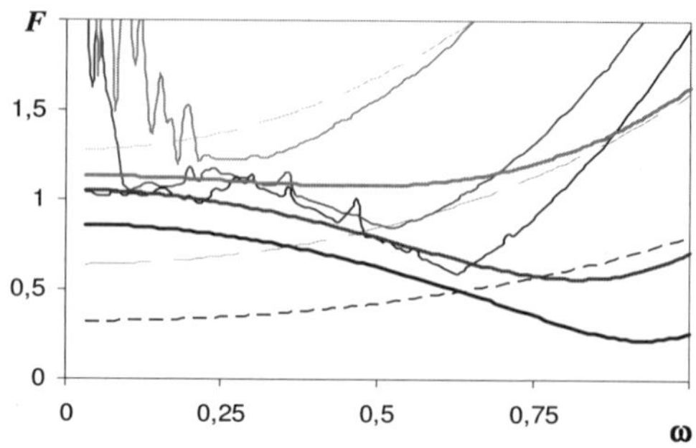

Fig. 2. Phase diagram of the pendulum system (12). The borders of Lyapunov stability areas are shown as heavy lines. They are compared to those obrained with Melnikov criterion (dashed lines) and actual position of chaotic areas obtained by direct numerical integration of Eq. (12) (lite lines). Blue, pink, and green lines correpond to $\delta = 0.25$, 0.5, and 1, respectively.

In order to check the efficiency of the proposed criterion and compare it to the results of previous works [10], we consider the case of harmonic excitation of type $f(t) = F\cos(\omega t)$ and use the following formula for an approximate solution of the Eq.(12)

$$x = A\cos(\omega t + \varphi) \tag{14}$$

where the amplitude A is as a root of the equation

$$\left[2J_1(A) - \omega^2 A\right]^2 + (A\omega\delta)^2 = F^2 \tag{15}$$

where $J_1(A)$ is Bessel function of the first kind [10]. By substituting the value of maximal stable amplitude (13) to Eq. (15), we obtain the borderline of the asymptotic stability area in explicit form $\gamma(\omega)$. Extensive numerical analysis of the oscillator (12) has been performed in [10] where the loci of chaotic areas on the plane of control parameters ω – F have been established at the value of $\delta = 0.25$. In Fig.2, we plot the borders of chaotic areas found numerically, Melnikov condition [11] and Lyapunov stability lines Eqs. (13a) and (15) at $\delta = 0.25$, 0.5, and 1. Apparently, the latter analysis provides a good prediction for the onset of chaotic oscillations, especially in the low-frequency part of the diagram where Melnikov's method strongly underestimates the chaos threshold. We also note that the latter provides an incorrect prediction of the position of chaotic area at high value of the dissipation parameter $(\delta = 1)$, whereas our analysis always provides a correct estimate for the threshold amplitude of the external force.

3.3. *Two coupled oscillators*

The last example is a system of two coupled resonance oscillators under the action of harmonic external signal that can be described as

$$\begin{aligned}\dot{x}_1 &= x_2\\ \dot{x}_2 &= -\delta x_2 - \omega_0^2 x_1 - \varepsilon x_1^3 + \kappa(x_3 - x_1) + F\cos(\omega t)\\ \dot{x}_3 &= x_4\\ \dot{x}_4 &= -\delta x_4 - \omega_0^2 x_3 - \varepsilon x_3^3 - \kappa(x_3 - x_1)\end{aligned} \tag{16}$$

Lyapunov analysis leads to the identical estimate of the stability threshold for each of the interacting oscillators

$$x_{1,2}^{\max} < \sqrt{\frac{\delta\omega_1\omega_2}{3\varepsilon(\omega_1+\omega_2)}}\ ;\ \text{where}\ \omega_{1,2}^2 = \kappa + \omega_0^2 - \frac{\delta^2}{4} \mp \kappa \tag{17}$$

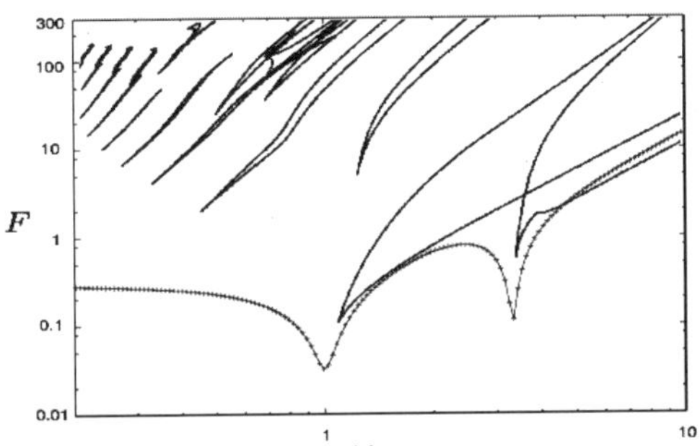

Fig. 3. Stability criterion for coupled oscillators (markers); saddle-node bifurcation (heavy lines) obtained by direct numerical integration of the system (16) (from Ref. [12]). Parameter values: $\varepsilon = 1; \kappa = 5;\ \delta = 0.1; \omega_0 = 1$.

Estimate of the amplitudes by, e.g., the method of harmonic balance requires solving the system of four nonlinear algebraic equations [13]. However, using the linear approximation approach described in Section 3.1, i.e. considering only linear terms in (16), we get

$$x_1^{\max} = F\sigma\sqrt{\left(\omega_0^2 - \omega^2\right)^2 + (\delta\omega)^2}\ ; x_3^{\max} = \kappa F\sigma \tag{18}$$

where

$$\sigma = \prod_{i=1}^{4}\sqrt{\left(\lambda_i^2 + \omega^2\right)^{-2}}\ ;\ \lambda_{1\div 4} = \frac{-\delta \pm \sqrt{\delta^2 - 4\left(\omega_0^2 \pm \kappa\right)}}{2}$$

In Fig. 3, we plot the lines of saddle-node bifurcation [12], together with the threshold line provided by our criterion. One can note a good quality of prediction given by Lyapunov stability analysis cobined with the linear response theory. Note that other instabilities, including chaotic motion are also located above the found threshold line [12].

4. Conclusions

The Second (direct) method of Lyapunov is shown to be an efficient technique for predicting the appearance of chaotic instability in nonlinear oscillatory systems. Due to the approximate character of additional methods that have to be used for estimating the maximal amplitude of the response signal, the method may over- or under-estimate the threshold value of controls where chaos occurs. This can be seen, e.g., in the high-frequency parts of Fig.2 where the predicted threshold lines lie well below actual chaotic areas, or in Fig. 3, where the threshold line lies above the saddle-node bifurcations. This means that special care should be taken in applying this criterion in combination with other approximate methods, especially in the systems possessing multiple coexisting attractors. However, compared to other methods like, e.g., Melnikov one, Lyapunov approach has an advantage of formulating a necessary condition for chaos, i.e. finding the maximal stable amplitude of motion, which is important for many applications.

References

[1] S. Wiggins. *Global Bifurcations and Chaos: Analytical Methods.* Springer, Berlin, 1988.

[2] G.A. Leonov. *Mathematical Problems of Control Theory, An Introduction.* World Scientific, Singapore, 2001.

[3] V.B. Ryabov. Using Lyapunov exponents to predict the onset of chaos in nonlinear oscillators. *Phys. Rev.E* 66(1): 016214(1–17), 2002.

[4] D.R. Merkin. *Introduction to the Theory of Stability.* Springer, New York, 1997.

[5] N.G. Chetaev. *The Stability of Motion.* Pergamon Press, 1961.

[6] G. Benettin, L. Galgani, A. Giorgilli, J.M. Strelcyn. Lyapunov exponents for smooth dynamical systems and Hamiltonian systems; a method for computing all of them, Part I: Theory. Meccanica, 15: 9–20, 1980.

[7] A.M. Lyapunov. *Stability of Motion.* Academic Press, New York, 1892.

[8] A. H. Nayfeh and D. T. Mook. *Nonlinear Oscillations.* John Wiley & Sons, New York, 1979.

[9] U. Parlitz and W. Lauterborn, Superstructure in the bifurcation set of the Duffing equation. Phys. Lett. A 107: 351-355, 1985.

[10] D. D'Humieres, M. R. Beasley, B. A. Huberman, and A. Libchaber. Chaotic states and routes to chaos in the forced pendulum. Phys. Rev. A 26: 3483-3496, 1988.

[11] A. H. Nayfeh and B. Balachandran. *Applied Nonlinear Dynamics. Analytical, Computational, and Experimental Methods.* John Wiley & Sons, New York, 1995.

[12] J. Kozlowski, U. Parlitz, W. Lauterborn. Bifurcation analysis of two coupled periodically driven Duffing oscillators. *Phys. Rev. E*, 51(3): 1861-1867, 1995.

[13] D.M. Vavriv, Yu.A. Tsarin, I.Yu. Chernyshov. Forced Oscillations of Two Coupled Passive Oscillators. Radiotekhnika I Elektronika, No. 10, 2015-2023, 1991 (in Russian)

Analysis of homoclinic bifurcation in Duffing oscillator under two-frequency excitation: Peculiarity of using Melnikov method in combination with averaging technique

Vladimir Ryabov and Kenta Fukushima

Future University Hakodate, 116-2 Kamedanakano-cho
Hakodate, Japan 041-8655
Email: riabov@fun.ac.jp

Abstract: We study Melnikov conditions predicting appearance of chaos in Duffing oscillator with hardening type of non-linearity under two-frequency excitation acting in the vicinity of the principal resonance. Since Hamiltonian part of the system contains no saddle points, Melnikov method cannot be applied directly. After separating the external force into two parts, we use a perturbation analysis that allows recasting the original system to the form suitable for Melnikov analysis. At the initial step, we perform averaging at one of the frequencies of the external force. The averaged equations are then analyzed by traditional Melnikov approach, considering the second frequency component of the external force and the dissipation term as perturbations. The numerical study of the conditions for homoclinic bifurcation found by Melnikov theory is performed by varying the control parameters of amplitudes and frequencies of the harmonic components of the external force. The predictions from Melnikov theory have been further verified numerically by integrating the governing differential equations and finding areas of chaotic behavior. Mismatch between the results of theoretical analysis and numerical experiment is discussed.

Keywords: Melnikov method, Duffing oscillator, chaos, averaging, global homoclinic bifurcation.

1. Introduction

Melnikov method, the original idea of which can be found in Melnikov [1], enables one to predict conditions of chaos by detecting the occurrence of global homoclinic bifurcation. Since the method is convenient to examine conditions of chaos in various dynamical systems, it has been often applied as the principal analytic tool [1-7]. In a typical setting [7], the analyzed system should possess a homoclinic (or heteroclinic) loop in the Hamiltonian part that allows studying the conditions for its intersection under the action of perturbation. If the system does not possess homoclinic trajectories in the unperturbed system, the conditions for chaos still can be studied in some cases by introducing a suitable coordinate transform, recasting the system to

Chaos Theory: Modeling, Simulation and Applications
C. H. Skiadas, I. Dimotikalis and C. Skiadas (Eds)

the form where Melnikov method becomes applicable. For example, in paper [2], the authors used the method of averaging prior to Melnikov analysis to obtain conditions for the onset of chaos in Duffing oscillator excited with a quasiperiodic external force. In [2], relation of Melnikov function to several attractors coexisting in the phase space was also addressed. Paper [3] studied a similar system subject to a combined action of a parametric and external resonances. In paper [4] the research has been focused on Duffing oscillator with external force, where Melnikov method and second-order averaging have been used to establish the conditions for homoclinic bifurcations near center-type points related to saddle-node bifurcations of subharmonics. In other works different coordinate transforms have been used for applying Melnikov method to the analysis of rotation in the ship model [5] and to Lorenz-type system (Lorenz-Robbins equation) [6].

However, it should be noted that in many works where Melnikov method was applied to the study of particular systems, the analysis usually ends up with a formula that establishes a border in the space of controls where one can expect homoclinic bifurcation to occur. Further analysis of chaos appearance conditions that would allow establishing links between physical properties of the system, homoclinic bifurcations, and chaos, has not been performed so far. It also remains unclear how well the criterion works for actual prediction of chaotic areas, since Melnikov condition cannot be considered neither necessary nor sufficient one for the existence of chaotic attractors in the phase space.

In this paper we study the properties of Melnikov condition in a quasiperiodically forced Duffing oscillator [2]. In particular, we are interested in the dependence of chaos threshold on amplitudes and frequencies of external forces and possible errors that utilization of a two-step perturbation technique can introduce. We also study interrelation between the well-known phenomenon of nonlinear resonance and Melnikov condition for chaos in this system.

2. Application of Melnikov method when unperturbed system does not contain separatrix

2.1. *Duffing oscillator under two-frequency excitation*

Quasiperiodically forced damped oscillator is one of the basic models in nonlinear physics. In the case of cubic nonlinearity it has the form of Duffing oscillator

$$\frac{d^2x}{dt^2}+\omega_0^2 x=\varepsilon\left[-2\alpha\frac{dx}{dt}-\gamma x^3+F_1\cos\omega_1 t+F_2\cos\omega_2 t\right] \tag{2.1}$$

The basic properties of the solutions of system (2.1) are

- quasiperiodic behavior and its bifurcations
- if $\gamma>0$, the nonlinearity is of "hard" type, and the potential of its Hamiltonian part is single-well function

- chaos is present in both the weakly $(\varepsilon << 1)$ and strongly $(\varepsilon \geq 1)$ nonlinear cases

Compared to the case of soft nonlinearity (γ<0), hardening potential has additional difficulty in the analysis, since Hamiltonian part does not contain saddle point and associated separatrix. The corresponding phase portraits of soft- and hard-type oscillators are given in Fig.1

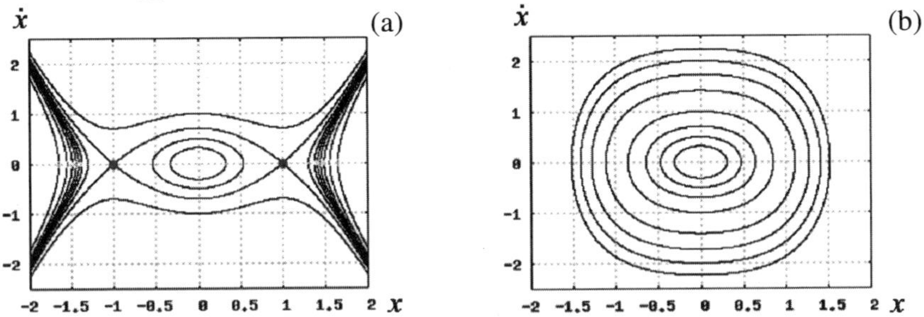

Fig. 1. Phase portrait of system (2.1) at $\alpha = F_1 = F_2 = 0$: (a) $\gamma < 0$:Two saddles and center; (b) $\gamma > 0$:Center only.

Melnikov method can be directly applied to the soft system, since it contains two heteroclinic trajectories connecting the saddle points (cf. Fig.1a). The case of hard nonlinearity defies analytical treatment, the corresponding phase portrait containing only one fixed point of center type. In the next section, we apply averaging method for recasting system (2.1) at $\gamma > 0$ to the form suitable for Melnikov analysis

2.2. *Averaged equations*

We consider here the case of the primary resonance, i.e. $\omega_1 \approx \omega_2 \approx \omega_0$. After introducing the new coordinates u, v via Van der Pol transformation

$$x = u(\tau)\cos\omega_1 t - v(\tau)\sin\omega_1 t\ , \tag{2.2}$$

parameters defining the frequency detuning $\Omega = \dfrac{\omega_2 - \omega_1}{\varepsilon}, \Delta = \dfrac{\omega_0 - \omega_1}{\varepsilon}$, and "slow" time $\tau = \varepsilon t$, averaged equations at the reference frequency ω_1 are obtained

$$\begin{aligned}\frac{du}{d\tau} &= -\alpha u - \left[\Delta + \beta(u^2 + v^2)\right]v + P_2 \sin(\Omega\tau) \\ \frac{dv}{d\tau} &= -\alpha v + \left[\Delta + \beta(u^2 + v^2)\right]u - P_1 - P_2 \cos(\Omega\tau)\end{aligned} \tag{2.3}$$

If to put $P_2 = 0$, we can find the equation for the frequency response curve

$$a^2\left[\alpha^2 + \left(\Delta + \beta a^2\right)^2\right] = P_1^2\ , \tag{2.4}$$

where $a^2 = u^2 + v^2$. The response curve (see Fig. 2) is the dependence of the roots of Eq. (2.4) (fixed points of Eqs. (2.3) at $P_2 = 0$) on Δ. Within certain range of frequencies, there is a hysteretic region, where multiple fixed points coexist in the phase space. One of those points is known to be of saddle type that, being considered in the Hamiltonian limit $(\alpha = 0)$, possesses a double homoclinic loop shown in Fig.3. From now on, we shall refer to the two loops as to "small" and "large", respectively.

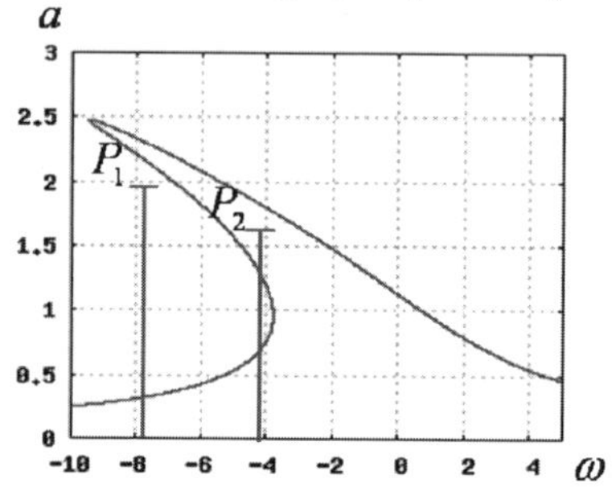

Fig. 2. Sketch of the frequency response curve of system (2.3).

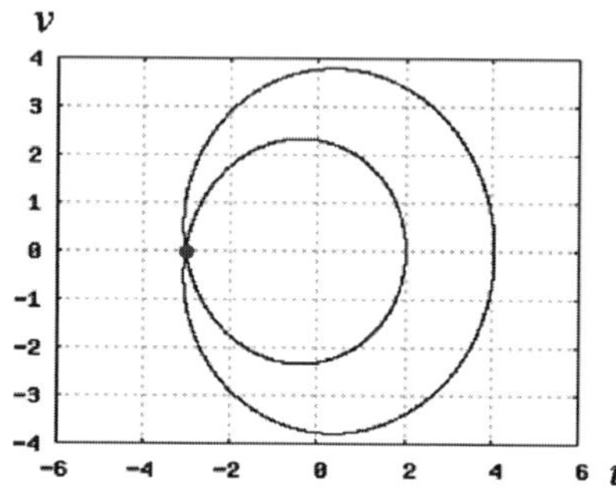

Fig. 3. Homoclinic orbits of system (2.3) at $\alpha = 0$.

Note that Melnikov method can be used for the analysis of system (2.3) only if certain conditions are met. First, it is necessary to choose the position of the frequency ω_1 within the area of multistability, i.e. where the curve shown in Fig.2 has three values at a single Δ-value. This ensures the existence of saddle point in both the unperturbed Hamiltonian system that is preserved after introducing dissipation. Second, the parameters α and P_2 should be considered small in order to validate the perturbation scheme used in Melnikov analysis.

In the case the parameter P_2 is not small compared to P_1, it is still possible to apply Melnikov method after the change of variables in Eq. (2.3)

$$u \to u\cos\Omega\tau + v\sin\Omega\tau; \; v \to v\cos\Omega\tau - u\sin\Omega\tau, \tag{2.5}$$

which is equivalent to performing averaging at the reference frequency ω_2. The system of averaged equations (2.3) in the new variables can be obtained from Eqs. (2.3) by substitution of parameters

$$\Delta \to \Delta - \Omega; \; \Omega \to -\Omega; P_1 \leftrightarrow P_2. \tag{2.6}$$

Although such change of variables looks like a trivial choice of notation

$\omega_1 \leftrightarrow \omega_2; P_1 \leftrightarrow P_2$,

we demonstrate below that Melnikov method is not invariant with respect to transformation (2.5), (2.6), i.e., depending on the choice of averaging frequency, it produces different predictions for the homoclinic bifurcation threshold.

2.3. *Melnikov criterion*

Melnikov analysis for the system (2.3) can be found in Ref. [2]. Here we reproduce some of its main results, with some corrections of the errors found in [2]. Introducing the parameters

$$\delta = -\Delta / \beta,\ f = 4P_1 / \beta,\ q = \pm 2\sqrt{-2fu_s},\ \eta^2 = -2fu_s - \frac{1}{4}b^2,\ b = -4(\delta - u_s^2),$$

where the upper sign corresponds to the small loop and the lower one – to the large loop, u_s stands for the u-coordinate of the saddle point found from

$$u_s^3 - \delta u_s - f/4 = 0\,, \tag{2.7}$$

we get the Melnikov distance $M(\tau_0)$ between stable and unstable manifolds in the Poincare section

$$M(\tau_0) = \alpha\left[\pm 8\delta \tan^{-1}\left(\sqrt{\frac{|q| \pm b}{|q| \mp b}}\right) - 3\eta\right] + \frac{2\pi P_2}{\sinh(\sigma\pi)}\left\{\sigma^2\eta^2 f^{-1}\exp\left(\mp\sigma\cos^{-1}\left(\mp\frac{b}{|q|}\right)\right)\right\}\sin(\Omega\tau_0) \tag{2.8}$$

where $\sigma = 2\Omega/\beta\eta$. The intersection of manifolds occurs if the function $M(\tau_0)$ has simple zeros at some values of the parameter $\tau_0 \in \left]-\infty;\infty\right[$. Noting that Eq. (2.8) can be represented in a simple form $M(\tau_0) = C_0 + C_1\sin(\Omega\tau_0)$, one can obtain the condition for intersection as $|C_0| = |C_1|$, or rewrite it as the explicit dependence of homoclinic bifurcation value $P_2^{(c)}$ on other control parameters

$$P_2^{(c)} = \alpha\left|\frac{\left[\pm 8\delta \tan^{-1}\left(\sqrt{\frac{|q| \pm b}{|q| \mp b}}\right) - 3\eta\right]\sinh(\sigma\pi)}{2\pi\left\{\sigma^2\eta^2 f^{-1}\exp\left(\mp\sigma\cos^{-1}\left(\mp\frac{b}{|q|}\right)\right)\right\}}\right| \tag{2.9}$$

Formula (2.9) thus represents the final result of applying Melnikov method to the system (2.3) and establishes a location of homoclinic bifurcation lines in the space of control parameters.

3. Properties of Melnikov criterion

3.1. *Homoclinic bifurcation condition*

We are now in a position of performing numerical analysis of Eq. (2.9) with the purpose of elucidating the general regularities in the positions of chaotic areas in the space of control parameters of system (2.3). In the following, we mainly focus the analysis on amplitudes and frequencies of the external force.

As a first step, we fix all the parameters except P_2 and calculate the values of $P_2^{(c)}$ where Eq. (2.9) holds. According to Melnikov theory, at larger values of the parameter P_2 the homoclinic structures should exist in the phase space. In Fig.4, we plot the dependence of the parameter $\mu_c = 20\lg\left[P_2^{(c)}/P_1\right]$ on frequency Ω, fixing other parameters at the values found in Ref. [2].

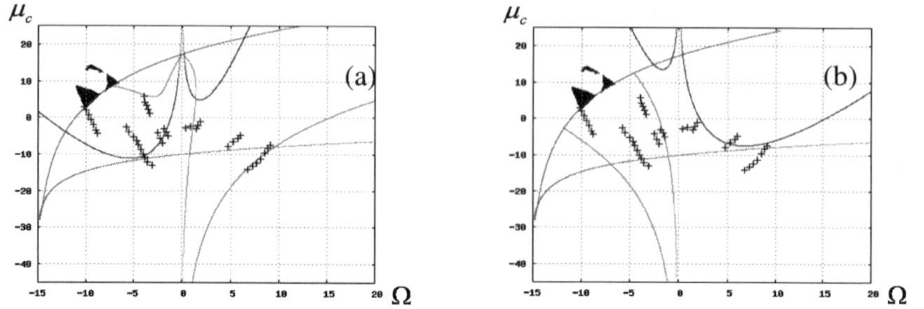

Fig. 4. Red lines: Melnikov condition (2.9) for the system (2.3) at $\alpha = 0.25, \beta = 1.54, P_1 = 2.48, \Delta = -15.05$ for small (a) and large (b) loops of separatrix. Green lines: same conditions as those shown in red, but calculated for the equations averaged at the frequency ω_2. Blue lines: position of saddle-node bifurcations calculated from the analysis of frequency response curves (saddle point exists in between). Black areas and crosses: chaos.

Our calculations show that, depending on Δ, from one to four lines of homoclinic bifurcations can appear in the diagram, corresponding to small and large homoclinic loops, positive or negative values of the parameter Ω. This means that bifurcation conditions for the two loops differ and depend on the position of the two components of external force on the frequency axis. We note also that bifurcation conditions for the both loops are not symmetric with respect to sign of the frequency, being lower for one sign than for the other, and possess certain "anti-symmetry" property with respect to the small and large loops.

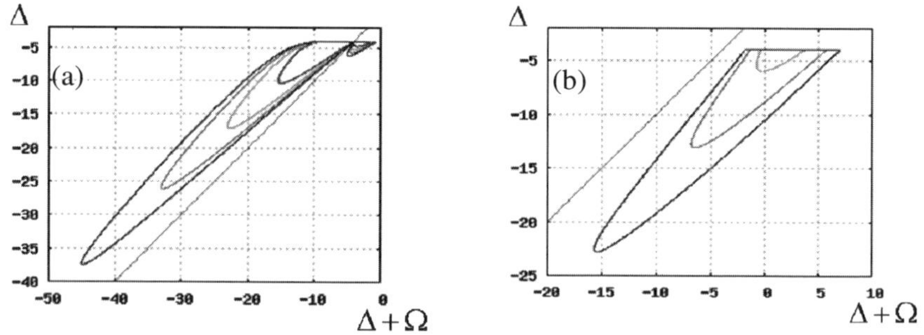

Fig. 5. Areas where homoclinic structures exist according to Melnikov criterion (2.9) for small (a) and large (b) homoclinic loops at $\alpha = 0.5, \beta = 1.54, P_1 = 2.48$. The line colors corresponds to the values of parameter $P_2 = 1.0$ (red); 1.5 (green); 2.0 (blue); 2.5 (purple). Gray line corresponds to the condition $\Omega = 0$.

In Fig.5 we plot the result of analyzing the homoclinic bifurcation conditions under the change in the parameters Δ and Ω at different levels of the parameter P_2. The closed loops delimit the areas where homoclinic structures exist in the phase space according to Melnikov criterion. The grey line indicates the position of origin in the Ω-axis, i.e. it is defined by the equation $\Omega = 0$. The asymmetry with respect to the sign of the parameter Ω is evident for both the small (Fig. 5a) and large (Fig. 5b) loops. In particular, the areas where homoclinic structures associated with the small loop exist are much larger at $\Omega < 0$, whereas homoclinic bifurcation of the big loop is absent at all for $\Omega < 0$ at the shown levels of P_2. This can be interpreted in terms of Fig. 2a as a result of higher system response to resonance excitation at $\Omega > 0$, since the response amplitude is larger at those frequencies.

3.2. Melnikov criterion and averaging

The red curves in Fig.4 are calculated under the assumption that averaging had been performed at the frequency ω_1. This choice of averaging frequency imposes a limitation on the parameters α and P_2, i.e. they should be small. If the value of one or both of those parameters increases, the estimate of homoclinic bifurcation condition provided by Melnikov method becomes incorrect. Approximate conditions of the validity of Melnikov criterion can thus be approximately established as

$$\alpha << 1; \quad P_2 \le P_1$$

The line $P_2 = P_1$ corresponds to the level $\mu = 0$ in Fig.4. Therefore, it can be expected that chaos areas could be found within the areas delimited by the Melnikov curves $P_2 = P_2^{(c)}$ and the line $\mu = 0$.

On the other hand, considering the averaging at the frequency ω_2 we can obtain Melnikov threshold in the areas, where $P_2 \ge P_1$ (green lines in Fig.4). In this case, it is also necessary to take into account the necessity of existence of saddle point in the unperturbed system. Saddle points are born in this system via saddle-node bifurcations denoted as blue lines in Fig.4. The locations of bifurcation lines have been calculated from Eq. (2.4).

Finally, we compare the predictions of Melnikov criterion to actual positions of chaotic areas in the plane of control parameters shown in Fig.4. For this purpose, we performed numerical integration of the system (2.3), marking the positions where we observed chaotic attractors by black dots and crosses. The general conclusion from this analysis is that Melnikov method provides a very approximate criterion for the existence of chaos in this system, since the mismatch between the predicted values and actual locations is not small. However, combining the predictions with the analysis of fixed points in the unperturbed system and performing averaging at different frequencies can substantially improve the precision of Melnikov criterion. Indeed, the combination of the areas between green lines in Fig.4b (averaging at ω_2), between blue lines (existence of saddle point) and above the line $\mu = 0$ (the condition $P_2 \geq P_1$) provides much better approximation to the chaotic area that simple utilization of the formula (2.9) (areas above red lines in Fig. 4).

4. Conclusions

Contrary to previous works that were not focused on the analysis of implications of the formulas obtained with Melnikov method, we perform an extensive study of the homoclinic bifurcation conditions in quasiperiodically forced Duffing oscillator under the variation of control parameters. The results of our analysis reveal that

- there is an asymmetry in the bifurcation conditions with respect to frequency detuning between the excitation frequencies;
- the prediction of Melnikov theory depends on the averaging procedure;
- there are two types of chaotic motions in the system caused by the presence of double homoclinic loop in the Hamiltonian part of the averaged system. The Melnikov conditions for the two loops are different and dependent on the sign of frequency detuning between the external force frequencies.

Finally, we verify our predictions with the results of numerical integration of the averaged equations and propose ways of improving the predictions made by Melnikov method in combination with averaging. We believe that similar approach can be also used in the analysis of other nonlinear oscillators under multi-frequency excitation.

References

[1] V.K.Melnikov, On the stability of the center for time periodic perturbations. *Trans. Moscow Math. Soc*, 12(3):1-57, 1963.

[2] V.B.Ryabov, D.M. Vavriv. Conditions of quasiperiodic oscillation destruction in the weakly nonlinear Duffing oscillator. *Physics Letters A*, 153(8, 9):431-436, 1991.

[3] K.Yagasaki et al. Dynamics of a Weakly Nonlinear System Subjective to Combined Parametric and External Excitation. *Journal of Applied Mechanics*, 57:209-217, 1990.

[4] K.Yagasaki. Second-Order Averaging and Melnikov Analysis For Forced Non-Linear Oscillators. *Journal of Sound and Vibration*, 190:587-609, 1996.

[5] Fritz Colonius et al. A Numerical Study Of Capsizing Comparing Control Set Analysis And Melnikov Method. *International Journal of Bifurcation and Chaos*, 18(5): 1503-1514, 2008.

[6] D.Huang et al. Dynamics of the Lorenz-Robbins system with control. *Physica D*, 218:131-138, 2006.

[7] S.Wiggins. *Introduction to Applied Nonlinear Dynamical Systems and Chaos.* Springer , 1990.

Exploring life expectancy limits: First exit time modeling, parameter analysis and forecasts

Christos H. Skiadas[1] and Charilaos Skiadas[2]

[1]Technical University of Crete, Data Analysis and Forecasting Laboratory, Chania, Crete, Greece
Email: skiadas@asmda.net
[2]Department of Mathematics and Computer Science, Hanover College, Indiana, USA
Email: skiadas@hanover.edu

Abstract: In this paper we explore the life expectancy limits by based on the stochastic modeling of mortality and applying the first exit or hitting time theory of a stochastic process. The main assumption is that the health state or the "vitality", according to Strehler and Mildvan, of an individual is a stochastic variable and thus it was introduced and applied a first exit time density function to mortality data. The model is used to estimate the development of mortality rates in the late stages of the human life span, to make better fitting to population mortality data including the infant mortality, to compare it with the classical Gompertz curve, and to make comparisons between the Carey med-fly data and the population mortality data estimating the health state or "vitality" functions. Furthermore, we apply the model to the life table data of Italy, France, USA, Canada, Sweden, Norway and Japan, and we analyze the characteristic parameters of the model and make forecasts. The case of female mortality in Sweden is extensively studied and forecasts to 2025 and 2050 are presented.

Introduction

The Health State of human beings is a term generally accepted as a crucial factor of the human life and is connected, even not verified quantitatively, to the life expectancy. The human health can be considered as a stochastic variable as it is strongly associated with uncertainties having to do with various factors both from the environment but also from the internal mechanism and information included into the DNA and the genes. The probability of sudden changes in the state of the human health due to illnesses or accidents is quite large, strengthening the assumption that the state of the human health could be considered as a stochastic variable. It is known that human characteristics vary around mean values, and the deviations seem to be the physical consequence of the probabilistic-stochastic character of the process. Moreover the probabilistic or better the stochastic character of the health state of individuals is generally accepted and measured at least qualitatively. Attempts to quantify the process go back to

Chaos Theory: Modeling, Simulation and Applications
C. H. Skiadas, I. Dimotikalis and C. Skiadas (Eds)

Strehler and Mildvan (1960) (1) who suggest the term "vitality" of a person, a stochastic function with a decreasing drift and changing randomly with age during the human lifetime. The end comes when the stochastic path crosses for the first time the zero line, representing the zero level of vitality or the zero health state.

The Stochastic Modeling

Based on a stochastic process for the health state function, we had explored the problem of modeling human mortality in 1995, when we developed and applied a dynamic model based on the first exit time or hitting time theory. That model is flexible enough as to account for expressing the human mortality distribution at all ages, including infant mortality (see the paper by Janssen and Skiadas, 1995 (2)). The proposed first exit or hitting time probability density function $g(t)$ includes 5 parameters. The two extra parameters account for the infant mortality, making it evident that a satisfactory model with fewer parameters could be obtained.

More recently (3, 4) we have developed more flexible stochastic models to express the human mortality. Though it is impossible to find the stochastic path for the state of health for every individual, it is possible to find the Mean Value of the health state or vitality of a population as a summation of the related health states of every individual. Starting with a model for the mean health state function $H(t)$ we can proceed in the construction of a stochastic model for the state of the human health, solve it, find the related probability density function $p(t)$ and then estimate the first hitting time probability density function $g(t)$ from a barrier located at zero for the stochastic process expressing the state of the human health. The parameters of the last function $g(t)$ are estimated by fitting it to the data collected from the Bureau of the Census for the yearly deaths of the population in a country.

Our approach was mainly to find models with explanatory power and able to fit the data directly instead of applying models to transformed data via logarithms as is the case in several applications starting from Gompertz (1825) (5). It was evident that the methods for nonlinear regression where not well developed when Gompertz introduced his model and the linearization via a logarithmic transformation was a reasonable substitute when working with mortality data. However, we should notice that by taking logarithms of the data and then fitting, the results highly underestimate the error term. The real error is of the order of the exponent of that estimated when taking logarithms.

Few years after the paper by Janssen and Skiadas (1995), Weitz and Fraser (2001) (6) proposed the inverse Gaussian as a model expressing the first hitting time probability density function for modeling the life time process of the med-fly based on the data collected and presented by Carey in a publication in Science (1992). The model proposed by Weitz and Fraser is nothing else but the classical model proposed by Schrödinger (1915) (7) and Smolukovsky (1915) (8) in the same journal issue. This model suggests a linear function for the mean value of the stochastic process thus assuming a decay process for the mean value of the health state. This is a correct argument in the case modeled by Schrödinger and Smolukovsky as they considered the spontaneous emission of particles from a radioactive material, but it is not appropriate for modeling the state of the human health. According to our experience the mean value of the human health could be stable or even increasing at the young ages and then decreasing slowly at the mid ages and fast decreasing at the old age. Even the simplest function should be non-linear. This is the underlying assumption in the theory proposed by Strehler and Mildvan in the 1960 publication in Science (1).

Properties of the Stochastic Model

However, the Weitz and Fraser approach gave good results in the case of the med-fly data, only differing from the nonlinear mean value by a small curvature as is presented in the paper by Skiadas and Skiadas (9). The data on med-flies are collected from an experiment (Carey et al. 1992 (10)) in which the life duration of 1.203.646 med-flies in cages was measured. As it was presented in the same comparative study by Skiadas and Skiadas (9) the inverse Gaussian is a right skewed model appropriate for some modeling approaches but not for the highly left skewed human life table data. The figure 1A for the mortality data for USA 2004 for females compared with the Carey med-fly data is evident. The age axis is rescaled as to account for both human and med-fly data. The rescaling is according to the method proposed by Robine and Ritchie (1993) (11).
The first exit time model used is of the following form (3)

$$g(t) = \frac{k(l + (c-1)(bt)^c)}{\sqrt{t^3}} e^{-\frac{(l-(bt)^c)^2}{2t}}$$

Where b, l, k and c are parameters. The key points in deriving this model where: 1) the proposal of a stochastic differential equation for the health state or the vitality (according to Strehler and Mildvan) of an individual, 2) the derivation of the transition probability density function $p(t)$ associated with the stochastic variable for the health state (vitality) by solving the appropriate Fokker-Planck partial differential equation and 3) the derivation of the first exit time probability density function $g(t)$ from the stochastic process having transition probability density function $p(t)$. These three steps where first achieved by Janssen and Skiadas (1995) and later Skiadas and Skiadas (10) suggested the model presented by the above hitting time density function. This model is already applied in mortality data for more than 35 countries and is quite flexible as to cover the infant mortality as well. The data are obtained from the Human Mortality Database (www.mortality.org) and the related applications along with the fitting programs appear in the webpage http://www.cmsim.net/id14.html .

The function used for the health state is: $H(t)=l-(bt)^c$. When $c=1$ the simple case studied by Schrödinger (1915) and Smolukovsky (1915) and applied by Weitz and Fraser (2001) results. As it is obvious from Figure 1A the Carey med-fly data is right skewed whereas the human mortality data, as seen in the case of USA females in 2004, is highly left skewed. The model is applied to the data with the estimates $b=0.2479$, $l=8.948$, $k=0.3803$ and $c=1.3$, $R^2=0.992$ for Carey med-fly data and $b=0.0194$, $l=12.705$, $k=0.3677$ and $c=5.3$, $R^2=0.988$ for females in USA 2004.

A very interesting case appears when we draw the tangent lines passing from the left inflection points of the curves fitted to med-fly and female (USA 2004) mortality data. Both lines, illustrated in Figure 1A, are almost parallel. The estimated values of the first derivative of the first exit time density functions are 0.00301 for the Carey med-fly data and 0.00327 for the USA 2004 female data, slightly deferring between each other. It is obvious that at least for this example the death rates at high ages are similar. Instead the behavior is different in the tails as the med-fly data present a very long tail compared to the USA 2004 female data.

The representation for the health state functions is given in the related graph in Figure 1B. The health state function is $H(t)=l-(bt)^c$. The data for males in USA 2004 are also fitted. The estimated parameters are $b=0.0218$, $l=10.936$, $k=0.362$ and $c=4.6$, $R^2=0.964$. The higher level accounts for the females in USA compared to USA males. The model suggests an almost linear development for the first decades of the human life span, a gradual decrease in the mid ages, and finally a fast decrease at old age. Instead for the Carey med-fly an almost linear decrease of the health state is estimated. The curvature expressed by the exponent $c=1.3$ slightly differs from the linear one ($c=1$) suggested by Weis and Fraiser. Instead for the human health states the exponent is $c=5.3$ and 4.6 for females and males in USA respectively, suggesting a fairly large curvature.

A very important point is to explore the properties of the Health State or Vitality function based on the curvature. It is clear that the curvature is larger as the exponent c takes higher values. Larger curvature is associated with a life span in the area of high vitality expressed by the first and almost linear part of the health state curves (see Figure 1B for males and females in USA). In the following we can estimate the point where the curvature $K(t)$ expressed by the following formula takes its maximum value.

$$K(t)=\frac{|\,c(c-1)b^c t^{c-2}\,|}{(1+c^2b^{2c}t^{2c-2})^{3/2}}$$

This maximum value for the curvature could be connected with the maximum deterioration of the physical repair mechanisms of the human organism (12). Estimating the maximum for the curvature we find that it is obtained at 77 years for the females and 74 years for males in USA the year 2004. Instead for the med-fly data the maximum curvature is obtained in the first year of the life span. We can assume that the mechanisms underlying the vitality of the med-fly are totally different from the related mechanisms of the human beings. Another interesting point resulting from the study of the curvature of the health state function is that it is smaller and smaller as we are moving at higher ages from its maximum value. This is in accordance with the assumption of slower mortality rates at higher ages (13, 14).

In the following we will examine how the mortality evolves at the late stages of the life span by using the first exit time density function. This will be checked by a comparative study (Sweden, females). The issue can be explored by observing the tangent lines at the right inflection point of the regression curve. This area accounts for the late years of the human life span that could be noted as the longevity area. We have drawn these lines for the case of females in Sweden from 1800 to 2000 illustrated in Figure 1C. It is clear that the tangent line is moving to the right part of the figure to the higher ages. This is in accordance with the improvement of the health state during the 200 years elapsed from 1800 to 2000. However, one should expect parallel lines moving to the right due to the significant improvement of the living conditions in our societies. To our surprise the tangent lines, while moving to the right, turn vertically and tend to consistently meet the horizontal axis just few a years after 100. The right part of the first exit time density function $g(t)$ is mainly expressed by the tangent line while a small part accounts for people included in the tail of the density function. The expectations for a longer life are mainly connected to the way the tangent line is moving to the right to the higher ages. The observations of the last decades show that the tangent line tends to gradually stabilize in a vertical position than to move significantly to the right. This can be due to the way our repair mechanisms included in the DNA and genes decide for the life expectancy. If it is so, the tangent line will be more and more vertical and the form of the first exit time density function at the maximum area more sharp following improvements in the living standards. Instead, if our destiny is not driven by the code, the tangent lines should move parallel to the right. The later may result following the improvement of biological sciences. However, in the following, we make parameter analysis of the mortality data for various countries to search for the future of human longevity.

As we have proposed and applied a model presenting a good fit to the data, including even the infant mortality we will now try an application to the data that perhaps Gompertz was aware at the time (1825) when he published his famous paper, and compare it with what Gompertz could see using his formula. The data used is the mean value of the females' deaths in Sweden from 1751 to 1825. The resulting mortality curves are illustrated in Figure 1D. As it was expected our model expresses the infant mortality as well, whereas the Gompertz model is expressed by the characteristic line. The data are transformed to account for 1000 people and the logarithm of the death rate is computed for every year of age. The estimated parameters of the first exit time model are b=0.0223, l=0.5878, k=0.552 and c=4.3 and for the Gompertz line b=0.062.

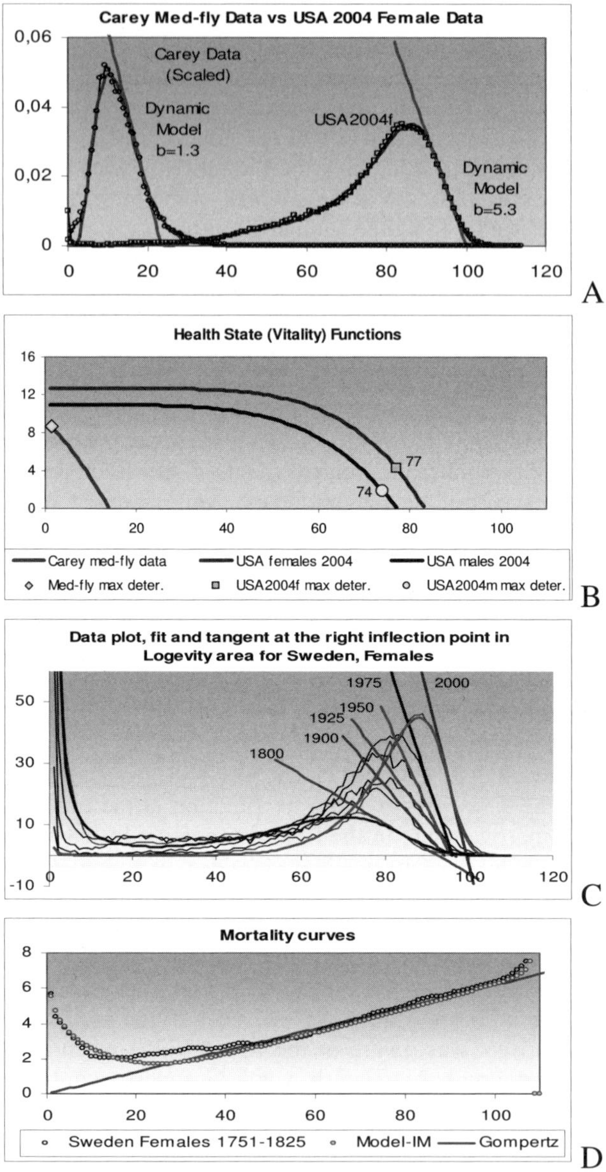

Figure 1. Applications of the First Exit Time Model.

A. Comparison of Carey med-fly data with the USA data for females, 2004.
B. Form of the Health State Functions for Carey data and USA data.
C. The characteristic tangent for the Sweden data for females.
D. The mortality curves.

Parameter Analysis

The parameters of the first exit time model proposed are estimated for several countries and for females for a quite large time period. The results are illustrated in Figure 2. Seven countries are selected: Italy, Sweden, USA, Japan, Canada, France and Norway. The parameters are estimated for as many larger periods as possible according to the data provided from the Human Mortality Database. In Figure 2A the parameter b is illustrated. The parameter values after 1950 form a group of almost parallel and slowly decreasing lines. In Figure 2B the parameter c is presented. The exponent c is growing in all the cases with Sweden, Italy, France and Norway to reach higher values. In figure 2C the parameter l is presented. This parameter accounts for the infant mortality data period and, as the infant mortality tends to decrease rapidly, the parameter l tends to zero. In this case we will have a simpler model of the form:

$$g(t) = \frac{k(c-1)(bt)^c}{\sqrt{t^3}} e^{-\frac{(bt)^{2c}}{2t}}$$

As the parameter k is a renormalization constant, the process is characterized by the parameters b and c. An illustrative example is given in Figure 2D. Two possible scenarios resulting from the projection of the parameters b and c to 2050 are presented: b=0.0125, c=14 and b=0.0115, c=14 are selected and the population is equal to 1000.

The maximum death rate is achieved when: $T_{max} = \left[\frac{2c-3}{2c-1}\right]^{1/(2c-1)} \frac{1}{b^{2c/(2c-1)}}$

For relatively large c an approximation is of the form: $T_{max} \approx \frac{1}{b^{2c/(2c-1)}}$

For the previous example the estimates are T=93.82 years for b=0.0125 and c=14 and T=102.3 years b=0.0115 and c=14.

Projection to 2050 of mortality for females in Sweden

An application of the previous theory and analysis to the mortality of the women in Sweden and a projection to 2050 is based on the following:

1. We draw the graphs for the maximum, left inflection point and right inflection point for the time period from 1751 to 2000 for the female mortality in Sweden. Every 25 years we compute the mean value of the next 5 years period and apply the model. The last period includes the years 2000-2004. So we have done 11 fittings to estimate the parameters of the model. The resulting values for the maximum, right inflection point and left inflection point are illustrated in Figure 3A. As the three resulting graphs show a systematic growth behavior we fit an exponential function of the form f(t)=a1+a2exp(a3t) to the data for the maximum and the right inflection point and a quadratic polynomial of the form $f(t)=a\text{-}+a2t+a3t^2$. The related curves are illustrated in Figure 3A. A projection of the curves of this figure to 2050 gives 95.1 years of age for the left inflection point, 95.1 years for the maximum and 100.7 years for the right inflection point.

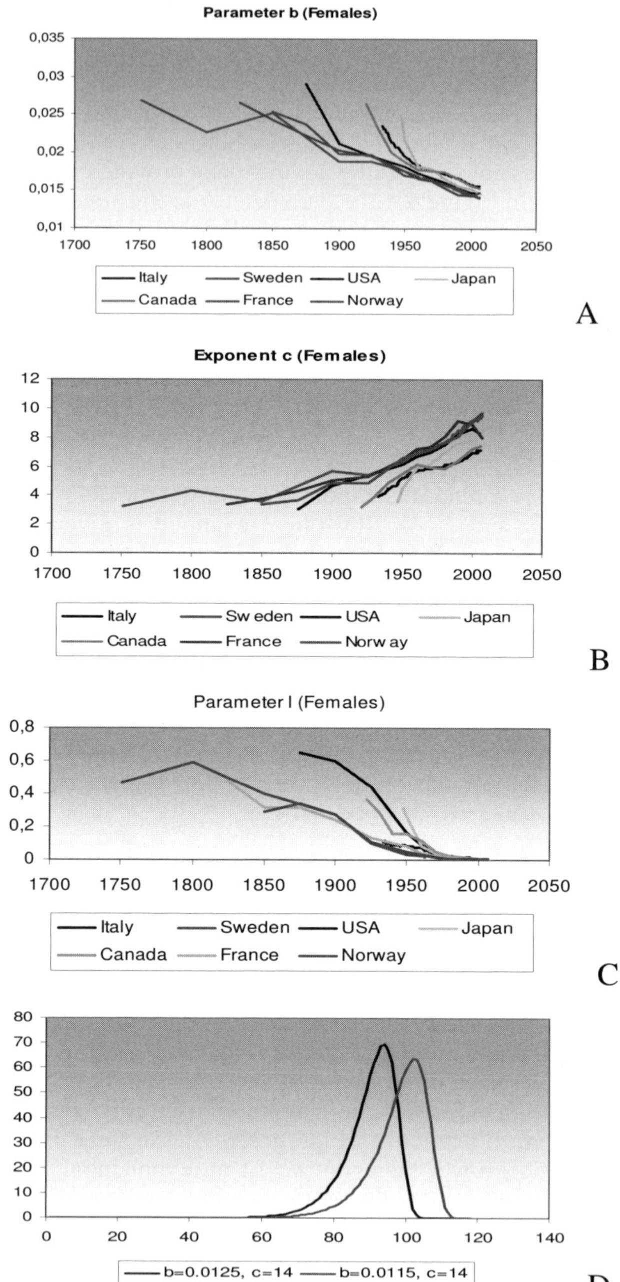

Figure 2. Parameter's values for various countries

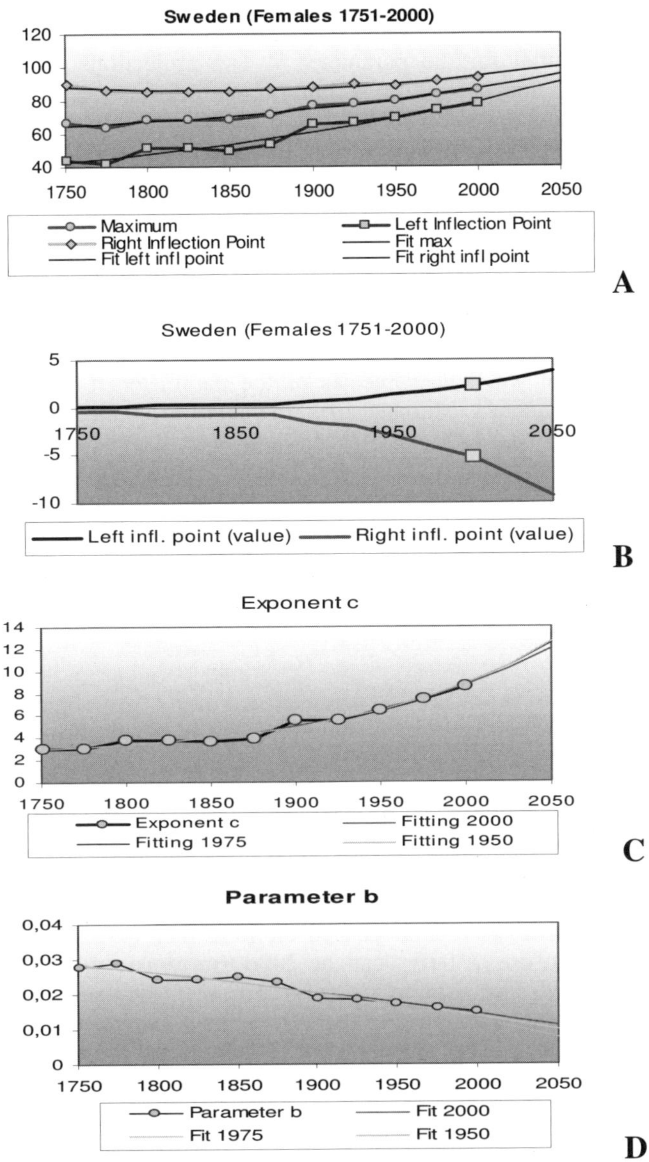

Fig. 3. Analysis of Sweden results for females and projections

2. We draw the graphs for the values of the first derivative of $g(t)$ at the left inflection point and at the right inflection point. As it is illustrated in Figure 3B the values for the left inflection point are positive and increasing and the values for the right inflection point are decreasing.

3. The behavior of the exponent c and the parameter b is illustrated in figures 3C and 3D. An exponential function is fitted to the data for the exponent c and a negative exponential for the parameter b. The projections of the resulting functions to 2025 and 2050 will give the needed information for the future of the mortality of females in Sweden.
4. Based on the previous projections for the age years for the characteristic points of the function $g(t)$, the values of the derivatives of $g(t)$ at the two characteristic points (the left and right inflection points) and the projections for the exponent c and the parameter b we can solve the inverse problem to find the parameters of the function $g(t)$ for the year 2050 and draw the curve of $g(t)$ as is illustrated in Figure 4 where the raw data for 2000 are also presented. The main problem is to explore for which are the more concrete projections from the parameters and the characteristic functions analyzed. To solve the problem we done forecasts of the parameters based on the data points for the parameters estimated from the data for females in Sweden from 1751 to 1950 and from 1751 to 1975. In both cases the estimates for the data for 2000 were done based on a combination of 2 parameters; the parameter l which accounts for the infant mortality was set equal to zero. The best results where achieved by using the projections for the exponent c (see Figure 3C) and for the point T_{max} of the maximum death rate (see Figure 3A). Instead the projections for the parameter b are quite sensitive. The values for b are estimated from the following formula:

$$b = \left[\frac{2c-3}{2c-1}\right]^{1/2c} \frac{1}{T_{\max}^{(2c-1)/2c}}$$

Table I

Base Year	1950	1975	2000	2000	2000
Forecasts Year	2000	2000		2025	2050
T_{max}	90.558	87.753	86	90.6	95.1
c	8.857	8.81	8.6	10.16	12.02
b	0.014139	0.014584	0.014887	0.0137	0.012661

5. Both cases of forecasts of 2000 based on 1950 and 1975 are illustrated in Figure 4. Now we use as base year the 2000 to estimate the 2025 and 2050 death distributions. The parameters estimated are presented in Table I. The parameter b is computed by using the values for c and T_{max} by applying the last formula. As can be seen from the above Table I and the related graphs for the parameters the estimates done from 1950 and 1975 slightly overestimate the direct estimates for 2000. However, the resulting distribution functions are quite well adapted to the raw data giving us a good tool and method to proceed to the forecasts for 2025 and 2050.
6. As the value of the parameter l is assumed to be negligible (l=0) the simpler form for $g(t)$ is selected and the results are illustrated in Figure 4. The model gives the following characteristic values for 2025: 84 years of age for the left inflection point, 90.6 years for the maximum and 97 years for the right inflection point and for 2050: 89 years of age for the left

inflection point, 95.1 years for the maximum and 101 years for the right inflection point. The values for the right inflection point are decreasing to an estimate of -7.124 for 2025 and -9.2555 for 2050 that is approximately 97.9 degrees for 2025 and 96.17 degrees for the 2050 for the slope of the tangent at the right inflection point, from 100.8 degrees at 2000. The results support our argument presented earlier that the tangent at the right inflection point (the longevity tangent) is continuously turning to a vertical position to the horizontal axis.

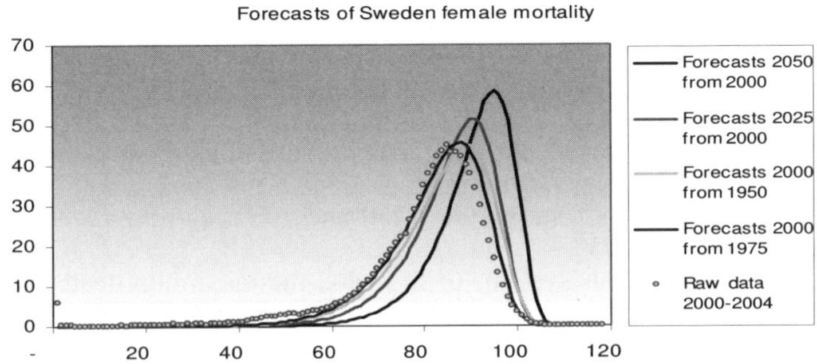

Figure 4. Forecasts for the years 2025 and 2050

Summary and Conclusions

In our studies on modeling and explaining the mortality laws we have used the first exit or hitting time theory for a stochastic process and we have applied the related formulas to human life table data and we have done comparisons with data as the Carey med-fly data that where presented, analyzed and compared with human mortality data.

There is a controversy regarding existing limits in life span or if it is prescribed through the DNA code and genes, and the influence of external factors affecting longevity. Perhaps the answer will come later following the advancement of biosciences. Today we can use the tools developed during the last centuries both from the analytic and also the applied point of view. We have mortality data that is quite well selected and provided by the bureau of the census of every country, models to apply starting from the famous Gompertz model, and theories regarding the life span and fitting techniques to data.

It was evident from the very early approaches that the stochastic character of the human life duration was inherent and underlying of every type of model that we could formulate, but the mathematic-analytic tools where less developed. For several studies the stochastic model applied was the inverse Gaussian, a model best suited to express a simple decay process, like for instance radioactive emission or the aging process of a simple organism. The decay process for the inverse Gaussian model is of the linear form $H(t)=l-bt$ instead of a nonlinear form of the type $H(t)=l-(bt)^c$ that we have suggested. In the later case it was found that very important results arise from the study of the curvature of this function.

Another point is that, due to the strong influence of the Gompertz technique in linearizing the mortality data by taking logarithms for the mortality rate, a large number of studies is directed to analyze the tail of the probability density function and the small amount of the history included in this part of mortality data. Instead the information included in the non-transformed mortality data and especially in the right part of the non-symmetric bell-shape form of the mortality data graph is more essential to understand the results of the unknown aging mechanisms. In the absence of any systematic aging mechanism (internal or external) the longevity tangent is expected to show no-characteristic behavior. Instead by analyzing the females' mortality data from Sweden of last two centuries we observe that the longevity tangent turns clockwise quite systematically. A further parameter analysis suggests a probability of increase of the life span in the future according to some systematic way. We have explored and modeled the variation of the parameters of the model, we have tested the forecasting ability of the model by making predictions for a recent year (2000) by based to two previous time points, 1950 and 1975, and then we have done forecasts for 2025 and 2050.

Now it looks that with our systems of life care, the maximum death rate in the short run will face a limit close to 100 years of age, at least for the case of females in Sweden studied. If it is so the tangent line in the longevity area will be more and more vertical during the years and the form of the first exit time density function at the maximum area more sharp following the improvement of the living standards. Instead, if our destiny is not driven by an internal mechanism the tangent lines should move parallel to the right. It remains to the improvement of the DNA and gene science and the related scientific techniques to improve our maximum life span.

Acknowledgments: 1) The data used can be downloaded from the Human Mortality Database at: http://www.mortality.org or from the statistical year-books of the countries studied. 2) The models, methods and fitting techniques and related programs can be downloaded from: http://www.cmsim.net

References

1. B. L. Strehler, A.S. Mildvan, *Science* 132, 14 (1960).
2. J. Janssen, C. H. Skiadas, C. H, *Appl. Stoc. Mod. Data An.* 11, 35 (1995).
3. C, Skiadas, C. H. Skiadas, *Com. Stat.*, 39, 444 (2010).
4. C. H. Skiadas, C. Skiadas, A modeling approach to life table data. in *Recent Advances in Stochastic Modeling and Data Analysis*, C. H. Skiadas, Ed. (World Scientific, Singapore, 2007), 350–359.
5. B. Gompertz, *Phil. Trans. Royal Soc. London A* 115, 153 (1825).
6. J. S. Weitz, H. B. Fraser, *Proc. Natl. Acad. Sci. USA*, 98(26),15383 (2001).
7. E. Schrödinger, *Phys. Zeit.*, 16, 289 (1915).
8. M. Smoluchowsky, *Phys. Zeit.* 16, 318 (1915).
9. C. H. Skiadas, C. Skiadas, Comparing the Gompertz Type Models with a First Passage Time Density Model in *Advances in Data Analysis*, C. H. Skiadas Ed. (Springer/Birkhauser, Boston 2010), 203-209.
10. J. Carey, P. Liedo, D. Orozco, J. Vaupel, *Science* 258, 457 (1992).
11. J. M. Robine, K. Ritchie, *Science* 260, 1665 (1993).
12. L. A. Gavrilov, N. S. Gavrilova, *Science*, 260, 1565 (1993).
13. S. Horiuchi, J. R. Wilmoth, *Demography*, 35, 391 (1998).
14. J. Oeppen, J. W. Vaupel, *Science*, 296, 1029 (2002).

Composing chaotic music from the letter m

Anastasios D. Sotiropoulos

School of Music, Composition-Theory Division
University of Illinois at Urbana-Champaign, USA
Email: asotiro2@illinois.edu

Abstract: Chaotic music is composed from a proposed iterative map depicting the letter m, relating the pitch, duration and loudness of successive steps. Each of the two curves of the letter m is based on the classical logistic map. Thus, the generating map is $x_{n+1} = r\, x_n(1/2 - x_n)$ for x_n between 0 and 1/2 defining the first curve, and $x_{n+1} = r\,(x_n - 1/2)(1 - x_n)$ for x_n between 1/2 and 1 representing the second curve. The parameter r which determines the height(s) of the letter m varies from 2 to 16, the latter value ensuring fully developed chaotic solutions for the whole letter m; r = 8 yielding full chaotic solutions only for its first curve. The m-model yields fixed points, bifurcation points and chaotic regions for each separate curve, as well as values of the parameter r greater than 8 which produce inter-fixed points, inter-bifurcation points and inter-chaotic regions from the interplay of the two curves. Based on this, music is composed from mapping the m- recurrence model solutions onto actual notes. The resulting musical score strongly depends on the sequence of notes chosen by the composer to define the musical range corresponding to the range of the chaotic mathematical solutions x from 0 to 1. Here, two musical ranges are used; one is the middle chromatic scale and the other is the seven- octaves range. At the composer's will and, for aesthetics, within the same composition, notes can be the outcome of different values of r and/or shifted in any octave. Compositions with endings of non-repeating note patterns result from values of r in the m-model that do not produce bifurcations. Scores of chaotic music composed from the m-model and the classical logistic model are presented.
Keywords: algorithmic composition, chaotic music, m-model music, double logistic map.

1. Introduction

Algorithmic music composition has been investigated in the literature for several decades. Examples are the monographs of Taube [1] and Miranda [2], the work of Tipei [3] and the paper by Serquera & Miranda [4]. The aesthetic aspects of algorithmic composition have been discussed by Garnett [5], Kretz [6] and Hagan [7], among others. Algorithmic chaotic music has also been dealt with, among others, by Dabby [8] who examined musical variations from a chaotic mapping, Rodet & Vergez [9] who related chaotic signals produced by numerical methods with true musical instrument models, and by Serquera & Miranda [4] who used

Chaos Theory: Modeling, Simulation and Applications
C. H. Skiadas, I. Dimotikalis and C. Skiadas (Eds)

coupled cellular automata to compose algorithmic sound. V. Sotiropoulos & A. Sotiropoulos [10] used a frequency-amplitude triangle mapping to compose chaotic music, while A. Sotiropoulos & V. Sotiropoulos [11] used a varying second order recurrence mapping. Further, V. Sotiropoulos [12] composed chaotic music using an elliptic iterative model and a mapping onto musical notes that resembles the colors of a single or a double rainbow.

In algorithmic music composition, each step is related to the preceding step or steps by some function. Each step involves the pitch, its duration, and its amplitude. To have chaotic music, the relating function must be nonlinear. The present paper deals with first order nonlinear iterative maps, so that each step is related directly only to the preceding step, and only the initial step needs to be picked. Such maps have been used in the literature not only in composing algorithmic music but also in simulating phenomena in social sciences. The most widely used has been the classical logistic map, $x_{n+1} = r\, x_n\, (1-x_n)$ with r a free parameter, introduced by May [13] as a discrete-time demographic model. Other examples of first order iterative maps generating chaos include the elliptic map, $x_{n+1} = r\, x_n^{1/2}\, (1-x_n)^{1/2}$, used by V. Sotiropoulos [12] to compose music, and the logistic-like maps, $x_{n+1} = r\, x_n^{\lambda}\, (1-x_n)^{\mu}$ with λ and μ positive, used by D. Sotiropoulos [14]. More on chaotic maps can be found in the book of Skiadas and Skiadas [15].

In the present study, a new first order iterative model is introduced to produce chaotic music. It is the m-model, resembling the letter m, with each of its two curves being the classical logistic map. The map is the analogue of the continuous double logistic map which has found applications in the field of biology, and, specifically in the growth of bones, e.g. Hunt and Hatch [16]. The map of the present paper is defined by the first step and a free parameter which determines whether chaos is produced. Values of the free parameter are obtained that produce fixed points, bifurcations or chaos either from each separate curve of the letter m or from the interplay of the two curves. Based on this, music is composed from mapping the m- iterative model solutions onto actual notes. The resulting musical score strongly depends on the sequence of notes chosen by the composer to define the musical range corresponding to the range of the chaotic mathematical solutions x from 0 to 1. Here, two musical ranges are used; one is the middle chromatic scale and the other is the seven- octaves range. At the composer's will and, for aesthetics, within the same composition, notes can be the outcome of different values of r and/or shifted in any octave. Compositions with endings of non-repeating note patterns result from values of r in the m-model that do not produce bifurcations. The effect of the choice of the initial pitch is also examined. Scores of chaotic music composed from the m-model and the classical logistic model are presented.

2. The mathematical chaotic pitches

In this study, a new first order iterative map is introduced as a proposition to generate a sequence of mathematical pitches. The map is defined as

$$x_{n+1} = r\, x_n\, (1/2 - x_n), \qquad 0 \leq x_n \leq 1/2, \quad r \leq 16 \tag{1a}$$

$$= r\, (x_n - 1/2)\, (1 - x_n), \qquad 1/2 \leq x_n \leq 1, \quad r \leq 16 \tag{1b}$$

where x_n is the mathematical pitch at step n, and x_{n+1} is the succeeding mathematical pitch so that only the initial pitch and the parameter r need to be picked by the composer. The correspondence between mathematical pitches and actual notes will be discussed in section 3. Similarly, the loudness and duration of each note can be obtained from the iterative model given by eqs. (1a) and (1b) with, generally, a different value of r. As seen in Figure 1 where eqs. (1a) and (1b) are plotted for r=16, the map chosen resembles the letter m; that's why it will be called the m-model. The largest value of the parameter r is 16 so that the largest mathematical pitch is 1. The maximum r is obtained by setting equal to one in eq. (1a), the pitch succeeding x=1/4. Full chaos will be achieved with r=16 since the curves are fully developed. Each of the two curves of the m-model is the classical logistic map, which is also plotted in Figure 1 fully developed (r=4) over the whole range [0,1].

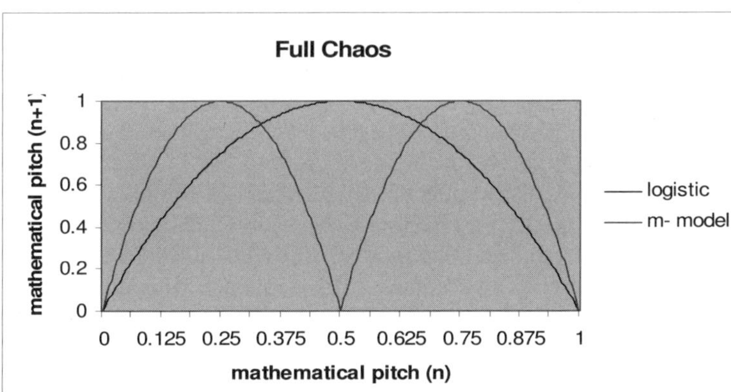

Fig. 1. The fully developed m-model (r=16) and logistic (r=4) iterative maps

It is remarked that the iterative m-model introduced here, is the analogue of the continuous double logistic model,

$$dx/dt = r\, x\, (1/2 - x), \qquad 0 \leq x \leq 1/2, \quad r \leq 16 \tag{2a}$$

$$= r\,(x - 1/2)\,(1 - x), \quad 1/2 \le x \le 1, \quad r \le 16 \tag{2b}$$

which yields as solution the double sigmoid curve, see for example, [16].

The value of r in the m-iterative model, determines whether the sequence of mathematical pitches is patterned and what the nature of the composition's ending is; repeated note(s) or chaotic notes. For each value of r, the mathematical pitch sequence x is computed from eqs. (1a) and (1b). It is noted that if we divide r by 2 and multiply x by 2 then the first curve of the m-iterative model becomes the classical logistic iterative model, $x_{n+1} = r\, x_n\, (1 - x_n)$, over the whole range of x. The behavior of the generated x's from the logistic iterative model is well known, see for example, [15]. It is also noted that if r of the m-model is smaller than 8, the resulting x is smaller than 1/2, meaning that only the first curve takes part in the iteration process. Therefore, when r of the m-model is between 2 and 4, the composition will quickly approach the note corresponding to the fixed point, $1/2\text{-}1/r$, independent of the initial pitch. When r is between 4 and 6, it will eventually approach the same note, fluctuating around it for some time. r=6 is the start of bifurcation. For r between 6 and $2(1+\sqrt{6})$, the composition will permanently oscillate between two notes, almost independent of the initial note. Values of r between $2(1+\sqrt{6})\approx 6.9$ and 7.08 will produce musical scores with endings oscillating between four notes. Increasing further the r value will also increase the number of oscillating notes towards the end of the composition. But for most values of r between 7.14 and 8, the composition will be chaotic and very much affected by the choice of initial note.

When r is larger than 8, the second curve of the m-model comes into play as well. If r is between 8 and $6+4\sqrt{2}\approx 11.66$, the minimum r value that yields a fixed pitch from the second curve as solution of $x=r(x-1/2)(1-x)$, the composition will be interchanging between chaotic and periodic with increasing r, because of the interplay of the two curves. For r between $6+4\sqrt{2}$ and $6+4\sqrt{3}\approx 12.93$, the minimum r value that yields bifurcated pitches from the second curve as solution of $x= f(f(x))$ with $f(x)=r(x-1/2)(1-x)$, the composition will tend to one repeating pitch. Between the r values of $6+4\sqrt{3}$ and 14.03 the composition will tend to an increasingly periodic pitch sequence with increasing r, while between the r values of 14.03 and 16, a chaotic pitch sequence will be generated. Conclusively, chaotic compositions will result from the iterative m-model introduced here for most values of r in the range (7.14, 11.66) and (14.03, 16).

For r=15.5, the chaotic pitch sequence generated from the iterative m-model is shown in Figure 2 for two different initial pitches, x_1=0.1, 0.05.

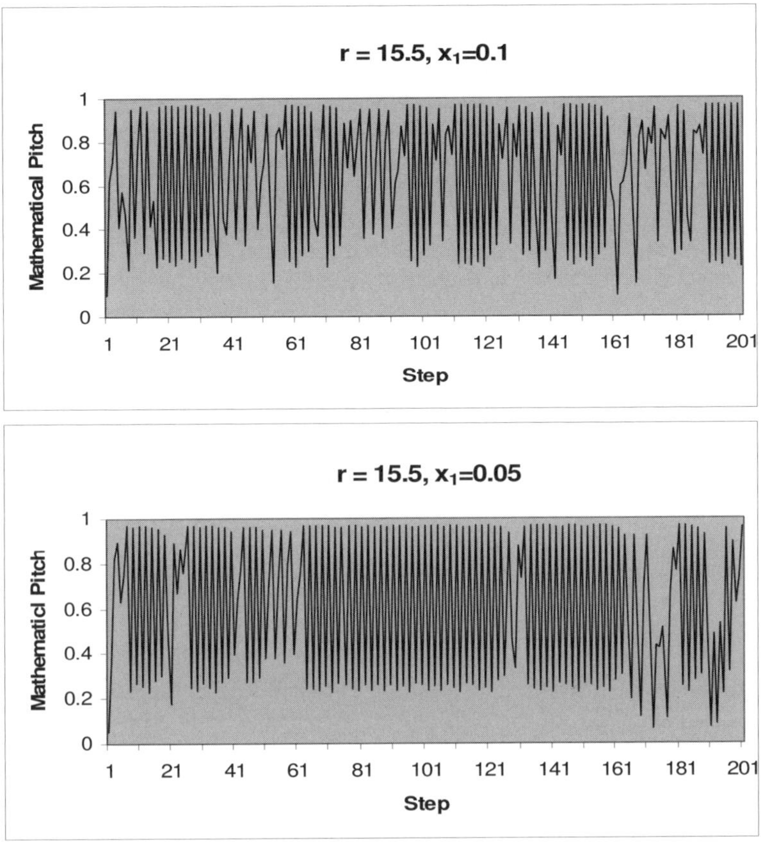

Fig. 2. The chaotic pitch sequence generated by the iterative m-model

The top graph of Figure 2 shows the generated pitch sequence for an initial pitch equal to 0.1, while the bottom sketch depicts the computed pitch sequence for an initial pitch equal to 0.05. As will be discussed in section 3, the 0.1 pitch is the musical note $C^{\#}/D^{b}$ or $G^{\#}/A^{b}$, depending on whether one octave or seven octaves correspond to the whole mathematical pitch range, [0,1]. It is seen that the pitch sequence generated from the iterative m-model is chaotic, as expected from the discussion above. Moreover, it is observed that the chaotic composition is very much affected by the choice of initial pitch.

3. The musical pitches

The mathematical pitch range of the m-model considered goes from 0 to 1. This range can be mapped in any fashion onto the actual musical pitch range producing different compositions. Here, in a straight forward manner, each ascending interval of the mathematical range is associated with an ascending note of either one octave, the middle chromatic scale, or seven octaves, making up the musical range. In the case of a single octave with 12 different notes, the range from 0 to 1 is divided into 12 equal intervals corresponding to the musical range [C,B]. For seven-octaves, the mathematical range is divided into 84 equal intervals to correspond to the musical range [C_1, B_7], the subscript denoting the octave number. An example of how the full chaotic (r=16) m-model affects the succeeding note of a composition for the two musical ranges, is shown in Figure 3.

one octave: **A** (x=0.75$^-$) or **G**$^\#$ (x=0.75$^-$) or **D**$^\#$ (x=0.25$^+$) or **D** (x=0.25$^-$) → **B** (x=1$^-$) → ...

seven octaves:
$\mathbf{C}^\#_5$ (x≈0.586) or $\mathbf{E}_7$ (x≈0.914) or $\mathbf{B}_4$ (x≈0.57) or $\mathbf{G}_1$ (x≈0.086) → $\mathbf{A}^\#_3$ (x≈0.414) → ...

Fig. 3. Two consecutive two notes of fully chaotic compositions in two musical ranges from the m-model

It is observed, as expected from the shape of the iterative m-model, that four different initial notes will generate the same second note within each musical range.

The iterative m-model introduced in this study, generates music that is distinctly different from the music of the classical logistic iterative map. Figure 4 shows two examples of the difference for two consecutive notes generated by the two iterative maps for the single octave musical range.

i) m-model: **A** (x=0.75$^-$) or **G**$^\#$ (x=0.75$^-$) or **D**$^\#$ (x=0.25$^+$) or **D** (x=0.25$^-$) → **B** (x=1$^-$) → ...
logistic model: **F** (x=0.5$^-$) or **F**$^\#$ (x=0.5$^+$) → **B** (x=1$^-$) → ...

ii) m-model: **F** (x=0.5-) or **F**$^\#$ (x=0.5+) or **B** (x=1$^-$) → **C** (x=0$^+$) → ...
logistic model: **B** (x=1$^-$) → **C** (x=0$^+$) → ...

Fig. 4. Two consecutive notes of fully chaotic compositions from two iterative models

It is seen that the effect of the initial note on the composition is strong for both iterative models and stronger for the m-model. The same note in the compositions from the two models, is generally preceded by a different note or notes.

Three exemplary compositions from the m-model are considered next, having a periodic sequence of a different number of notes after the first tens of notes. In Figure 5, the repeating notes and their sequence involved in the oscillations are shown in the two musical ranges for each of the three r's.

i) r=9.15:	**D$^{\#}$** (x=0.251) → **F$^{\#}$** (x=0.572) → **D$^{\#}$** (x=0.282) → **F$^{\#}$** (x=0.563)	one octave
	A_2 (x=0.251) → **C_5** (x=0.572) → **B_2** (x=0.282) → **B_4** (x=0.563)	seven octaves
ii) r=13:	**A** (x=0.769) → **A** (x=0.808)	one octave
	E_6 (x=0.769) → **G_6** (x=0.808)	seven octaves
iii) r=14:	**G** (x=0.657) → **A** (x=0.754) → **A$^{\#}$** (x=0.875)	one octave
	G_5 (x=0.657) → **$D^{\#}_6$** (x=0.754) → **$C^{\#}_7$** (x=0.875)	seven octaves

Fig. 5. The sequence of repeating notes in two musical ranges from periodic m-models

It is observed that for a fixed m-model (same r), the number of repeating notes and the notes themselves in the ending of the composition depend on the musical range picked. When r=9.15, there are either two notes repeating, D$^{\#}$, F$^{\#}$ or four, A_2, C_5, B_2, B_4, depending on the musical range. When r=13, there is one repeating note, A, in the single octave range but two repeating notes, E_6, G_6, in the seven octaves range. For r=14, the composition ends with three repeating notes, G, A, A$^{\#}$ or G_5, $D^{\#}_6$, $C^{\#}_7$, in either musical range.

4. The musical scores

Based on the iterative m-model with r=15.5 and initial mathematical pitch equal to 0.1, whose resulting mathematical pitches were shown in the top diagram of Figure 2, I composed two musical scores by mapping the mathematical pitches onto musical notes with a range of one octave and seven octaves, respectively. The duration, and loudness of each note I picked at will. In the case of the single octave range, I also freely shifted the octave of the resulting note, for aesthetic purposes. Next, I transcribed the notes for each composition in the Finale software program. The two scores are shown below with the top score corresponding to the *free* 'one' octave range and the bottom score to the seven octaves range.

Score 1

Score 2

5. Conclusions

A new iterative model, the m-model (double logistic), was proposed in this study to generate chaotic music. It was shown that the composition is different from that generated by the classical logistic map and it very much depends on the choice of the musical range that the mathematical range is mapped onto. It was also shown that the composition depends on the initial pitch chosen by the composer. The musical scores produced in this study were only intended as exemplary parts of complete compositions and to demonstrate the abundance of scores that the proposed algorithmic model can generate. At the composer's will, parts of the same composition may be the outcome of a different height (r value) of the m-model and/or initial pitch. The aesthetic intervention of the composer plays a significant role in the final outcome of a complete musical score, as in the case of any algorithmically generated music.

6. References

[1] H. K. Taube. *Notes from the metalevel: an introduction to algorithmic music composition. Studies in New Music Research*, Routledge, N. York, 2004.
[2] E. Miranda. *Composing music with computers.* Bath Press, Bath, UK, 2004.
[3] S. Tipei. Manifold compositions: formal control, intuition, and the case for comprehensive software. *Proceedings 2005 International Computer Music Conference*, Copenhagen, Denmark, 2007.
[4] J. Serquera and E. Miranda. Algorithmic sound composition using cellular automata. In *Chaos Theory: Modeling, Simulation and Applications*, C. H. Skiadas, I. Dimoticalis and C. Skiadas, eds. World Scientific, 2011.
[5] G. E. Garnett. The aesthetics of interactive computer music. *Computer Music Journal,* v. 25, 21-33, 2001.
[6] J. Kretz. Freedom and necessity in computer aided composition: a thinking framework and its application. In *Chaos Theory: Modeling, Simulation and Applications*, C. H. Skiadas, I. Dimoticalis and C. Skiadas, eds. World Scientific, 2011.
[7] K. Hagan. Aesthetic consideration in algorithmic and generative composition. In *Chaos Theory: Modeling, Simulation and Applications*, C. H. Skiadas, I. Dimoticalis and C. Skiadas, editors. World Scientific, 2011.
[8] D. S. Dabby. *Musical variations from a chaotic mapping*. PhD thesis. MIT, 1995.
[9] X. Rodet and C. Vergez. Nonlinear dynamics in physical models: from basic models to true musical instrument models. *Computer Music Journal*, v. 23, no.3: 35-49, 1999.
[10] V. D. Sotiropoulos and A. D. Sotiropoulos. Music composition from the cosine law of a frequency-amplitude triangle. In *Chaotic Systems: Theory and Applications*, ed. by C. H. Skiadas and I. Dimotikalis, World Scientific, 2010.
[11] A. D. Sotiropoulos and V. D. Sotiropoulos. Composing chaotic music from a varying second order recurrence equation. In *Chaotic Systems: Theory and Applications*, ed. by C. H. Skiadas and I. Dimotikalis, World Scientific, 2010.
[12] V. D. Sotiropoulos. The rainbow effect in composing chaotic algorithmic music. In *Chaos Theory: Modeling, Simulation and Applications*, C. H. Skiadas, I. Dimoticalis and C. Skiadas, eds. World Scientific, 2011.
[13] R. M. May. Simple mathematical models with very complicated dynamics." *Nature*, v. 261: 459-467, 1976.
[14] D. A. Sotiropoulos. On logistic-like iterative maps. In *Chaos Theory: Modeling, Simulation and Applications*, C. H. Skiadas, I. Dimoticalis and C. Skiadas, eds. World Scientific, 2011.
[15] C. H. Skiadas and C. Skiadas. *Chaotic Modelling and Simulation*. CRC Press, N. York, 2009.
[16] E. E. Hunt and J. W. Hatch. The estimation of age at death and ages of formation of transverse lines from measurements of human long bones. *American J. Physical Anthropology*, v. 54: 461-469, 1981.

On the timbre of chaotic algorithmic sounds

Dimitrios A. Sotiropoulos, Anastasios D. Sotiropoulos* and Vaggelis D. Sotiropoulos*

Department of Sciences, Technical University of Crete, Greece
Email: dimsotirop@yahoo.com

*School of Music, Composition-Theory Division, University of Illinois at Urbana-Champaign, USA
Emails: asotiro2@illinois.edu, sotirop1@illinois.edu

Abstract: Chaotic sound waveforms generated algorithmically are considered to study their timbre characteristics of harmonic and inharmonic overtones, loudness and onset time. Algorithms employed in the present work come from different first order iterative maps with parameters that generate chaotic sound waveforms. The generated chaotic sounds are compared with each other in respect of their waveforms' energy over the same time interval. Interest is focused in the logistic, double logistic and elliptic iterative maps. For these maps, the energy of the algorithmically synthesized sounds is obtained numerically in the chaotic region. The results show that for a specific parameter value in the chaotic region for each one of the first two maps, the calculated sound energy is the same. The energy, though, produced by the elliptic iterative map is higher than that of the other two maps everywhere in the chaotic region. Under the criterion of equal energy, the discrete Fourier transform is employed to compute for the logistic and double logistic iterative maps, a) the generated chaotic sound's power spectral density over frequency revealing the location (frequency) and relative loudness of the overtones which can be associated with fundamental frequencies of musical notes, and b) the generated chaotic sound's frequency dependent phase, which together with the overtones' frequency, yields the overtones' onset time. It is found that the synthesized overtones' loudness, frequency and onset time are totally different for the two generating algorithms (iterative maps) even though the sound's total generated power is equal. It is also demonstrated that, within each one of the iterative maps considered, the overtone characteristics are strongly affected by the choice of initial loudness.
Keywords: chaotic sound, sound synthesis, sound analysis, algorithmic music, chaotic maps, music timbre, sound spectrum, power density, phase, overtones, partials.

1. Introduction

The present work is concerned with the timbre characteristics of algorithmically generated chaotic sounds as a stepping stone towards synthesizing music with pre-determined features. It is well known that timbre is that characteristic of sound which distinguishes sounds of the same duration, pitch and loudness (see, for example, Fletcher & Rossing [1]). Timbre describes the quality of sound,

Chaos Theory: Modeling, Simulation and Applications
C. H. Skiadas, I. Dimotikalis and C. Skiadas (Eds)

distinguishing the same notes produced by different musical instruments. In the time domain, timbre may be thought of as the sound's waveform shape. However, the timbre characteristics are revealed in the sound's Fourier spectrum and are associated with partial harmonic or inharmonic overtones, their location (frequency), loudness and onset time. Overtone loudness and frequency are included in the magnitude of the sound's Fourier transform, while overtone onset time is calculated from the phase of the sound's Fourier transform.

Several works in the literature have dealt with a sound's timbre and with synthesizing music of set timbre characteristics. When playing a note, the panflute comes very close to producing a pure sine waveform, thus, having only a fundamental frequency (pitch) and no partial overtones. Wind instruments produce square waveforms which include only odd harmonics, while string instruments produce sawtooth waveforms whose timbre includes both odd and even harmonics (see, for example, [2]). Algorithmic music composition and sound generation has been of wide interest for several decades; see, for example, the monographs of Taube [3] and Miranda [4], the work of Tipei [5] and the paper by Serquera & Miranda [6]. The aesthetic aspects of algorithmic composition have been discussed by Garnett [7], Kretz [8] and Hagan [9], among others. The frequency content of sounds produced by musical instruments was analyzed by Alm & Walker [10] using Fourier analysis, spectrograms and scalograms. Yasuda & Hama [11] worked on synthesizing the timbre of stringed instruments with a set structure.

Chaotic sound or music produced algorithmically has also been researched in the literature. Dabby [12] examined musical variations from a chaotic mapping, Rodet & Vergez [13] related chaotic signals produced by numerical methods with true musical instrument models, while Serquera & Miranda [6] used coupled cellular automata to compose algorithmic sound. To compose chaotic music, V. Sotiropoulos & A. Sotiropoulos [14] used a frequency-amplitude triangle mapping, while A. Sotiropoulos & V. Sotiropoulos [15] used a varying second order recurrence mapping. Further, V. Sotiropoulos [16] used an elliptic iterative mapping and the rainbow effect whereas A. Sotiropoulos [17] used a double logistic iterative mapping, to compose chaotic music. Chaotic signals, in general, were analyzed by Liu & Liao [18] using dual-lifting wavelet transform. Very recently, Zech et al [19] presented the magnitude of the power spectrum of chaotic signals generated from the logistic iterative map, based on the complex network approach of Marwan et al [20] for the chaotic signal.

In the present paper, the timbre overtone characteristics of pitch, loudness, and onset time, are examined for chaotic sound waveforms generated by algorithms of first order iterative maps. Particular interest is focused in the logistic, double logistic and elliptic iterative maps. The generated chaotic sounds from the different maps are compared with each other in respect of their waveforms' energy over the same time interval. The energy is obtained numerically in the chaotic region. For chaotic sounds with equal energy from the logistic and double logistic iterative maps, the Discrete Fourier Transform is used to compute and compare the generated sound's power spectral density over frequency revealing the location (frequency) and relative loudness of the overtones. The fundamental frequency (pitch) of musical notes is then associated with the frequency of partial overtones. We also compute and compare the generated sound's frequency dependent phase, which together with the overtones' frequency, yields the overtones' onset time. Last, the effect of initial loudness on the overtone characteristics, within an iterative map, is considered.

2. The sound generating maps

The chaotic sounds considered are generated from first order iterative maps. In the present study, particular attention is paid on the chaotic sound signals produced from three specific iterative maps, the logistic, the double logistic, and the elliptic, defined by the following recurrence equations:

$$\text{logistic:}\quad x_{n+1} = r\, x_n\,(1 - x_n),\quad 0 \le x_n \le 1,\ \ r \le 4 \tag{1}$$

$$\text{double logistic:}\quad x_{n+1} = r\, x_n\,(1/2 - x_n),\quad 0 \le x_n \le 1/2,\ \ r \le 16 \tag{2a}$$

$$= r\,(x_n - 1/2)\,(1 - x_n),\quad 1/2 \le x_n \le 1,\ \ r \le 16 \tag{2b}$$

$$\text{elliptic:}\quad x_{n+1} = r\, x_n^{1/2}\,(1 - x_n)^{1/2},\quad 0 \le x_n \le 1,\ \ r \le 2 \tag{3}$$

where x_n denotes the sound loudness at time step n. The maximum value of loudness at any step is set equal to one limiting the value of the free parameter r as 4, 16, and 2 respectively for the three maps. Depending on the value of r, the above iterative maps may produce steady state or chaotic signals. Full chaos is generated by the corresponding maximum r value for each map, for which the three maps are depicted graphically in Figure 1 as

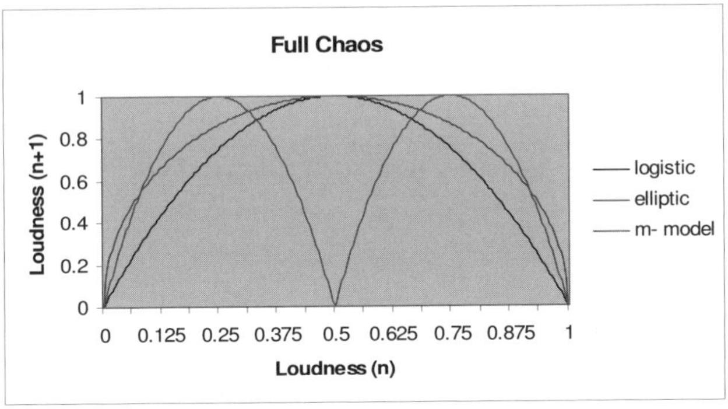

Fig. 1. The fully developed iterative maps

In Figure 1, the double logistic map is named the m-model, after one of the authors of the present paper A. Sotiropoulos [17], as its shape resembles the letter m. It is known that the logistic map produces chaotic signals for most values of r above 3.6 (see, for example, Skiadas and Skiadas [21]). For the double logistic map, A. Sotiropoulos [17] found that chaotic signals are generated for most values of r above 14, while one of the authors of the present paper V. Sotiropoulos [16] found that the elliptic map produces chaotic signals for values of r above 1.95. Examples of chaotic signals have been given in [21], [17] and [16] for the logistic, double logistic (m-model) and elliptic map respectively.

3. The sound's energy

One of the goals of the present study is to compare chaotic sounds generated by different maps. The comparison can be made between arbitrary maps. However, here the generated sound signals will be compared on physical grounds, having the same energy. Inversely, this criterion will define the generating maps. The energy, E, of the generated sound is given by

$$E = \sum_{n=1}^{N} |x_n|^2 \tag{4}$$

as the sum of the squared loudness at each time step, n, where N denotes the total number of time steps. Since the energy depends on the duration of the sound signal, dividing it by the duration, T, gives

$$P = E / T \tag{5}$$

the power (energy per unit time) of the sound signal to be used for comparison.

For the three iterative maps defined by eqs. (1), (2) and (3) and for 11 values of their parameter r ranging from 3.5 to 4 for the logistic, from 15 to 16 for the double logistic, and from 1.9 to 2 for the elliptic, the chaotic sound signals are computed for 512 time steps. This range of r includes values for which the maps generate chaotic sound signals, as discussed in section 2. The energy per unit time (power), of the generated sound is subsequently calculated using eq. (4). The results are shown collectively in Figure 2 by using a single integer index r, varying from 1 to 11, to correspond to the 11 values of the actual parameter r for each of the three maps under consideration.

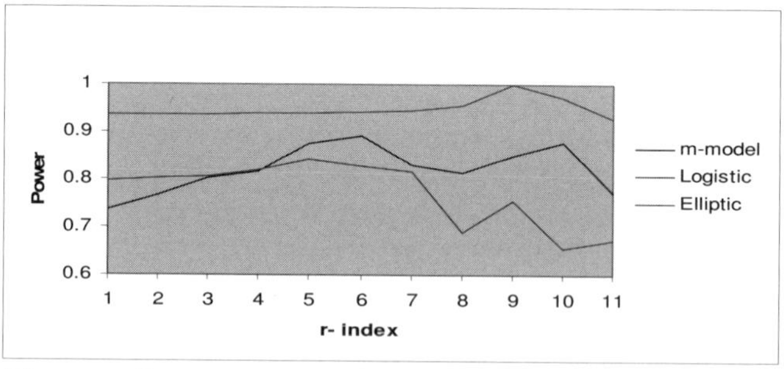

Fig. 2. The sound's energy per unit duration vs. the mapping parameter r

It is noted that the initial loudness picked has no effect in the computed power, as is predicted theoretically. This was tested by using as initial loudness in the second run that of the 513th time step of the first run, initial loudness of the third run that of the 513th time step of the second run, and so on. It is observed that for all the values chosen for the parameter r, the power of the sound generated from the elliptic map is higher than that of the sound generated from the logistic or the m-model (double logistic) map. The latter two maps produce chaotic sound signals that have the same energy per unit duration when r=3.65 in the logistic map and r=15.3 in the double logistic map. For these two specific maps, the generated chaotic sound's overtone characteristics will be examined below by computing the magnitude and phase of the sound's power spectrum.

4. The sound's overtones

For the logistic and double logistic (m-model) maps, the overtones of the generated chaotic sound signals are identified as the frequency peaks in the magnitude of the sound's power spectrum. The magnitude of the power spectrum is given by the squared absolute value of the discrete Fourier transform of the sound signal in time, that is, the squared loudness at each frequency. The overtone loudness at each frequency step, k, is given as

$$|X_k| = |\sum_{n=1}^{N} x_n \exp(-2\pi ikn / N)|, \quad k=1, 2, \ldots, N \tag{6}$$

The frequency, f_k, at each step k is given by

$$f_k = k / (N \Delta t), \quad k = 1, \ldots, N \tag{7}$$

where Δt is the time interval between successive steps in time. The frequency interval between successive frequencies is $1/(N\Delta t)$ and the bandwidth is $1/\Delta t$. A typical time interval, Δt, for synthesized musical sounds is 1 msec. Then, the spectrum's bandwidth is 1 kHz and, for N=512, the frequency interval is 1.953 Hz.

For the logistic and double logistic (m-model) maps with r=3.65 and r=15.3 respectively, the loudness of the generated sound at each frequency is computed using the Fast Fourier Transform method on eq. (6), once the discrete sound signal, x_n, has been obtained in time. The chaotic sound signals in time generated from the two iterative maps are shown in Figure 3 where Δt=1msec, N=512 and the initial loudness is 0.1.

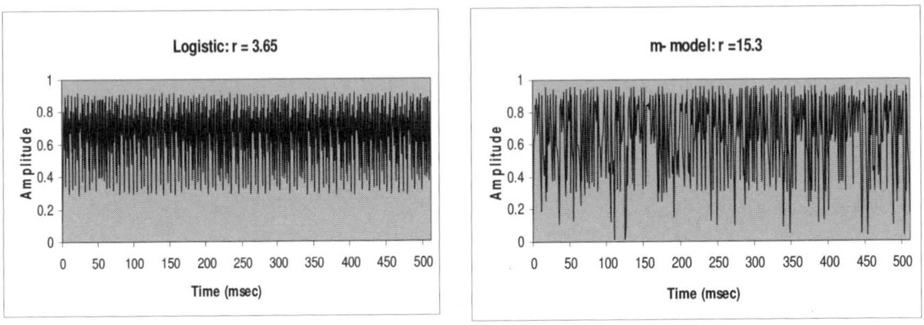

Fig. 3. Equal power chaotic sound signals from the logistic and m iterative maps

It is observed that the generated sound signals from the two iterative maps, the logistic and m-model, have a distinctly different chaotic shape.

The spectral density, defined as the ratio of the squared overtone loudness to the signal power, is shown in Figure 4 where the fundamental frequency occurring at the lowest frequency of almost 2 Hz is not shown as this frequency has no application in musical sounds; the human ear not been able to pick up frequencies below 16 Hz. The squared overtone loudness is scaled relative to the largest partial overtone, whose squared loudness is then equal to one.

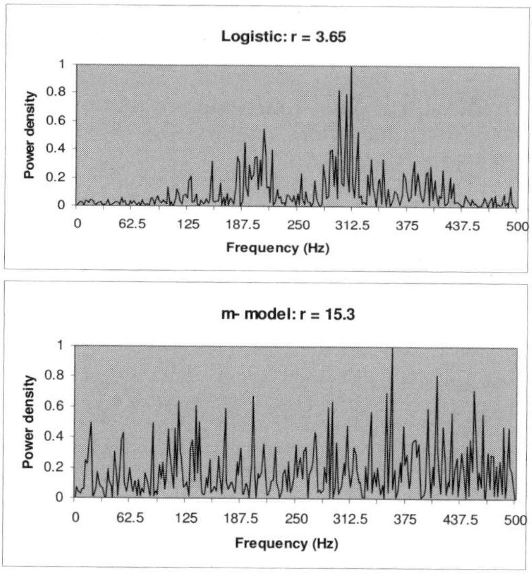

Fig. 4. The squared overtone loudness relative to the largest overtone

It is observed that the chaotic overtone content of the two equal power sounds is very different. The number of overtones from the double logistic (m-model) map with similar loudness is much larger than that from the classical logistic map. The loudest overtone from the logistic map occurs at 311 Hz while that from the double logistic map at 359 Hz. These partial overtones can be associated with the fundamental frequency (pitch) of the musical notes $D^{\#}_4$ and F_4 respectively, as can also be the other overtones.

5. The sound's phase

The sound's frequency dependent phase is important in determining the overtone onset time, since the overtone onset time, t_k, is given by

$$t_k = \varphi_k / (2\pi f_k) \tag{8}$$

where the overtone phase, φ_k, is calculated from the X_k of eq. (6) as

$$\varphi_k = \tan^{-1} [\mathrm{Im}(X_k) / \mathrm{Re}(X_k)] \tag{9}$$

For the equal power chaotic sound signals generated by the logistic map with r=3.65 and the m-model map with r=15.3, the overtone phase is computed from eqs. (9) and (6) with Δt=1msec, N=512, and initial loudness x_n=0.1. The phase's principal value between -π and π obtained as the atan2 function from computer programming, is shown in Figure 5.

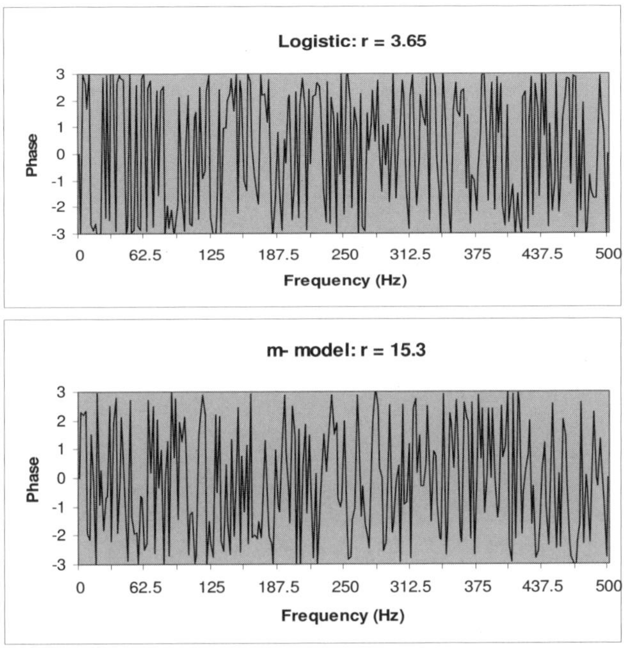

Fig. 5. The overtones' frequency dependent phase

The frequency dependent phases from the two maps are very different and they exhibit strong variations with frequency, implying that particular care must be taken in calculating the overtones' onset time from eq. (8) where the actual phase must be used and not its principal value. For the loudest overtones at 311 Hz for the logistic map and at 359 Hz for the double logistic map, the results of Figure 5 together with eq. (8) after carefully considering the jumps in phase values, yield the onset times of 124 msecs and 140 msecs, respectively.

6. Effect of initial loudness

The effect of initial loudness on the partial overtones of the chaotic sounds generated by the two iterative maps, the logistic and the double logistic (m-model) is now examined. In the first run, the initial loudness is the same as before, equal to 0.1. The computed squared overtone loudness over the frequency spectrum has been presented above in Figure 4, scaled with respect to the largest overtone excluding the fundamental pitch. For comparison with a different initial loudness, it is depicted again with the dark (blue) line in Figure 6. The initial loudness picked for the second run is the 513th loudness of the corresponding iterative map with initial loudness equal to 0.1. This loudness is equal to 0.906 for the logistic map and 0.889 for the m-model map. The computed squared overtone loudness over the frequency spectrum, scaled with respect to the largest overtone excluding the fundamental pitch, is drawn with a light (yellow) line in Figure 6 below.

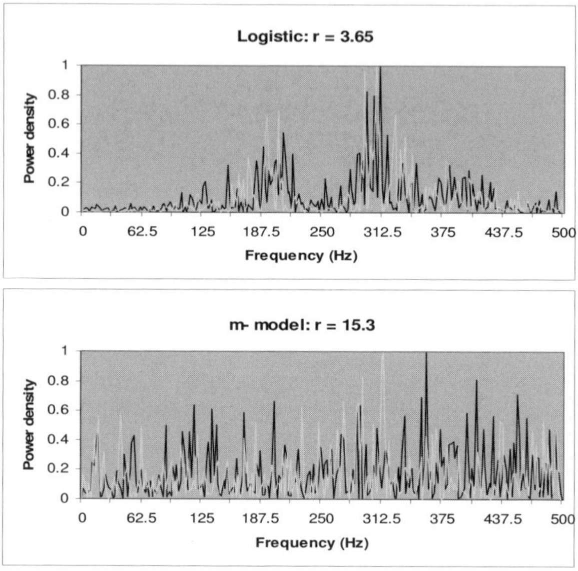

Fig. 6. Effect of initial loudness, x_1, on overtones' scaled squared loudness.

Dark (blue) line: $x^{(1)}_1 = 0.1$, light (yellow) line: $x^{(2)}_1 = x^{(1)}_{513}$

The generated overtones exhibit a dependence on the initial loudness of the iterative map, even though the energy per unit time (power) of the generated sound is the same in both cases. It is observed that the overtones' dependence on initial loudness is stronger for the m-model than for the classical logistic model.

7. References

[1] N. H. Fletcher and T. D. Rossing. *The Physics of Musical Instruments*. Springer Science and Business Media, Inc., 1998.

[2] *http://www.allaboutcircuits.com/vol_2/chpt_7/4.html*

[3] H. K. Taube. *Notes from the metalevel: an introduction to algorithmic music composition. Studies in New Music Research*, Routledge, N. York, 2004.

[4] E. Miranda. *Composing music with computers*. Bath Press, Bath, UK, 2004.

[5] S. Tipei. Manifold compositions: formal control, intuition, and the case for comprehensive software. *Proceedings 2005 International Computer Music Conference*, Copenhagen, Denmark, 2007.

[6] J. Serquera and E. Miranda. Algorithmic sound composition using cellular automata. In *Chaos Theory: Modeling, Simulation and Applications*, C. H. Skiadas, I. Dimoticalis and C. Skiadas, eds. World Scientific, 2011.

[7] G. E. Garnett. The aesthetics of interactive computer music. *Computer Music Journal,* v. 25, 21-33, 2001.

[8] J. Kretz. Freedom and necessity in computer aided composition: a thinking framework and its application. In *Chaos Theory: Modeling, Simulation and Applications*, C. H. Skiadas, I. Dimoticalis and C. Skiadas, eds. World Scientific, 2011.

[9] K. Hagan. Aesthetic consideration in algorithmic and generative composition. In *Chaos Theory: Modeling, Simulation and Applications*, C. H. Skiadas, I. Dimoticalis and C. Skiadas, eds. World Scientific, 2011.

[10] J. F. Alm and J. S. Walker. Time-frequency analysis of musical instruments. SIAM Review, v. 44, no. 3: 457-476, 2002.

[11] K. Yasuda and H. Hama. Analysis/Synthesis of stringed instruments using formant structure. *Int. J. Computer Science & Network Security*, v. 7, no. 9: 192-197, 2007.

[12] D. S. Dabby. *Musical variations from a chaotic mapping*. PhD thesis. MIT, 1995.

[13] X. Rodet and C. Vergez. Nonlinear dynamics in physical models: from basic models to true musical instrument models. *Computer Music Journal*, v. 23, no.3: 35-49, 1999.

[14] V. D. Sotiropoulos and A. D. Sotiropoulos. Music composition from the cosine law of a frequency-amplitude triangle. In *Chaotic Systems: Theory and Applications*, ed. by C. H. Skiadas and I. Dimotikalis, World Scientific, 2010.

[15] A. D. Sotiropoulos and V. D. Sotiropoulos. Composing chaotic music from a varying second order recurrence equation. In *Chaotic Systems: Theory and Applications*, ed. by C. H. Skiadas and I. Dimotikalis, World Scientific, 2010.

[16] V. D. Sotiropoulos. The rainbow effect in composing chaotic algorithmic music. In *Chaos Theory: Modeling, Simulation and Applications*, C. H. Skiadas, I. Dimoticalis and C. Skiadas, eds. World Scientific, 2011.

[17] A. D. Sotiropoulos. Composing chaotic music from the letter m. In *Chaos Theory: Modeling, Simulation And Applications*, C. H. Skiadas, I. Dimoticalis and C. Skiadas, eds. World Scientific, 2011.

[18] Y. Liu and X. Liao. Adaptive noise reduction method for chaotic signals using dual-lifting wavelet transform. *IEEE Computer Society. Int. Workshop on Chaos-Fractals Theories & Applications*, 2009.

[19] A. Zech, J. F. Donges, N. Marwan, and J. Kurths. Power spectrum of the logistic map. The Wolfram demonstration project. *http://demonstrations.wolfram.com/PowerSpectrumOfTheLogisticMap,* 2010.

[20] N. Marwan, J. F. Donges, Y. Zou, R. V. Donner, and J. Kurths. Complex Network Approach for Recurrence Analysis of Time Series. *Physics Letters A* 373 (46): 4246–4254, 2009.

[21] C. H. Skiadas and C. Skiadas. *Chaotic Modelling and Simulation*. CRC Press, N. York, 2009.

The rainbow effect on composing chaotic algorithmic music

Vaggelis D. Sotiropoulos

School of Music, Composition-Theory Division, University of Illinois at Urbana-Champaign, USA; Email: sotirop1@illinois.edu

Abstract: The effect of rainbow color sequence on composing chaotic algorithmic music is examined. The mathematical range of the chaotic algorithm is mapped onto musical notes whose sequence follows the sequence of the seven main rainbow colors and their in-between five auxiliary colors. Each musical note is identified with the frequency of a color by a frequency shift. As a result, for a single rainbow, the scale of the chaotic music comprises an ascending chromatic F major scale without the thirteenth note, followed by its corresponding descending chromatic scale, for a total of twenty four notes. For aesthetic purposes, a note can be placed in any octave at the composer's will. The effect of a double rainbow on composing chaotic music is also studied. It is known from nature that the outer bow has its color sequence reversed. Thus, in this case, the double rainbow musical scale comprises forty eight notes on a repeated reversed full chromatic F major scale without the thirteenth note in the ascent or the first note in the descent, resembling in shape the letter w. Colorless regions in the rainbow or dark (Alexander's bands) regions in a supernumerary rainbow are included in the musical range as rests. With the musical scale based on the described rainbow mapping, chaotic music is composed from an algorithm defined by a semi-elliptical first order iterative map. The minor axis of the ellipse is defined by the range of the mathematical pitch from 0 to 1 while the semi-major axis by that of the succeeding pitch from 0 to r/2; r is a free parameter that varies from 1 to 2 to be chosen by the composer. The lower limiting value of the free parameter r corresponds to a circle of radius 1/2 yielding steady state music whereas all the other values of r correspond to ellipses. Chaotic compositions result from r values between 1.95 and 2, the latter value yielding full chaos from an ellipse with its major axis double its minor axis. Fixed notes are obtained for all r's, i.e., notes to be avoided for a prematurely ending composition. Chaotic musical scores are composed from the semi-elliptical iterative map with different initial pitches, on a single or a double rainbow musical scale.

Keywords: chaotic composition, algorithmic music, rainbow musical scale, chromatic music, rainbow music, color spectrum, elliptical music.

1. Introduction

The relation between musical notes and the colors of the visible spectrum was a scientific subject of discussion for more than two millennia since Aristotle. At Renaissance set, Newton in his Opticks (1704) divided the color wheel into seven wedges in the order of the color spectrum, including the color *indigo*, starting from *red* and ending in *violet*, identifying each color with the first seven natural

Chaos Theory: Modeling, Simulation and Applications
C. H. Skiadas, I. Dimotikalis and C. Skiadas (Eds)

musical notes of the ascending C major scale, from D to C. About forty years later, Castel (1742) correlated the colors of the visible spectrum with the descending F major scale. Castel's idea with the sharps or flats corresponding to the auxiliary, in-between, colors found many fans among composers well into the twentieth century, see for example, Scholes [1]. Physical and non-physical based relations have been used between color and music since the 1950's. Examples are Xenakis' compositions *Polytope* (1967) and *Le Diatope* (1978) and Gubaydulina's *Alleluja* (1990) in which color is a basic rhythmic element in the development of the music. A comprehensive historical account is found in Jewanski [2].

In the present study, the seven colors of the visible spectrum as well as their in-between five colors are placed on twelve equal arcs on the circumference of a color wheel in the order of their ascending frequency, that is, from red to red-violet. Based on the frequency of musical notes which is obtained by shifting down the frequency of colors, each color of the visible spectrum is identified with a note. Starting with the color red, the resulting musical range is an ascending chromatic F major scale without its 13th note, G, for a total of twelve notes. However, the purpose of the present paper is to have the musical range correspond to the sequence of rainbow colors. For a single rainbow the colors go from red on the outside from the left and right, to a colorless region in the middle with violet on either side of it. The resulting rainbow musical scale is a full (ascending-descending) chromatic F major scale, from G to $F^{\#}$ and from $F^{\#}$ to G, without the thirteenth note in the ascent or the first note in the descent, for a total of twenty four notes. The colorless region of the rainbow is represented by rest. A double rainbow is also considered since double rainbows do exist in rainy days. In this case, the rainbow musical scale is comprised of forty eight notes on a repeated reversed full chromatic F major scale, without the thirteenth note in the ascent or the first note in the descent, resembling in shape the letter w. It should be remarked that the notes can be in any octave depending on the shift of the colors' frequency. Supernumerary rainbows also exist in nature whose dark regions are known as Alexander's bands, correspond to musical rest. An analogous procedure to the aforementioned one, maps the colors of the supernumerary rainbow onto musical notes.

The goal of this paper is to examine the rainbow effect on composing chaotic algorithmic music. Algorithmic music composition has been investigated in the literature for several decades. Examples are the monographs of Taube [3] and Miranda [4], the work of Tipei [5] and the paper by Serquera & Miranda [6]. The aesthetic aspects of algorithmic composition have been discussed by Garnett [7], Kretz [8] and Hagan [9], among others. Algorithmic chaotic music has also been dealt with, among others, by Dabby [10] who examined musical variations from a chaotic mapping, V. Sotiropoulos & A. Sotiropoulos [11] used a frequency-amplitude triangle mapping to compose chaotic music, and A. Sotiropoulos & V. Sotiropoulos [12] used a varying second order recurrence mapping.

The algorithm used here is based on a first order iterative map that represents a semi-ellipse whose semi-major axis is chosen by the composer while its minor axis defines the mathematical pitch range from 0 to 1. Depending on the length of the major axis, the generated music is either chaotic or patterned. The mathematical pitch range is associated with the single or the double rainbow range of colors which, in turn, is correlated with the musical range as described above. The most widely used first order iterative map to compose chaotic music represents a parabola and is the well known classical logistic map, introduced by May [13]. A. Sotiropoulos [14] used a double logistic map, resembling the letter m, to generate chaotic music. Other first order iterative maps producing chaos are found in the monograph of Skiadas & Skiadas [15] and in D. Sotiropoulos [16].

2. The rainbow musical scale

The visible spectrum comprises seven main colors which, in ascending order of frequency, are: red, orange, yellow, green, blue, indigo (blue violet), and violet. This order is identical to the order of colors in a rainbow, from the outside to the inside. Each color is identified by its frequency interval, which is (e.g. [17]) respectively, 430-480 THz, 480-510 THz, 510-540 THz, 540-610 THz, 610-650 THz, 650-690 THz, and 690-750 THz. Since we have twelve musical notes, to associate each color with a note's pitch, the five in-between (auxiliary) colors must also be considered; red orange, yellow orange, yellow green, blue green, and red violet. The last auxiliary color may be thought of as the last color if all the colors are placed in a wheel, after Newton. Analogously, the five auxiliary colors have frequency intervals in-between the frequency intervals of the main colors given above.

The pitch of a musical note is associated with the frequency of a specific color by dividing it by two to a power of an integer, such as to reach nearest the fundamental frequency of the note. For example, if the frequency of the orange color, taken to be 490 THz, is divided by 2^{40}, we obtain 445 Hz which is near the frequency of A_4. Thus, the note A will be associated with the orange color. Following this procedure, one note corresponds to one color as shown in Figure1.

Red	Redorange	Orange	Yelloworange	Yellow	Yellowgreen	Green	Bluegreen	Blue	Indigo	Violet	Redviol
↕	↕	↕	↕	↕	↕	↕	↕	↕	↕	↕	↕
G	$G^{\#}/A^{b}$	A	$A^{\#}/B^{b}$	B	C	$C^{\#}/D^{b}$	D	$D^{\#}/E^{b}$	E	F	$F^{\#}/G^{b}$

Fig. 1. The correspondence between colors and notes

The horizontal sequence of colors in a single rainbow forms, from left to right, a visible spectrum followed by its mirror image with a colorless region separating them. Therefore, mapping the color range of a single rainbow onto a musical scale, yields the ascending chromatic major F scale without its thirteenth note as shown in Figure 1, followed by the corresponding descending chromatic major F scale with a pause in-between to represent the rainbow's colorless region. This single rainbow musical scale resembles the shape of the Greek capital letter Λ.

A double rainbow consists of an outer bow with its color sequence reversed and an inner bow with the same order of colors as the single rainbow. The two bows are separated by a dark region which will be represented by musical rest. Therefore, the range of colors of a double rainbow will correspond to a musical scale comprised of a repeated reversed full (descending-ascending) chromatic F major scale, resembling the letter w, as the double rainbow musical scale. This double rainbow musical scale comprises forty eight notes, as the thirteenth note is missing form the ascent and the first note from the descent. For supernumerary rainbows consisting of more than two bows with several dark regions between them known as Alexander's bands, the musical scale will comprise a number of repetitions of the full F major scale, without its two middle notes.

3. Mapping the mathematical pitch onto the rainbow musical scale

Chaotic music will be composed from an elliptic algorithmic model, to be presented in section 5. The algorithm generates mathematical pitches with a range from 0 to 1. The mathematical pitch is mapped onto a musical note on the rainbow musical scale, described in section 2. So, for the *single* rainbow, the mathematical pitch range is divided into 24 intervals with each interval in ascending order corresponding to a musical note in increasing order on the rainbow musical scale. The result is shown in Figure 2.

(0,1/24] → G, (1/24, 2/24] → $G^{\#}/A^{b}$, … (10/24, 11/24] → F, (11/24, 12/24] → $F^{\#}/G^{b}$, (12/24, 13/24] → $F^{\#}/G^{b}$, (13/24, 14/24] → F, (14/24, 15/24] → E, … (23/24, 1] → G

Fig. 2. The mapping of the mathematical pitch onto the single rainbow scale

The mapping of the mathematical pitch onto the *double* rainbow scale involves matching forty eight intervals of the mathematical pitch from 0 to 1 to the forty eight notes of the scale at hand. Figure 3 shows this mapping.

(0,1/48] → $F^{\#}/G^{b}$, (1/48, 2/48] → F, (2/48, 3/48] → E, … (11/48, 12/48] → G, (12/48, 13/48] → G, (13/48, 14/48] → $G^{\#}/A^{b}$, … (10/24, 11/24] → F, (23/48, 24/48] → $F^{\#}$/C (24/48, 25/48] → $F^{\#}/G^{b}$, (25/48, 26/48] → F, (2/48, 3/48] → E, … (35/48, 36/48] → G, (36/48, 37/48] → G, (37/48, 38/48] → $G^{\#}/A^{b}$, … (46/48, 47/48] → F, (47/48, 1] → $F^{\#}/G^{b}$

Fig. 3. The mapping of the mathematical pitch onto the double rainbow scale

4. Duration of the chaotic rainbow notes

The duration of the chaotic notes, which is also chaotic in sequence, is computed from the elliptic iterative map described in section 5. However, the algorithmic model generates mathematical durations whose range is from 0 to 1. The actual note duration is obtained from mapping the mathematical duration onto the duration of notes in the rainbow musical scale described in sections 3 and 4. So, for the single rainbow scale each of the twenty four divisions of the mathematical duration will be assigned a conventional note duration, by approximating the fraction by its closest conventional fraction for a note value. The result, shown in Figure 4, is a rainbow duration scale resembling in shape the Greek capital letter Λ, whose lowest end (the minimum note duration) is the note value 1/32 and its highest end (the maximum note duration), the note value 1/2.

(0,1/24] →1/32 (G), (1/24, 2/24] → 1/16 ($G^\#/A^b$), (2/24, 3/24] →1/8 (A), (3/24, 4/24] → 1/8+1/32 ($A^\#/B^b$),
(4/24, 5/24] → 1/8+1/16 (B), (5/24, 6/24] → 1/4 (C), (6/24, 7/24] → 1/4+1/32 ($C^\#/D^b$),
(7/24, 8/24] → 1/4+1/16 (D), (8/24, 9/24] → 1/4+1/8 ($D^\#/E^b$), (9/24, 10/24] → 1/4+1/8+1/32 (E),
(10/24, 11/24] → 1/4+1/8+1/16 (F), (11/24, 12/24] → 1/2 ($F^\#/G^b$),
(12/24, 13/24] → 1/2 ($F^\#/G^b$), (13/24, 14/24] → 1/4+1/8+1/16 (F), … (23/24, 1] → 1/32 (G)

Fig. 4. The mapping of the mathematical duration onto the single rainbow duration scale

The mapping of the mathematical duration onto the double rainbow duration scale is picked so that each note's duration is the same as in the single rainbow duration scale. The resulting double rainbow duration scale is represented by Figure 3 with each interval being assigned the conventional duration of its note as given by the duration of the same note in the single rainbow duration scale of Figure 4. Thus, the double rainbow duration scale resembles in shape the letter w, whose depth (the largest note duration difference) is 1/2-1/32=15/32, equal to the height of the letter Λ of the single rainbow duration scale.

5. The elliptic chaotic model

The mathematical pitches and their duration are generated by an elliptic iterative map defined by

$$x_{n+1} = r\, x_n^{1/2} (1 - x_n)^{1/2}, \qquad 0 \leq x_n \leq 1, \quad 1 \leq r \leq 2 \tag{1}$$

where n denotes the time step and x_n represents the pitch or duration at step n. Eq. (1) yields all the pitches or their duration if the first one is given by the composer. The equation states that successive pitches or durations fall onto a semi-ellipse whose height is given by the semi-major vertical axis and is equal to r/2, while its minor axis is equal to 1. The lower limiting value of r, r=1, makes the semi-ellipse a semi-circle, while the upper limiting value of r, r=2, gives the fullest developed semi-ellipse with pitches or durations that do not exceed the value 1.

All the other r values between 1 and 2 yield a semi-ellipse of different height. Eq. (1) for the two extreme values of r, 1 and 2, is shown graphically in Figure 5.

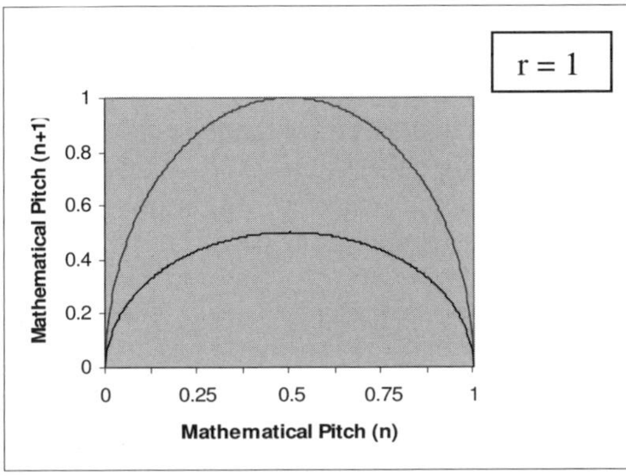

Fig. 5. Successive pitches from the semi-elliptical model (r=2: top curve, r=1: bottom curve)

The effect of the fullest developed semi-ellipse (r=2) on the generated sequence of six notes preceding the note G, for both the single and the double rainbow scales, is demonstrated in Figure 6. It is remarked that the sequence will depend on the starting note and that it is different for the two rainbows scales.

E (x=0.386) → **G** (x=0.974) → **D** (x=0.32) → **G$^{\#}$/A^{b}** (x=0.933) → **F$^{\#}$/G^{b}** (x=0.5^{+}) → **G** (x=1^{-}) → **G** (x=0.02)
F (x=0.973) → **A$^{\#}$/B^{b}** (x=0.322) → **D$^{\#}$/E^{b}** (x=0.935) → **F$^{\#}$/G^{b}** (x=0.494) → **F$^{\#}$/G^{b}** (x=1^{-}) → **F$^{\#}$/G^{b}** (x=0.018) → **G** (x=0.262)

Fig. 6. Sequence of generated notes from the semi-elliptical model on the single rainbow scale (first row) and the double rainbow scale (second row)

The choice of the free parameter, r, in the elliptic algorithmic model affects the pitch and duration characteristics of the composition. In this respect, four separate intervals of r are found:

i) $1 < r \leq \sqrt{3}$. For these r values, the pitch and duration after a few steps tend to one repeating value, given by the solution of the equation, $x=f(x)=rx^{1/2}(1-x)^{1/2}$, which is,

$$x_f = (1 + r^{-2})^{-1} \tag{2}$$

For example, when r=1.5, the generated composition will end in a repeated D note of duration 1/4+1/16 on the single rainbow scale or in a repeated A note of value 1/8 on the double rainbow scale. It is remarked that x_f of Eq. (2) is the fixed point for *any* r, meaning that x_f is a point to be avoided if the composition is not to end prematurely, for $r>\sqrt{3}$, in a repeated note corresponding to this x_f. However, when $r>\sqrt{3}$ the fixed point is repelling, $|f'(x_f)|>1$, meaning that points near x_f will not cause the composition to end prematurely. The way to avoid the fixed point if generated by the algorithm, is to set the generated (fixed) point equal to a number slightly smaller or larger than the exact fixed point.

ii) $\sqrt{3} < r \leq 1.9$. For each value of r in this interval, the generated composition will end in two notes (with their duration) repeating, corresponding to the solutions of the bifurcation equation, x=f(f(x)), which are

$$x_{1,2}=[1 \pm \{(r^2 - 3)/(r^2+1)\}^{1/2}]/[2\,(1-r^{-2})] \qquad (3)$$

For example, when r=1.9, in the single rainbow scale the composition will end in F, $G^{\#}/A^{b}$ repeating, of durations 1/4+1/8+1/16 and 1/16 respectively, while in the double rainbow scale the composition will end in E, E repeating, of duration 1/4+1/8+1/32. However, for each value of r in this interval, also exists the repelling fixed point of Eq. (2), as discussed above in i.

iii) $1.9 < r < 1.95$. In this range of r, the generated composition will end also periodically, with a larger number of notes than two within each cycle. When $1.9< r\leq 1.93$, there are four repeating notes, when r is larger than 1.93 and smaller than 1.94 there are eight notes repeating, and so on. Moreover, the fixed point of Eq. (2) exists for each r in the whole interval.

iv) $1.95 \leq r \leq 2$. For each of these values of r, the elliptic map will generate a composition which will have no pattern as the generated notes are chaotic no matter how long the composition is. Full chaos is developed when r=2. The fixed point of Eq. (2) is to be avoided also here, if the composition is not to end prematurely.

6. Fully chaotic pitches and durations

The fully chaotic pitches and their durations generated by the semi-elliptical algorithmic model, with r=2 and two different initial steps, are shown in Figure 7. The top diagram shows the pitch variation (or duration) with increasing step for the first 101 steps, with an initial pitch (or duration) equal to 0.05. This initial step corresponds to the note $G^{\#}/A^{b}$ of duration 1/16 in the single rainbow scale, while in the double rainbow scale it corresponds to the note E of duration 1/4+1/8+1/32. In the bottom diagram, the pitch (or duration) sequence is shown with increasing step for also the first 101 steps, but with an initial pitch (or duration) equal to 0.1. This initial step corresponds to the note A of duration 1/8 in the single rainbow scale, and to the note D of duration 1/4+1/16 in the double rainbow scale. An overall observation is that the initial step chosen has a strong effect on the generated, from the elliptic map, chaotic pitches and their durations. It is further observed that for both cases, there is an interval of about six steps in which the pitch stays near the value 0.8, because it is the fixed point of Eq. (2) for r=2. However, because the fixed point is repelling, as discussed in section 5, the generated pitch escapes from the neighborhood of this value after a few steps. The interval between the values of 0.7917=19/24 and 0.8333=20/24 corresponds to the note B of duration 1/8+1/16 in the single rainbow scale, and either the note A of duration 1/8 in the interval (0.7919=38/48, 0.8125=39/48] or the note $A^{\#}/B^{b}$ of duration 1/8+1/32 in the interval (0.8125=39/48, 0.8333=40/48], in the double rainbow scale. Therefore, the generated composition will repeat the same note for a few steps before it changes note. However, this repetition occurs at a completely different time, depending on the initial pitch chosen by the composer. It is seen from the top diagram of Figure 7, where the initial note is either $G^{\#}/A^{b}$ of duration 1/16 or E of duration 1/4+1/8+1/32 depending on the single or double rainbow scale, that the multiple repetition of the note starts at the 58th step. While in the bottom diagram, where the initial note is either A of duration 1/8 or D of duration 1/4+1/16, the multiple repetition of the note starts at the 25th step.

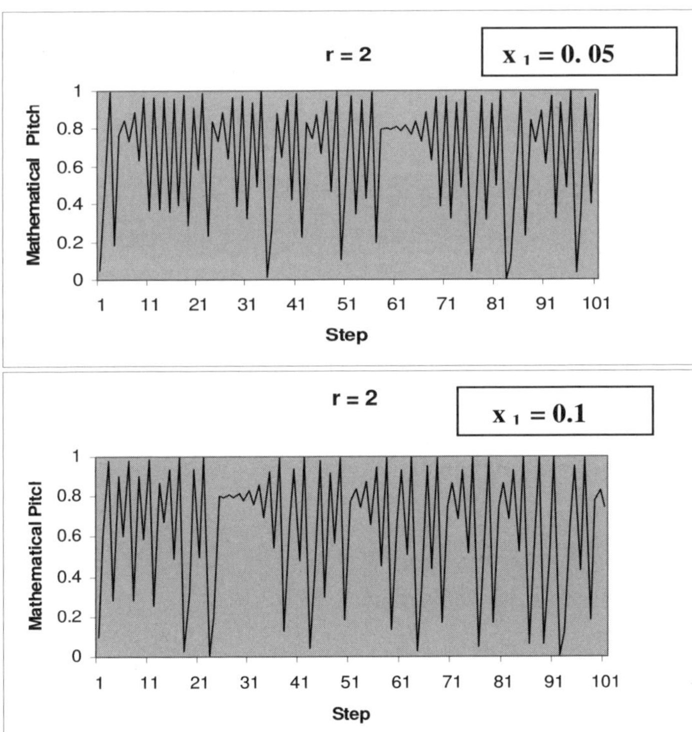

Fig. 7. The fully chaotic pitches (r=2) from the elliptic model for two initial values

7. The musical scores

For the case that is depicted in the top diagram of Figure 6, the corresponding musical pitches and durations were transcribed in the Finale notation program for both the single and the double rainbow scale. The single rainbow score shown below first, is mostly comprised of *two-note* chords with each note of the chord not appearing simultaneously as its other note. These two-note chords were mostly achieved by assigning every other iteration to each voice. It should also be remarked that, for the sake of practicality, all the 32nd notes were turned into grace notes. There are three sections in the short composition. First, there is a rising section. The second section is static with much octave displacement. The final section is a falling section. The prevailing "color" of the harmonies changes rapidly. The double rainbow score was composed in a similar fashion and is shown below, second. This composition lasts longer than its single rainbow counterpart because the algorithmically generated note durations are longer. Moreover, its harmonies seem more consonant.

The *single* rainbow fully chaotic score from the elliptic map:

The *double* rainbow fully chaotic score from the elliptic map:

8. References

[1] P.A. Scholes. Colour and Music. In *The Oxford Companion to Music*. 10th edition, London, 1970.

[2] J. Jewanski. Ist C = rot?: Eine kultur- und wissenschaftsgeschichte zum problem der wechselseitigen bezeihung zwischen ton und farbe. Von Aristoteles bis Goethe. *Berliner Musik Studien*, Band 17. Sinzig: Studio, Berlin, 1999.

[3] H. K. Taube. *Notes from the metalevel: an introduction to algorithmic music composition. Studies in New Music Research*, Routledge, N. York, 2004.

[4] E. Miranda. *Composing music with computers.* Bath Press, Bath, UK, 2004.

[5] S. Tipei. Manifold compositions: formal control, intuition, and the case for comprehensive software. *Proceedings 2005 International Computer Music Conference*, Copenhagen, Denmark, 2007.

[6] J. Serquera and E. Miranda. Algorithmic sound composition using cellular automata. In *Chaos Theory: Modeling, Simulation and Applications*, C. H. Skiadas, I. Dimoticalis and C. Skiadas, eds. World Scientific, 2011.

[7] G. E. Garnett. The aesthetics of interactive computer music. *Computer Music Journal,* v. 25, 21-33, 2001.

[8] J. Kretz. Freedom and necessity in computer aided composition: a thinking framework and its application. In *Chaos Theory: Modeling, Simulation and Applications*, C. H. Skiadas, I. Dimoticalis and C. Skiadas, eds. World Scientific, 2011.

[9] K. Hagan. Aesthetic consideration in algorithmic and generative composition. In *Chaos Theory: Modeling, Simulation and Applications*, C. H. Skiadas, I. Dimoticalis and C. Skiadas, eds. World Scientific, 2011.

[10] D. S. Dabby. *Musical variations from a chaotic mapping*. PhD thesis. MIT, 1995.

[11] V. D. Sotiropoulos and A. D. Sotiropoulos. Music composition from the cosine law of a frequency-amplitude triangle. In *Chaotic Systems: Theory and Applications*, ed. by C. H. Skiadas and I. Dimotikalis, World Scientific, 2010.

[12] A. D. Sotiropoulos and V. D. Sotiropoulos. Composing chaotic music from a varying second order recurrence equation. In *Chaotic Systems: Theory and Applications*, ed. by C. H. Skiadas and I. Dimotikalis, World Scientific, 2010.

[13] R. M. May. Simple mathematical models with very complicated dynamics." *Nature*, v. 261: 459-467, 1976.

[14] A. D. Sotiropoulos. Composing chaotic music from the letter m. In *Chaos Theory: Modeling, Simulation and Applications*, C. H. Skiadas, I. Dimoticalis and C. Skiadas, eds. World Scientific, 2011.

[15] C. H. Skiadas and C. Skiadas. *Chaotic Modelling and Simulation*. CRC Press, N. York, 2009.

[16] D. A. Sotiropoulos. On logistic-like iterative maps. In *Chaos Theory: Modeling, Simulation and Applications*, C. H. Skiadas, I. Dimoticalis and C. Skiadas, eds. World Scientific, 2011.

[17] H. Rossotti. *Color*. Princeton University Press, 1983.

A highly chaotic attractor for a dual-channel single-attractor, private communication system

Banlue Srisuchinwong and Buncha Munmuangsaen

Sirindhorn International Institute of Technology, Thammasat University
131 M.5, Tivanont Road, Bangkadi, Muang, Pathum-Thani, Thailand 12000
Email: banlue@siit.tu.ac.th

Abstract: A one-parameter highly chaotic attractor is presented and its application to a dual-channel, single-attractor, private communication system is demonstrated based on self-synchronization and chaotic masking techniques. Only a single attractor is required for a dual-channel transmitter or receiver, and can be either the well-known Lorenz attractor, the Lorenz-like attractor, or the one-parameter highly chaotic attractor developed in this paper. The latter is particularly well suited for an application to private communications due to the relatively high values of both the maximum Lyapunov exponent of 2.6148 and the maximum Kaplan-Yorke dimension of 2.1921. An advantage of the dual channel is the possibly twice increase in higher speed.
Keywords: highly chaotic attractor, self-synchronization, dual-channel single-attractor private communications.

1. Introduction

Since the discovery of the eminent Lorenz chaotic attractor in 1963 [1], the study of chaotic behaviour in nonlinear systems has attracted great attention due to possible applications in various fields of science and technology. In particular, the pioneer work of Pecora and Carroll [2-4] based on chaos synchronization has attracted special interests in applications of chaos to various communications including private communication systems. A certain class of chaotic systems possesses a self-synchronization property [2-4], if it can be decomposed into at least two subsystems, i.e. a drive system (transmitter) and a stable response subsystem (receiver), that synchronize when coupled with a common drive signal. The ability of self-synchronization in the Lorenz systems and others for communications has been found experimentally to be highly robust to perturbations in the drive signal leading to the chaotic signal masking techniques [5,6]. Such techniques, however, have been applied mainly to a single channel of communication systems.

Most existing chaotic attractors with five terms [7-10], six terms [e.g. 11] or seven terms [e.g. 1] in three first-order ordinary differential equations (ODEs) have been suffered from relatively low values of either the Lyapunov exponent (L), or the Kaplan-Yorke dimension (DKY). As the former (L) is a measure of chaoticity whilst the latter (DKY) is a measure of complexity (or

Chaos Theory: Modeling, Simulation and Applications
C. H. Skiadas, I. Dimotikalis and C. Skiadas (Eds)

strangeness) [8], high values of both are desirably expected for applications of chaos to private communication systems.

In this paper, a one-parameter highly chaotic attractor is presented and its application to a dual-channel, single-attractor, private communication system is demonstrated. Only a single attractor is required for a dual-channel transmitter or receiver, and can be either the well-known Lorenz attractor [1], the Lorenz-like attractor [11], or the one-parameter highly chaotic attractor developed in this paper. The latter is an improved version of [11] and is particularly well suited for an application to private communications as a result from the highest value of the Lyapunov exponent and the relatively high value of the Kaplan-Yorke dimension. As an example, robust self-synchronization of only the one-parameter highly chaotic attractor is demonstrated for a dual-channel single-attractor private communication system using the signal masking technique. An advantage of the dual channel is the possibly twice increase in higher speed.

2. A One-Parameter Highly Chaotic Attractor

An existing Lorenz-like chaotic attractor [11] has been described as a set of three first-order, autonomous, ODEs of the form :

$$x=a(y-x),\ y=bx-dxz,\ z=xy-cz. \tag{1}$$

where a = 5.8, b = 16, c = 1.8. The parameter d simply scales the size of the attractor and therefore d can be any factor, e.g. d = b/c. A trivial simplification is to set one or more of the four parameters to 1. One way to do this without destroying the chaos is to set the parameters d = c = 1. Therefore, the system (1) becomes a simpler ODE as follows:

$$x=a(y-x),\ y=bx-xz,\ z=xy-z. \tag{2}$$

In an attempt to adjust the parameters a and b for the maximum Lyapunov exponent, the parameter a in (2) increases to more than a hundred which is undesirable because the solution of state variables x, y, z in (2) occupies a wide dynamic range and the parameter b must also be adjustable to be a very large number in order to maintain its chaotic behaviour.
An alternative approach is to scale down the variables x, y and z to be smaller whilst a relatively high value of the maximum Lyapunov exponent can be maintained. For example, let a = 10, u = x/20, v =y/20 and w = z/40.
Therefore, u=10v-u, v=bu-40uw, w=10uv-w. Consequently,

$$x=10(y-x),\ y=bx-40xz,\ z=10xy-z. \tag{3}$$

It can be seen from (3) that b is now the only single controllable parameter, and therefore (3) is a one-parameter chaotic attractor. Figure 1(a) illustrates a bifurcation diagram exhibiting a period-doubling route to chaos of the peak of z of (3) versus b which is varied from 0 to 600. It is apparent that there are some periodic windows in the chaotic region. The chaotic behavior disappears quickly for $b \leq 0$. Figures 1(b) and (c) shows the corresponding Lyapunov exponent spectrum and the Kaplan-Yorke dimension DKY of (3), respectively. It can be observed from Figs. 1 (a-c) that, at b near 296.5, the maximum Lyapunov exponent L(max) = 2.6148 and the maximum Kaplan-Yorke dimension DKY(max) = 2.1921.

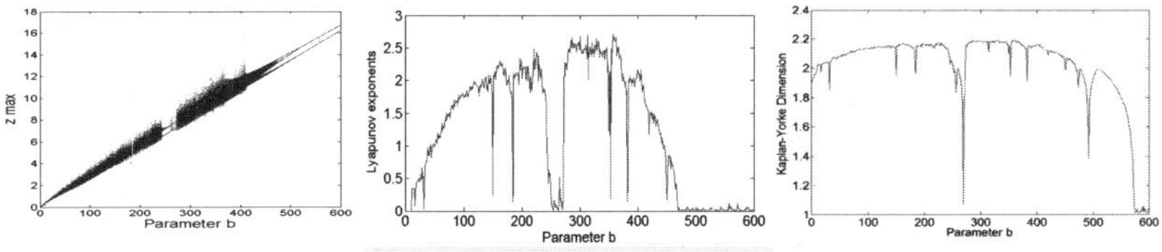

Figure 1: For the one-parameter highly chaotic attractor, (a) A bifurcation diagram, (b) Lyapunov exponent spectrum, (c) Kaplan-Yorke dimension DKY.

Table 1 : Lyapunov exponent (L) and the Kaplan-Yorke dimension (DKY) of the system (3) developed in this paper compared to those of others.

Attractors	Lyapunov Exponents L	DKY
MCLS [8]	(0.2101, 0, -1.2101)	2.2354
Lorenz-Like [11]	(0.5907, 0, -8.3907)	2.0701
Lorenz [12]	(0.9056, 0, -14.5723)	2.0621
Lu[12]	(1.2890, 0, -20.2890)	2.0635
Five Terms [7]	(1.4913, 0, -6.4939)	2.2301
Chen [12]	(2.0836, 0, -12.0836)	2.1724
This Paper	(2.6148, 0, -13.6148)	2.1921

Table 1 summarizes the Lyapunov exponent (L) and the Kaplan-Yorke dimension (DKY) of the system (3) developed in this paper compared to those of others. It is apparent from Table 1 that the value of L=2.6148 of the system (3) is the highest, although the value of DKY=2.1921 of the system (3) is the third highest, which is only 2% smaller than that of the maximally complex Lorenz system (MCLS) [8] where DKY= 2.23542. Consequently, the one-parameter attractor described in (3) is highly chaotic at $b \cong 296.5$, and is therefore particularly well suited for an application to a private communication system.

By using the Fourth-order Runge-Kutta method to solve the system (3) with time step size of 0.001, Figure 2 displays the resulting strange attractor of (3) using the initial values x(0) = 1, y(0) = 1 and z(0) = 6. It appears from Fig. 2 that the new attractor exhibits abundantly complex behaviour of chaotic dynamics.

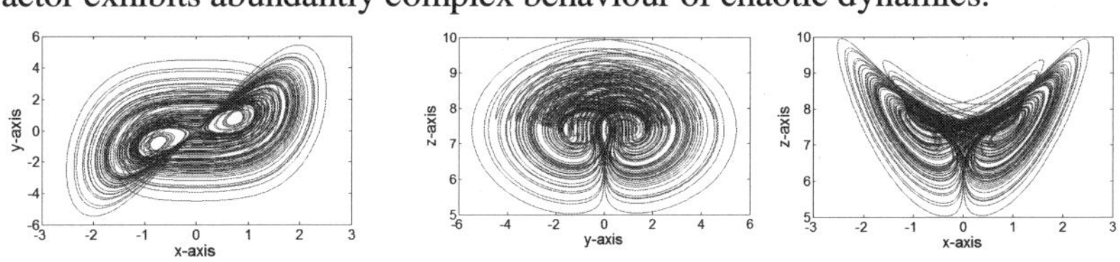

(a) (b) (c)

Figure 2: A one-parameter highly chaotic attractor, (a) x-y plane, (b) y-z plane, (c) x-z plane.

3. Dual-Channel Single-Attractor Private Communications

An existing chaotic masking technique [5,6] can be modified using the one-parameter highly chaotic attractor described in (3), as a drive system at the transmitter and a response subsystem at the receiver, for a dual-channel single-attractor private communication system, as shown in Fig. 3.

3.1. A Dual-Channel Single-Attractor Transmitter

At the transmitter, the one-parameter highly chaotic attractor described in (3) can be used as a single drive system for a dual-channel transmitter independent of its response subsystem at the receiver as follows :

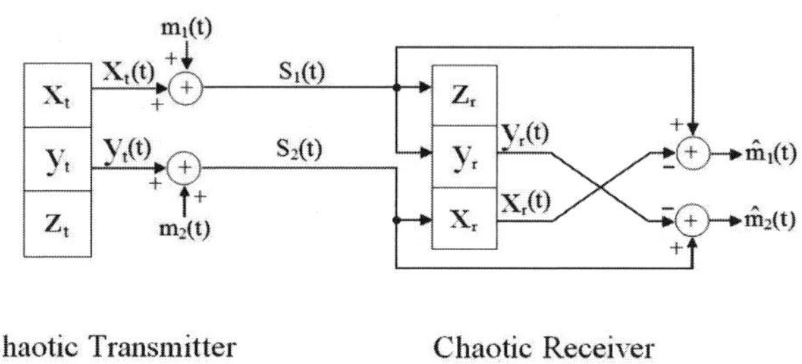

Figure 3 : A dual-channel single-attractor private communication system based on the one-parameter highly chaotic attractor.

xt=10(yt-xt), yt=296.5xt-40xtzt, zt=10xtyt-zt. (4)

As shown in Fig. 3, the dual-channel transmitter consists of two parallel transmitted signals. The first transmitted signal is s1t=xtt+m1t, where xtt is a chaotic masking signal and m1t represents the first original message to be transmitted. The second transmitted signal is s2t=ytt+m2(t), where ytt is a chaotic masking signal and m2t represents the second original message to be transmitted. As suggested in [5], the power spectra of the chaotic masking signals should be much higher than those of the messages, i.e. they should be highly overlapping with an average signal-to-masking ratio of approximately 20 dB.

3.2. A Dual-Channel Single-Attractor Receiver

At the receiver, the one-parameter highly chaotic attractor described in (3) can be used as a single response subsystem for a dual-channel receiver as follows :

xr=10s2(t)-xr,
yr=296.5s1(t)-40s1(t)zr, (5)
zr=10s1(t)yr-zr.

As shown in Fig. 3, the dual-channel receiver consists of two parallel received signals s1t and s2t, each of which regenerates a clean masking signal yrt and xrt, respectively. When the receiver synchronizes with s2(t), then xrt≅ xtt. As shown in Fig. 3, the message signal m1(t) can be recovered as m1t=s1t-xrt=xtt+m1t-xrt≈m1(t). Similarly, when the receiver synchronizes with s1(t), then yrt≅ytt. As shown in Fig. 3, the message signal m2(t) can be recovered as m2t=s2t-yrt=ytt+m2t-yrt ≈m2(t).

3.3. Numerical Results

As simple examples, the first transmitted message m1t=0.2sin($2\pi f1t$), where the frequency f1 = 1 kHz, and the second transmitted message m2t is a pulse-train rectangular waveform with an amplitude = 0.4 and the frequency f2 = 2 kHz. Self-synchronization can be achieved over a wide range of initial conditions, e.g. at the transmitter xto, yto, zto=2, 2, 6, and at the receiver xro, yro, zro=-2, 4, 3.

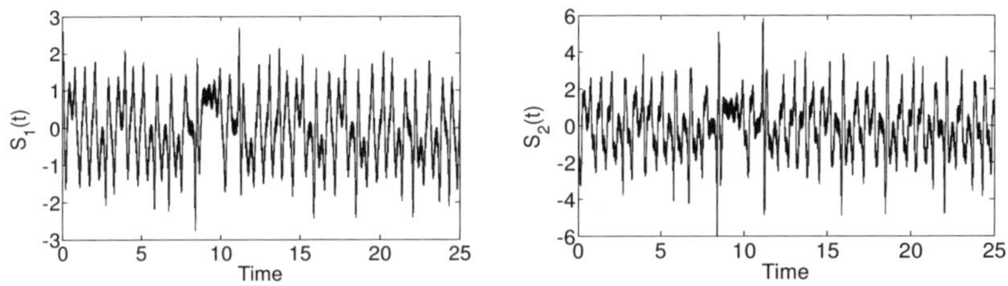

Figure 4 : (a) A transmitted signal s1t, (b) A transmitted signal s2t.

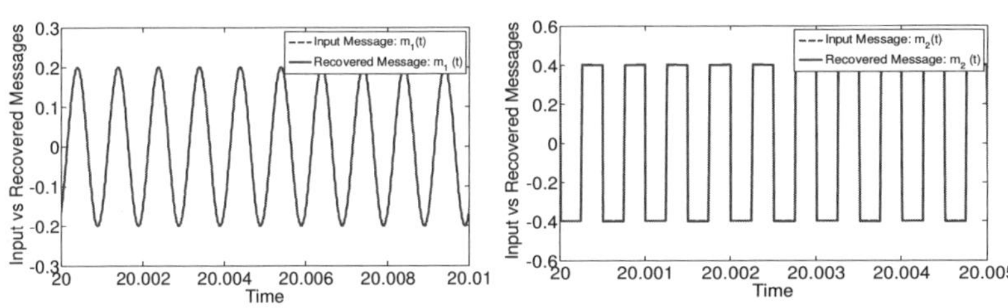

Figure 5: (a) Recovered message m1t, (b) Recovered message m2t.

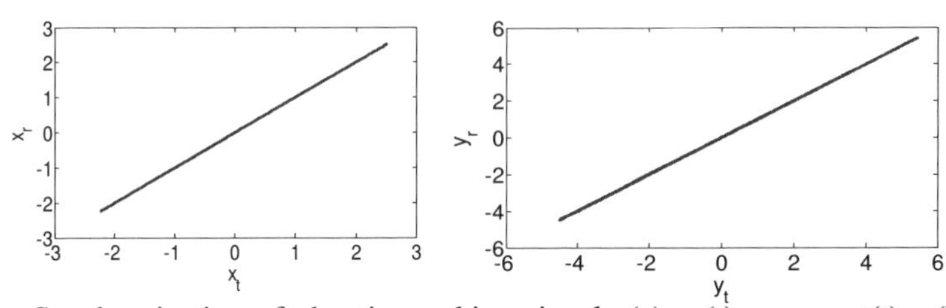

Figure 6 : Synchronization of chaotic masking signals (a) xr(t) versus xt(t), (b) yr(t) versus yt(t).

Figures 4(a) and 4(b) show the transmitted signal s1t and s2t, respectively. Figures 5(a) and 5(b) depicts the recovered messages m1t and m2t, respectively. Figure 6(a) displays the synchronization of the clean chaotic masking signals xr(t) versus xt(t) whilst Fig. 6(b) shows the synchronization of the clean chaotic masking signals yr(t) versus yt(t). Figure 7(a) illustrates a transient prior to the successful synchronization of signals xt(t) and xrt. Plots of errors xr-xt, yr-yt and zr-zt are depicted in Figs 7(b), 7(c) and 7(d), respectively.

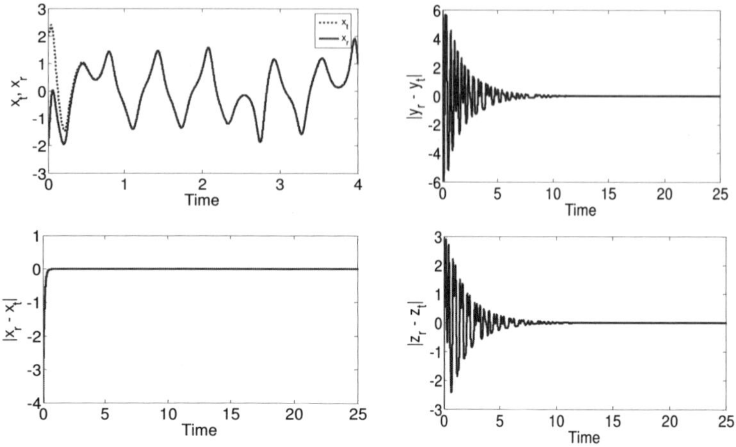

Figure 7 : Transients of (a) signals xt(t) and xrt, (b) errors xr-xt, (c) errors yr-yt and (d) errors zr-zt.

3.4. An Alternative Dual-Channel Single-Attractor Receiver

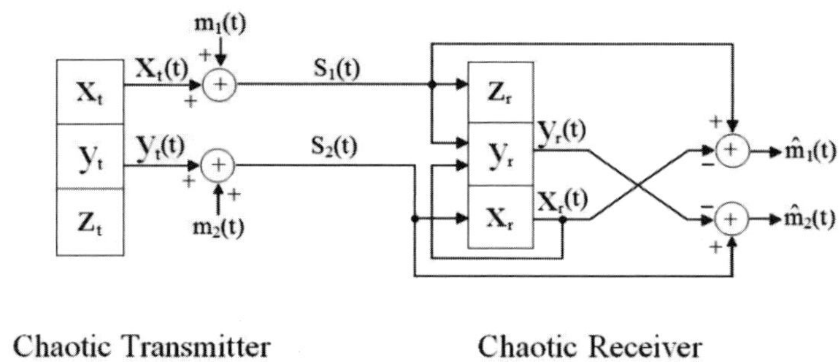

Figure 8 : An alternative dual-channel single-attractor receiver based on the one-parameter highly chaotic attractor.

As an alternative to the receiver shown Fig. 3, Fig. 8 shows another possible dual-channel single-attractor receiver whilst the similar transmitter remains the same. The one-parameter highly chaotic attractor described in (3) can be used as a single response subsystem for an alternative dual-channel receiver as follows

$$
\begin{aligned}
xr &= 10s2(t)-xr, \\
yr &= 296.5s1(t)-40xr(t)zr, \\
zr &= 10s1(t)yr-zr.
\end{aligned}
\tag{6}
$$

A major disadvantage of the system (6), however, is the required longer transient time prior to the steady synchronization.

4. Conclusions

A one-parameter highly chaotic attractor has been presented and its application to a dual-channel, single-attractor, private communication system has been demonstrated based on self-synchronization and chaotic masking techniques. Only a single attractor is required for a dual-channel transmitter or receiver, and can be either the well-known Lorenz attractor, the Lorenz-like attractor, or the one-parameter highly chaotic attractor developed in this paper. The latter is particularly well suited for an application to private communications due to the highest value of the maximum Lyapunov exponent of 2.6148 and the relatively high value of the maximum Kaplan-Yorke dimension of 2.1921. Twice increase in higher speed can be expected due to the dual channel.

References

[1] E. N. Lorenz, Deterministic nonperiodic flow, J. Atmos. Sci., 20, 130–141, 1963.
[2] L. M. Pecora and T. L. Carroll, Synchronization in chaotic systems, Phys. Rev. Lett, 64(8):821–5, 1990.
[3] L. M. Pecora and T. L. Carroll, Driving system with chaotic signals, Phys. Rev. A, 44:2374–2383, 1991.
[4] T. L. Carroll and L. M. Pecora, Synchronizing chaotic circuits, IEEE Trans. Circuits Syst., 38:453-456, 1991.
[5] K. M. Cuomo and A. V. Oppenheim, Synchronization of Lorenz-based chaotic circuits with applications to communications, IEEE Trans. Cir. Sys., 40(10): 626-633, 1993.
[6] I. Pehlivan and Y.Uyaroğlu, Simplified chaotic diffusionless Lorentz attractor and its application to secure communication systems, IET Commun., 2007, 1(5):1015-1022, 2007.
[7] B. Munmuangsaen and B. Srisuchinwong, A new five-term simple chaotic attractor, Phys. Lett. A., 373(44):4038-4043, 2009.
[8] J. C. Sprott, Maximally complex simple attractors, CHAOS, 17:033124-1-6, 2007.
[9] G. van der Schrier and L. R. M. Maas, The diffusionless Lorenz equations; Shil'nikov bifurcations and reduction to an explicit map, Physica D 141:19-36, 2000.
[10] J. C. Sprott, Some simple chaotic flows, Phys. Rev. E 50(2):R647-650, 1994.
[11] B. Munmuangsaen and B. Srisuchinwong, A New Lorenz-Like Chaotic Attractor and Its Synchronization, Proceedings of 2009 Chinese Control and Decision Conference (CCDC 2009), Guilin, China, 17-19 June, 1508-1512, 2009.
[12] Q. Yang, G. Chen, T. Zhou, A unified Lorenz-type system and its canonical form, Int. J. Bifur. and Chaos, 16(10), 2855–2871, 2006.

Manifestation of chaos in collective models of nuclei

Pavel Stránský[1,2], Michal Macek[1], Pavel Cejnar[1], Alejandro Frank[2], Ruben Fossion[2] and Emmanuel Landa[2]

[1]Institute of Particle and Nuclear Physics, Charles University in Prague, Czech Republic
[2]Institute of Nuclear Sciences, UNAM, Mexico City, Mexico
Email: pavel.stransky@nucleares.unam.mx

Abstract: We present results of an extensive analysis of classical and quantum signatures of chaos in the geometric collective model (GCM) and the interacting boson model (IBM) of nuclei. Apart from comparing the regular fraction of the classical phase space and the Brody parameter for the nearest neighbor spacing distribution in the quantum case, we also adopt (i) the Peres lattices allowing one to distinguish ordered and disordered parts of spectra and to reveal main ordering principles of quantum states, (ii) the geometrical method to determine the position where the transition from order to chaos occurs, and (iii) we look for the $1/f^{\alpha}$ power law in the power spectrum of energy level fluctuations. The Peres method demonstrates the adiabatic separation of collective rotations in the IBM.
Keywords: collective model of nuclei, Hamiltonian system, Peres lattice, classical-quantum correspondence, $1/f$ noise, Riemannian geometry, adiabatic separation, quasi-dynamical symmetry.

1. Introduction

It has been shown in recent years that simple models of nuclear collective dynamics – the geometric collective model (GCM) and the interacting boson model (IBM) – exhibit a high degree of variability in regular and chaotic features with energy and control parameters [1,2,3]. The complex dynamics encoded in relatively simple Hamiltonians well position these systems to be laboratories for detailed investigation into classical-quantum correspondence and for testing different approaches of measuring and visualizing chaos. In addition, it appears that the study of chaos can help in understanding the dynamical structure of the systems and allows for observation of new phenomena, such as the quantum phase transitions [4] and the quasi-dynamical symmetry (QDS) [5].

Classical chaos is commonly studied by means of the visual method of Poincaré sections and numerically characterized by quantities derived from Lyapunov exponents. This approach requires the tedious solution of differential equations of motion for a large set of trajectories in order to obtain an overall image of the degree of regularity. To bypass these difficulties, Horwitz *et al.* [6] have proposed recently another way to distinguish between regular and chaotic dynamics. When the system,

Chaos Theory: Modeling, Simulation and Applications
C. H. Skiadas, I. Dimotikalis and C. Skiadas (Eds)

evolving in a flat Cartesian space with some potential, is transformed to be in a free motion on a special curved manifold, the Riemannian geometry gives a very simple condition of stability: a negative eigenvalue of a matrix obtained from the curvature tensor appearing inside the kinematically accessible area of the system implies instability, and thus chaoticity of the motion.

Quantum measures of chaos are based on certain statistical properties of the energy levels, for example on the nearest-neighbor-spacing distribution (and the related Brody parameter ω) [7]. It is clear that the statistics describe only some bulk features of a sufficiently large portion of the spectrum. The method introduced by Peres [8], however, makes it possible to assign regular, chaotic, or a mixed type of dynamics to *individual states*. It uses lattices formed from the expectation values $A_i = \langle\psi_i|A|\psi_i\rangle$ of an arbitrary operator A plotted against the energies $E_i = \langle\psi_i|H|\psi_i\rangle$. Due to arguments based on the semiclassical EBK quantization, the lattices of points (E_i, A_i) show regular patterns in integrable systems, whereas in the chaotic case the resulting images are formed by a combination of ordered and disordered patterns, whose relative size depends on the degree of chaoticity. Note that the Peres method is entirely visual and can be considered as a quantum analogue of the classical Poincaré sections.

Long-range correlations of eigenlevels are usually studied by the spectral rigidity Δ_3 or the number variance Σ^2, from which, however, it is difficult to quantify the degree of regularity of a system. Incorporating long-range correlations into quantitative studies has been done by Relaño *et al.* [9], considering spectral fluctuation of a time series of deviations between the unfolded eigenenergies and uniformly distributed levels. The ensemble average of the power spectrum obeys the power law $\langle S(f)\rangle = 1/f^{\alpha}$ where α, bounded between 1 and 2 (for a totally chaotic and an integrable system, respectively), can serve as a measure of regularity.

The aim of this contribution is to establish the standard methods of classical and quantum chaos in the GCM and IBM and then make a step beyond; we use Peres lattices to present the manifestation of QDS by well-pronounced rotational bands for *high*-energy states in the IBM with symmetry breaking interaction switched on.

2. Geometric collective model

We use the GCM in the nonrotating regime, *i.e.* considering only the vibrations with zero angular momentum $\vec{J}$. The Hamiltonian $H = T + V$ describes quadrupole motions of nuclei. Quantizing the system and separating the vibrational and rotational degrees of freedom by moving into the intrinsic frame we obtain the kinetic term in the form [10]

$$T^{5D} = -\frac{\hbar}{2K}\left(\frac{1}{\beta^4}\frac{\partial}{\partial\beta}\beta^4\frac{\partial}{\partial\beta} + \frac{1}{\beta^2\sin 3\gamma}\frac{\partial}{\partial\gamma}\sin 3\gamma\frac{\partial}{\partial\gamma}\right) + T_{rot}$$

The system has in total 5 degrees of freedom which is indicated by the label 5D. K is a mass parameter and (β,γ) are the so called Bohr coordinates, describing the shape of the deformed nucleus. For states with $J=0$ the rotational part T_{rot} vanishes. The potential (with adjustable parameters A,B,C) reads as

$$V = A\beta^2 + B\beta^3 \cos 3\gamma + C\beta^4$$

There is another way how to deal with the kinetic term in the case of the principal frame at rest. If the 3 rotational degrees of freedom are frozen *before* the quantization is carried out, we get the familiar 2D kinetic term in polar coordinates (β,γ)

$$T^{2D} = -\frac{\hbar}{2K}\left(\frac{1}{\beta}\frac{\partial}{\partial\beta}\beta\frac{\partial}{\partial\beta} + \frac{1}{\beta^2}\frac{\partial^2}{\partial\gamma^2}\right)$$

Both kinetic terms T^{5D} and T^{2D} have the same classical limit and offer a possibility to study the influence of the quantization method on the quantum chaotic features of the spectrum, i.e. test the famous Bohigas conjecture [11]. Let us emphasize that only the 5D case represents the standard (nuclear) GCM. Note that the system is integrable for $B=0$.

We diagonalize both types of the GCM Hamiltonian in the respective 2D or 5D oscillator bases. Due to scaling properties, only one of the parameters (A,B,C,K) in the classical case, or two in the quantum case, determine the qualitative features of the system [3,12]. We fix $C=1$ and move along the path of three connected lines (each described by only one principal control parameter), passing all possible configurations of the system: (i) $B\in(0,1)$ with $A=-1$, (ii) $A\in(-1,1)$ with $B=1$ where the deformed-spherical ground-state shape phase transition is crossed at $(A,B)=(1/4,1)$, and (iii) $B\in(1,0)$ with $A=1$. In the quantum case we consider the classicality constant $\kappa=\hbar/K$ as the second fundamental parameter, which influences only the density of energy levels [12]. We choose two independent Peres operators identified with (i) the square of the angular momentum operator L^2 connected with rotations about angle γ, which is the $J=0$ projected O(5) invariant in the 5D case or the O(2) invariant in the 2D case, and (ii) the nonintegrable perturbation $H'=\beta^3\cos 3\gamma$.

Figure 1 presents the complete map of the classical regularity quantified by the regular fraction of the classical phase space f_{reg}. Leaving the leftmost and the rightmost $B=0$ integrable limits, we observe a progressive retreat of regularity with increase of the perturbation B. In the intermediate domain the structure is extremely complex, forming sharp branching valleys and curved edges. At $B\approx 0.6$ the regularity achieves a well-pronounced maximum, which is tightly connected with the IBM semiregular arc [13]. The shape-phase transition region (the white vertical dashed line) is characterized by accentuated chaoticity. The dashed black line encircles the region where the kinematically accessible region in the $(x \equiv \beta\cos\gamma, y \equiv \beta\sin\gamma)$ plane has concave shape. The lower part of the line corresponds exactly with the curve obtained by the Horwitz's geometrical method, which should separate regions with regular and chaotic dynamics. However, it is evident from the figure that this criterion provides only an estimation.

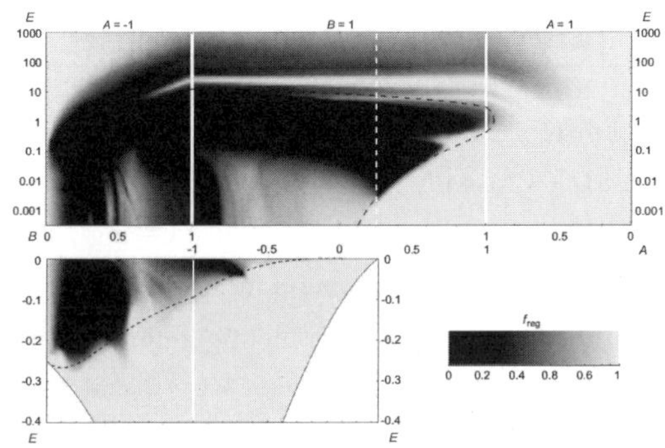

Figure 1. Complete map of the classical regular fraction f_{reg} as a function of control parameters and energy E. The degree of chaos is coded in shades of gray, with light gray (black) corresponding to complete order (chaos). The plot is divided by the thick white lines into three regions of different scaling ($A=-1$, $B=1$, and $A=1$, respectively) with free parameters being varied along the horizontal axis. The white dashed line indicates the deformed-spherical shape phase transition. The black dashed line corresponds to the convex-concave transition of the border of kinematically accessible area.

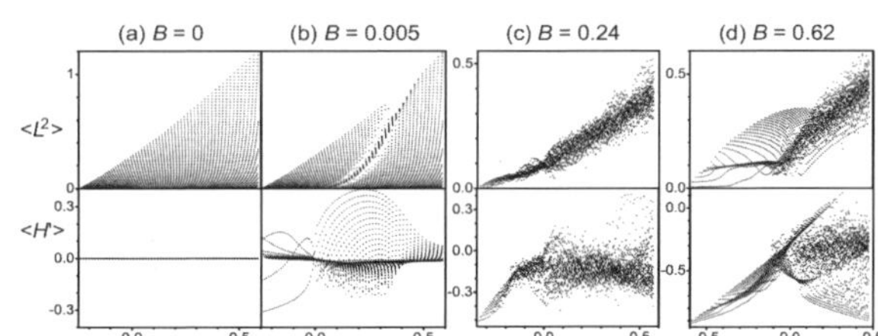

Figure 2. Peres lattices of the 2D GCM for $\kappa = 2.5\cdot 10^{-5}$ and $A=-1$ with 3000 levels plotted. The transition between integrable (panel a) and nonintegrable (panels b – d) regimes are observed.

The decay of regular pattern in the quantum Peres lattices when introducing nonintegrability is demonstrated in Figure 2. Starting from a totally ordered case (panel a), small nonintegrable perturbation causes disturbances only in a part of a lattice (panel b). Maximal disorder is reached at $B=0.24$. For higher B the progression reverses and new regular structures start to appear, which is observed in both $\langle L^2 \rangle$ and $\langle H' \rangle$ lattices. The regular islands can be identified with remnants of classical tori.

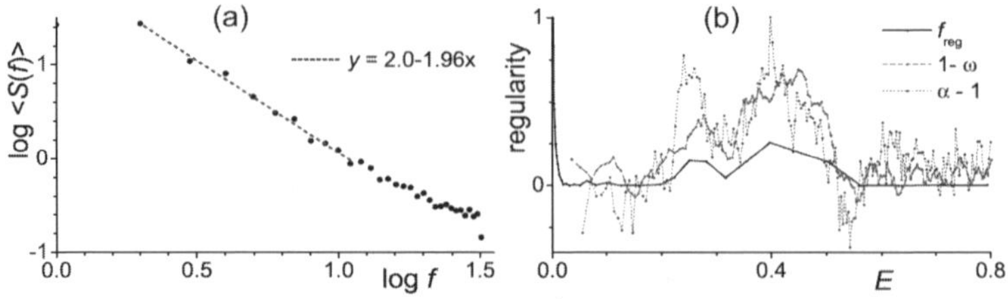

Figure 3. (a) Average power spectrum of the $1/f^{\alpha}$ noise method in the integrable case $(A=-1, B=0)$ of the 2D GCM with $\kappa = 4\cdot 10^{-6}$, using 500 sets of 64 levels, starting from level 8000. We use the simplified notation $x \equiv \log f$ and $y \equiv \log\langle S(f)\rangle$. (b) Comparison of the classical f_{reg} and quantum measures of regularity $1-\omega$ and $\alpha-1$ for the 2D GCM in the mixed dynamics configuration with $A=0.25, B=1$ and $\kappa = 1\cdot 10^{-6}$. The standard errors of the measures are 3%, 5% and 20%, respectively.

Figure 3 presents the power spectrum for the method of $1/f^{\alpha}$ noise in an integrable case of the model. The value of the exponent $\alpha = 2$ is in full agreement with the prediction of the theory [9].

In the quantum case we use two measures of regularity: (i) the adjunct $1-\omega$ of the Brody parameter, which is estimated by the χ^2 fit applied to the nearest-neighbor spacing distribution of the unfolded levels, and (ii) the quantity $\alpha-1$ of the $1/f^{\alpha}$ noise in the averaged power spectrum of energy level fluctuation. In order to obtain the energetic dependence of the regularity, we divide the calculated spectra (usually 30 000 well-converging levels) into windows of 1000 levels. Both quantum measures can then be directly compared with f_{reg} [see Figure 3 (b)]. The correspondence is apparent, though the quantum measures tend to overemphasize the regularity.

Figure 4 confronts the Peres lattices and the measures of chaos for both 2D and 5D quantization schemes. The lattices are displayed in panels (a) and (b), showing small differences in structure, but the distribution of regular and chaotic areas are equal. This is in agreement with the negligible difference between the 2D and 5D cases of $\omega(E)$ (panel d), which remains within the limits of the standard error in our calculations. Finally, observe the influence of the classicality parameter κ on the level density by comparing panel (a) with Figure 2 (d).

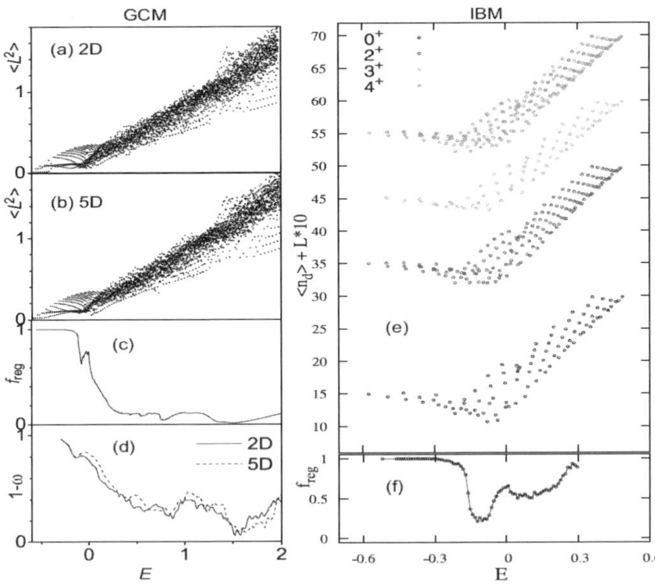

Figure 4. First column: Classical and quantum chaos of the GCM for $A=-1$ and $B=0.62$. Peres lattices for $\kappa=1\cdot10^{-4}$ are compared for 2D (a) and 5D (b) quantization. (c) Classical regular fraction of the phase space f_{reg} and (d) corresponding quantum counterpart – Brody parameter ω calculated for $\kappa=2.5\cdot10^{-5}$. Second column: IBM dynamics at $(\eta,\chi)=(0.5,-0.9)$. (e) Peres lattices of the n_d operator for different angular momenta ($N=30$ bosons), and (f) corresponding classical fraction of regularity f_{reg}.

3. Interacting boson model

To demonstrate a different application of the Peres method we consider the Hamiltonian [14]

$$H = a\left[\frac{\eta}{N}n_d - \frac{1-\eta}{N^2}Q_\chi \cdot Q_\chi\right]$$

with n_d the *d*-boson number operator (N is a conserved total number of bosons), Q_χ the quadrupole operator, and (η,χ) two external parameters defining the symmetry triangle. Its vertices $(\eta,\chi)=(0,0)$, $(1,0)$, and $\left(0,-\sqrt{7}/2\right)$ correspond to the O(6), U(5), and SU(3) dynamical symmetries, respectively. The system is integrable in the triangle vertices and along the side with $\chi=0$. The Peres operator we can take any of the Casimir invariants connected with the symmetry groups, but our results are best demonstrated by the choice the U(5) operator n_d.

Recently, some low-energy rotational bands populating the spectrum below the saddle point of the classical energy surface were explained analytically using specific SU(3)-like boson condensates [15]. In this contribution we turn to the highly excited states. Figure 6(e) shows the eigenstates of IBM in the region of enhanced regularity lying in the interior of the triangle (at the so called arc of regularity), distributed into 4 Peres lattices according to their angular momentum. Apart from a few ordered low-lying states (present also in other parts of the parametric triangle), we may observe two distinct regular regions: one at $E \approx 0$ and the second reaching the highest accessible energy. Surprisingly, the lattices for different angular momentum L form the same pattern there. The analyses based on wave function correlation, $E2$ transitions and moments of inertia reveal that these "regular" energies associate into bands, showing signatures of SU(3) QDS in the $L(L+1)$ excitation energy pattern [16]. This behavior is strongly accentuated along the regular arc. While the low-energy dynamics are almost completely regular in the entire symmetry triangle, the partially increased regularity of the arc (relative to the neighboring regions) is brought about essentially by the regular high-lying modes, which has already been shown [1] [this is also supported by the classical fraction of regularity f_{reg} for states with $L=0$, drawn in the panel (f)]. It suggests a connection between quasi-dynamical symmetry and regularity.

4. Conclusions

We have calculated some measures of chaos for the classical and quantum versions of the geometric collective model (we dispose of two different quantum GCM with the same classical limit) and the interacting boson models in the wide range of external parameters and energy, observing good correspondence between the classical fraction of regularity and the Brody parameter that quantifies chaos in the quantum case. This correspondence is supported by the visual observation of the regular and chaotic structures in the Peres lattices.

In addition, we adopted the geometric method by Horwitz *et al.* that estimates the position of regular-chaotic transition and follows the convex-concave shape change of the kinematically accessible area border in the GCM, which has been proven analytically [17]. We also adopted the method of $1/f^{\alpha}$ noise by Relaño *et al.*, obtaining expected $\alpha = 2$ in an integrable configuration and a correspondence in the energetic dependence of the measures $1-\omega$ and $\alpha - 1$ in a configuration with mixed dynamics.

In IBM we have observed a strong correlation between the variation of classical and quantum degrees of regularity and the occurrence rate of the rotational bands. This suggests that the separation of rotational motion from intrinsic vibrational dynamics is enhanced by the intrinsic regularity, while on the other hand, the chaotic intrinsic states are more prone to mixing, if the rotation comes into play; the regular states seem to be protected by their symmetry, in contrast. Different methods of rotational band identification are described in [16], being in full agreement with the results obtained by Peres method.

This work was supported by the Czech Science Foundation (202/09/0084), by the Czech Ministry of Education (MSM 0021620859), by PAPIIT-UNAM (grant IN113808), by DGAPA, UNAM, and by Conacyt, Mexico.

References

[1] Y. Alhassid, and N. Whelan. *Phys. Rev. Lett.* 67:816, 1991; N. Whelan, and Y. Alhassid. *Nucl. Phys.* A 556:42, 1993.

[2] P. Cejnar, and P. Stránský. *Phys. Rev. Lett.* 93:102502, 2004.

[3] P. Stránský, M. Kurian, and P. Cejnar. *Phys. Rev.* C 74:014306, 2006.

[4] P. Cejnar, and P. Stránský. *Phys. Rev.* E 78:031130, 2009.

[5] D. J. Rowe. *Nucl. Phys.* A 745:47, 2004; G. Rosenstell, and D. J. Rowe. *Nucl. Phys.* A 759:92, 2005

[6] L. Horwitz, Y.B. Zion, M. Lewkowicz, M. Schiffer, and J. Levitan. *Phys. Rev. Lett.* 98:234301, 2007.

[7] M.C. Gutzwiller. *Chaos in Classical and Quantum Mechanics.* Springer-Verlag, New York, 1990.

[8] A. Peres. *Phys. Rev. Lett.* 53:1711, 1984.

[9] A. Relaño *et al. Phys. Rev. Lett.* 89:244102, 2002; E. Faleiro *et al. Phys. Rev. Lett.* 93:244101, 2004; J.M.G. Gómez *et al. Phys. Rev. Lett.* 94:084101, 2005.

[10] G. Gneuss, U. Mosel, and W. Greiner. *Phys. Lett.* B 30:397, 1969.

[11] O. Bohigas, M. J. Giannoni, and C. Schmit. *Phys. Rev. Lett.* 52:1, 1984.

[12] P. Stránský, P. Hruška, and P. Cejnar. *Phys. Rev.* E 79:046202, 2009; *Phys. Rev.* E 79:066201, 2009.

[13] M. Macek, P. Stránský, P. Cejnar, S. Heinze, J. Jolie, and J. Dobeš. *Phys. Rev.* C 75:064318, 2007.

[14] F. Iachello, and A. Arima. *The Interacting Boson Model.* Cambridge Univ. Press, Cambridge, 1987.

[15] M. Macek, J. Dobeš, and P. Cejnar. *Phys. Rev.* C 80:014319, 2009.

[16] M. Macek, J. Dobeš, and P. Cejnar. Phys. Rev. C in press.

[17] P. Stránský *et al.* in preparation.

Importance of the chaos for computational processes of collective intelligence in social structures

Tadeusz (Ted) Szuba

AGH University, Dept. of Control, Cracow, Poland
Email: szuba@ia.agh.edu.pl

Abstract: Today, computers are based on automata model of computations i.e. Turing Machine and are designed to be deterministic; chaotic behavior is undesirable when considering the stability of algorithms for example. However, when attempting to build the theory of the phenomena of Collective Intelligence (*CI*), it appears that molecular models of computations relying on chaotic behavior of its components must be used as the computational model. Moreover, Chaos emerges as the essential component for Collective Intelligence (*CI*) computational processes, providing some required computational mechanisms and computational properties. The paper attempts to define Collective Intelligence and describe relations between *CI* and Chaos.
Keywords: Collective Intelligence (*CI*), models of computations, parallel distributed and unconscious computations, Chaos.

1. Introduction

Collective Intelligence (*CI*) phenomena are widely observed among beings at all biological levels [5]: starting from Collective Intelligence of bacterial colony fighting against an applied drug, through the Collective Intelligence of social insects (e.g. ants, honeybees), wolves hunting for deer, up to the Collective Intelligence of human social structures. The power of Collective Intelligence as a computational process (for social structures' problem solving) is well known; see for example ant algorithms [1]. Recent research has shown that the computational model of Collective Intelligence is quite different comparing to digital computers. Single beings, weather a small ant (considered by researchers as automata) or a highly intelligent human, ultimately face the same problem – a small accidental event considered as the initial condition for their period of activity, can dramatically change expected final results, causing a butterfly effect. In the case of a small ant, a random turn to the left when exploring the range around a nest, can result in the discovery of a huge food deposit, and consecutively in an avalanche response from all its worker nest mates gathering there, to transport the food towards the nest. A similar situation could be experienced by sir Branson for example – a small piece of information in the morning newspaper or on the internet can result in Branson's activity and thus huge shifts of capital - e.g. in abandoning his space program in favor of say, a new anti-cancer program that promises fantastic benefits. The ant and Sir Branson doing his business, in fact both function in the frame of their global social structure activity

Chaos Theory: Modeling, Simulation and Applications
C. H. Skiadas, I. Dimotikalis and C. Skiadas (Eds)

(global means: from nest perspective or from humankind perspective). Analyzing the social structure from a higher formal level, and applying the proper model of computations to the social structure, allows us to find stability level, because Chaos of individual existence summarized over the social structure, converts into a kind of determinism of Collective Intelligence. Despite the chaotic behavior of single beings, the Collective Intelligence of given social structure is predictable, can be formally described and Collective Intelligence Quotient (IQS) for this social structure can be defined and estimated [2], [3], [4]. However, it must be highlighted that Chaos is most important for mergence of Collective Intelligence in the social structure. The paper will give a short overview of the theory of computational Collective Intelligence, as is presented in [4] and will discuss relations between the two phenomena: Chaos of individual existence and Collective Intelligence of a social structure.

2. Computational model of Collective Intelligence

The entry point for all discussion about Collective Intelligence phenomena is based on the following observations:

1. The individuals inside a social structure usually cooperate or interact in chaotic, often non-continuous ways. In a social structure, beings move quasi-randomly because needs and opportunities of real life force them to do so.
2. Predicate calculus, where facts, rules and goals are used to build inferences, seems to be the most suitable mathematical formalism to describe behavior of beings.
3. Inference processes performed by beings are made randomly, starting when there is a need, when higher-level needs are temporarily not disturbing a given being, or when there is a chance to rendezvous and make inference(s). Most inferences are never finished. This makes the similarity to Brownian movements almost striking, and suggests using quasi-Brownian movements[1] for modeling behavior in a social structure.
4. Individual inferences are also accidental and chaotic.
5. Resources for inferences are distributed in space, time, and among beings.
6. The performance of a given social structure is highly dependent on its organization.
7. Facts and rules in an inference system can create inconsistent systems. Multiple copies are allowed.
8. Probability over the domain of problems must be used as an IQ measure for a social structure.

Collective Intelligence is defined [4] as the unconscious, nondeterministic, parallel and distributed inference process run by social structure. Thus, mathematical logic (not Boolean algebra) is used as the basic computational formalism in the *CI* model of computations. To run this process, beings "are used" in an unconscious way as computational elements, data and software transferring elements and data and software storage elements. The brains of

---

[1] http://en.wikipedia.org/wiki/Fractional_Brownian_motion

beings necessary for everyday life, mutual communication and displacements activity of beings allow these Collective Intelligence functions to be accomplished. The molecular model of computations allows us to formalize this i.e. to give precise definitions and to run simulations to achieve final results on e.g. how intelligent a given social structure is against a specific problem. Let us look at some basic concepts and definitions.

1st level Computational Space (*CS*) with inside quasi-random traveling Clause Molecules (*CMs*) of facts, rules, and goals c_i is denoted as the multiset $CS^1 = \{c_1,...,c_n\}$. Thus, clauses of facts, rules, and goals are themselves 0-level *CS*. For a given *CS*, we define a membrane denoted by $|\bullet|$ which encloses inherent facts, rules, and goals. It is obvious that $CS^1 = \{c_1,...,c_n\} \equiv \{|c_1,...,c_n|\}$. For a certain kind of membrane $|\bullet|$ its type p_i is given, which will be denoted $|\bullet|_{p_i}$ to define which *CMs* can pass through it. Such an act is considered as Input/Output for the given *CS* with a given $|\bullet|$. It is also allowable to define degenerated membranes marked with or $|\bullet$ i.e. a collision-free (with membrane) path can be found going from the exterior to the interior of an area enclosed by such a membrane, for all types of *CMs*. The simplest possible application of degenerated membranes in the *CS* simulating a given social structure is to make, e.g. streets or other boundaries. If the *CS* contains clauses as well as other *CSs*, then it is considered to be one of a higher order, depending on the level of internal *CS*. Such internal *CS* will be also labeled with $\hat{v}_j$ e.g.

$$CS^2 = \{|c_1,...CS^1_{\hat{v}_j},...c_n|\} \quad iff \quad CS^1_{\hat{v}_j} \equiv \{|b_1,...,b_n|\}$$

$$where \quad b_i \quad i=1...m \quad and \quad c_j \quad j=1...n \quad are \quad clauses$$

Every c_i can be labeled with $\hat{v}_j$ to denote characteristics of its individual quasi-random displacements. The general practice will be that higher level *CSs* will take fixed positions, i.e. will create structures, and lower level *CSs* will perform displacements. For a given *CS* there is a defined position function *pos*:

$$pos: O_i \rightarrow \langle position\ description \rangle \cup undefined \quad where \quad O_i \in CS$$

If there are any two internal *CS* objects O_i, O_j in the given *CS*, then there is a defined distance function $D(pos(O_i), pos(O_j)) \rightarrow \Re$ and a rendezvous distance d. We say that during the computational process, at any time t or time period Δt, two objects O_i, O_j come to rendezvous iff $D(pos(O_i), pos(O_j)) \leq d$. The rendezvous act will be denoted by the rendezvous relation ®, e.g. O_i ® O_j which is reflexive and symmetric, but not transitive. The computational process for the given *CS* is defined as the sequence of frames F labeled by t or Δt, interpreted as the time (given in standard time units or simulation cycles) with a well-defined *start* and *end*, e.g. $F_{t_0},...,F_{t_e}$. For every frame its multiset $F_j \equiv (|c_1,...,c_m|)$ is explicitly given, with all related specifications: *pos(.)*, membrane types *p*, and movement specifications *v* if available. The system is initialized for the start of the inference process when the set of clauses, facts, rules, and goals (defined by the programmer) is injected into this *CS*. When modeling the *CI* of certain closed social structures, interpretations in the structure will be given for all CS_n^m, i.e. "this *CS* is a message"; "this is a single human"; "this is a village, a city", etc. The importance of properly defining $\hat{v}_j$ for every CS_j^i should be emphasized. As has been mentioned, the higher level CS_j^i will take a fixed position to model substructures like villages or cities. If we model a single human as CS_j^1, then $\hat{v}_j$ will reflect displacement of that being. Characteristics of the given $\hat{v}_j$ can be purely Brownian or can be quasi-random, e.g. in lattice, but it is profitable to subject it to the present form of CS_j^i. Proper characteristics of $\hat{v}_j$ provide the following essential tools:

- The goal clause, when it reaches the final form, can migrate toward the defined *Output* location (membrane of the main *CS* or even a specific *CS*). Thus, the appearance of a solution of a problem in the *CS* can be observable.
- Temporarily, the density of some *CMs* can be increased in the given area of *CS*. After the given CS_j^i reaches the necessary form, it migrates to specific area(s) to increase the speed of selected inferences.

Let there be given a set *S* of individuals $indiv_1, \ldots, indiv_n$ existing in any environment *Env*. No specific nature is assumed for the individuals nor for their environment. It is necessary only to assume the existence of a method to distinguish $indiv_i$ $i=1,\ldots,n$ from the *Env*. Let there be also given a testing period $t_{start} - t_{end}$ to judge/evaluate the property of *CI* of *S{…}* in *Env*. Let there now be given any universe *U* of possible problems $Probl_i$ proper for the environment *Env,* and be let there be given a complexity evaluation for every problem $Probl_i$ denoted by $f_o^{\text{Probl}_i}(n)$. *CI* deals with both *formal* and *physical* problems; thus we should write the following:

$$f_O^{\text{Probl}_i}(n) \overset{def}{=} \begin{cases} \text{if } Probl_i \text{ is a computational problem, then apply a standard} \\ \text{complexity definition with } n \text{ defining the size of the problem;} \\ \text{if } Probl_i \text{ is a physical problem, then use as a complexity} \\ \text{measure any proper physical units, e.g. weight, size, time,} \\ \text{etc to express } n. \end{cases}$$

Let us also denote in the formula the ability to solve the problems of our set of individuals *S* over *U* when working/thinking without any mutual interaction (absolutely alone, far from each other, without exchange of information):

$$Abl_U^{all\ indiv} \overset{def}{=} \bigcup_{\text{Probl}_i \in U} \max_S \left(\max_n f_O^{\text{Probl}_i}(n) \right)$$

This set defines the maximum possibilities of *S* when individuals are asked one by one to e.g. display their abilities through all the problems. Observe that if any problem is beyond the abilities of any individual from *S*, this problem is not included in the set.

DEFINITION 1. COLLECTIVE INTELLIGENCE AS A PROPERTY

Now assume that individuals coexist together and interact in some way. We say that *CI* emerges because of cooperation or coexistence in S, iff at least one problem Probl' can be pointed to, such that it can be solved by a lone individual but supported by the group, or by some individuals working together, such that

$$f_O^{\text{Probl}_i}(n') \overset{significantly}{>} f_O^{\text{Probl}_i}(n) \in Abl_U^{all\,indiv}$$

or

$$\exists Probl'\, such\ that\ \left(\forall n Probl' \notin Abl_U^{all\,indiv}\right) \wedge \left(Probl' \in U\right)$$

The basic concept of the definition is that the property *CI* emerges for a set of individuals *S* in an environment *U* iff there emerges a quite new problem $\in U$, which can be solved from that point, or similar but even more complex problems can be solved. Even a small modification in the structure of a social group, or in its communication system, behavior, or even in, e.g. the education of some individuals, can result in *CI* emergence or increase. This is one among many cases found when working on *CI* theory, of key importance of the Chaos for this phenomena. An example is when the shoemakers moved their shops from remote villages into the City. This can be the result of a king's order or creation of a "free trade zone". The important thing is that the distance between them has been reduced so much that it triggers new communication channels of some nature (e.g. spying). Defining *CI* seems simple but measuring it is quite a different problem. The difficulty with measuring *CI* lies in the necessity of using a specific model of computations, which is not based on the DTM Turing Machine.

The pattern of inference generalized for any CS has the form:

DEFINITION 2. GENERALIZED INFERENCE IN CS^N (abbreviated *GI)*

Assuming that $CS = \{...CS_j^i...CS_l^k...\}$, we give these definitions:

CS_j^i ® CS_l^k and $U(CS_j^i, CS_l^k)$ and C(one or more CS_n^m of conclusions)

one or more CS_n^m of conclusions, $R(CS_j^i \text{ or } CS_l^k)$

where: $CS^i_j \circledR CS^k_l$ denotes rendezvous relation; $U(CS^i_j, CS^k_l)$ denotes unification of the necessary type successfully applied; *C(one or more* CS^m_n *of conclusions*) denotes that CS^m_n are satisfiable. $R(CS^i_j \text{ or } CS^k_l)$ denotes that any parent *CMs* are retracted if necessary. Later, when discussing N-element inference, we will only be interested in "constructive" inferences, i.e. when a full chain of inferences exists. Thus the above diagram will be abbreviated as $CS^i_j ; CS^k_l \xrightarrow{GI} \sum_n CS^m_n$ without mentioning the retracted *CMs* given by $R(CS^i_j$ or $CS^k_l)$. In general, successful rendezvous can result in the "birth" of one or more *CMs*. All of them must fulfill a *C(...)* condition; otherwise, they are aborted. Because our model of computations is designed to evaluate the *CI* of closed social structures, simplifying assumptions based on real life can be made. It is difficult to find cases of direct rendezvous and inference between two CS^m_i *and* CS^n_j *if* $m, n \geq 1$ without an intermediary involved $CS^0_k \quad k = 1,2,...$ (messages, pheromones, observation of behavior, e.g. the bee's dance, etc.). Even in *Genetic Algorithms*, the crossover of genes can be considered the inference of the two genomes CS^0_i *and* CS^0_j. Only if we consider CS^n on the level of nations, where exchange (migration) of humans takes place, can such a case be considered an approximation to such high level rendezvous and inferences. This is, however, just approximation, because finally, this exchange is implemented at the level of personal contact of humans, which are just rendezvous and inferences of two CS^0_i *and* CS^0_j with the help of $CS^0_k \; k = 1,2...$. Thus, rendezvous and direct inference between two CS^i_j *if* $i \geq 1$ are rare. We will only make use of a single CS^n_{main} *for* $n > 1$ as the main *CS*. Single beings (humans, ants) can be represented as $CS^1_{individual}$. Such beings perform internal brain-inferences, independently of higher, cooperative inferences inside CS_{main} and exchange of messages of the type CS^0. It will be allowable to have internal CS^k inside the main *CS*, as static ones (taking fixed positions) to define sub-structures such as streets, companies, villages, cities, etc. For simplicity, we will try to approximate beings as CS^0; otherwise, even statistical analysis would be too complicated. It is also important to assume that the results of inference are not allowed to infer recursively. They must immediately disperse after inference; however, after a certain time, inferences between them are allowed again (this is called *refraction* in Expert Systems).

The two basic definitions for *CI* and its measure IQS have the form:

DEFINITION 3: N-ELEMENT INFERENCE IN CS^N

There is a given *CS* at any level $CS^n = \{CS_1^{a_1}, \ldots CS_m^{a_m}\}$, and an allowed Set of Inferences *SI* of the form $\{set\ of\ premises\ CS\} \xrightarrow{I_j} \{set\ of\ conclusions\ CS\}$, and one or more CS_{goal} of a goal. We say that $\{I_{a0}, \ldots, I_{a_{N-1}}\} \subseteq SI$ is an N-element inference in CS^n, if for all $I \in \{I_{a0}, \ldots, I_{a_{N-1}}\}$ the premises $\in$ the present state of CS^n at the moment of firing this inference, all $\{I_{a0}, \ldots, I_{a_{N-1}}\}$ can be connected into one tree by common conclusions and premises, and $CS_{goal} \in \{set\ \text{of conclusions for } I_{a_{N-1}}\}$.

DEFINITION 4: COLLECTIVE INTELLIGENCE QUOTIENT (IQS)

IQS is measured by the probability *P* that after time *t,* the conclusion CM_{goal} will be reached from the starting *state of* CS^n, as a result of the assumed N-element inference. This is denoted $IQS = P(t, N)$.

Please note that IQS is not scalar like human IQ – it is a function (surface). For evaluating *CI* the last two definitions fulfill these requirements: • N-element inferences can be allowed to be interpreted as any problem-solving process in a social structure or inside a single being, where N inferences are necessary to get a result; or any production process, where N-technologies/elements have to be found and unified into one final technology or product. Therefore, in the same uniform way, we model inferring processes or production processes within a social structure. This is very important because some inference processes can be observed only through resultant production processes or specific logical behavior (of ants, bees, bacteria). • Simulating N-element inferences allows us to model the distribution of inference resources between individuals, dissipation in space or time, or movements (or temporary concentration) in the *CS*. This reflects well the dissipated, moving, or concentrated resources in a social structure of any type. • Cases can be simulated where some elements of the inference chain are temporarily not available, but at a certain time *t*, another inference running in the background or in parallel will produce the missing components. This is well known in human social structures, e.g. when a given research or technological discovery is blocked until missing theorems or sub-technology is discovered. • Humans infer in all directions: forward, e.g. improvements of existing technology; backward, e.g. searching how to manufacture a given product going back from the known formula; and also through generalization, e.g. two or more technologies can be combined into one more general and powerful technology or algorithm. The N-element inference simulated in the RPP reflects all these cases clearly.

3. Importance of Chaos in Collective Intelligence systems

When working on phenomena of Collective Intelligence and later on during process of building its computational model, and finally when constructing basic definitions, is has been found that Chaos as a property of the considered systems is one of the fundamental components for *CI* and must be "of proper form" to keep *CI* working properly. A good example is bacteria, where small mutation immunizing some bacteria against an applied drug, results in the butterfly effect (positive) for this species of bacteria (and negative for us). This happens frequently because specific Collective Intelligence [6] based on mutual observation (of metabolism products) and the exchange of DNA emerges in bacterial colonies. However, for this kind of *CI* , a proper form of Chaos must be maintained in the bacterial colony, otherwise the *CI* phenomena may vanish. Please note, that on this basis, a research problem of key importance can probably be proposed; "what should the measure and form of Chaos be in the given social structure, e.g. making a company immune to market fluctuations". Another important issue when regarding relations of Chaos and *CI* is that in most cases, researchers working with Chaos theory, by default think about "one process" whereas in case of *CI* we should think in terms of "parallel Chaos". Only parallel chaotic existence of beings can produce convincing, strong global *CI* results. In most cases, (especially when individual intelligence of a being is low) it must be massive parallelism. Increasing the organization of a social structure reduces chaotic behavior of individuals, awarding with individual safety, lower cost of daily life, higher global performance but on the other hand, dramatically reduces ability to solve unexpected, new problems, which a social structure can face. Please note that built-in behavioral strategy of single being (at all intelligence levels) in case of any catastrophe is "panic behavior" which has a chaotic nature. In case of bacteria, this is forced mutation, while in case of more complex beings it can take any form of escape. Keep in mind that *CI* easily devotes single beings when whole social structure profit from this – which implies cancelation of given chaotic local (individual) process.

It is interesting that social structures frequently add a "chaotic component" to their existence in an artificial way. Good examples are: slave & aristocrat: a slave's life is deterministic, whereas the aristocrat, on the basis of resources produced by slaves, is allowed to lead a more chaotic life. Other examples are; a regiment on a battle field with rangers around, the concept of lottery giving an illusory chance to simple people, to change their life, etc. Summarizing the discussion, the ability of the social structure to solve specific class of problems, expressed through *CI* parameter IQS becomes non-Chaotic; despite a lot of individual Chaos inside the social structure. Perhaps the most convincing example is phenomena called "A. Smith Invisible Hand of Market". Parameters describing the free market, display all symptoms of a chaotic system [7], however the Collective Intelligence of the free market makes it stabilize, even robust against actions trying to destabilize it.

References

[1] M. Dorigo, T. Stuetzle: Ant Colony Optimization . *The MIT Press.* 2004.

[2] Szuba T.: A Molecular Quasi-random Model of Computations Applied to Evaluate Collective Intelligence. *Future Generation Computing Journal. volume/issue: 14/5-6.* 1998.

[3] T.Szuba: A formal definition of the phenomenon of collective intelligence and its IQ measure. *Future Generation Computing Journal. 17, Elsevier,* 2001.

[4] T. Szuba: Computational Collective Intelligence. *Wiley bookseries on parallel and distributed computing. Hardcover, 410 pages*, 2001.

[5] T. Szuba: "Was there Collective Intelligence Before Life on Earth? (Considerations on the Formal Foundations of Intelligence, Life and Evolution)". *World Futures - The Journal of General Evolution. Gordon and Breach Publishing. 58/1,* 2002.

[6] Szuba T., Polański P., Schab P., Wielicki P. : On efficiency of Collective Intelligence Phenomena. *Submitted in 2009 to Information Sciences Journal, Elsevier.*

[7] Baumol W. J, Benhabib J.: Chaos - Significance, Mechanism, and Economic Applications. *Journal of Economic Perspectives, American Economic Association. 1989 vol. 3(1).*

Complex signal generators based on capacitors and on piezoelectric loads

Horia-Nicolai L. Teodorescu[1,2] and Victor P. Cojocaru[3]

[1] Faculty of Electronics, Telecommunications and Information Technology, 'Gheorghe Asachi' Technical University of Iasi, Romania
[2] Institute of Computer Science of the Romanian Academy, Iasi, Romania
Email: hteodor@etc.tuiasi.ro
[3]The "Gheorghe Asachi" Technical University of Iasi, Iasi, Romania
Email: vcojocaru@etti.tuiasi.ro

Abstract: We briefly analyze and demonstrate several nonlinear signal generators, according to the principles proposed in [1]. These systems apparently should not oscillate, because they include only capacitors in the positive feedback loops. However, the design attempted to make use of the well-known parasitic elements of the capacitors to create a selective feedback loop with physical capacitors only. The use of the resonant (piezoelectric) load adds a new potential mode of oscillation. The design aimed to make achievable as many as possible frequency points where oscillation modes ay exist, against the rather common belief that a single oscillation mode can be supported at one time in electronic oscillators. For this purpose, variable, independent gains have been provided to the positive feedback loops that correspond to the various modes in the circuit. We describe the schemes, briefly analyze their behavior, show simulation and experimental results, and discuss potential uses. The paper is largely based on [1].
Keywords: nonlinear capacitor, chaotic electronic circuit, multimodal oscillation, parasitic elements, Warburg model, SPICE simulation

1. Introduction

Recently, a large number of chaotic circuits have been proposed, for example [1-6], with applications ranking from secure communications [7] to predicting the market evolutions [8-9]. Here, we provide some experimental results on the new class of chaotic systems proposed in [1]. Also, we provide a few more details on the operation of these circuits. The analysis is performed in the frames of the linear theory and of the general Barkhausen condition; the treatment in the framework of nonlinear differential equations will be provided elsewhere. In the linear approximation, we draw the equivalent scheme of the amplifiers and of the main electric components in the circuit and focus on the frequency dependence of the gain and phase in the open loop.

Chaos Theory: Modeling, Simulation and Applications
C. H. Skiadas, I. Dimotikalis and C. Skiadas (Eds)
 (pp. 423 - 430)

The proposed circuit is basically a loop composed of two inverting amplifiers, one of them having apparently an essentially capacitive load. The capacitive load adds one pole and is known to produce the instability of the amplifier. However, the scheme proposed uses an *electrolytic* capacitor, moreover uses *a double capacitor* (preferably) *in parallel.* The ideas presented in [1] are (i) the electrolytic capacitor is a *non-linear* capacitor, in fact, it is a *fracpol* (fractionary pole) which boosts feedback in a whole band instead of a single frequency; (ii) the double capacitor, one electrolytic and one non-electrolytic, introduces an almost flat region in the phase plot of the coupling system, thus allowing a broader possible region for the chaotic regime; (iii) using two or more loops helps extending the range where the Barkhausen condition might be satisfied and helps increasing the number of bifurcation points.

2. The basic circuit and its equivalent scheme

The basic circuit consists of a single loop composed of two OA coupled through a capacitor (see Fig. 1). An extra element is a piezoelectric load to one of the OAs.

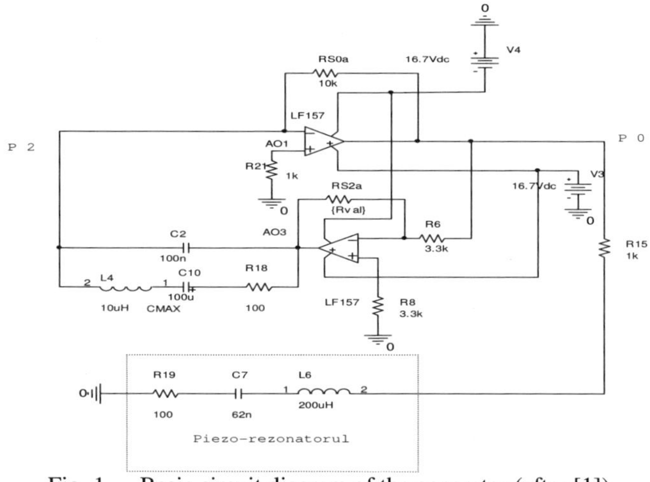

Fig. 1. Basic circuit diagram of the generator (after [1])

We used an operational amplifier type LF 157 -- actually, a noisier version, 574UD3, as higher noise favors the starting of oscillations in the circuit and may add to the overall chaotic process. The LF 157 circuit has two main poles, represented in the equivalent scheme by two RC circuits (see Fig. 2 a). The equivalent scheme is obtained by replacing the amplifiers and the capacitors by their equivalent circuits. The result will be shown, after a brief discussion, in Fig 3, for the circuit in Fig. 1.

The scheme may oscillate as a relaxation oscillator if we admit saturation nonlinearity for the operational amplifiers (OA), according to (see simplified equivalent circuit next to the equations):

$$U_2 = AU_1 \ or \pm V_{CC} = \begin{cases} \max\left(-AU_1, V_{CC}\right) \ if \ U_1 \le 0 \\ \min\left(-AU_1, -V_{CC}\right) \ if \ U_1 > 0 \end{cases}, \quad U_3 = -AU_2 \ or \pm V_{CC}$$

$$\frac{dU_C}{dt} = \frac{1}{C}\frac{U_3 - U_1}{R_1} = \frac{1}{C}\frac{\max\left(AU_2, V_{CC}\right)}{R_1}, \quad U_3 - U_1 = U_C$$

$$\begin{cases} \max\left(-AU_2, V_{CC}\right) - U_1 = U_C \\ \min\left(-AU_2, -V_{CC}\right) - U_1 = U_C \end{cases}.$$

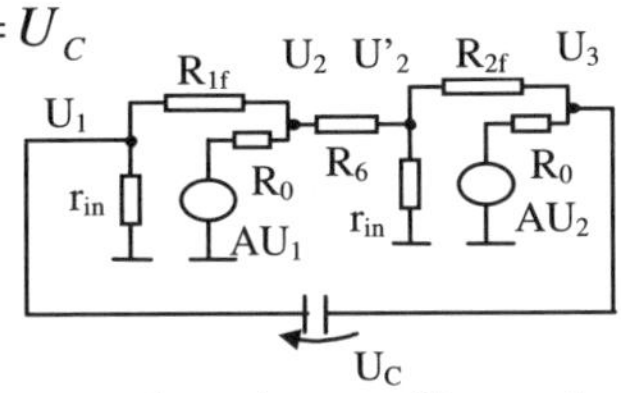

The operation in the saturation case corresponds to a relaxation oscillator, hence chaos should not appear. The nonlinear dynamics is favored by the key element of the scheme, the electrolytic capacitor, which is a reputed nonlinear element. In fact, the class of chaotic circuits discussed is best characterized by and is based on the use of the electrolytic capacitors as nonlinear elements and by the use of two parallel capacitors for extending the range where the (generalized) Barkhausen relation may be satisfied. Because an electrolytic capacitor includes an ionic conductor, its equivalent scheme includes a nonlinear element, beyond the proper capacitance. Actually, the ionic and the mixture conductors have an intricate behavior, which is translated in the equivalent scheme to a distributed series and parallel RC circuits with pole frequencies up to infinity (see [1]). The nonlinear element is characterized by a so called *fracpol*, a pole with fractional power, $1/s^{\alpha}$, where $\alpha \in \mathbf{Q}$, typically $\alpha \in (-1,0)$ for capacitive elements. The most frequently used scheme of an electrolytic capacitor contains the so-called *Warburg element*, with $\alpha = -0.5$. A few details pertaining to the operation are shown in Fig. 2. Correlating the first and second panel in Fig. 2, we derive that several points may satisfy the Barkhausen condition at the same time, which means instability.

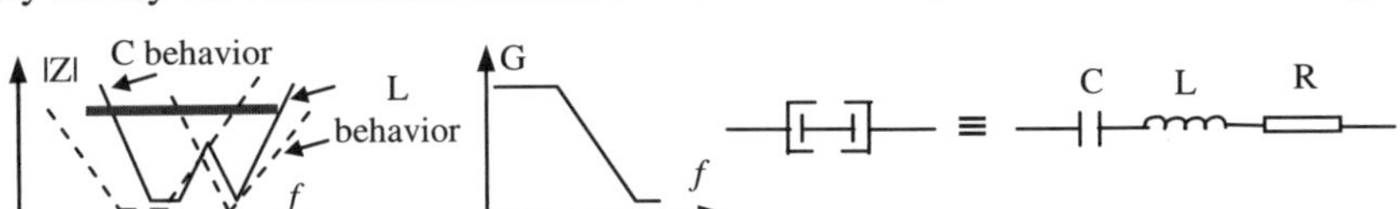

Fig. 2. Approximate variation of the impedance of the two parallel capacitors with frequency (left). The two poles of the OA (middle). The basic (linearized) equivalent scheme of a capacitor

The equivalent scheme of the circuit is shown in Fig. 3. The scheme is built taking into account only three poles, for the operational amplifier, but the LF 157 circuit used is better represented by a four pole scheme. In the equivalent circuit, which stands for a single capacitive loop, the electrolytic capacitor has been replaced by a nonlinear scheme with a Warburg element, while a linear RLC scheme replaces the ceramic capacitor.

The system is oscillating for any frequency value satisfying the Barkhausen condition, $\underline{G}\cdot\beta = 1 \in \mathbf{C}$. This condition is satisfied for several frequencies, according to the scheme and to the gain and phase plots of the amplifiers and of the group of the capacitor. Recall that the impedance of the Warburg element is $Z_W = \dfrac{a}{(j\omega)^{0.5}}$, with a basic phase of – 45°.

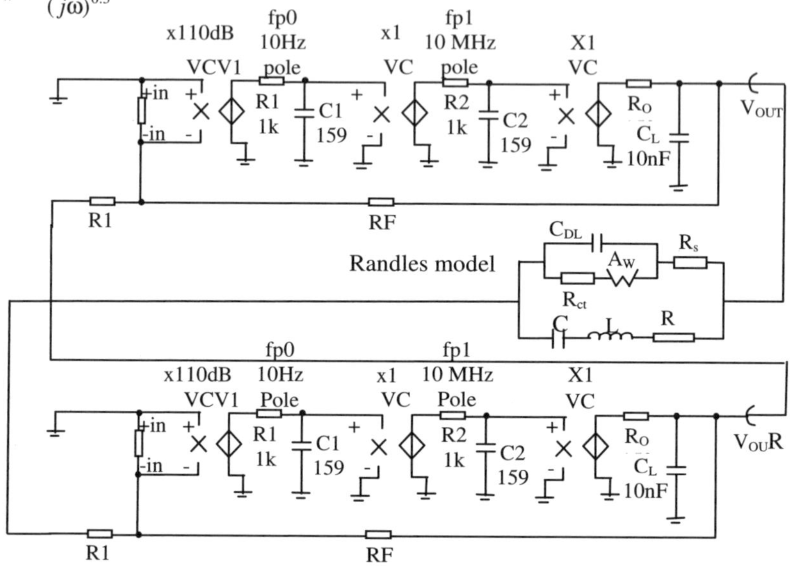

Fig. 3. Equivalent scheme, assuming a model with three poles for the OpAmp, and a Warburg model for the electrolytic capacitor; single loop configuration

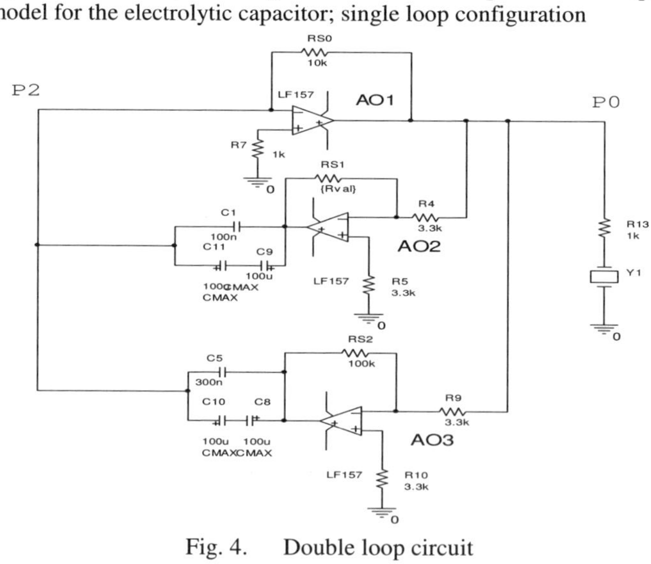

Fig. 4. Double loop circuit

3. Double loop circuit

We extended the circuit presented in the previous sections by a second loop, motivated by the fact that the new loop will add more intricate (in)stability conditions and will produce a larger family of chaotic regimes, due to new modes of self-oscillations allowed. This more complex scheme uses two feedback loops decoupled through two operational amplifiers, see Fig. 4. The double loop circuit further extends the range where the Barkhausen condition may be satisfied and creates new bifurcation conditions with respect to the simple loop. Thus, it is not surprising that the double loop circuit has an even more complex nonlinear dynamic.

4. Simulation and experimental results

4.1. Simulations

While SPICE simulation can not be expected to produce correct results because of the lack of appropriate models for the capacitors and because of the partly correct model of the OAs, it is still useful to grasp a view of the potential instability regions of operation. The results of the simulations are exemplified in Fig. 5, showing incept of the instability.

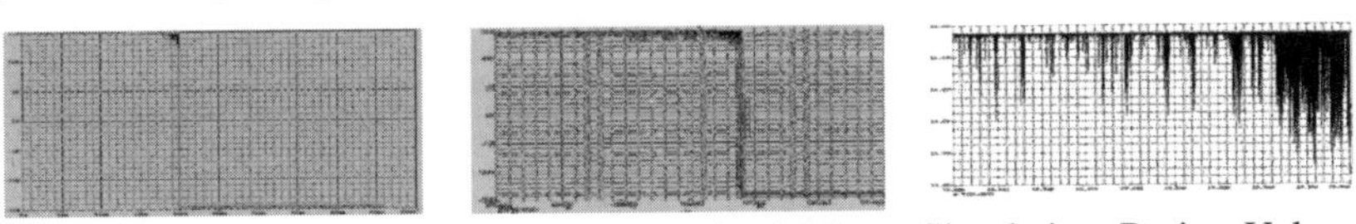

Simulation, PSpice. Values for RS1- 500 Ω RS2 100 Ω time 0 - 50 ms

Simulation, Pspice. Values for RS1-500 Ω RS2 100 Ω time 18 - 21ms

Simulation, Pspice. Values for RS1-500 Ω RS2 100 Ω time 19.0 - 1.9ms

Fig. 5. Example of SPICE simulation, for three time frames, showing the initiation of a chaotic regime. Double loop circuit

4.2. Measuring results

The measuring scheme used in the experiments is shown in Fig. 6. The measurements were performed at the temperature t = +25°, VDD = 16V.

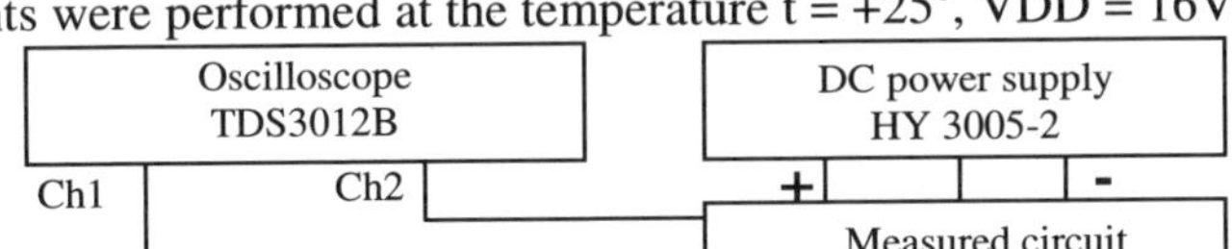

Fig. 6. The setup for the measurements

The experimental evidence is shown in Figures 7,8. In these figures, several waveforms obtained at the output of the circuit are shown, along with the corresponding spectra obtained with by FFT algorithm implemented on the oscilloscope type TDS3012B. In Fig. 7, the phase diagrams are also shown, using the signals in the measuring points P1 and P2.

Notice in the above figures the $1/f^{\gamma}$ type of spectrum at low frequencies, and the prevalence of the white noise at higher frequencies. The overall noise spectrum is well described for these circuits, for large ranges of the values of the resistors in the circuit, by the formula $S(f) = a/f^{\gamma} + b$.

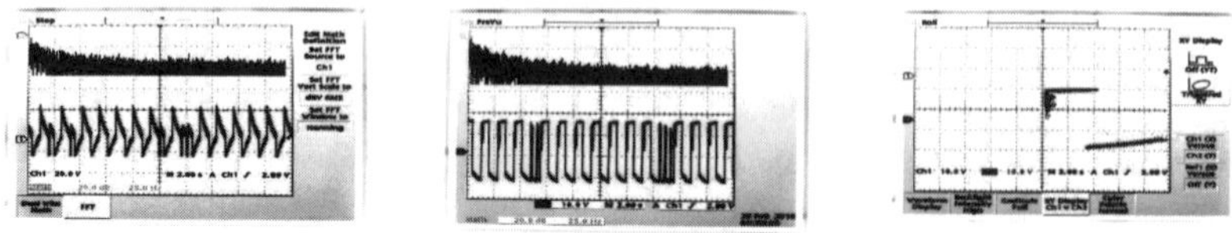

RS1-1.5k, RS2-100k, Top: FFT P2, Bottom: P2

RS1-1.5k, RS2-100k, Top: FFT P0, Down: P0

RS1-1.5k, RS2-100k. The XY diagram

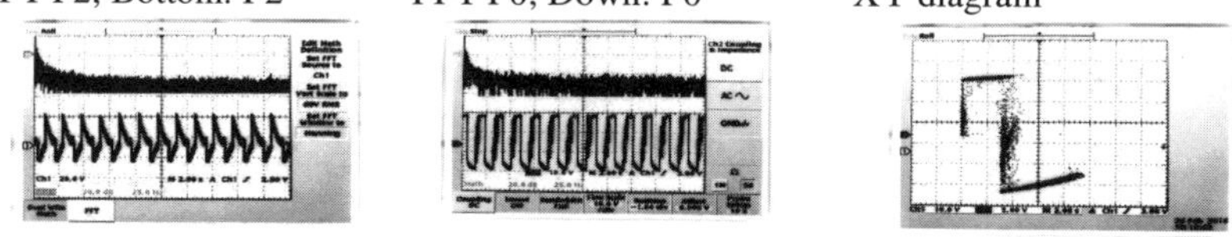

RS1-680Ω, RS2-100, Top: FFT P2, Bottom: P2

RS1-680Ω, RS2-100k, Top: FFT P0, Bottom: P0

RS1-680Ω, RS2-100k. The XY diagram

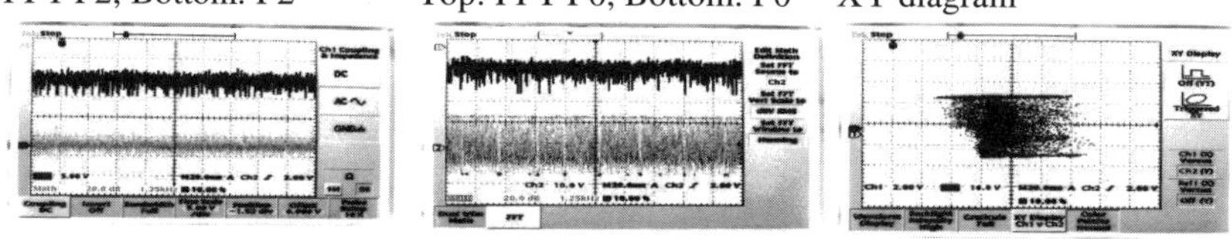

RS1-100k, RS2-200Ω, Top: FFT P2, Bottom: P2

RS1-100k, RS2-200Ω, Top: FFT P0, Bottom: P0

RS1-100k, RS2-200Ω. The XY plot

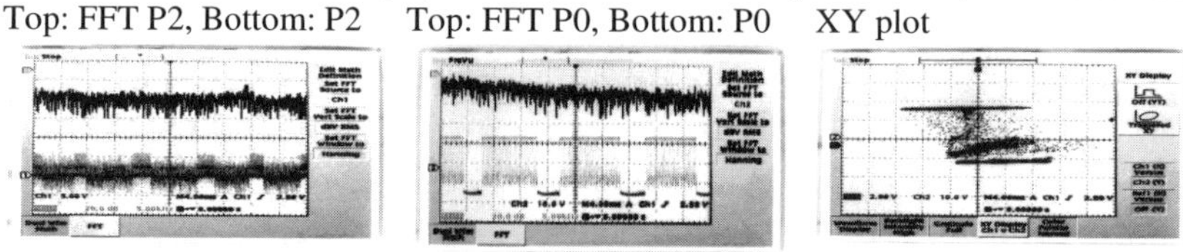

RS1-100k, RS2-330Ω, Top: FFT P2, Bottom: P2

RS1-100k, RS2-330Ω, Top: FFT P0, Bottom: P0

RS1-100k, RS2-330Ω. The XY plot

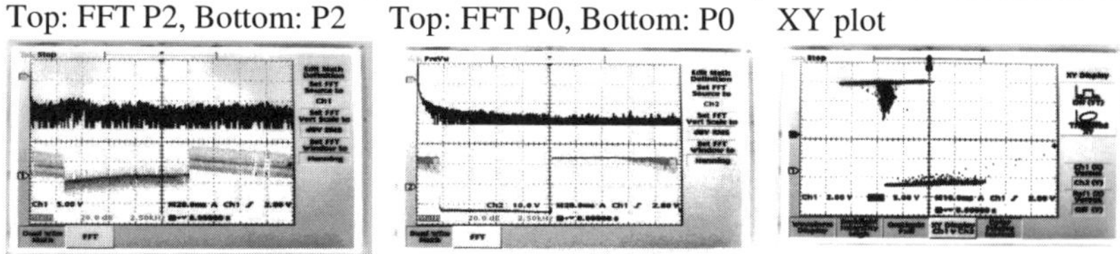

RS1-100k, RS2-470Ω, Top: FFT P2, Bottom: P2

RS1-100k, RS2-470Ω, Top: FFT P0, Bottom P0

RS1-100k, RS2-470Ω. The XY plot

Fig. 7. Examples of experimental results, for various values of the resistors in the scheme. Double loop circuit

Notice in the first two rows of Fig. 7 the existence of two competing regimes and the transition from one to the other with the instability in the middle (see XY plot in the second row of Fig. 7).

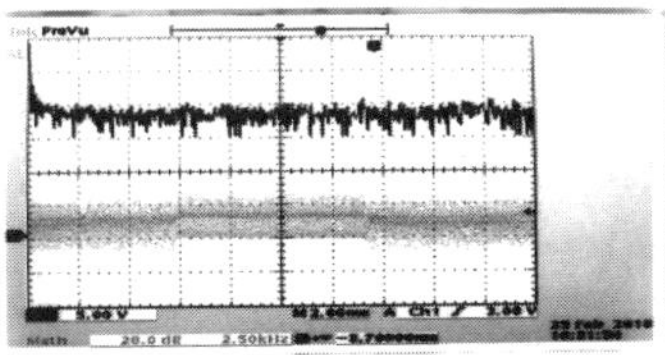

RS2-68Ω. Top: FFT P2, Bottom: P2

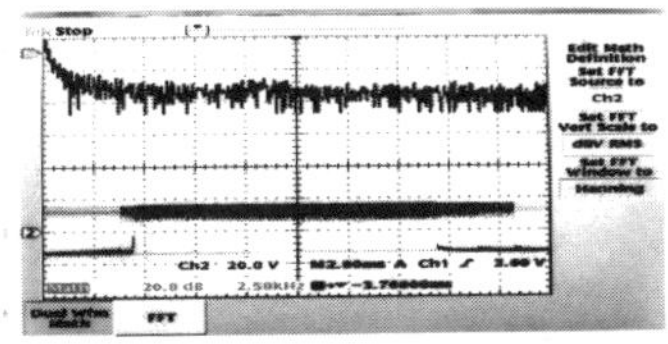

RS2-68Ω. Top: FFT P0, Bottom: P0

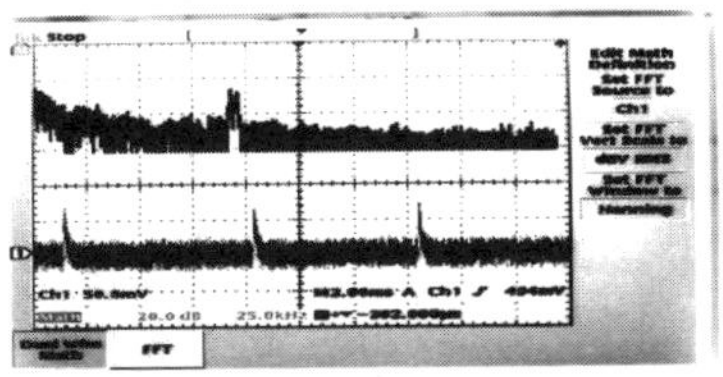

RS2-200Ω. Top: FFT P2, Bottom: P2

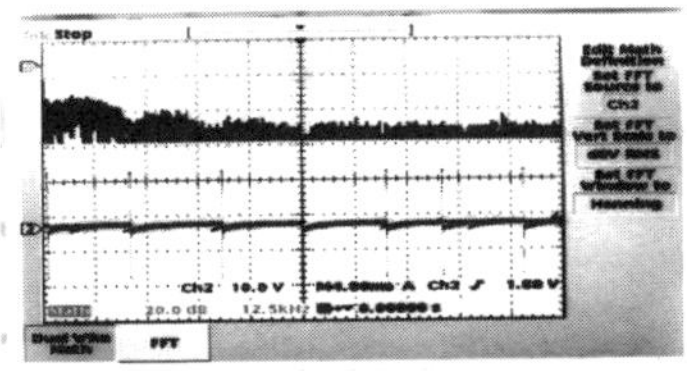

RS2-200Ω. Top: FFT P0; Bottom: P0

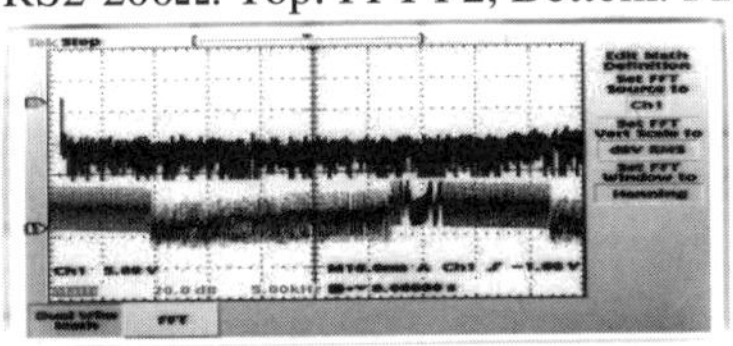

RS2-220Ω. Top: FFT P2, Bottom: P2

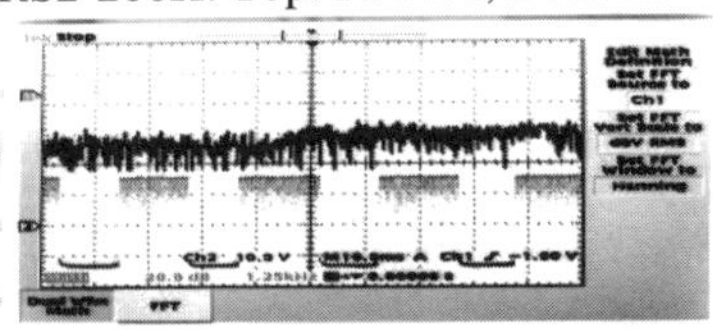

RS2-220Ω. Top: FFT P0, Bottom: P0

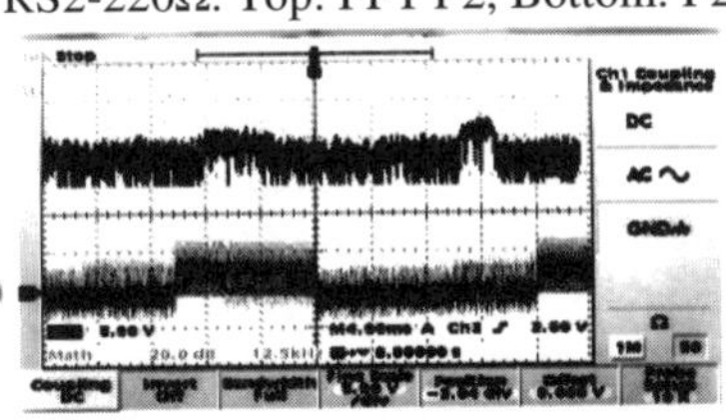

RS2-270Ω. Top: FFT P2, Bottom: P2

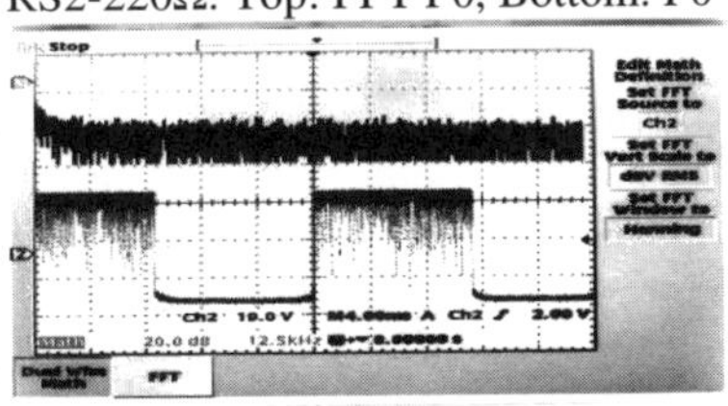

RS2-270Ω. Up: FFT P0, Down: P0

Fig. 8. Examples of experimental results, for various values of the resistors in the scheme. Single loop circuit

Notice in the Figures 8 that an almost periodic relaxation oscillation is superposed over a chaotic behavior.

5. Conclusions

We implemented and measured under various conditions two schemes proposed in [1], demonstrating the chaotic behavior obtained under broad conditions. Essential to the operation of the schemes under chaotic regime are the use of two capacitors in parallel and the nonlinear behavior of the electrolytic capacitor. The scheme is very simple, cheap, and works reliably with no special devices and with little trimming (one or two variable resistors) to find a convenient regime.

The schemes have already been used for increasing the security of a system based on a microcontroller. That application is described in [10]. Another application where we consider applying the circuit is driving a piezoelectric emitter in an echo-location system for robotic applications. The circuits described are also intended for the generation of large bandwidth ultrasounds for a bio-mimetic system.

Authors' contributions. The circuits were proposed by the first author who also carried out the planning of the work and the method of analysis and implementation. The second author implemented and simulated the circuits under the supervision and with the support of the second author. Both authors contributed to the measurements. The first author wrote the paper. Both authors recorded the photographs. The copyright remains with the first author.

Acknowledgment. The work of the second author was supported by the doctoral project grant BRAIN; the first author was partly supported by the research grant AdBioSonar (CNMP, Romania).

References

[1]. H.N. Teodorescu, *A New Class of Chaotic Circuits based on Capacitive Feedback.* Proc. *ITEI*, Kishnew, R. Moldova, May 20-22, 2010.

[2]. Y. Kobayashi, H. Nakano, T. Saito, *A Simple Chaotic Circuit with Impulsive Switch Depending on Time and State.* Nonlinear Dynamics (2006) 44: 73–79, Springer 2006.

[3]. S. Mandal, S. Banerjee, *An Integrated CMOS Chaos Generator*, Proc. 1st Indian National Conf. on Nonlinear Systems & Dynamics, 2003; also S. Mandal, S. Banerjee, *Analysis and CMOS Implementation of a Chaos-Based Communication System*, IEEE Trans. Circ. & Systems, 51(9):1708-1722, 2004

[4]. K. Mitsubori and T. Saito, *Dependent Switched Capacitor Chaos Generator and Its Synchronization.* IEEE Transactions Circuits and Systems-L: Fundamental Theory and Applications, Vol. 44, No. 12, pp. 1122-1128, December 1997.

[5]. A. Algaba, E. Freire, E. Gamero, A.J. Rodríguez-Luis, *Analysis of Hopf and Takens–Bogdanov Bifurcations in a Modified van der Pol–Duffing Oscillator.* Nonlinear Dynamics 16: 369–404, 1998. Kluwer Academic Publ.

[6]. A.H. Sato, H. Takayasu, Y. Sawada, *Power law fluctuation generator based on analog electrical circuit.* arXiv.org, Condensed Matter, http://arxiv.org/PS_cache/cond-mat/pdf/0007/0007326v1.pdf (20 Jul 2000).

[7]. K.G. Mishagin, V.V. Matrosov, L.V. Kuzmin, A.V. Kletsov, *Multi-band Chaotic Oscillator with Phase-locked Loop.* Piers Online, 5, 6, 531-535, 2009.

[8]. H. Fanchiotti, C.A. Garcıa Canal, N. Martınez, *Critical Analysis of Electronic Simulation of Financial Market Fluctuations.* http://arxiv.org/PS_cache/cond-mat/pdf/0203/0203067v1.pdf (February 1, 2008).

[9]. A.H. Sato, H. Takayasu, *Market price simulator based on analog electrical circuit.* arXiv.org > cond-mat > arXiv:cond-mat/0104318. http://arxiv.org/abs/cond-mat/0104318 (18 Apr 2001).

[10]. H.N. Teodorescu, F. Iftene, *Increasing the Operation Security of a Microsystem with a Chaotic Oscillator.* Accepted, ISIIE Conference Galati, Sept. 2010.

Drift waves' synchronization by using an external signal. The stabilization of a chaotic plasma turbulence

C. L. Xaplanteris[1, 2] and E. Filippaki[2]

[1]Hellenic Military Academy, Vari Attikis
[2]Plasma Physics Lab, IMS, NCSR "Demokritos", Athens, Greece
Email: lfilip@ims.demokritos.gr

Abstract: In a cylindrical cold rf plasma, a variety of drifts and other sorts of waves are usually observed; when a turbulence is created, the state becomes chaotic and then the plasma turns out to be more unstable. In the present work, an external signal is enforced on the plasma's waves (or turbulence), which strongly affects the physical magnitudes of the plasma instabilities. The final result is that plasma stabilization occurs when plasma waves are synchronized with the external signal. Moreover, nonlinear phenomena occur, such as a vigorous coupling among the waves' frequencies, which affect the Hall conductivity. Another significant observation is the influence of boundaries on the interaction waves.

Keywords: plasma instabilities, wave-wave interaction, synchronization, stabilization of a chaotic state.

1. Introduction

The unstable behavior of the plasmas and the physical tendency to shape nonlinear dynamics, which appear by oscillations or waves form [1], are well known. At many cases the plasma's inconstancy raises so much, that a turbulence is formed and the state becomes chaotic [2]. During the early 60's plasma instabilities and the interactions among them, were considered as a serious obstacle in the thermonuclear fusion process, and since that time the plasma's chaotic configuration has been put under strict consideration.The purpose is to acquire the knowledge and consequently the ability to control and suppress the chaotic states by using feed-back process or applying simple external signals [3-5]. This successful intervention on oscillatory unstable plasma gives the expectation that the chaotic states of plasmas can be stabilized by imposing suitable external signals.
The last idea was capable to give a reason for the experiments and the researches that have been carried out at the Plasma Laboratory of 'Demokritos' during the recent decades [6-8]. In these researches drift waves have appeared and explained as originally, caused by the rf power gradient

Chaos Theory: Modeling, Simulation and Applications
C. H. Skiadas, I. Dimotikalis and C. Skiadas (Eds)

[6,7], or as collisional instability [9] and a thorough experimental study is carried out, by using an external signal on the drifts. For the first time Lashinsky managed to synchronize drift waves by using an external driver perturbation to an antenna placed into the plasma column [10].

In the present work, the drift waves' synchronization with a periodic driving signal is shown; in a cylindrical symmetry, an rf magnetized argon plasma is produced, since drift waves proragate azimouthally. As the magnetic field's intensity B crosses the upper hybric resonance ($B_0 \cong 8{,}35.10^{-2} T$), the drift changes the propagating direction on the opposite way [6]. In this way, copropagating and counterpropagating electric fields are easily formed to achieve the wave-wave interaction [2]. Another ability in the cylindrical device is the placing of the external field antenna, radially [6] or axially [6,7] so that the direction of the driving wave can be affected. Phenomena of frequency coupling were observed by the resonance, between the external wave and any harmonic of the drift wave.

In Section 2, the experimental set-up is presented, as well as the results are shown too. An interaction among waves produced into plasma accompanied by theoretical interpretative model is presented at Section 3. Finally, in Section 4 the interpretations and conclusions are given.

2. Experiment's description

2.1. Set-up description

In a quasi Q-machine device, a cold, magnetized rf argon plasma is produced. The plasma production keeps the following factors and the physical parameters unchanged: the plasma cavity is always constituted by a metallic closed cylinder, with fixed inner radius ($r = 3cm$) and its axis is placed along the external homogeneous magnetic field B. The cylinder length is taken according to the experiment purpose; however, its value varies between $40 - 140cm$. The rf power for the plasma's maintenance is produced by a magnetron generator operating at $2{,}45GHz$; this wavy power is entered in the cavity by a metallic antenna fastened on the circular base center, along the cylinder axis, so that a coaxial line is formed with the externar conductor. The antenna's diameter is $0.6cm$ and its length is adapted to the cylinder's length. The described set-up has a complete cylindrical symmetry and consequently, the mathematical elaborations are always carried out on cylindrical system. Suitable electrical probes have been adapted, in order to control every interesting point of the plasma column; these probes move radially, azimouthally and axially, providing the ability to meassure the plasma parameters (density, temperature, electric potential,e.t.c.). Specifically, the disk probe is moving radially and around its axis and it is used to measure the electron drift current; moreover, it may operate as an emissive antenna directing the driving signal on the azimouthal drift wave.

Likewise, the coaxial antenna 2 is used for the driving wave entering. Figure 1 shows the outline of the device described above.

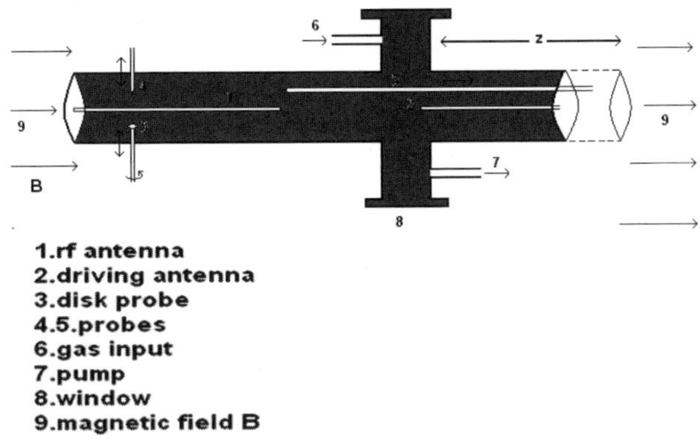

Figure 1. The experimental set-up placed at the external magnetic field.

2.2. Experimental results

A drift wave's existence is persistent and an extensive research has been carried out in Plasma Laboratory of 'Demokritos' [6-9]. These waves fall into three (regions) ranges of frequency, propagate azimouthally but have an axial component sometimes. The first group with the lower frequency ($\approx 17KHz$) has been examined firstly and listed as drift wave, caused by the rf electric fild gradient [6,7]. As it was experimentally found and theoretically interpreted, the wave propagates azimouthally and changes direction as the external magnetic field intensity crosses the upper hybrid resonance B_0. Figure 2a shows a typical spectrum of this drift wave accompanied by its' harmonics. In Figure 2b a turbulence is shown when the plasma becomes more unstable.

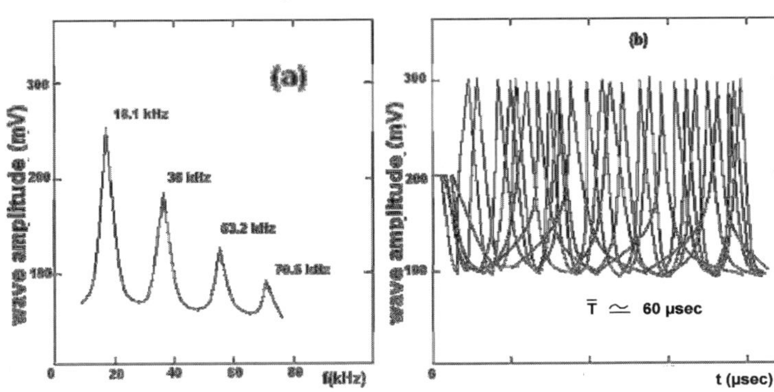

Figure 2a. A typical wave spectrum is shown, while in Figure 2b a turbulence is presented.

2.2.1. Waves' coupling

The drift wave synchronization is initially attempted by using the coaxial antenna 2 for the external signal entering. The main concern in this phase of the experiment was to change the driving wave frequency and amplitude. The drift wave (its fundamental) frequency was $f_0 \cong 18{,}1KHz$, since the driving frequency f_d started from a lower value ($f_d \cong 6KHz$). As Figure 3a shows, when the f_d approaches the fundamental frequency f_0, a strong coupling takes place and the otherwise constant drift wave frequency is attached to the driving wave one and follows it from value $f_d \cong 14KHz$ to $f_d \cong 22KHz$. Enough attention has to be paid to the way by which the driving wave affects the drift wave's harmonics. The same research has been repeated as the driving frequency crosses one harmonic of the drift wave and examines how this influences the fundamental f_0; although the results are obscure, we are able to say that the affected drift's harmonic (from the driving wave) affects the whole drift spectrum. However, this influence is powerless, and more evident only on the frequency, than on the wave amplitude. Just as it is expected, an increase of wave amplitude is only observed on the syntonized harmonic, since the other harmonics'amplitude remains uninfluenced. Figure 3b shows the driving influence on plasma wave's harmonics.

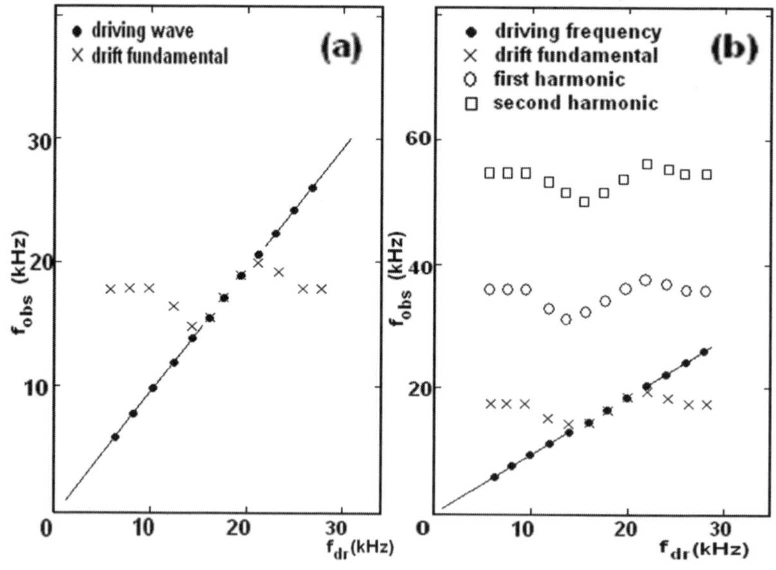

Figure 3a. The coupling between driving and fundamental is presented, while in Figure 3b the influence on harmonics appears.

2.2.2. Drift wave's synchronization

To synchronize the azimouthally moved drift wave easily, the disk probe is used due to its ability to turn around itself. By setting the plane surface perpendicularly to the azimouthally propagating drift, the coincidence of two waves direction is achieved; these waves may copropagate or counterpropagate and this becomes possible in two ways: firstly, with the disk probe surface rotating by 180^0 and secondly, when the external magnetic field is crossing the upper hybrid resonance B_0. It is remarkable that the driving wave follows an azimouthal orbit as if the drift wave was operating as guidewave. Figure 4 presents the cut of the plasma production device, where the antenna, the disk probe and the drift wave representation are shown.

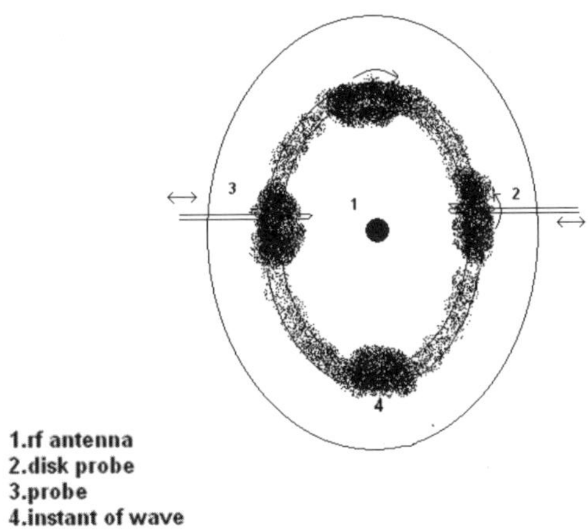

Figure 4. A cut of the plasma column shows the azimouthal wave's propagation.

In both cases, with the copropagation or counterpropagation of the azimouthal waves, the result depends on the exciter amblitude, mode number and much more on the driver frequency. If the experiment is extended in phase space, then the complicated state may end in chaotic and the drift's synchronization demands painful effort. Figure 5 shows the copropagation of the two waves when the $E_0 = E_0^1$, and $\omega \approx \omega^1$ conditions are valid.

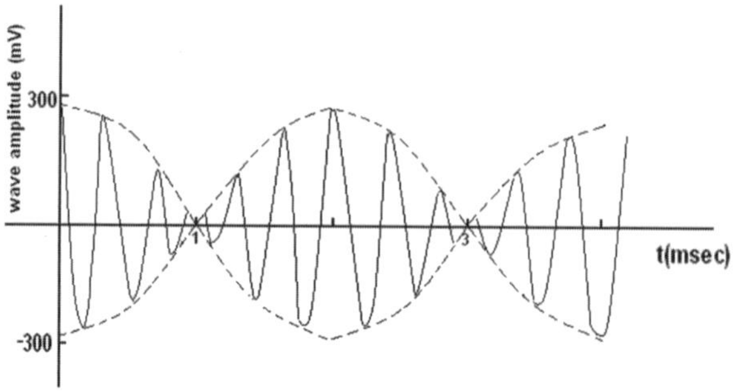

Figure 5. Drift's synchronization by the waves' copropagation is shown

It must be noted that the waves' interaction behaves as a pure beat.
Figure 6 presents the waves' counterpropagation under the same conditions ($E_0 = E_0^1$, and $\omega \approx \omega^1$), and the standing wave's configuration is evident.

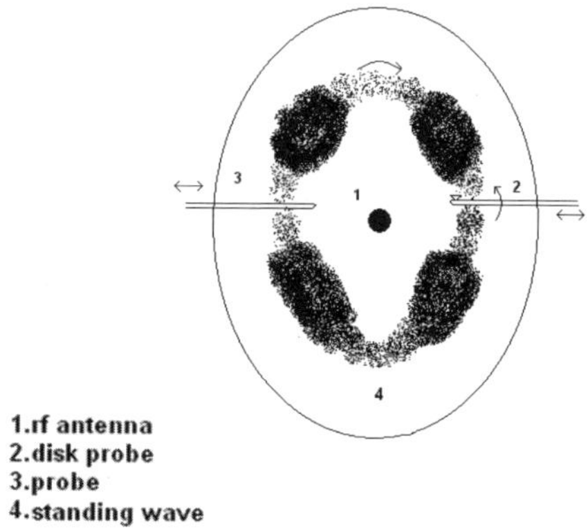

Figure 6. A standing wave's construction by the waves' counterpropagation is presented

3. Theoretical model

A drift wave in the lower frequency range is caused by rf field gradient [6] and obeys the following dispersion relation:

$$\omega_l \cong l\Omega_i + \frac{l}{2}(\Omega_R - \Omega_D) - j\frac{\nu_i}{2} + j|s|C_s\sqrt{\frac{U_R}{U_D} - 1} \qquad (1)$$

Where, $\Omega_i, \Omega_R, \Omega_D$ the ions,the rf field gradient and density gradient cause respectivelly gyrofrequency, ν_i the ion-neutral collisions, $s \equiv {}^{1}\!/_{n_0}\left({}^{dn_0}\!/_{dr}\right)$

the density radial gradient, C_s the ion sound velocity and, U_R, U_D the drift velocities.

Eq. (1) justifies directly the drift wave spectrum (Figure 2) and the fundamental f_0 as the upper harmonics. Moreover, it interprets the wave excitement as eq. (1) gives growth rate.

Then, the drift wave propagates azimouthally with the form $E_0.e^{j(\omega t-l\theta)}$ or $E_0.e^{j(\omega t+l\theta)}$, according to the propagation direction. If the same mode driving wave meets the plasma wave by the same phase shift, but its frequency and amplitude have different values ω^1 and E_0^1 respectively, then the composition depends on the waves' direction:

i) in copropagation, the addition is,

$$E_{tot} = E_0.e^{j(\omega t-l\theta)} + E_0^1.e^{j(\omega^1 t-l\theta)} \tag{2}$$

By taking the case $E_0 = E_0^1$, Eq. (2) becomes,

$$E_{tot} = E_0.e^{-jl\theta}.(e^{j\omega t} + .e^{j\omega^1 t})$$

or

$$E_{tot} = E_0.e^{-jl\theta}.2\cos\frac{\omega-\omega^1}{2}t.e^{j\frac{\omega+\omega^1}{2}.t} \tag{3}$$

If it is taken that $\omega \approx \omega^1$, Eq. (3) gives,

$$E_{tot} = 2E_0.e^{-jl\theta}.\cos\frac{\omega-\omega^1}{2}t.e^{j\omega .t} \tag{4}$$

Eq. (4), reveals that for the fixed azimouthal angle θ, the resultant wave has a beat_behavior.

When $\omega = \omega^1$, Eq. (4) is written,

$$E_{tot} = 2E_0.e^{-jl\theta}.e^{j\omega .t} = 2E_0.e^{j(\omega t-l\theta)}$$

That is simply the doubling of wave amplitude.

In case $E_0 \neq E_0^1$, the problem is aproached with the separation of the bigger amplitude, but the solution is compound and difficult to confirm experimentally.

ii) in counterproragation the addition is,

$$E_{tot} = E_0.e^{j(\omega t - l\theta)} + E_0^1.e^{j(\omega^1 t + l\theta)} \tag{5}$$

By aproaching with $E_0 \cong E_0^1$,Eq. (5) becomes,

$$E_{tot} = E_0.e^{j\omega t}(e^{-jl\theta} + e^{jl\theta}) = E_0.e^{j\omega t}(\cos l\theta + \cos l\theta)$$

or

$$E_{tot} = 2E_0.\cos l\theta.e^{j\omega t} \tag{6}$$

It must be noted that Eq. (6) represents a standing wave.

4. Experimental data interpretation

Eq. (1) has the ability to interprete the drift wave existence (Figure 2a) and the fundamental frequency f_0, as well as the upper harmonics [6]. If non-linearity is taken into estimation when Eq.(1) is produced, then the turbulence of Figure 2b may be justified. Figure 3 shows the waves' coupling when the driving signal is placed away (antenna 2). It is evident that the synchronization is strongly affected by the frequencies' value. Moreover, Figure 5 and Figure 6 results can be interpreted with Eq. (4) and Eq. (6) respectively. Temporaly, only the simple conditions are considered into the theoretical model; when the waves' magnitudes are different, the phenomena become much complicated to be interpreted mathematically, although they are evident during the experiment (optically and acoustically). The formation of a suitable model capable to give explanation, is the following challenge.

References

[1] J. Wesson, Tokamaks, 2nd edn.Oxford: Clarendon Press, 1997.

[2] T. Klinger, in Handbook of Chaos Control, edited by H.G. Schuster (Wiley VCH, Weinheim, 1998) pp. 513-562.

[3] K.D. Weltmann, M.E.K. 0epke,and C.A. Selcher, Phys. Rev. E **62**, 2773, 2000.

[4] E. Gravier, X. Caron, and G. Bonhomme, Phys. Plasmas **6,** 1670, 1999.

[5] D. Block et al, Phys. Rev. E **63,** 056401, 2001.

[6] A. Anastassiades, and C.L. Xaplanteris, J. Phys.Soc. of Jpn **52,** 492 1983.

[7] C.L. Xaplanteris, Astrophys. Space Science **136,** 171-181, 1986.

[8] C.L. Xaplanteris and E. Filippaki, Topics on Chaotic Systems, Selected Papers from CHAOS 2008 International Conference, pp. 406-415, 2009.

[9] C.L. Xaplanteris, J. Plasma Physics, Vol. **75,** part 3, pp. 395-406, 2009.

[10] H. Lashinsky, in Symposium on Turbulence of Fluids and Plasmas, edited by J. Fox (Polytechnic Press, New York), pp. 29-46, 1968.

Chaos game technique as a tool for the analysis of natural geomorphological features

G. Žibret[1] and T. Verbovšek[2]

[1]Geological Survey of Slovenia, Ljubljana, Slovenia
Email: gorazd.zibret@geo-zs.si
[2]Faculty of Natural Sciences and Engineering, University of Ljubljana, Slovenia
Email: timotej.verbovsek@ntf.uni-lj.si

Abstract: The paper presents a novel method for evaluating the different natural processes. The method is based on the chaos game technique, and evaluation of the results is based on the measurements of the fractal dimension of the obtained Sierpinski triangles. Pathways of the natural rivers channels case study has been used for the demonstration of the methodology, despite the method can be applied to many natural phenomena. Authors would like to encourage others to make similar analysis, because more case studies are needed to draw more solid conclusions.
Keywords: chaos game, Sierpinski triangle, fractal dimension, river.

1. Introduction

Fractal properties of many natural systems are known and thoroughly described. Among them, river networks and channels were first natural objects analysed by using fractal methods, and some well-known laws has been discovered, such as bifurcation and stream-length order or power law of length and basin area. In this study a different and novel approach is used. The question addressed was: can the topological 2–D properties of natural objects, like coastal lines, river channels pathways, mountain top ridges etc. be described by equations or are they mainly influenced by random or, even better by pseudorandom[1] processes? In the presented case, the river channel pathways have been chosen as a case study of the used methodology.

2. Methods

The methodology is based on the chaos game technique. If played with a sufficiently long sequence, created by a random process like throwing a perfect dice, than this technique generates a complete Sierpinski triangle. The theory goes further – any sequence, obtained by a non-random process

[1] Authors have to state at this point, that term "random process" is used here for a process, influenced by very large number of factors – measuring and taking all of them into account is impossible with the current technology (for example, like a calculation of a trajectory of every sand particle in a sandstorm, despite it might be, at least theoretically, possible in the future).

Chaos Theory: Modeling, Simulation and Applications
C. H. Skiadas, I. Dimotikalis and C. Skiadas (Eds)

(chaotic, periodic or similar sequences) produces an incomplete triangle [1]. The question addressed in this study was, what happens if the chaos game is played with the sequences, obtained from the natural systems. In this case the sequences describing the pathways of river channels have been used as an example. Prior to that, the sequences had to be generated first. For the demonstration, the four large rivers have been chosen: Amazon, Mississippi, Danube and Sava River. The procedure begins with (a) digitalization of the river channel in Euclidean space with Autodesk AutoCAD application. (b) On the obtained course line, n numbers of points P_i; $i=1..n$ has been superimposed, dividing the line into $n-1$ equally long segments. The elements of the sequence obtained in this way (denoted as *Seq*) are the angles between different segments. Element i of the sequence *Seq* thus equals the angle between two lines, connecting the points P_{i-1} and P_i and points P_i and P_{i+1}. Angle has a positive value, if the course of a river turns in clockwise direction and vice versa. The obtained sequence has been then (c) converted to the sequence of three classes (denoted as *Seq**), which can be used as an input for the chaos game technique. To balance the obtained Sierpinski triangle, the limit values for each class have been set as 33.333^{th} and 66.666^{th} percentile values (P^{33} and P^{66}) of the elements of the *Seq*. If the element i of the sequence *Seq* was smaller than the value of P^{33}, then the value of the element i of the sequence *Seq** was set to 1, if it fell between P^{33} and P^{66}, it was set to 2 and if was greater than the value of P^{66} then it has been assigned with a value of 3. The procedure (d) continues with playing the chaos game with the sequence *Seq**, with the initial point set at the origin of the coordinate system. Finally, as a quantitative measure of the obtained result, (e) the fractal dimension of the generated image, based on the box-counting method, has been calculated.
The mathematical calculations and specially-written computer programs have been firstly tested on different mathematical sequences to evaluate and detect possible errors of the used method. Tested sequences were: digits of constants π and e and internal rand() function of MS Excel application. All three of them produced almost complete triangle and measured fractal dimension of the triangle was different not more than 2% of the expected theoretical value. The methodology was thus recognized to be correct and satisfactory.

The aforementioned methodology was applied to 4 large rivers: Amazon, Mississippi, Danube and Sava with values of n = 1000, 2000, 4000, 8000, 16000 and 32000. To make a comparison of the natural river pathways with the mathematical curves, which try to simulate them, the same methodology was then applied to them. Curves were named as *sinus* (simple sine wave), *supersin* (superposition of 4 sine waves with different wavelengths and amplitudes) and *complex* (superposition of 4 waves with alteration of amplitudes and wavelengths – curve with the most similar appearance as the natural rivers, used in this study). Moreover, two types of very small amount of noise have been introduced to the *complex* function, producing two additional functions. The first function, named *external* was generated as the addition of noise to the *complex* function (*external(x)=complex(x)+noise*) has been introduced. Second function, named *internal*, noise was introduced to the variable (*internal(x)=complex(x+noise)*).

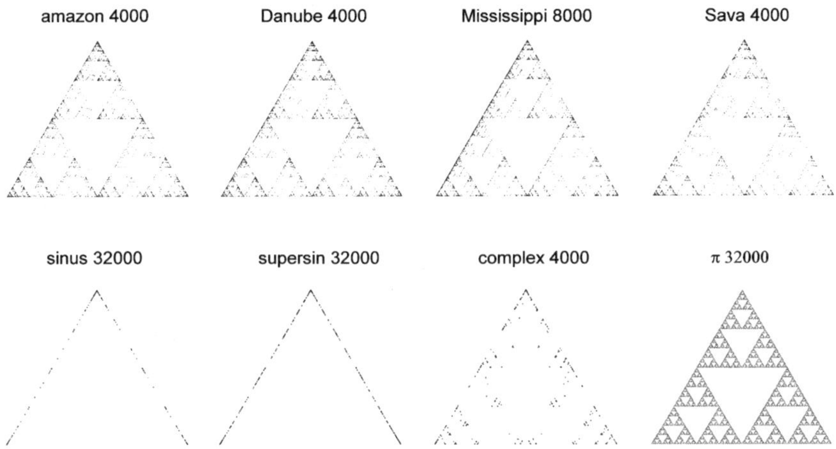

Figure 1: Images of the incomplete Sierpinski triangles, obtained by the chaos game technique. Upper images were generated with the sequences obtained from natural rivers and lower triangles with the sequences, obtained from different superposition of sinus waves. A more detailed description of functions can be found in the text. π represents a triangle obtained by playing a chaos game technique using decimals of π number. Number beside the name of the function represents the number of points used. Pixels in the upper row are artificially enhanced by factor 4, and pixels in the lower row (except for the π case) by a factor 8 to increase visibility on hardcopy.

3. Results

Images of the obtained Sierpinski triangles at the selected number of segments, where biggest fractal dimension has been found, are presented on Figure 1. Chaos game has been played with sequences, representing four natural rivers, aforementioned mathematical functions and the number π (test sequence). Figure 2 shows how the fractal dimension of the obtained triangle varies in relation to the number of segments used.

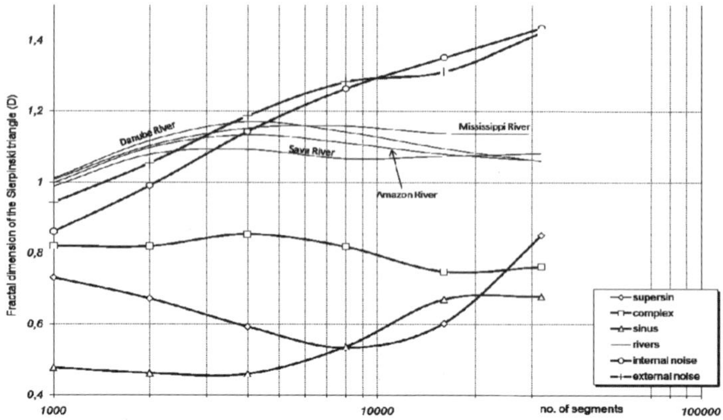

Figure 2: Fractal dimensions of the obtained triangle in the relation to the number of the segments, used for the generation of the sequence, needed for the chaos game.

Due to the fact that 1000 or 2000 points are not enough to construct a complete triangle using imaging resolution of 1024x1024 pixels (as in our study), all values from Figure 2 have been corrected. Correction was made on the basis of the measurement of the dimension of the Sierpinski triangle, obtained with the same procedure as the analysis of the rivers. The only difference is that a random sequence was used to play the chaos game. The correction factor was then calculated with the formula r=t/D, where r = correction factor, t = mathematical dimension of the complete Sierpinski triangle (log(3)/log(2)=1.585), D = dimension of the obtained triangle using methodology described in chapter 2. Results are shown in the Table 1. Figure 3 shows the values from Figure 2, which are corrected by the correction factors from Table 1. With this procedure the limitations of the methodogy used in our study are bypassed.

Table 1: Comparison of the measure fractal dimension of the Sierpinski triangle (D) using random sequences of different length (N) with the theoretical value of the dimension (t). Correction factor (r) is calculated according to the equation provided below.

N	D	t	r=t/D
1000	1.0143	1.585	1.5627
2000	1.1338	1.585	1.3980
4000	1.2490	1.585	1.2690
8000	1.3518	1.585	1.1725
16000	1.4357	1.585	1.1040
32000	1.5057	1.585	1.0527
64000	1.5567	1.585	1.0182

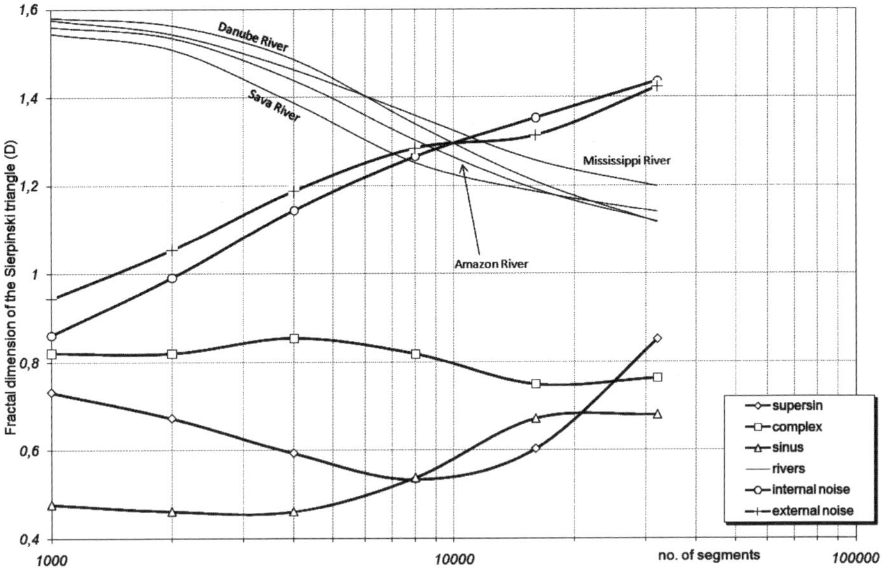

Figure 3: Corrected fractal dimensions of the natural rivers and generated curves for all rivers and artificial functions.

4. Discussion

One of the findings of the study is that natural rivers can be most accurately described by a complex superposition of sinus waves as the shape of the curve from Figure 2 of *complex* function corresponds best to the natural rivers, despite they exhibits higher degree of randomness in their pathways. Noise introduction, neither internal not external, does not completely solve this problem. When comparing the rivers, Danube River exhibits the highest degree of randomness in its main channel pathway, and Sava River the smallest. Rivers with smaller gradients, like Danube or Mississippi, have a higher degree of randomness in their pathways than those with larger gradient [2].

The results in Figure 3 suggest that the rivers, observed on the larger scale, behave as almost completely random, because the dimensions of the Sierpinski triangles are very close to the theoretical ones. But when "zooming in" or, in other words, observing the river pathway with a higher precision, the influence of the random factors decreases (there is a decrease of the obtained dimension D) and a more predictable behaviour is observed (meanders). So the following conclusion can be made; that the regional geological and tectonical conditions are the "random" driving force in the behaviour of the river course, while meanders, despite being chaotic in their nature (sensitive to the initial conditions), are much more predictable and can be better described with mathematical equations or simulations.

However the major aim of the study was not only to discuss the results of the river behaviour, but also to present a methodology and to encourage others to try similar research on any natural phenomenon which can be described with three different classes which fits as input for the chaos game technique. The problem is, that despite as much of the whole process of the analysis has been automated (up to the authors' skills and possibilities), it is still a long-lasting and time-consuming. And more analyses, based on this method, are needed to be able to successfully draw any further conclusions.

5. Remark

The methodology, results and the discussion, along with the references, has been thoroughly described in the article entitled "Quantitative analysis of randomness exhibited by river channels using chaos game technique: Mississippi, Amazon, Sava and Danube case studies" [2], published in the journal Nonlinear Processes in Geophysics. It is published under Creative Common licence and is freely available to download from the NPG journal web site (http://www.nonlin-processes-geophys.net/16/419/2009/npg-16-419-2009.pdf). The readers are invited to test their results with the method proposed.

This contribution has been also presented on the conference *Chaos 2010* (01-04 June 2010, Chania, Greece). Despite the fact that some new outcomes are presented in this paper, the readers are kindly requested to refer to the original paper [2].

References

[1] HO. Peitgen, H. Jurgens and D. Saupe. *Chaos and fractals – New Frontiers of Science*. Springer-Verlag, New York, 2004.

[2] G. Žibret and T. Verbovšek. Quantitative analysis of randomness exhibited by river channels using chaos game technique: Mississippi, Amazon, Sava and Danube case studies. *Nonlin. Processes Geophys.*, 16:419–429, 2009

Dynamics of a steel turning process

Grzegorz Litak and Rafał Rusinek

Lublin University of Technology, DepartNadbystrzycka 36, 20-618 Lublin, Poland
Email: g.litak@pollub.pl

Abstract: This paper explores the cutting force oscillations. Forces have been measured during the stainless steel turning. We provide the results of standard statistical analysis of the corresponding time series together with their recurrence properties. We claim that the system, which is initially in a regular vibration region for some fairly larger cutting depth, is unstable to chaotic oscillation appearance. This could have the important implication to the process control procedure.
Keywords: turning process, nonlinear vibrations, recurrence plots, recurrence quantification analysis.

1. Introduction

The machining technology is the most important component in the modern massive production. Over the past years, its fast development gave way to a reliable high-speed cutting procedure. Consequently, elimination and stabilization of the associated chatter oscillations have become a high interest in science and technology [1]. The plausible adaptive control concept, based on relatively short time series, has been studied to gain deeper understanding.

We investigate dynamics of a turning process by the recurrence plot technique [2]. The experimental time series of a stainless steel cutting process enable us to distinguish different types of the system response. This method, supplemented by recurrence quantification analysis (RQA), was used to analyze relatively short time series [2-4]. Changing the single system parameter, cutting depth, we observed qualitative change of the system response. The special attention is focused on high speed cutting [1,5-6]. The knowledge about dynamics of this process can be used to optimize the process to obtain surfaces of the best quality.

2. Experimental Procedure and Results

Our turning experiment has been conducted using the workpiece circular shaft of stainless steel (EZ6NCT25) with the diameter of 22mm and the tool cutting edge angle of 45 degrees. The shaft angular velocity was fixed to 710 rpm while the corresponding feeding ratio was 0.25 mm/rev. Experiments have been done for three different cutting depths. The obtained results, the three orthogonal force components (feed F_x, thrust F_y, and cutting F_z components, respectively) measured with the sampling frequency was 2 kHz are presented in the 3D force diagram (Fig. 1). In Tab. 1 including the statistical (standard deviation-*SD*, skewness-*SK*, and kurtosis-*KU*) rend recurrence (*RR*-recurrence rate) information.

Chaos Theory: Modeling, Simulation and Applications
C. H. Skiadas, I. Dimotikalis and C. Skiadas (Eds)

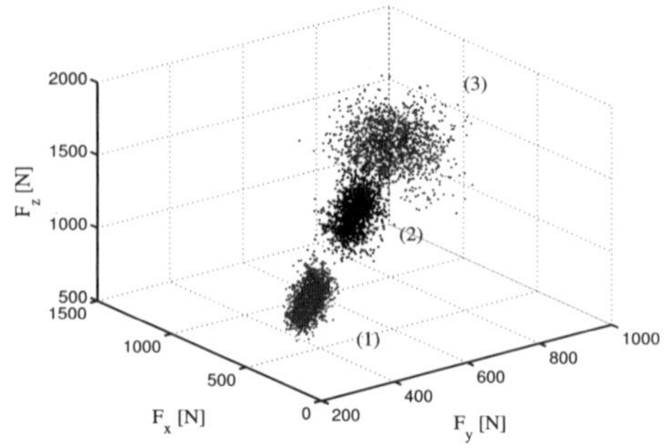

Figure 1. The 3D force diagram. Points (1), (2), and (3) denote the cutting force measurement points for the cutting depth h=1.00, 1.75, and 2.30 mm, respectively. The sampling frequency was 2 kHz, while the number of considered measurement points N=1500.

In this note we show the results of this method limiting to the fairly small time interval of 1500 measurement points. The aim of our consideration is to follow the force fluctuations evolution with changing the cutting depth h. Figure 1 illustrate the distribution of measured points for the cutting depth h=1.00, 1.75, and 2.30 mm, respectively. Note that the all three averages of the corresponding forces components are increasing monotonically with increasing h (Fig. 1). Furthermore, the fluctuation range of cutting force components are also increasing. The summary of their statistical properties (Tab. 1) confirms this trend in terms of the standard deviations SD_i (where i=x,y,z). Interestingly for the smaller h (1.00 and 1.75 mm) the dominant fluctuations is z-component (see the DS_i in Tab. 1) while for the largest h (2.30mm) y-force component. Skewness SK_i is relatively small for all cases indicating the left-to-right mirror symmetry of distribution. Finally, kurtosis values KU_i are relatively close to that of the Gaussian distribution 3.0 for i=z force components. On the other hand for the feed force component (i=x) we observe the typical leptokurtic density (KU_i<3). However, the last case of the thrust force component (i=y) is the most interesting. For the cutting depth h=1.00 and 2.30 the distribution is leptokurtic (KU_y=2.591 and 2.661, respectively) while for h=1.75 we observe the platykurtic (KU_i > 3) distribution with kurtosis KU_y reaching 3.40. This is the largest discrepancy from the monotonic tendencies and may signal intermittency. To tell more about this transition and the difference between dynamical response of the h=1.00 and 2.30 oscillations one need to do more extensive analyze. In this paper we propose the recurrence analysis by recurrence plots quantified by recurrence rate RR. In this aim we reconstruct the embedding space including the current forces measured at given sampling event and time delayed quantities:

$$x(i) = [F_x(i), F_y(i), F_z(i), F_x(i-\Delta i), F_y(i-\Delta i), F_z(i-\Delta i)],$$

where Δi is the characteristic time delay. Note, basing on the system response main periodic behaviour, for our system we assumed Δi=45 in the units of sampling intervals and the embedding space dimension has been fixed to 6. The corresponding recurrence matrix is defined [2]

$$R_{ij} = \Theta(\varepsilon - \| x(i) - x(j) \|),$$

where ε is the threshold value and simultaneously the tolerance for states identification. Note that the matrix elements are taking 0 and 1 values depending on the $\mathbf{x}(i)$ and $\mathbf{x}(j)$ states similarities in the embedding space. $\Theta(.)$ is the step Heaviside function.

Tabele 1. The summary of statistical (mean value-*MV* standard deviation-*SD*, skewness-*SK*, and kurtosis-*KU*) rend recurrence (*RR*-recurrence rate) information. ϵ has been chosen as 0.6*SD*.

h [mm]	SD_x, SD_y, SD_z [N]	SK_x, SK_y, SK_z	KU_x, KU_y, KU_z	MV_x, MV_y, MV_z [N]	RR [10^{-3}]
1.00	32.6 45.0 86.5	-0.041 0.013 0.240	2.016 2.591 2.821	403.5 581.2 794,9	9.3
1.75	36.2 68.8 97.3	0.037 0.253 0.089	2.102 3.400 2.847	594.6 740.9 1189.1	7.8
2.30	44.9 151.1 131.9	0.157 0.153 0.204	2.628 2.661 3.055	743.5 869.6 1501.8	6.9

The ratio of the nonzero matrix elements number to the total number of elements, excluding the diagonal ones R_{ij}-δ_{ij}, defines the recurrence rate parameter *RR*. Their values for examined cases defines the recurrence properties of the system response and are strictly connected with the correlation sum [2-4]. We estimated this parameter for the examined cases (Fig. 1) and included in the Tab. 1. For equal approach to different cases we expressed ϵ in units of the three dimensional standard deviation *SD* obtained as the geometrical sum of the corresponding deviations related to *x, y,* and *z* force components:

$$SD = \sqrt{SD_x^2 + SD_y^2 + SD_z^2}$$

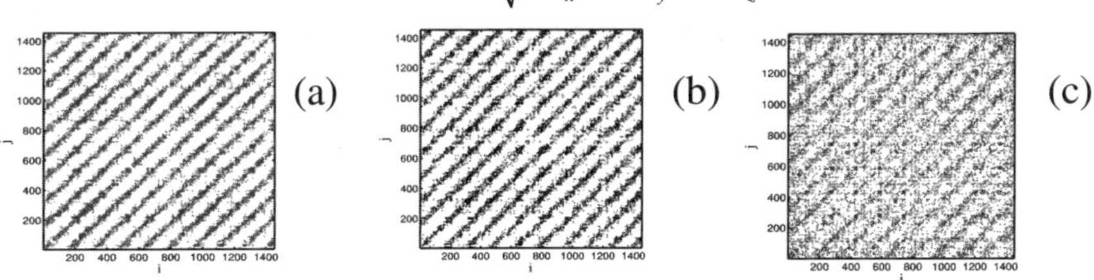

Figure 2. Recurrence plots for different cutting depths h=1.00, 1.75, and 2.30 mm for (a), (b), and (c), respectively. ϵ has been chosen as 0.6*SD*. Statistical and recurrences parameters are summarized in Tab. 1

Interestingly, one can see the tendency: the larger cutting depth *h*, the smaller recurrence rate *RR*. To study the distributions of recurrence we show the corresponding recurrence plots in Fig. 2. The coloured regions in the three figures (Fig. 2a, b, and c) are noticeably different. For instance the diagonal lines patterns correspond to periodic nature of oscillations. Thus, from the above plots one can easily see that increasing cutting depth *h* leads to less periodic or even chaotic behaviour.

3. Conclusions

This work presents experimental investigations of cutting process in which the stainless steel was cut. During the experiment three component cutting forces are measured as a function of the cutting depth. The obtained signals were analyzed using statistical and recurrence methods. Our results show that the increasing cutting depth leads to higher values of all cutting forces components. Furthermore, the fluctuations are also increasing showing, however, different distributions for different force components. The most interesting contribution appeared in the *y* component. Its local non-monotonous change of kurtosis signals an interesting change in the system dynamic response. The recurrence plots are clearly indicating that the response is less periodic (or more chaotic) for larger cutting depth *h*. The obtained results can be used to identify types of vibrations and select the most proper cutting parameter [5-6]. Note that the recurrence plots and recurrence quantification (by the *RR* parameter) are tools suitable to analyze fairly short time series [2-4].

To draw more conclusions we will study the fluctuations more systematically by including intermediate cutting depths and consider longer time series to apply additional dynamical tests as Fourier and wavelet analysis.

Acknowledgment

The financial support of Structural Funds in the Operational Programme - Innovative Economy (IE OP) financed from the European Regional Development Fund - Project "Modern material technologies in aerospace industry", No. POIG.01.01.02-00-015/08-00 is gratefully acknowledged.

References

[1] Y. Altintas, *Manufacturing Automation: Metal Cutting Mechanics, Machine Tool Vibrations, and CNC Design*, Cambridge University Press, Cambridge, 2000.

[2] N. Marwan, M.C. Romano, M. Thiel, and J. Kurths, Recurrence plots for the analysis of complex systems, *Physics Reports,* 438: 237-329, 2007.

[3] G. Litak, M. Wiercigroch, B.W. Horton , and X. Xu, Transient chaotic behaviour versus periodic motion of a parametric pendulum by recurrence plots, *Z. Angew. Math. Mech.* 90: 33-41, 2010.

[4] G. Litak, J.T. Sawicki, R. Kasperek, Cracked rotor detection by recurrence plots, *Nondest. Test. Eval.* 4: 347-351, 2009.

[5] R. Rusinek, Vibrations in cutting process of titanium alloy, *Eksploatacja i Niezawodność – Maitenance and Reliability* 3: 48-55, 2010.

[6] G. Litak, A.K. Sen, A. Syta, Intermittent and chaotic vibrations in a regenerative cutting process, Chaos, Solitons & Fractals 41: 2115-2122, 2009.

Author Index

Adzhemyan L.Ts., 89
Aidanpää Jan-Olov, 97
Aleixo Sandra M., 309, 333
Aniszewska Dorota, 105
Axenides Minos, 110
Aybar O. O., 127, 231

Baba Yuya, 281
Barbot Jean-Pierre, 143
Bauch Szymon, 265
Bhattacharjee Jayanta K., 120
Budyansky M.V., 3
Busawon Krishna, 207

Cakar O., 127
Caneco Acilina, 135
Cejnar Pavel, 406
Chakraborty Sagar, 120
Chatziioannou Aristotelis, 239
Chrousos George P., 239
Cojocaru Victor, 423
Conde Luis, 151

Daniels S, 255
Datcu Octaviana, 143
Del Rio Ezequiel, 151
Dick O.E., 159
Donoso Jose M, 151
Dowling D P, 255

Elaskar Sergio, 151

Filippaki E., 431
Fossion Ruben, 247, 406
Frank Alejandro, 247, 406
Fukushima Kenta, 349
Fyman P.A., 3

Georgaki Anastasia, 167
Ghassemlooy Z., 207
Grácio Clara, 135
Grigoras Carmen, 175
Grigoras Victor, 175

Hacinliyan A. S., 127, 131
Hagan Kerry L., 183
Hnatich M., 89
Honkonen J., 89
Hul Oleh, 265

Iliopoulos A.C., 297

Jevtic N., 191

Kalashnikov Vladimir L., 199
Karakatsanis L.P., 297
Kharel Rupak, 207
Kharkov Yaroslav A., 59
Khavroshkin O. B., 14
Khots Boris, 215
Khots Dmitriy, 215
Kretz Johannes, 223
Kugiumtzis Dimitris, 289
Kusbeyzi I., 127, 231

Lambrou George I., 239

Landa Emmanuel, 247, 406
Law V J, 255
Ławniczak Michał, 265
Leonov Gennady A., 21
Lindkvist Göran, 97
Litak Grzegorz, 445
Lopez Vieyra J.C., 247

Macek Michal, 406
Miranda Eduardo, 273
Morales I., 247
Moschovi Maria, 239
Munmuangsaen Buncha, 399
Nwankire C E, 255

Onishi Ryo, 281
Orman Gabriel V., 30

Papana Angeliki, 289
Pavlos E.G., 297
Pavlos G.P., 297
Pestana Dinis D., 309, 333
Pikulin Dmitry , 317
Ponomarev V.I., 3
Pound Eleri A. , 325
Prants S.V., 3

Ramm Alexander G., 41
Rocha J. Leonel, 135, 309, 333
Rusinek Rafał, 445
Ryabov Vladimir B., 341, 349
Rybaczuk Marek, 105

Sarkar Amartya, 120
Schweitzer J.S., 191
Serquera Jaime, 273
Sirko Leszek, 265
Skiadas Charilaos, 357
Skiadas Christos H., 53, 357
Sokolov Valentin V., 59
Sotiropoulos Anastasios, 369, 379
Sotiropoulos Dimitrios A, 77, 379
Sotiropoulos Vaggelis, 379, 388
Srisuchinwong Banlue, 399
Stine P.,191
Stránský Pavel, 247
Szuba Tadeusz (Ted), 414

Takahashi Keiko, 281
Teodorescu Horia-Nicolai, 423
Tsolakis Christos, 167
Tsoutsouras V.G., 297
Tsyplakov V. V., 14

Uleysky M.Yu., 3

Velazquez V., 247
Verbovšek T., 439
Vlad Adriana, 143
Vlahopoulos Spiros, 239

Xaplanteris C. L., 431

Zhirov Oleg V., 59
Žibret G., 439